D1368442

PLANT BIOCHEMISTRY

Plant
Biochemistry

BY JAMES BONNER

Kerckhoff Laboratories of Biology,
California Institute of Technology,
Pasadena, California

ACADEMIC PRESS INC., PUBLISHERS
NEW YORK 10, N. Y.
1950

This book is affectionately
dedicated to my first
teacher of chemistry

WALTER D. BONNER
Professor of Chemistry, Emeritus
The University of Utah

PREFACE

The remarkable advances achieved in several branches of plant biochemistry in recent years have gone far toward establishing the basic similarity of the metabolic processes of higher plants to the metabolic processes of other groups of organisms. It is in fact becoming apparent that much of biochemistry is common to many kinds of living things and that metabolic pathways elucidated for, perhaps, a microorganism apply equally well to the mammal and to the higher plant. A familiar example is the way in which yeast studies have increased our knowledge of the steps involved in the glycolytic transformation of hexose to pyruvate—a metabolic pathway which we now know to be ubiquitous. More recently investigations of bacterial systems have held a key position in the discovery and study of heterotropic CO_2 fixation and have contributed greatly to our understanding of the role of energy-rich phosphate in energy transfer, both processes of the widest general significance in living organisms. The pathways of amino acid synthesis and interconversion, discovered and worked out with the fungus Neurospora, appear to have their counterparts also in other organisms and to be of value as a guide to studies with higher plants as well as with mammals.

Although living things now appear to operate along lines that are basically similar, still each group has its own special metabolic features superimposed on this basic metabolic pattern. In this discussion of the biochemistry of the plant we shall be concerned, not only with metabolic pathways which are well understood in other groups of organisms, but also with matters which peculiarly concern plants, such as the metabolism of starch and of the cell wall, the formation of certain typical secondary metabolic products, and the biochemistry of photosynthesis. It is the purpose of this book to bring together the scattered work on general biochemistry as it applies to plants as well as to summarize those fields of biochemistry pertaining to the plant.

This book is written for the student whose interest in biochemistry is especially directed toward the plant. It is based upon a course in plant biochemistry which the author has given for the past twelve years to students of the plant sciences as well as to students of the chemistry of natural products, and which in the absence of any appropriate text has been conducted solely by lectures and by readings in the original literature.

This volume is suitable as a text for courses in plant biochemistry given on the senior and early graduate level, provided that the student has a background of some organic chemistry, although not necessarily any previous course in biochemistry. Selections from the general reading appended to each chapter may serve as additional material for such courses, while the specific references may prove helpful to the advanced student or research worker who wishes to initiate a detailed study of a particular field.

The author believes that a book of this nature, in order to be useful, should not consist merely of a collection of facts or of an annotated bibliography, but should rather present a series of as well ordered and consistent pictures as are possible with the present state of our knowledge. Every effort has been made therefore to present the facts of each subject in such a manner as to bring some order out of the whole. This is by no means easy in the present unorganized state of our knowledge concerning the biochemistry of plants. Future work and future concepts will no doubt revise our notions as to the proper way in which to present and organize the information on this subject.

There is much work to be done in plant biochemistry. Our understanding of many basic metabolic pathways in the higher plant is lamentably fragmentary. While the emphasis in this book is on the higher plant, it will frequently be necessary to call attention to conclusions drawn from work with microorganisms or with higher animals. Numerous problems of plant biochemistry could undoubtedly be illuminated by the closer application of the information and the techniques which have been developed by those working with other organisms.

Biochemistry today is a rapidly expanding discipline and new facts as well as new concepts are being continuously brought forward. It is inevitable therefore that certain of the material contained in the present book will be superseded by new facts and broader concepts. In certain particularly active fields this has indeed occurred even during the preparation of the manuscript.

Certain important aspects of biochemistry have been entirely omitted from the present volume simply because of the lack of pertinent information from the domain of higher plants.

The author wishes to express thanks to his colleagues who have so generously helped him in the preparation of this book. To A. J. Haagen-Smit in particular the author wishes to express his appreciation not only for his unfailing and enthusiastic assistance with the present volume but also for his counsel over the years. The author is indebted to the students in his successive classes for aid in collection of much of the original material. Thanks are especially due to those who diligently checked the original references cited in the text: Francis Haskins, Herbert Hull, Leonard Jansen, Kenneth Paigen, Carl Price, Bernard Strauss, Albert Siegel, and Emanuel Windsor.

Critical reading of various portions of the manuscript has been shared by A. A. Benson, W. H. Burris, Ray Dawson, Jacob W. Dubnoff, A. J. Haagen-Smit, W. Z. Hassid, James Henderson, Norman H. Horowitz, K. Markley, John Nance, Walton B. Sinclair, Emil Smith, Adrian Srb, W. Stepka, Paul K. Stumpf, Samuel G. Wildman and Laszlo Zechmeister. To each the author wishes to express his appreciation. Daniel E. Atkinson and Gerald Fling have contributed by their critical reading of the proof. It is improbable that this book could have been brought to completion without the untiring and expert editorial assistance of Betty Jean Wood, who has helped in all matters. The author wishes to also thank Harriet Bonner, who has done much of the proof reading and Shigeru Honda who has aided in the preparation of the illustrations. For the task of typing the successive versions of the manuscript the author is particularly indebted to Delores Craton and Kay Asakawa. A final note of appreciation is due to the directors of the Herman Frasch Foundation for Agricultural Chemistry, who in their policy of supporting basic work in plant biochemistry have done so much to promote study in this field, both at the California Institute of Technology and elsewhere.

JAMES BONNER

Pasadena, California
July 1950

CONTENTS

PART I

CARBOHYDRATES AND CARBOHYDRATE METABOLISM

PART II

THE CELL WALL AND CELL WALL METABOLISM

PART III

PLANT ACIDS AND PLANT RESPIRATION

PART IV

METABOLISM OF NITROGENOUS COMPOUNDS

PART VI

CERTAIN ASPECTS OF PLANT GROWTH

Chapter 1

AN INTRODUCTION TO PLANT BIOCHEMISTRY

Biochemistry is concerned with the chemical operation of living things, and this book treats therefore the chemical operation of plants. We will discuss the manner in which plants synthesize some of their many chemical constituents, the functions of these constituents in the plant, and the ways in which these constituents may be again broken down or altered into still other constituents; in fact, we will discuss, in so far as it is possible, the metabolism of the plant. Biochemistry merges imperceptibly into the related field of the organic chemistry of natural products. It can readily be sensed that this must be so since we must evidently be aware at least of some of the exact chemical compounds present in a plant before we are able to initiate any study of their metabolism. A study of the biochemistry of plant growth for example must first entail finding out what chemical substances constitute the growth increment. It is not surprising therefore that the great bulk of the chemical work thus far carried out with plants has concerned the isolation, identification, and structure determination of some of the vast number of compounds synthesized by and occurring in higher plants. This book is not intended to be a general catalog of the chemical compounds found in plants. On the contrary, such information is tabulated in several excellent works which are listed at the end of this chapter and which should be consulted for details concerning the phylogenetic distribution, amounts, isolation, and structure determinations of particular plant products. Since biochemistry is, however, the study of metabolic changes in chemical structure it will be necessary for us to discuss in some detail the chemistry of the particular plant materials with which we shall be concerned. In this discussion of individual plant components we should not however lose sight of the relationship which these individual components bear to the structure of the plant cell as a whole.

Biochemistry of the Cell. The generalized plant cell is made up of a thin layer of protoplasm which surrounds the large vacuole and which is in turn surrounded by the characteristic cell wall. From the point of view of cell volume, the vacuole is all important. Viewed physiologically, the vacuole is also doubtless of great significance, particularly as to its osmotic role in the water relations of the plant. From the standpoint of dry weight, however, the vacuole is relatively insignificant since its contents are made up so largely of water. Metabolically also the vacuole is not of great interest since few if any important metabolic processes take place within this portion of the cell. The wall of the plant cell ordinarily makes up a much smaller portion of the total volume than the vacuole, although in certain cases very massive secondary walls may occupy the bulk of the cell lumen. Even in the generalized cell, however, the cell wall tends to make up a quarter to a half or more of the total dry weight, and is then the largest single item in the overall cellular constitution. The remaining dry matter of the cell is shared among the components which go to make up the cytoplasm and its inclusions, the nucleus, starch grains, fat droplets, the various granules and, particularly in the case of the cells of green leaves, the chloroplasts. What now can we say about the chemical composition and the metabolic functions of these different portions of the plant cell?

The cell wall consists primarily of polysaccharides and polysaccharide derivatives (Part II), and although there are a few special cases in which the cell wall is made up largely of a single polysaccharide, still in the more usual case, several distinct and different compounds are intimately intermingled in its structure. Pores in this structure permit the ready exchange of smaller molecules and it is undoubtedly the protoplasmic membrane rather than the cell wall which constitutes the semipermeable barrier in the plant cell. Cell wall constituents are evidently formed by processes which take place in the protoplasm, although we have but little evidence on this point. In any case, the cell wall once laid down is relatively inert and its constituents ordinarily reenter the metabolism of the plant sluggishly if at all. The cell wall is, however, subject to at least two striking and important transformations, related respectively to cell elongation (Chapter 7) and to fruit ripening (Chapter 9). In both of these cases as in cell wall growth itself, the secondary transformations of the wall constituents are undoubtedly mediated by metabolic processes carried out within the protoplasm of the cell.

The nonaqueous constituents of the vacuole appear to consist principally of substances of low molecular weight including water soluble pigments (Chapter 28) and possibly other glycosides, in addition to inorganic ions, sugars (Chapters 2 and 3), and organic acids (Chapter 14). Proteins if they occur in the vacuole at all are present only in small amounts, and enzymes have been located with certainty in the vacuole only in latex vessels where vacuolar and protoplasmic contents are inextricably mixed.

The full complexity of the plant cell is achieved only in the protoplasm, of which proteins are the principal and characteristic components. Of the protoplasmic protein, which may constitute a fourth to one-half of the total dry weight of a typical leaf cell, approximately one-half goes to make up the soluble cytoplasmic fraction while the remaining half is contained in various particulate inclusions, the nucleus, chloroplasts, and other granules. Many different and chemically individual proteins are undoubtedly concerned in the composition of these different cellular components and further cytochemical differentiation between the various cellular structures is found in the distribution of still other materials. Thus chlorophyll and the carotenoid pigments are restricted in higher plants to the plastid, desoxyribonucleic acid to the nucleus, and ribonucleic acid largely to the cytoplasm. This cytochemical differentiation is associated with important biochemical differentiation in metabolic function. Thus we know that the synthesis of certain proteins, the desoxyribonucleoproteins, occurs in the nucleus and there are indications that the nucleus may also directly synthesize other proteins which are then liberated into the cytoplasm. It is probable also that the synthesis and degradation of the starch grains and of the fat droplets which occur as particulate inclusions in the cytoplasm are reactions which take place in or on the surface of these particles. It is the chloroplast, however, which most clearly carries on chemical processes quite different in kind from those of the cytoplasm in which the chloroplast is imbedded. Thus the typical reaction of photosynthesis, the light-induced splitting of water, is a process confined in the higher plant to chloroplasts. The terminal stages of chlorophyll synthesis as well as the synthesis of the associated carotenoid pigments appear also to occur only within the chloroplast structure. Recent work indicates that much of the active respiratory system of the plant cell may be included in the particulate structure. What then is left to the cytoplasm proper by way of biochemical function? What synthetic and what degradative reactions are typically cytoplasmic processes? We cannot yet answer this with certainty, although it seems probable that there are biochemical reactions which are cytoplasmic in occurrence and which may include among other processes the formation and transformation of the simple carbohydrates and the synthesis of particular amino acids.

The point to be stressed is this. Biochemists have long tended to treat the cell as a homogeneous mixture of catalysts and substrates, the whole capable of carrying out the manifold activities of living things. Biologists on the contrary have been deeply concerned with the description of the structures which may be observed within the cell. We now know that each cellular structure, the nucleus, the chloroplast, and the other particulate inclusions, is concerned with its own typical and characteristic aspect of the overall metabolism. The concept of the cellular granule or particle as a metabolic unit has emerged with particular clarity from work on the animal enzyme systems responsible for such processes as pyruvate or fatty acid oxidation. Respiratory pyruvate oxidation takes place (Chapter 15) through an orderly sequence of reactions involving at least fifteen different kinds of enzymes. These enzymes are not distributed randomly among the cytoplasmic proteins of the cell but are combined into one large highly organized particle. It has been known for many years that certain types of biochemical reactions, including respiration, are dependent on cellular structure and are lost when cellular structure is destroyed. Just what aspect of cellular structure is it which is essential to such a process? It now seems probable that the answer to this question is somewhat as follows. Biochemical reactions in which an orderly sequence of reactions are required frequently take place only in the presence of highly organized particles made up of the required array of enzymes. The proper functioning of this metabolic particle may also require the presence of components of the cytoplasm such as coenzymes, energy acceptors, or other enzymes. The proper functioning of a cell as a whole undoubtedly requires the appropriate functioning of many different kinds of these metabolic units, all suspended in and interacting through the same cytoplasm. In this book we shall in general talk about metabolic processes as occurring in a particular kind of cell or tissue and will not be able from sheer lack of information to define more closely the exact nature of the cellular units involved in the particular reaction. It should be borne in mind however that any full elucidation of a biochemical process will necessarily include definition of the cellular units concerned.

Enzymes. Cellular metabolism consists essentially in the chemical alteration of substrate molecules by reactions which take place to an appreciable extent only in the presence of individual and specific enzymes. It is appropriate therefore to include in this introduction some discussion of the general nature and properties of enzymes. It will be assumed that the reader of this book is acquainted with the fact that the majority of biological reactions are catalyzed by enzymes and that all enzymes thus far isolated in pure form are protein in nature. Since each enzyme specifically catalyzes only one reaction or one type of reaction and since the cells of many tissues are known to be able to carry out many different kinds of reactions, it is clear that each cell must contain a great many different kinds of enzymes. Sumner and Somers have, in fact, calculated that the animal cell may contain of the order of 1000 different enzymes. While nowhere near this number have as yet been detected in any single kind of cell or tissue, a total of several hundred different enzymes is known in all. Although in a few cases a particular enzyme may make up a moderate proportion of the total dry weight of a cell, for example phosphatase of leaves (several per cent) or urease of jack bean (ca. 0.1%), still individual enzymes are ordinarily present in exceedingly low concentrations such as 0.01% or less of the total cellular dry matter. It is clear then that the isolation of an enzyme or its identification by a property other than its catalytic activity is ordinarily a difficult task and for this reason relatively few plant enzymes have actually been isolated in pure form.

Modern nomenclature makes use of the substrate acted upon with the suffix *ase* for the names of enzymes. Thus we have amylase (starch hydrolyzing), protease (protein hydrolyzing), and phosphorylase (phosphate transferring) as general groups of enzymes. There are however several amylases which differ in the exact way in which they attack starch. Two such different starch hydrolyzing enzymes are then called respectively α- and β-amylase. Similarly the phosphate transferring enzyme concerned with starch metabolism is known as starch phosphorylase, while that concerned with sucrose metabolism is known as sucrose phosphorylase. In earlier work enzymes were named rather more in accord with the spirit of free enterprise and we still have such names as papain and ficin in common use today.

It cannot be assumed that the comparable enzymes of different species are necessarily identical from a chemical point of view. Thus the leaf phosphatases of spinach, tobacco, Chinese cabbage and *Xanthium*, although generally similar in nature, do differ in their characteristic electrophoretic mobilities (Chapter 17). Amylases, as well as proteases, differ among themselves as to pH optima, inactivation characteristics, and other qualities, indicating also a measure of species specificity which is superimposed upon the enzymatic character of the individual enzyme.

Each different enzyme is, so far as known, a definite compound or molecular species, and those which have thus far been isolated in pure crystalline form are all proteins. The molecular weights of the known plant enzymes range from the 13,000 of cytochrome C to the 483,000 of urease, with the majority lying in the lower half of this range. Their protein nature confers on enzymes as a class two of their most typical characteristics, namely sensitivity to pH and susceptibility to heat inactivation. Proteins are amphoteric substances containing both acidic carboxyl and basic amino groups and capable then of ionizing cationically in acid solutions and anionically in alkaline solutions. At some intermediate pH, at which the number of cations of the protein just balances the number of anionic groups, the protein is effectively neutral and is said to be at its isoelectric point. It is not surprising therefore that the catalytic activity of enzymes likewise changes with pH. The pH optimum does not ordinarily reveal any simple relation to the isoelectric point, and it is probable that in part the pH optimum reflects a change in stability as well as a change of ionic nature of the enzyme with pH. With but few exceptions, the catalytic activity of enzymes is readily destroyed by heat. The effect of this treatment is to denature the enzyme protein, that is, the protein becomes insoluble and loses most if not all of its biological specificity. Protein denaturants other than heat also inactivate enzymes; they may for example be destroyed by the action of ultraviolet light, by heavy metal ions, especially silver, mercury, and lead, and by the action of concentrated acid or alkali.

Certain groups of enzymes consist wholly of protein, and in such cases it appears that the protein itself is solely responsible for the catalytic power of the enzyme. Enzymes of this group include the amylases and the proteases. Enzymes of other groups consist of protein combined with non-protein moieties or prosthetic groups, and in such cases it is the prosthetic group which is commonly the active catalytic center of the enzyme. The prosthetic groups of certain enzymes are readily removable from the protein and the prosthetic group free protein moiety is then referred to as the apoenzyme. This is true of the dehydrogenases which are described in detail in Chapter 15, and which contain di- or triphosphopyridine nucleotides as their loosely bound prosthetic groups. With other enzymes the prosthetic group may be simply a metal ion, copper in the case of several of the oxidases or magnesium in certain peptidases, phosphatases, and enzymes of the carbohydrate metabolism. Finally there are enzymes containing complex prosthetic groups which are very tightly bound and which can be removed only by such drastic treatment as denaturation of the protein moiety. This is true of the magnesium porphyrin protein of the photosynthetic system as well as of certain iron porphyrin enzymes including catalase. A striking fact concerning the prosthetic groups of enzymes is the occurrence in them of such chemical compounds as thiamine, pyridoxine, niacin, adenine, riboflavin, and possibly pantothenic acid and biotin, which we also know as the vitamins of the B group. Many of these substances were recognized as essential dietary factors for animals, microorganisms, and even for certain higher plant tissues before the biochemical bases of the requirements became known. Similarly, the minor elements, copper, iron, and manganese, are required by plants as prosthetic groups or portions of prosthetic groups of specific enzymes, and it may be suspected that other essential minor elements of plant nutrition, molybdenum, zinc, and boron, may also function as the active groups of particular enzymes.

Enzymes catalyze reactions which are spontaneous from a thermodynamic standpoint but which take place spontaneously at a low or imperceptible rate. In the presence of the proper enzyme a reaction will then merely proceed more rapidly toward the attainment of its true thermodynamic equilibrium. In many enzymatic reactions, as for example the hydrolysis of starch by amylase or of sucrose by β-fructosidase, the reaction catalyzed is one in which the thermodynamic equilibrium is so far toward one side that it is not possible to measure the equilibrium constant or to enzymatically synthesize any detectable amount of the original substrate from the reaction products. It is probable, none the less, that most if not all enzymes catalyze reactions in both directions. The reversibility of enzymatically catalyzed reactions has been demonstrated for a wide variety of cases, beginning with the reversal of the hydrolysis of ethyl butyrate to ethyl alcohol and butyric acid in the presence of the enzyme lipase. In aqueous media this reaction proceeds largely to the right, as written below. If lipase is allowed to act on ethyl alcohol and

$$\text{Ethyl butyrate} + H_2O \rightleftharpoons \text{Ethyl alcohol} + \text{butyric acid}$$

butyric acid in reaction media containing but little water, as for example in carbon tetrachloride, a good yield of ethyl butyrate may be obtained. Similarly glucosides may be synthesized from glucose and an appropriate alcohol in the presence of the hydrolytic glucosidases. In this instance also, the synthetic reaction depends upon reduction of the water concentration in the reaction medium, in this case by the alcohol added. More detailed information concerning the reversibility of enzymatic reactions has been obtained with the oxidative enzymes which are discussed in

$$\beta\text{-Ethyl glucoside} + H_2O \underset{\text{In 60–85\% alc.}}{\overset{\text{In water}}{\rightleftharpoons}} \text{Ethyl alcohol} + \text{glucose}$$

Hydrolysis and synthesis of β-ethyl glucoside in the presence of the enzyme β-glucosidase (emulsin)

Chapter 15. Despite the fact that enzymatic reactions are in the main reversible, it should be borne in mind that in many cases the reaction may proceed in one direction only in living tissue. This is strikingly the case with the hydrolysis of starch by amylase. The reverse reaction, synthesis of starch, takes place by an entirely different route, involving a different enzyme and a different substrate.

Mode of Enzyme Action. It is commonly believed that not all molecules of a substance or reactant are equally reactive and that only those molecules which possess more than a certain minimal energy are in fact capable of reaction. This minimal energy which the molecule must have in order to insure reaction is called the activation energy and it may be thought of as an energy barrier which the molecule must surmount before it can rearrange its electronic structure, react, and descend to the new and more stable state. The rate of chemical reaction of a substance is proportional then not to the total number of molecules of the substance but to the number of such molecules which possess the required activation energy. The energy of activation may be supplied by thermal collisions among molecules. Thus reactions ordinarily proceed more rapidly as the temperature is raised and as thermal movements become accordingly more intense. The relation between reaction rate, energy of activation and temperature is given by the Arrhenius equation

$$\text{Velocity constant of reaction} = A e^{-E_a/RT}$$

in which A is a constant, E_a the energy of activation, e the base of natural logarithms, R the gas constant (1.99 cal. per mol) and T the absolute temperature. The reaction rate is hence an exponential function of activation energy and reciprocally an exponential function of temperature. This means that small changes in either activation energy or in temperature exert great effects on the rate of chemical reactions.

Enzymes increase reaction rate by lowering the apparent activation energy. Table 1-1 summarizes data for four reactions in which the acti-

TABLE 1-1. Activation energies of enzymatically catalyzed and non-catalyzed reactions. (After Lineweaver)

Reaction	Enzyme	Activation energy E_a in cal./mol	
		No catalyst or H^+ ions alone	Enzyme as catalyst
H_2O_2 decomposition	Catalase	18,000	5,000
Sucrose hydrolysis	β-Fructosidase	26,000	11,500
Casein hydrolysis	Trypsin	20,600	12,000
Ethyl butyrate hydrolysis	Lipase	13,200	4,200

vation energy of an uncatalyzed reaction or of a simple hydrogen ion induced hydrolysis can be compared directly with the activation energy of the corresponding enzymatic reaction. In each case the activation energy of the enzymatic reaction is lower, in some instances much lower, than the activation energy of the nonenzymatic reaction. Because of the exponential relation between reaction velocity and activation energy, these decreases in activation energy correspond to very large increases in reaction rate. Thus a decrease of the activation energy from 20,000 cal./mol to 10,000 cal./mol corresponds to an increase in reaction velocity of roughly 500,000 times.

Although the exact manner in which an enzyme decreases the activation energy is not known, still it is clear that the critical point about the process is the formation of a complex between enzyme and substrate. Such enzyme-substrate complexes have been demonstrated by several indirect methods and can be detected directly in the case of certain metal-containing enzymes in which the absorption spectrum of the enzyme is altered as the result of complex formation. The early belief that enzymes merely accelerate but do not enter into a reaction is therefore not strictly correct. The enzyme does enter into the reaction but may be regenerated in its original form as the complex decomposes to yield the final reaction product.

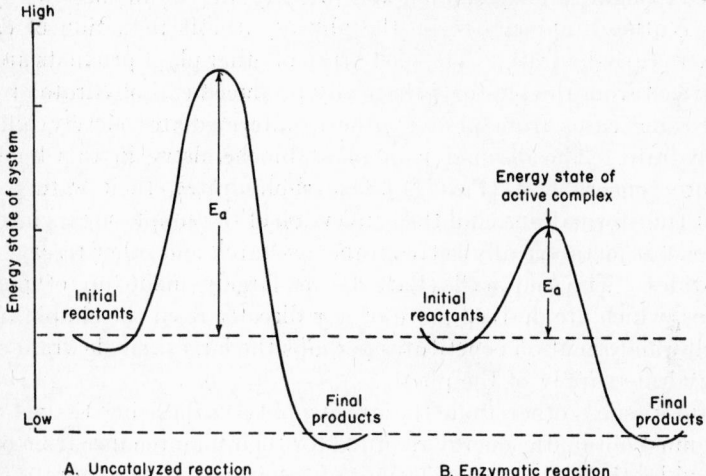

FIG. 1-1. Diagrammatic representation of the activation energy as related to the energies of the initial reactants and final products (A) in an uncatalyzed reaction and (B) in an enzymatic reaction involving formation of an enzyme-substrate complex. E_a represents energy of activation.

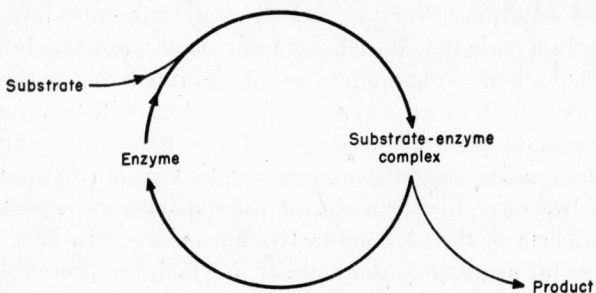

FIG. 1-2. Substrate-enzyme complex formation followed by decomposition of the complex into reaction product with regeneration of the enzyme.

A great deal of valuable information on the kinetics of enzymatic reactions and on other specialized aspects of enzyme lore has been gathered together in several works which are cited at the end of this chapter. The balance of this book will not be so much concerned with enzymology *per se* as with the nature of the enzymatically catalyzed transformations which take place in the higher plant.

Pathways of Plant Metabolism. The most striking feature of the metabolism of higher green plants is their ability to synthesize a wealth of complex organic compounds from simple substrates, e.g., carbon dioxide, water and certain inorganic ions, the energy required for the process being supplied as light. The common and primary step in all these many and varied syntheses appears to be the photosynthetic reduction of carbon dioxide to carbohydrate. The vast array of other plant products are then synthesized from the photosynthetically produced carbohydrate or possibly in some cases from photosynthetic intermediates closely allied to carbohydrate. The discussion of plant biochemistry in this book will therefore concern first (Part I) the carbohydrates, their nature, their simpler transformations, and the conversion of the simple sugars into such complex but metabolically active forms as starch and other reserve polysaccharides. The cell wall (Part II) is largely made up of polysaccharides, which are derived more or less directly from the simple sugars, and cell wall formation constitutes perhaps the largest single drain on the carbohydrate supply of the plant.

Plant tissues, other than the green photosynthesizing tissues themselves, must obtain the energy required for their maintenance from energy liberated by the biological oxidations of respiration. In respiration, carbohydrate originally formed in photosynthesis is again metabolized to carbon dioxide and water, but the organism is provided with mechanisms for the conservation of the energy liberated by this highly exothermic process. The energy thus conserved is then available in forms suitable for the transformation of further carbohydrate molecules into other still more energy-rich compounds such as amino acids, proteins, fats, and isoprenoids. The plant acids are important intermediates in the oxidative breakdown of carbohydrates in respiration and constitute in fact a major component of many plant tissues. Part III is therefore concerned firstly with the plant acids and secondly with plant respiration as a whole.

Thus far we have tacitly assumed the existence of a protoplasm in which metabolism of the carbohydrates takes place. In Part IV we will consider in so far as possible the manner in which the proteins and other nitrogenous constituents of the protoplasm are synthesized and otherwise metabolized. Although nitrogen content is only one of the characteristic features of the proteins and allied materials, it will nevertheless be convenient to refer to this section as the metabolism of nitrogenous compounds. The plant is able to fulfill its nitrogen requirement with inorganic forms such as nitrate, which are taken up by the root, reduced, and incorporated into the various amino acids which are the essential units of the more complex proteins (Chapter 16). Chapters 17 and 18 concern the nature and metabolism of the plant proteins themselves, and this is followed by discussion of other important nitrogenous compounds, including the purines and the pyrimidines, which are of particular significance as components of the active groups of many proteins. The alkaloids which are of so much pharmacological and chemical interest are treated in this book only in so far as the scanty knowledge of their metabolism would seem to justify. The present section on nitrogen metabolism closes with brief discussion of the biochemistry of nitrogen in nature as a whole and concerns particularly the conversion of molecular nitrogen of the air into forms available to higher plants.

From the carbohydrate formed in photosynthesis and with the energy liberated in respiration, the plant synthesizes and accumulates a host of secondary products of which three general groups are considered in Part V. The first of these groups to be taken up is that of the lipids or fats and allied materials. These lipids are secondary only in sense of biochemical derivation since quantitatively they make up substantial proportions of many seeds, fruits, and even of particular vegetative tissues. Fats, like starch, serve as forms for storage of food reserves and are readily mobilized by the plant when conditions require. The isoprenoid compounds, simple terpenes, carotenoids, and rubber, have no such obvious general function in the plant. Even though the simple terpenes or rubber itself may be formed in large amounts by particular species, still these materials perform no known role and appear at present largely to represent a reckless squandering of resources by the plant. Of the isoprenoids only the carotinoids and the chlorophyll component, phytol, can reasonably be demonstrated to fulfill vital plant functions. This does not mean that the biochemistry of the isoprenoids is not of interest. On the contrary, both the simple terpenes represented by the essential oils and the polyterpenes represented by rubber are of great economic value to man, and it is clearly of interest to understand the steps by which the plant synthesizes these materials in which it has apparently a purely platonic interest. Still another group of products of uniquely botanical interest are the water soluble plant pigments, the anthocyanins and related compounds, which go to make up the great bulk of flower as well as of many fruit and leaf pigments.

From the metabolism of the individual plant constituents we pass on to the biochemistry of the processes which regulate and integrate the growth of the plant as a whole. It is now clear that internal harmony in the higher plant is brought about by a complex system of hormones, organic substances which are synthesized in one organ or tissue and translocated to other organs or tissues where they bring about specific physiological responses. These responses in so far as they have been studied have very largely to do either with the extent or quantitative aspect of the growth of the organ affected or, in other instances, with the differentiation of the tissue into new kinds of tissues. It has been an outstanding accomplishment of plant physiology to recognize, separate, and study the processes thus hormonally regulated in the plant and a large and steadily increasing literature on the subject already exists. Only a small beginning has been made, however, in the biochemistry of hormonal growth regulation, and it can hardly be said that more than the general nature of the processes involved is now understood.

This book closes with a summary of the presently known facts concerning the biochemistry of photosynthesis. From a logical standpoint this subject could better be the first taken up, but so complex is the subject that it seems better to consider it in the light of what has earlier been developed concerning protein structure and metabolic pathways in the plant.

Like so many of the metabolic pathways here described, the subject matter of this book is cyclical rather than linear in arrangement. We begin with the carbohydrates, pass on to respiration, proteins, various secondary products and finally through photosynthesis we return to the carbohydrates. Strictly speaking each aspect of biochemistry must be considered in the light of and as related to all the other aspects. It is merely an unfortunate geometric necessity which confines the textbook of biochemistry to a linear sequence in its presentation.

General Reading

The works listed in sections 1 to 6 are those which cover the whole field indicated. Specialized references to particular topics in plant biochemistry are given at the end of each chapter.

1. Distribution and chemistry of plant products.
   Klein, G., Handbuch der Pflanzenanalyse (6 vols.). Springer, 1931–1933.
   Wehmer, C., Die Pflanzenstoffe. Fischer, 1931, Supplement, 1935.
   Czapek, A., Biochemie der Pflanzen. Fischer, 3rd ed., 1922.
   Gisvold, O., and Rogers, C. H., The Chemistry of Plant Constituents, Burgess, 2nd ed., 1938.
   Haas, P., and Hill, T. G., Chemistry of Plant Products. Longmans, Green and Company, 4th ed., 1928.
2. General analytical methods.
   Klein, G., Handbuch der Pflanzenanalyse. Springer, 1931–1933.
   Glick, D., Techniques of Histo- and Cytochemistry. Interscience Publishers, 1949.
   Association of Official Agricultural Chemists. Official and tentative methods of analysis. (Pub. by The Assoc. of Official Agricultural Chemists, Washington, D. C.), 5th ed., 1940.
   Piper, C., Soil and Plant Analysis. Interscience Publishers, 1944.
   Tunmann, O., and Rosenthaler, L., Pflanzenmikrochemie. Borntraeger, 2nd ed., 1931.
   Molisch, H., Microchemie der Pflanze. Fischer, 3rd ed., 1923.

3. Earlier works bearing largely on plant biochemistry.
    Thatcher, R., Chemistry of Plant Life. McGraw-Hill, 1921.
    Kostytschev, S., Chemical Plant Physiology. Blakiston, 1931.
    Onslow, M., Practical Plant Biochemistry. Cambridge, 2nd ed., 1923.
    Onslow, M., Principles of Plant Biochemistry. Cambridge, 1931.
    Tottingham, Wm. E., Plant Biochemistry. Burgess, rev. ed., 1937.
    Snyder, H., Chemistry of Plant and Animal Life. Macmillan, 4th ed., 1920.
4. General works on enzymes and enzymatic reactions.
    Sumner, J. B., and Somers, G. F., Chemistry and Methods of Enzymes. Academic Press, 2nd ed., 1947.
    Northrop, J. H., Kunitz, M., and Herriott, R. V., Crystalline Enzymes. Columbia University Press, 2nd ed., 1948.
    Glasstone, S., Laidler, K. J., and Eyring, H., The Theory of Rate Processes. McGraw-Hill, 1941.
    Bamann, E., and Myrbäck, K. Die Methoden der Fermentforschung. Georg Thieme, 1941 (Academic Press, 1945).
    Sumner, J. B., and Myrbäck, K., The Enzymes. Two volumes, Academic Press, in press.
5. Important review series which occasionally contain articles on subjects relevant to plant biochemistry.
    Chemical Reviews
    Cold Spring Harbor Symposia on Quantitative Biology
    Annual Review of Biochemistry
    Physiological Reviews
    Quarterly Review of Biology
    Biological Reviews
    Ergebnisse der Enzymforschung
    Advances in Enzymology
    Advances in Protein Chemistry
    Advances in Carbohydrate Chemistry
    Vitamins and Hormones
    The Botanical Review
    Bacteriological Reviews
6. Journals which at the present time contain numerous original contributions to plant biochemistry, together with the abbreviations used in this book.

    U.S.A.

    Archives of Biochemistry (Arch. Biochem.)
    Journal of Biological Chemistry (J. Biol. Chem.)
    Journal of the American Chemical Society (J. Am. Chem. Soc.)
    American Journal of Botany (Am. J. Botany)
    Botanical Gazette (Botan. Gaz.)
    Plant Physiology (Plant Physiol.)
    Soil Science (Soil Sci.)
    Journal of Agricultural Research (J. Agr. Research)

    Britain

    Biochemical Journal (Biochem. J.)
    Journal of the Chemical Society, London (*J. Chem. Soc.*)
    New Phytologist
    Proceedings of the Royal Society (London), Series B (*Proc. Roy. Soc. (London)*)

Australia

Australian Journal of Experimental Biology and Medical Science (Australian J. Exptl. Biol. Med. Sci.)
Australian Journal of Science (Australian J. Sci.)

Switzerland

Helvetica Chimica Acta (Helv. Chim. Acta)

Sweden

Annals of the Royal Agricultural College of Sweden (Ann. Roy. Agr. Coll. Sweden)

France

Bulletin de la société chimique de France (Bull. soc. chim.)
Bulletin de la société de chimie biologique (Bull. soc. chim. biol.)

U.S.S.R.

Biokhimiya (Biokhim.)
Bulletin of the Academy of Sciences, Moscow

Germany

Planta
Biochemische Zeitschrift (Biochem. Z.)
Berichte der deutschen chemischen Gesellschaft (Ber.), now Chemische Berichte
Hoppe-Seyler's Zeitschrift für physiologische Chemie (Z. physiol. Chem.)
Liebig's Annalen der Chemie (Ann.)

Netherlands

Biochimica et Biophysica Acta (Biochim. Biophys. Acta)

PART I.   CARBOHYDRATES AND CARBOHYDRATE METABOLISM

Chapter 2

THE MONOSACCHARIDES

The Hexoses.   Four hexose sugars, D-glucose, D-mannose, D-galactose, and D-fructose, are commonly found in higher plants although other sugars including sorbose may also be found occasionally.   Of the four principal sugars, glucose and fructose occur both in combined forms and in the free uncombined state, while the others are found mainly, perhaps exclusively, in various types of chemical combination as, for example, in the polysaccharides and in other derivatives including the glycosides. The chemical structures of the principal hexoses are given below.

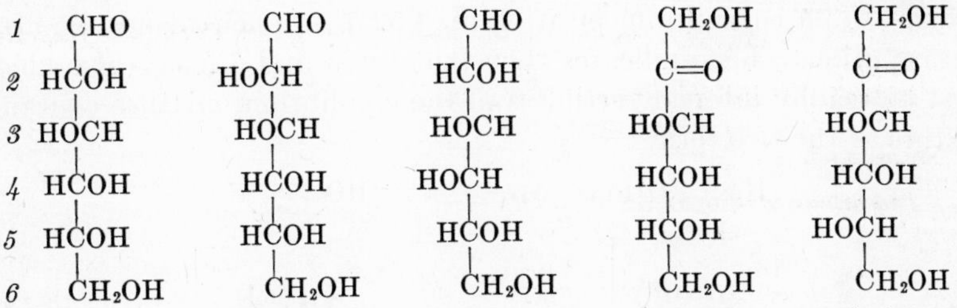

1	CHO	CHO	CHO	CH₂OH	CH₂OH

Glucose and fructose differ from one another only in the groups present at carbon atoms 1 and 2, the carbonyl group of glucose being terminal (position 1) and hence aldehydic, whereas that of fructose is at the second carbon atom and is hence a ketonic group.   The other hexoses differ from glucose or from fructose by their configurations about the asymmetric carbon atoms, which in glucose are four in number.   Thus mannose differs from glucose in orientation about carbon atom 2, whereas galactose differs from glucose in orientation about carbon atom 4.

   The classical work of Tollens and others has shown that glucose and other hexoses exhibit chemical properties which indicate that they exist in solution in the cyclic form as well as in the straight chain structures shown above.   Good evidence for the existence of such cyclic forms follows from the work of Fischer[1] who found that when glucose is treated

15

with methyl alcohol and HCl under acetal-forming conditions, one methyl group enters the molecule at carbon atom 1 with formation of the methyl glucoside. Such treatment of D-glucose gave rise, however, to two isomeric glucosides differing from one another in optical rotation. To these two compounds, α- and β-methyl glucoside, Fischer assigned the structures shown below.

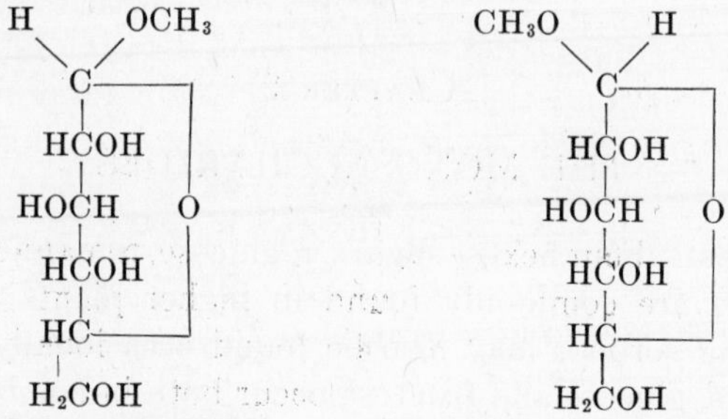

Methyl α-D-glucoside Methyl β-D-glucoside

D-Glucose itself also exists in two forms, an α form having a rotation in water of $+113°$, and a β form having a rotation of $+19°$. When either of these forms is dissolved in water the rotation of the solution gradually changes to an equilibrium position of $+52.5°$, a phenomenon known as mutarotation. This indicates then that the α and β forms of D-glucose must be readily interconvertible and the equilibrium mixture contains a mixture of the two forms.

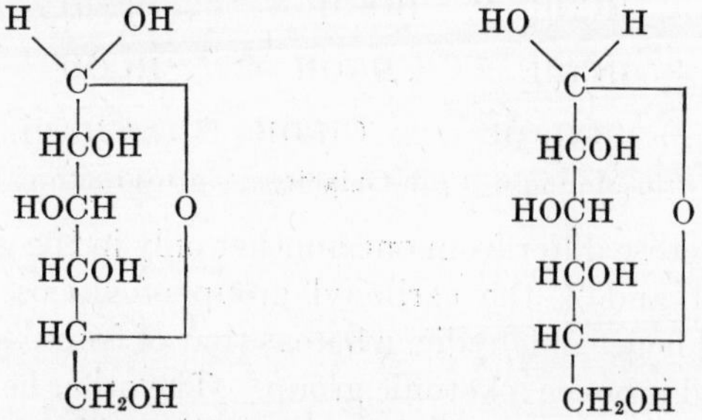

α-D-Glucose β-D-Glucose

Correlation of the α and β forms of D-glucose with the α- and β-methyl glucosides of Fischer was accomplished by Armstrong[2] who showed that by treatment of α-methyl glucoside with the enzyme maltase, α-D-glucose is liberated, while treatment of β-methyl glucoside with the enzyme emulsin results in the liberation of β-D-glucose. These facts establish then that glucose does in fact exist in a ring or cyclic form. The closing of the ring results in formation of a new asymmetric carbon atom at position 1, and the possibility of isomerism at this point is responsible for the existence of the α and β forms of D-glucose. The point of closure of the ring in the cyclic form of D-glucose was originally assigned by Fischer to carbon atom 4, which would involve the formation of a five-membered or furanose ring. Proof that the cyclic form of D-glucose is in fact a six-membered or pyranose ring was brought forth only in much more recent work. The methods used in the establishment of the pyranose ring structure of D-glucose are basic to structure determinations in other more complicated cases and will therefore be described in some detail. The theory of the structure determination is based upon the complete methylation of the free hydroxyl groups of the sugar, a method first described by Purdie and Irvine.[3] D-Glucose is first treated with methyl alcohol and HCl with formation of the methyl glucoside; this is followed by methylation of the free hydroxyl groups with methyl iodide and silver oxide to form the tetramethyl methylglucoside. The glucosidic methyl group at carbon atom 1 is less stable to hydrolysis than the methyl ether groups and may be readily hydrolyzed off to yield a tetramethyl-D-glucose. This substance

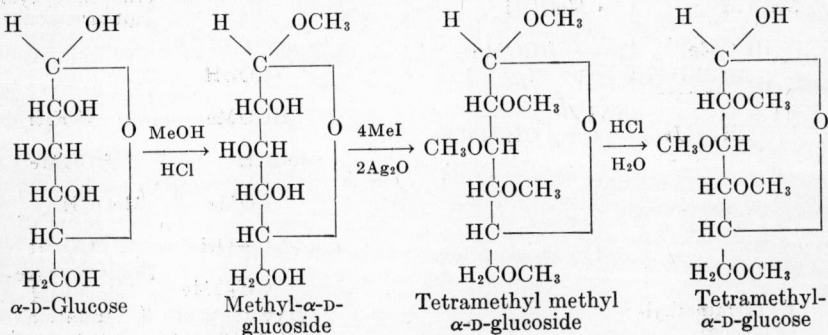

FIG. 2-1. Steps involved in the exhaustive methylation of D-glucose.

is much more stable than the original glucose and may be subjected to treatments which are not applicable to glucose itself. Alternative methylation methods for the sugars have been proposed, notably that of Haworth[4] in which dimethyl sulfate in the presence of sodium hydroxide is employed. Tetramethylglucose may be oxidized and from the oxidation products the point of ring closure, that is, the carbon atoms which lack a methyl ether group, may be ascertained. Final proof that glucose is in fact D-glucopyranose was obtained in this way by Hirst,[5] who subjected tetramethylglucose to oxidation with nitric acid. From the oxidation products the appropriate derivatives of glutaric and succinic acids were isolated, indicating that in the oxidation the molecule is first attacked adjacent to carbon atom 5, which must therefore lack the methyl ether group and is then the site of the ring closure.

Subsequent to the identification of the ring structure of glucose by the degradation of the methylated sugar, a simpler method was found in periodate oxidation of the methyl glycoside.[6] This reagent has the property of cleaving a carbon chain between any two adjacent carbon atoms bearing hydroxyl groups. In the case $R \cdot CHOH \cdot CH_2OH$, the primary alcohol group yields formaldehyde; the secondary alcohol yields an aldehyde. If it is adjacent on both sides to other carbon atoms bear-

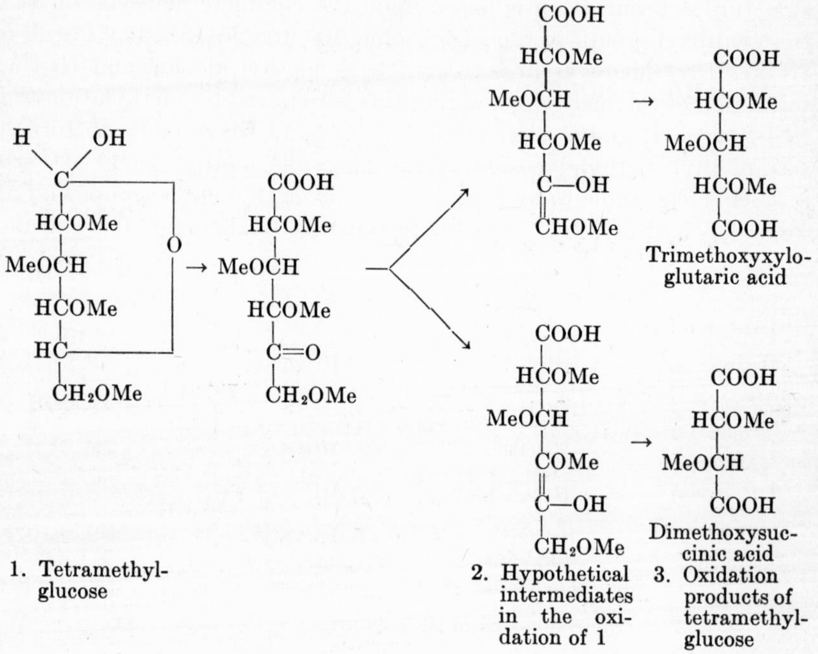

FIG. 2-2.   Proof of structure of tetramethyl-D-glucose pyranoside by oxidative method of Hirst.[5]

ing hydroxyl groups ($R \cdot CHOH \cdot CHOH \cdot CH_2OH$), the central CHOH group yields formic acid. Thus the nature and quantities of the oxidation products formed yield information as to the structure of the compound oxidized. Oxidation of the methyl α-D-glycosides of furanose and pyranose structures yield the oxidation products as shown in FIG. 2-3. The glucofuranose derivative requires two mols of periodic acid for oxidation and yields one mol of formaldehyde, while the glucopyranoside uses up two mols of oxidant and yields one mol of formic acid. Application of the periodate method confirms the conclusion arrived at above that D-glucose exists in solution in the pyranose ring form.

A pyranose ring structure is common to the aldose hexoses, including glucose, galactose, and mannose, and this is true both as to the free sugars and as to the same sugars in their combinations as disaccharides and the more complex polysaccharides. The furanose forms of these sugars do not appear to be of common occurrence in nature, a fact which may be related to the relative instability of the furanose as compared with the pyranose sugars.

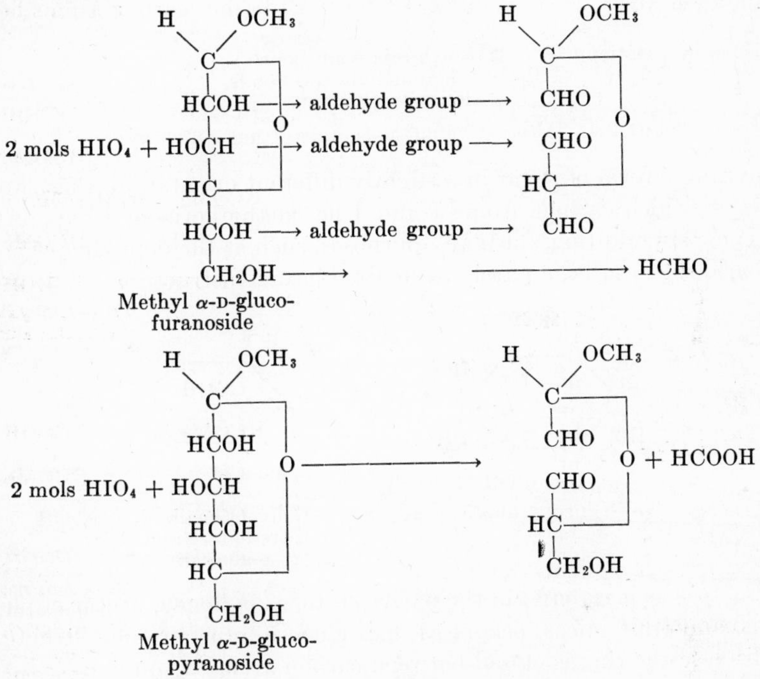

Methyl α-D-glucofuranoside

Methyl α-D-glucopyranoside

FIG. 2-3. Oxidation with periodic acid.

The spatial relationships of the atoms in glucose are not accurately shown in the type of formulas used above but are best depicted by a hexagonal formula, and Haworth[7] has suggested that in such formulas the orientation of the hydrogen atoms and hydroxyl groups be indicated by their position with respect to the ring. Thus the groups pictured above the atom to which they are attached are those which are understood to be above the plane of the formula (or in a position to the right in the Fischer formula), while those pictured below the atom to which they are attached are understood to be below the plane. The Haworth representation may be made clearer by heavy shading of the lines connecting atoms of the hexagonal ring to indicate that the observer is looking down on a projection of the hexagon. In this book it will be convenient to write

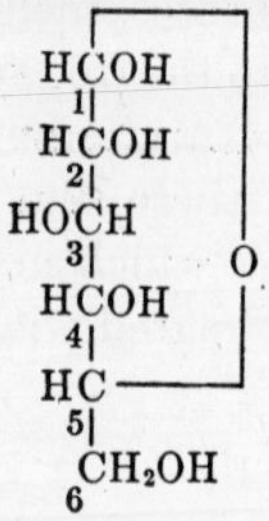

HCOH
1
HCOH
2
HOCH
3
HCOH
4
HC
5
CH₂OH
6

Fischer representation

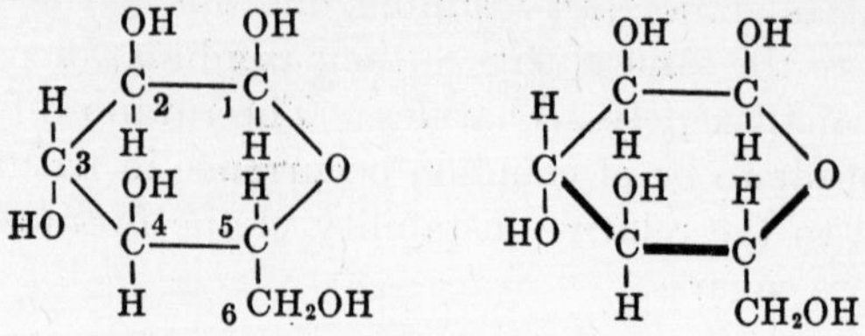

OH   OH
| |
H   C   C
 \ / 2 1 \
  C   H   H   O
 / 3 OH   H
HO   C   C
     4   5
     | |
     H   CH₂OH
         6

OH   OH
| |
H   C   C
 \ /   H   H   \
  C   OH   H   O
 /
HO   C   C
     | |
     H   CH₂OH

Haworth representations
(heavy shading indicates bonds
closest to observer)

Fɪɢ. 2-4.   Alternative methods of depicting α-ᴅ-glucose.

the pyranose form of sugar in a slightly different orientation, that shown
below, in which carbon atoms 1 and 4 lie on the horizontal axis of the
hexagon.   In addition, for many purposes, such as the depicting of polysaccharide structures, a purely formalized hexagon structure will be used.

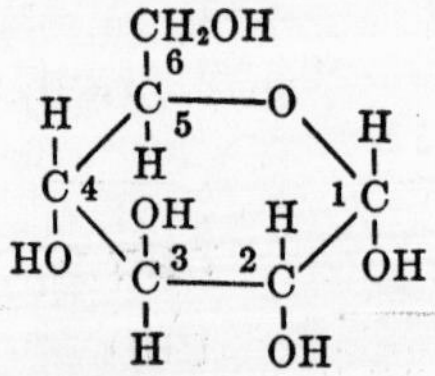

CH₂OH
6
C   O
H   5
C 4 OH   H   1 C
HO   C 3 2 C   OH
     | |
     H   OH

α-ᴅ-Glucopyranose

6
5   O
4   1
3   2

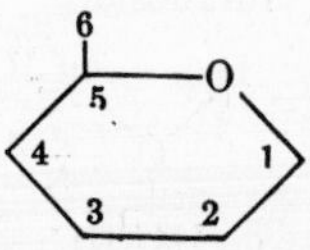

Conventionalized
representation of
ᴅ-glucose

Fructose as it is found in the native state, as in honey, nectar of plant
glands, and fruit juices, occurs as does glucose in a pyranose structure.
In this case the ring is closed between carbon atoms 2 and 6.

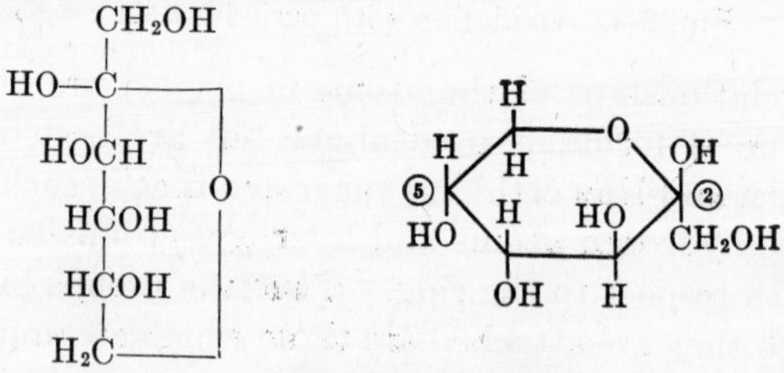

CH₂OH
|
HO—C
|
HOCH
| O
HCOH
|
HCOH
|
H₂C

H
H   O
H   H   OH
⑤   ②
HO   H   HO CH₂OH
     OH   H

ᴅ-Fructopyranose

Fructose in the bound state as in sucrose and in various polysaccharides
is found only in a furanose form in which the ring is formed between
carbon atoms 2 and 5.   The two cyclic forms of fructose are however
probably in equilibrium in aqueous solution, with the more stable
pyranose form predominating.

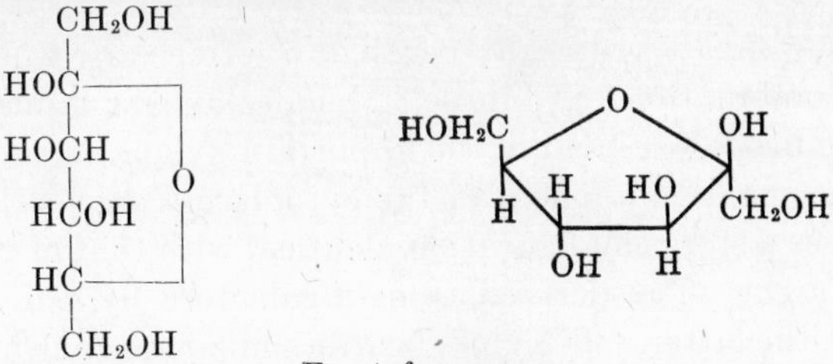

D-Fructofuranose

D-Galactose, as noted above, occurs only in combined forms, notably in galactans, gums, and mucilages. It is an interesting fact that galactose and the sugars sterically related to it, fucose and arabinose, are found, although rarely, in both the D and L forms, occasionally even in the same plant product. Thus L-galactose is a constituent of agar, a polysaccharide which consists principally of D-galactose.

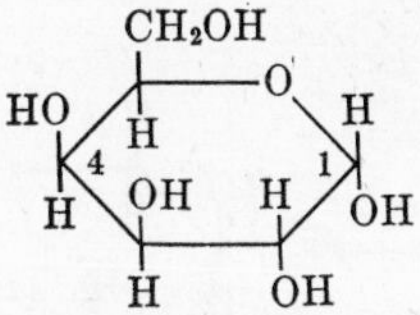

α-D-Galactopyranose 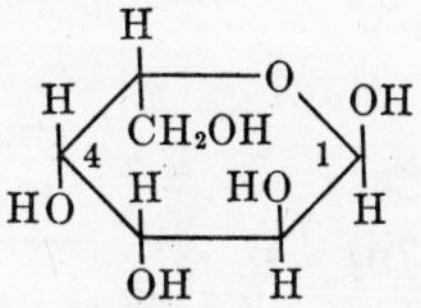   α-L-Galactopyranose

D-Mannose like galactose occurs in the plant principally in the form of polysaccharides, including particularly the mannans and various gums. With this sugar as with the other aldohexoses it is the pyranose form which is found exclusively in natural products.

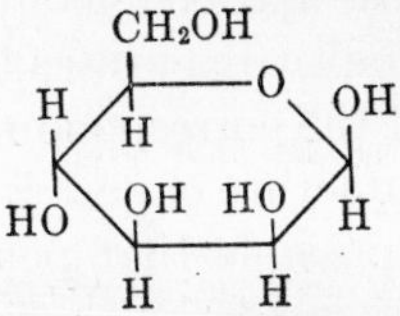

β-D-Mannopyranose 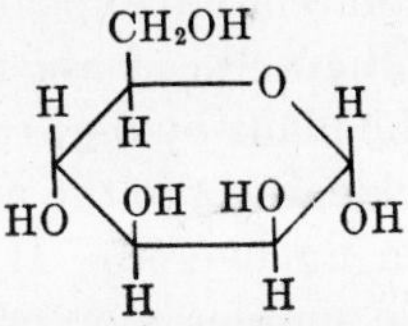   α-D-Mannopyranose

The Pentoses and Heptoses. Three pentoses, or five-carbon sugars, may be considered as being general and major constituents of the higher plant. These include the isomeric aldoses L-arabinose, D-xylose, and D-ribose.

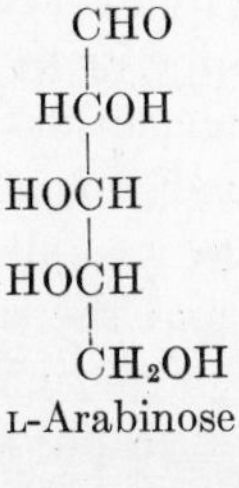

L-Arabinose 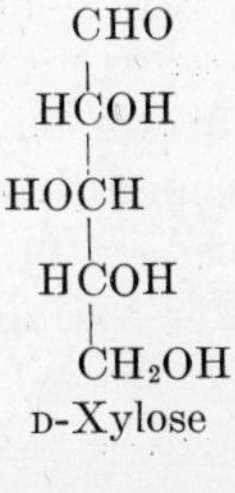   D-Xylose 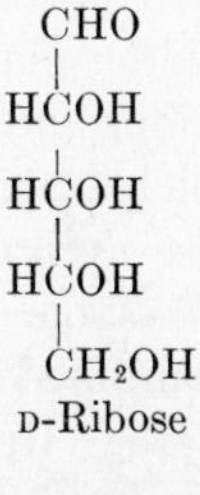   D-Ribose

None of these sugars ordinarily occur in the free state but all are found, rather, in combination: arabinose in hemicelluloses, gums, and arabans; xylose in hemicelluloses and in the important xylans of woody tissue, and ribose as the sugar component of the cytoplasmic nucleotides. L-Arabinose possesses a steric configuration identical with that of D-galactose and may be thought of as derived from D-galactose by removal of carbon atom 6 from the latter. D-Xylose bears a similar relationship to D-glucose.

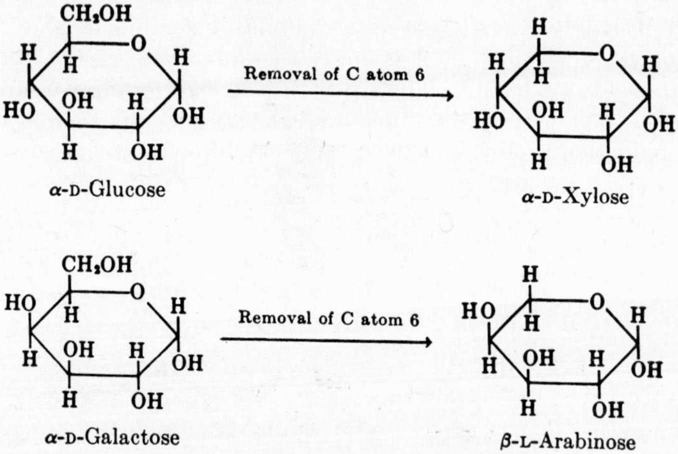

FIG. 2-5. Configurational relationships between glucose, galactose, and the corresponding pentoses, xylose and arabinose.

The structural relationships between glucose and xylose or between galactose and arabinose need not imply that the two pentoses are necessarily derived in the plant by decarboxylation of the corresponding hexoses, and in fact no evidence as to the mode of manufacture of pentoses in the plant has yet been uncovered. It is entirely possible that pentoses may be built up from simpler compounds as, for example, by the union of a two-carbon with a three-carbon unit and that pentoses may represent slightly aberrant products of reverse glycolysis (Chapter 30) rather than direct hexose degradation products.

D-Ribose is not intimately related to any other natural sugar but is nevertheless probably a universal constituent of living cells, including those of higher plants. The occurrence of ribose appears to be restricted to nucleotides and it constitutes the sugar of the nucleotides of the cytoplasmic nucleic acids as well as of at least three broad groups of enzymes. In all these derivatives, ribose is attached through its aldehydic group to a carbon atom of a purine or pyrimidine base, and the sugar is additionally combined through an hydroxyl group with one or more molecules of phosphate. Ribose like fructose is always found in its derivatives in the furanose form.

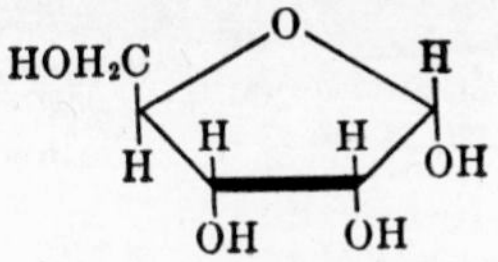

α-D-Ribofuranose

The methyl pentoses may be considered as hexose reduction products in which the primary alcohol group of carbon atom 6 has been reduced to a terminal methyl group. Typical methyl pentoses are L-rhamnose, widely distributed in glycosides and found in certain gums; L-fucose, a component of the algal polysaccharide, fucan; and D-fucose, or rhodeose, a component of a few rare glycosides. Although D-fucose is structurally related to D-galactose, the other methyl pentoses are, from a structural view, reduction products of hexoses which do not occur in nature. It is

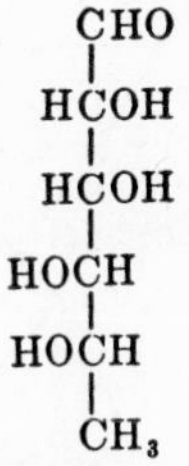

L-Rhamnose

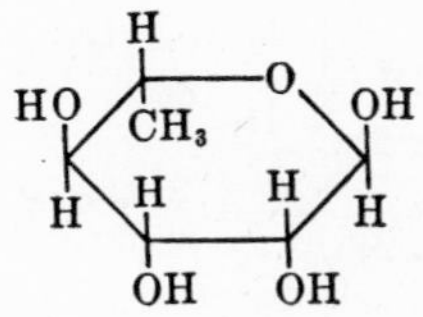

α-L-Rhamnopyranose

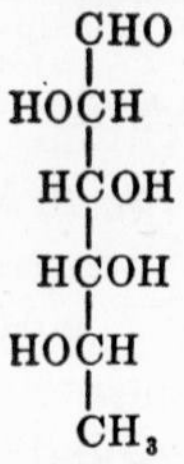

L-Fucose

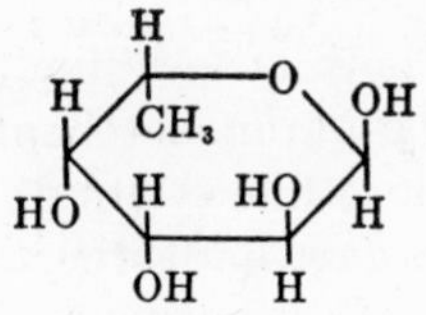

α-L-Fucopyranose

therefore entirely possible that the methyl pentoses are synthesized in the plant by mechanisms other than the seemingly obvious route through reduction of the appropriate hexose.

A sugar which is reduced at carbon atom 2 is known in the form of 2-desoxy-D-ribose, a component of the nucleic acids of the nucleus. The distribution of 2-desoxyribose in the cell is therefore to be contrasted with that of ribose which is, as pointed out above, largely confined to cytoplasmic nucleic acids and other extra-nuclear structures. A 2-desoxy-6-methyl pentose, digitoxose, has been found among the sugars liberated by hydrolysis of the *Digitalis* glycosides, sterols combined with sugars which have striking effects on the heart. The rare methylated sugar digitalose also occurs in these glycosides.

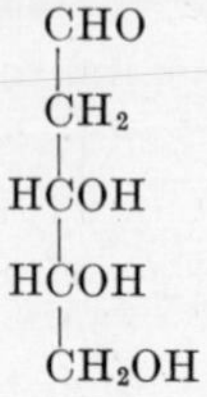

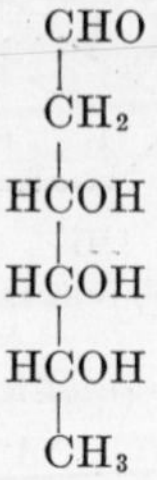

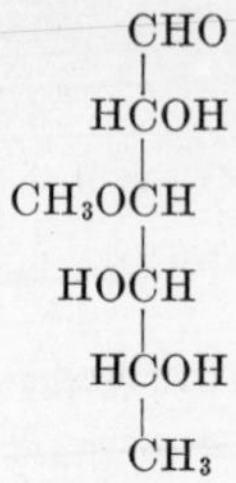

$$
\begin{array}{ccc}
\text{CHO} & \text{CHO} & \text{CHO} \\
| & | & | \\
\text{CH}_2 & \text{CH}_2 & \text{HCOH} \\
| & | & | \\
\text{HCOH} & \text{HCOH} & \text{CH}_3\text{OCH} \\
| & | & | \\
\text{HCOH} & \text{HCOH} & \text{HOCH} \\
| & | & | \\
\text{CH}_2\text{OH} & \text{HCOH} & \text{HCOH} \\
& | & | \\
& \text{CH}_3 & \text{CH}_3 \\
\text{2-Desoxy-D-ribose} & \text{Digitoxose} & \text{Digitalose}
\end{array}
$$

Attention should be further called to a branched-chain pentose, hamamelose, known only from *Hamamelis*, the witch hazel. The sugar is of particular interest in that it could conceivably be built up by aberrant condensation of two molecules of a well-known sugar breakdown product, glyceraldehyde.

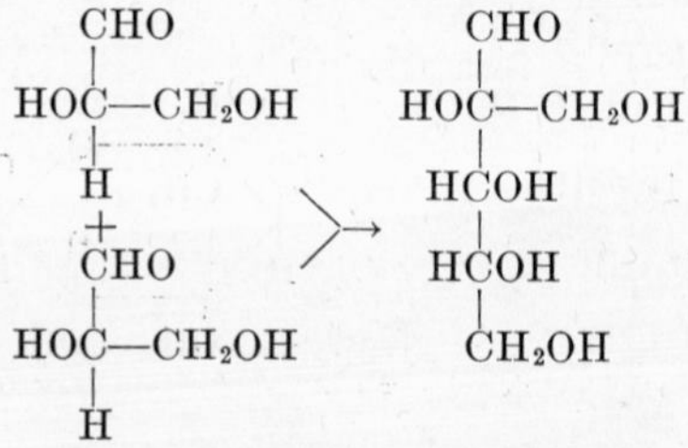

$$
\begin{array}{cc}
\text{CHO} & \text{CHO} \\
| & | \\
\text{HOC—CH}_2\text{OH} & \text{HOC—CH}_2\text{OH} \\
| & | \\
\text{H} & \text{HCOH} \\
+ & | \\
\text{CHO} & \text{HCOH} \\
| & | \\
\text{HOC—CH}_2\text{OH} & \text{CH}_2\text{OH} \\
| & \\
\text{H} &
\end{array}
$$

Glyceraldehyde   Hamamelose
Possible, but hypothetical, production of hamamelose

Two heptoses, both of very limited distribution, are known to occur in higher plants. D-Mannoketoheptose is found in the free form in the fruit of the avocado, while D-altroketoheptose or sedoheptose occurs free in the leaves of *Sedum* and other succulents. D-Mannoketoheptose is structurally related to fructose, while sedoheptose, a pyranose sugar in which the ring is closed between carbon atoms 2 and 6, is not referable to any other natural sugar.

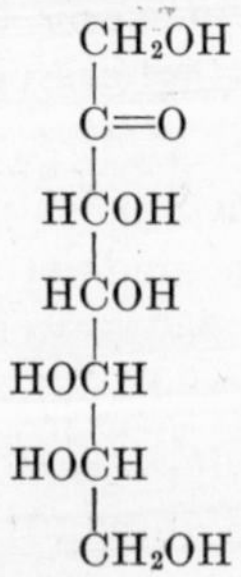

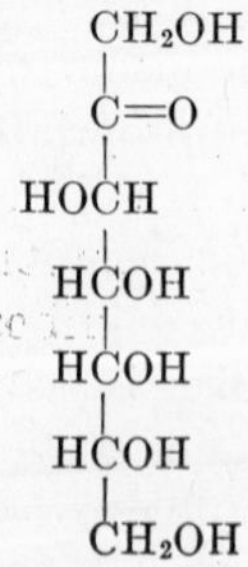

$$
\begin{array}{cc}
\text{CH}_2\text{OH} & \text{CH}_2\text{OH} \\
| & | \\
\text{C=O} & \text{C=O} \\
| & | \\
\text{HCOH} & \text{HOCH} \\
| & | \\
\text{HCOH} & \text{HCOH} \\
| & | \\
\text{HOCH} & \text{HCOH} \\
| & | \\
\text{HOCH} & \text{HCOH} \\
| & | \\
\text{CH}_2\text{OH} & \text{CH}_2\text{OH} \\
\text{D-Mannoketoheptose} & \text{D-Altroketoheptose} \\
& \text{(sedoheptose)}
\end{array}
$$

TABLE 2-1. The occurrence of various pentoses and pentose derivatives in the plant kingdom

Type of sugar	Name of sugar	Species in which sugar is found	Compounds in which sugar is found
Pentose	L-Arabinose	General	Gums, arabans, hemicellulose
	D-Xylose	General	Gums, xylans, hemicellulose
	D-Ribose	General	Cytoplasmic nucleotides
Methyl pentose	L-Rhamnose	Common, not general	Gums, glycosides as quercitrin, naringin
	L-Fucose	*Fucus*	Fucan
	L-Rhodeose (D-fucose)	*Convolvulaceae, Tubera jalapae*	Glycoside jalapin
Desoxy pentose	2-Desoxy ribose	General	Nuclear nucleic acid
Desoxy methyl pentose	Digitoxose	*Digitalis*	*Digitalis* glucosides

General References

Haworth, W. N., The Constitution of Sugars. Longmans, Green and Co., 1929.

Pigman, W. W., and Goepp, R. M., Jr., Carbohydrate Chemistry. Academic Press, 1948.

Wolfrom, M. L., Chapters on Carbohydrate Chemistry in Gilman *Organic Chemistry.* Wiley, 2nd ed., 1943.

References

1. Fischer, E., *Ber.*, **28**, 1145 (1895).
2. Armstrong, E. F., *J. Chem. Soc.*, **83**, 1305 (1903).
3. Purdie, T., and Irvine, J. C., *J. Chem. Soc.*, **83**, 1021 (1903).
4. Haworth, W. N., *J. Chem. Soc.*, **107**, 8 (1915).
5. Hirst, E. L., *J. Chem. Soc.*, **129**, 350 (1926).
6. Jackson, E. L., and Hudson, C. S., *J. Am. Chem. Soc.*, **59**, 994 (1937); **61**, 1530 (1939).
7. Haworth, W. N., The Constitution of Sugars. Longmans, Green & Co., 1929.

Chapter 3

THE DISACCHARIDES

The disaccharide sucrose is of central importance in the carbohydrate metabolism of the plant, while the disaccharides maltose and cellobiose are of quantitative significance as the building blocks of the two important polysaccharides, starch and cellulose. The chemistry of these and other disaccharides is hence of great importance, and determination of the structures involved has attracted much interest. Structure determination in the disaccharide series entails a whole series of chemical problems, including identification of the monoses involved, identification of the stereochemical configuration (α or β) of the glycosidic bond between the two components, identification of the sugar involved as the alcohol in the glycosidic linkage, identification of the carbon atom involved in the glycosidic linkage, and finally determination of the ring structures of the component sugars. The component monoses involved in sucrose or in any disaccharide are identified by isolation of the pure compounds after hydrolysis of the disaccharide. Thus maltose yields only D-glucose on hydrolysis. The nature of the glycosidic linkage can frequently be determined enzymatically, a method originally developed by Armstrong.[1] Alpha glycosides, such as α-methyl glucoside or maltose, are hydrolyzed to D-glucose by maltase, the enzyme complex of sprouting barley, or by α-glycosidase of yeast. Beta glycosides on the contrary are hydrolyzed by emulsin, an enzyme complex containing β-glycosidase and obtained from seeds of the *Rosaceae*. This enzymatic specificity holds throughout in the hydrolysis of the α or β glycosidic linkages in the disaccharides.

In the determination of the carbon atoms involved in the glycosidic and ring structures it is necessary to make use of the technique of methylation of the disaccharide, followed by hydrolysis of the methylated product and identification of the methylated monoses obtained. Methylation of the reducing disaccharides requires milder treatment than that used in methylation of the monoses. Thus methyl alcohol and HCl cannot be used to introduce a methyl group into the reducing part of the molecule since the acid causes hydrolysis of the glycosidic linkage. The methyl iodide-silver oxide method cannot be used since the reagent acts as an oxidizing agent on any free sugar reducing group. The dimethyl sulfate procedure, however, has been used with success by Haworth.[2] In this technique the methyl glycoside of the sugar is first formed by treatment with methyl sulfate in alkali at low temperature. The temperature is then raised and methylation of the hydroxyl groups of the glycoside accomplished. As a final safeguard, a treatment with methyl iodide and silver oxide may be added to insure complete methylation. The methylated disaccharide may then be hydrolyzed with acid to its two constituent methylated monosaccharides and these may in turn be separated by fractional distillation. Through the work of many investigators, the properties of the several tetra-, tri-, di-, and monomethyl sugars are known. Thus the identities of the methylated sugars derived from hydrolysis of any oligo (di-, tri-, etc., saccharides) or polysaccharide may be determined by the comparison of boiling points, melting points, and the properties of derivatives such as the anilides, with those recorded in the literature.

Maltose. Maltose as we know today is 4-D-glucose-α-D-glucopyranoside; that is, it consists of two glucopyranose residues linked through carbon atoms 1 and 4 and is a reducing disaccharide since one of the two aldehyde groups is not involved in linkage. Maltose as such is not found in nature but occurs only as a constituent of polysaccharides, particularly of starch from which it is liberated as a product of the action of amylase.

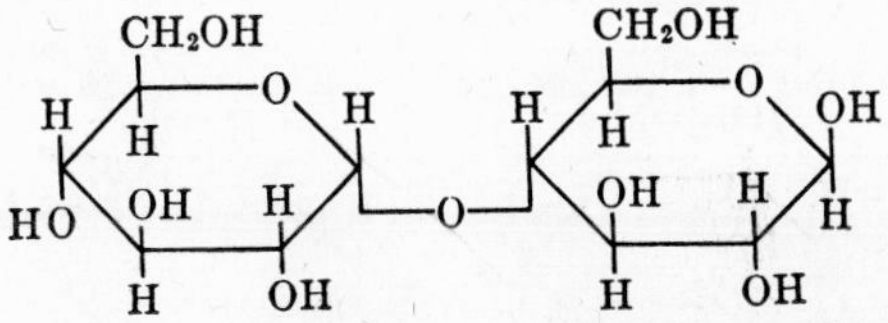

Maltose: 4-D-Glucose-α-D-glucopyranoside

Hydrolysis of maltose by maltase or by acid yields two molecules of D-glucose and attack of the compound by maltase establishes the compound as an α-glucoside. Completely methylated maltose may be hydrolyzed with acid to yield equimolecular amounts of 2,3,4,6-tetramethyl-D-glucopyranose and 2,3,6-trimethyl-D-glucose. (It should be noted again that although the free glucosidic carbon atom 1 is methylated during the methylation treatment, this methyl group is lost during hydrolysis of the methylated disaccharide.) The isolation of 2,3,4,6-tetramethylglucose clearly establishes D-glucopyranose as a constituent of maltose. The presence of 2,3,6-trimethylglucose might however represent either of two possibilities, namely, glucopyranose bound through a glycosidic linkage at carbon atom 4, or glucofuranose bound through a glycosidic linkage at carbon atom 5. This question can be settled by oxidation of the free aldehydic group to a carboxyl group with bromine. The ring is opened by this treatment. Methylation of the acid disaccharide derivative (maltobionic acid) followed by hydrolysis

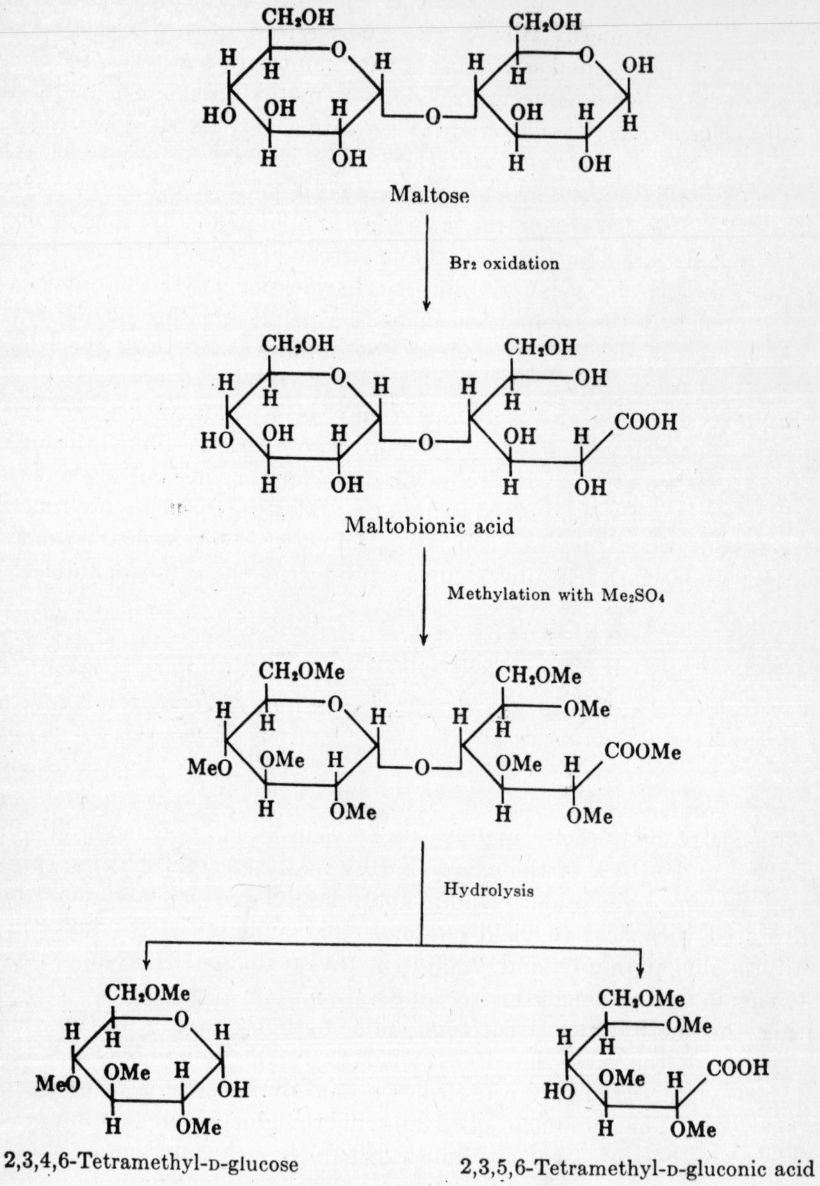

FIG. 3-1.   Steps in structure determination of maltose.

results in the isolation of a tetramethyl aldonic acid, 2,3,5,6-tetramethyl gluconic acid. This leaves only carbon atom 4 for involvement in the glycosidic bond of the second residue.

Other Disaccharides of Glucose. The disaccharide cellobiose resembles maltose in that it is a reducing disaccharide which yields two molecules of D-glucose on hydrolysis. Cellobiose also does not occur as such in nature but may be obtained as the result of carefully controlled hydrolysis of cellulose. Methylation and hydrolysis of the disaccharide yields 2,3,4,6-tetramethyl-D-glucopyranose and 2,3,6-trimethyl-D-glucose. The pyranose nature of the second residue can be established through bromine oxidation as with maltose. Cellobiose, like maltose, hence consists of two D-glucopyranose units linked through carbon atoms 1 and 4. In contrast to maltose, however, cellobiose is a β-glucoside since it is hydrolyzed by emulsin and other β-glucosidases.

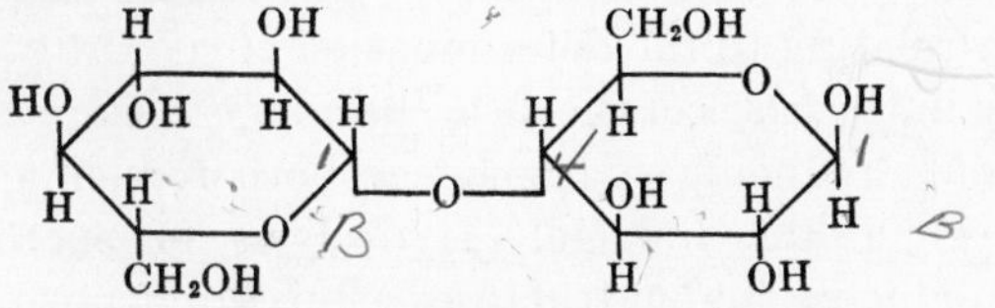

Cellobiose: 4-D-Glucose-β-D-glucopyranoside

Gentiobiose, a reducing β-disaccharide, is found in glucosides including the amygdalin of the bitter almonds and the crocin of *Crocus*. On hydrolysis the disaccharide yields two molecules of β-D-glucose, while methylation followed by hydrolysis yields 2,3,4,6-tetramethylglucopyranose and 2,3,4-trimethylglucopyranose. This establishes gentiobiose as 6-glucose-β-D-glucopyranoside.

The nonreducing disaccharide trehalose is produced by many fungi and stored by them as a reserve sugar and occurs also in *Selaginella*. Trehalose yields two molecules of D-glucose on hydrolysis. Since the sugar is nonreducing the two constituent molecules must be linked

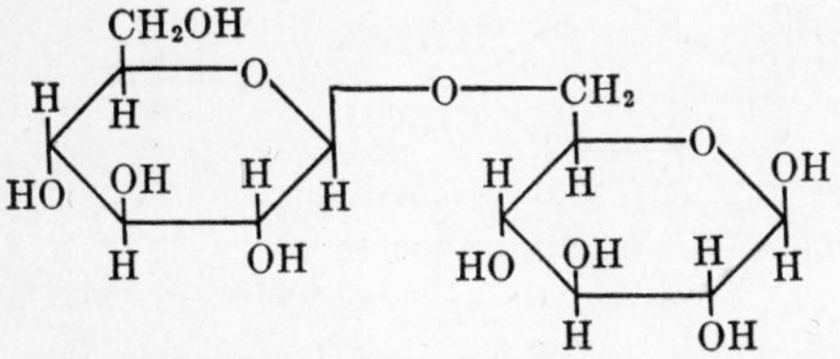

Gentiobiose

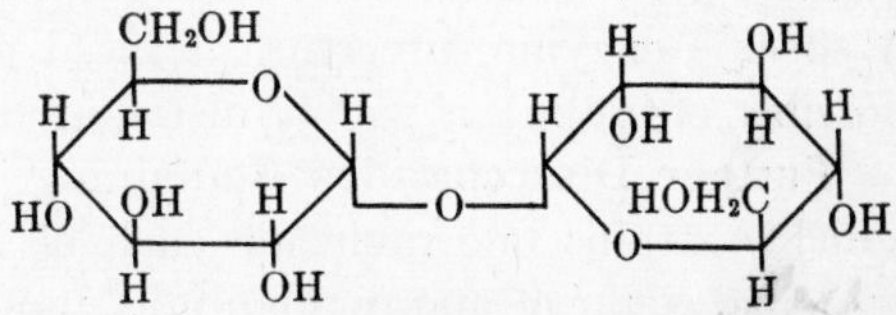

Trehalose

through the two carbon atoms 1 to form α-D-glucopyranosyl-α-D-gluco-pyranoside. This structure is confirmed by the results of periodate oxidation which yields two mols of formic acid per mol of trehalose, indicating the presence of three adjacent OH groups in each glucose residue.

Sucrose. Of all the disaccharides, sucrose is of by far the most significance both because of its wide distribution and its intimate metabolic role in the plant. Sucrose is not only a major photosynthetic product and a general constituent of higher plant tissues but is also one of the principal forms of carbohydrate storage and accumulation. In addition, carbohydrates appear to be largely transported within the plant in the form of sucrose. It is of interest to note that, important though it is, the complete structure of sucrose was worked out only in the investigations of Haworth and co-workers in 1927[3] while final enzymatic synthesis of the molecule was achieved by Hassid and others only in 1944.[4] The problems relating to the determination of sucrose structure are more involved than in the cases of the disaccharides considered up to this point. Sucrose is not a reducing sugar, hence both reducing groups must be involved in glycosidic linkage. Hydrolysis of sucrose yields equal amounts of D-glucose and D-fructose, a mixture known as invert sugar from the fact that the dextrorotation of a sucrose solution changes to a levorotation during hydrolysis, due to the liberation of the strongly levorotatory D-fructose. Methylation of sucrose followed by hydrolysis yields equal amounts of 2,3,4,6-tetramethylglucopyranose, and 1,3,4,6-tetramethylfructofuranose, while enzymatic investigations have established that sucrose is both an α-glucoside and a β-fructoside. These facts taken together establish sucrose as 1-α-D-glucopyranose-β-D-fructofuranoside. A striking property of sucrose is its remarkable ease of hydrolysis in acid solution, its rate of hydrolysis in acid being of the order

Sucrose: 1-α-D-glucopyranose-β-D-fructofuranoside

of 1000 times greater than the comparable rates for maltose or lactose. A specific enzyme, invertase, limited to the hydrolysis of sucrose (and possibly of inulin), is widely distributed in higher and lower plants.

Further Disaccharides Containing Two Different Sugars. Lactose, which contains one molecule each of D-glucose and D-galactose, is the principal sugar of mammalian milk but appears to be absent from higher plants. Lactose is a reducing sugar and since it is hydrolyzed by emulsin it must possess a β-glycosidic linkage. Methylation and subsequent hydrolysis yield 2,3,6-trimethylglucose and 2,3,4,6-tetramethylgalactose while oxidation to the aldobionic acid, followed by methylation and hydrolysis yields 2,3,5,6-tetramethylgluconic acid as does maltose. The structure of lactose is hence analogous to that of cellobiose with the substitution of a galactosidal residue for the glucosidal residue. Melibiose, a sugar found as a constituent of various bark exudates and formed also by partial hydrolysis of raffinose, is similarly a glucose-galactose disaccharide structurally related to gentiobiose. In melibiose, a D-glucose residue is linked through carbon atom 6 to the aldehydic group of a D-galactose residue.

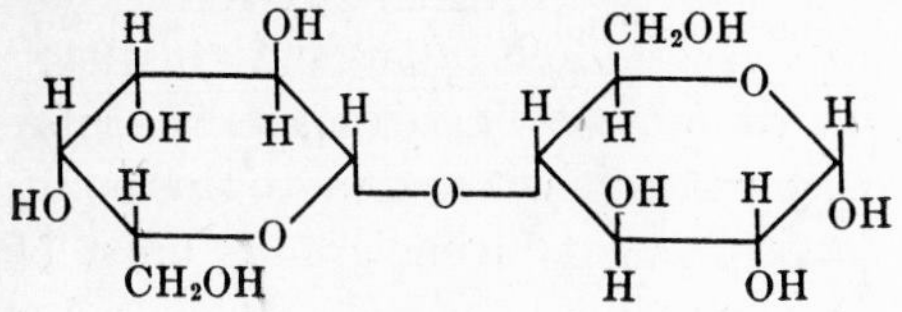

Lactose

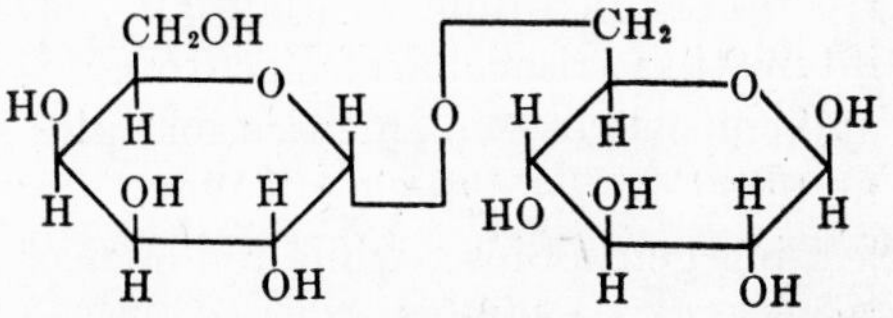

Melibiose

There are known in nature several disaccharides in which a glucose residue is linked to a pentose molecule. Thus the glucoxylose found in leaves of *Daviesia* contains glucose and xylose linked through their two reducing groups to form a nonreducing disaccharide. In primeverose, found in leaves of *Primula officinalis*, glucose is linked through its reducing group to carbon atom 6 of xylose.

Trisaccharides. Of the trisaccharides only one, raffinose, is known to occur widely in nature, cottonseed meal and sugar beet juice being usual sources. Raffinose on hydrolysis yields D-glucose, D-fructose, and D-galactose in equimolecular amounts. Enzymatic hydrolysis with emulsin yields sucrose and galactose while mild acid hydrolysis or hydrolysis with the yeast enzyme raffinase yields melibiose and fructose. The component monosaccharides are therefore arranged in raffinose as follows:

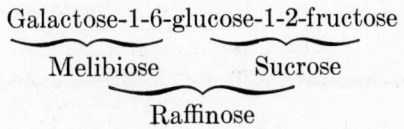

Galactose-1-6-glucose-1-2-fructose

Melibiose   Sucrose

Raffinose

TABLE 3-1. Occurrence and constituent sugars of trisaccharides in plant material

Trisaccharide	Constituent Sugars	Occurrence
Nonreducing tri-saccharides		
Raffinose	D-Galactose-D-glucose-D-fructose 1-[6-α-galactose-α-D-gluco-pyranose]-β-D-fructofuranoside	Widespread, sugar beet roots, cottonseed
Gentianose	D-Glucose-D-glucose-D-fructose 1-[6-D-glucose-α-D-glucose]-β-D-fructofuranoside	Roots of *Gentiana*
Melezitose	D-Glucose-D-fructose-D-glucose 2-[3-(α-glucose)-β(?)D-fructo-furanose]-α-D-glucoside	Exudate of conifers as *Larix*, *Pseudotsuga;* exudates of *Populus*, lime
Reducing trisac-charides		
Rhamninose	D-Galactose-L-rhamnose-L-rham-nose	Glycoside xanthorhamnin from *Rhamnus*
Robinose	L-Rhamnose-L-rhamnose-D-galactose	Glycoside robinin from flowers of *Robinia*

An isomer of raffinose, planteose, occurs in *Plantago* species. A second interesting trisaccharide, gentianose, is found in roots of *Gentiana lutea* and consists of two glucose molecules joined to a fructose molecule. The material is nonreducing. Weak acid hydrolysis removes the fructose, leaving gentiobiose, while removal of the terminal glucose residue yields sucrose as a product. Other trioses of limited distribution are listed in Table 3-1.

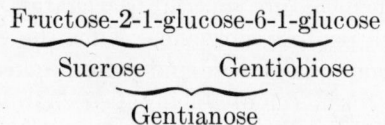

Fructose-2-1-glucose-6-1-glucose

Sucrose　　　Gentiobiose

Gentianose

Stachyose. But one tetrasaccharide, stachyose, is known at present as a natural product and has been reported from the roots of *Stachys* as well as from seeds of *Lupinus luteus, Soja hispida,* and *Ervum lens.* According to older authors this sugar is composed essentially of raffinose containing an added galactose unit.

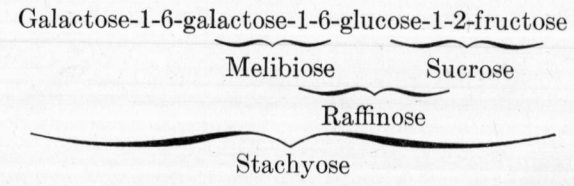

Galactose-1-6-galactose-1-6-glucose-1-2-fructose

Melibiose　　　Sucrose

Raffinose

Stachyose

More recent work has indicated however that the galactose-glucose linkage may be through carbon atoms 1 and 4, in which case stachyose would be unrelated to raffinose.

General References

Hudson, C. S., Advances in Carbohydrate Chemistry. Academic Press, 1946. Vol. 2, p. 1.

Pigman, W. W., and Goepp, R. M., Jr., Carbohydrate Chemistry. Academic Press, 1948.

References

1. Armstrong, E. F., *J. Chem. Soc.*, **83**, 1305 (1903).
2. Haworth, W. N., Long, C., and Plant, J., *J. Chem. Soc.*, 2809 **(1927)**.
3. Avery, O. T., Haworth, W. N., and Hirst E. L., *J. Chem. Soc.*, 2308 **(1927)**.
   Haworth, W. N., Hirst, E. L., and Learner, A., *J. Chem. Soc.*, 2432 **(1927)**.
4. Hassid, W. Z., Doudoroff, M., and Barker, H. A., *J. Am. Chem. Soc.*, **66**, 1416 (1944).

CHAPTER 4

TRANSFORMATION OF THE SUGARS

Phosphorylation of Hexoses. The initial step in many of the transformations to which the hexoses are subject is esterification with phosphoric acid. Four phosphorylated hexoses, two of which are glucopyranose derivatives and two fructofuranose derivatives, appear to occur univer sally in plant material and have been found whenever carefully sought The concentrations of these substances found in tissue vary from exceedingly minute to concentrations of several per cent of the dry weight of plant, as is shown in Table 4-1.

TABLE 4-1. Amounts of phosphorylated hexoses found in various plant materials

Species	Part	Treatment	Hexose phosphate content		
			Compound	%	Basis
Pisum sativum[6a]	Seed, meal	None	Hexose mono-phosphate	0.17	Air dry
			Fructose di-phosphate	0.14	" "
Pisum sativum[6a]	Seed, meal	Incubated 70 hr. with phosphate	Hexose mono-phosphate	1.18	" "
			Fructose di-phosphate	4.47	" "
Pisum sativum[6b]	Leaves	None	Hexose mono-phosphate	0.046	Fresh
Beta vulgaris[6c]	Leaves	None	Hexose mono-phosphate	0.034	"
Solanum tuber-osum[6d]	Tuber	Common storage	Glucose-6-phosphate	3.50	Dry
			Fructose-6-phosphate	0.17	"
Avena sativa[6e]	Coleoptile	None	Fructose-1,6-diphosphate	0.077	Fresh
			Fructose-6-phosphate	0.048	"

The phosphate esters of the hexoses are frequently referred in the older literature to the names of the biochemists who first discovered each. These names together with the chemically identifying terms used today are shown in Fig. 4-1. The role of phosphate in the transformations of hexoses was first appreciated by Harden and Young in 1905.[1]

These workers showed that during the fermentation of glucose by yeast preparations inorganic phosphate is used up and fructose diphosphate appears. The steps in the production of fructose diphosphate have since been studied in elaborate detail in yeast and in muscle by Emden, Meyerhof, Neuberg, Lohmann, Cori, and many others,[2] and the sequence of reactions discussed below appears to be essentially identical in all living tissues. The principal substrate for phosphorylation is glucose. In the presence of the enzyme hexokinase glucose is converted to glucose-6-

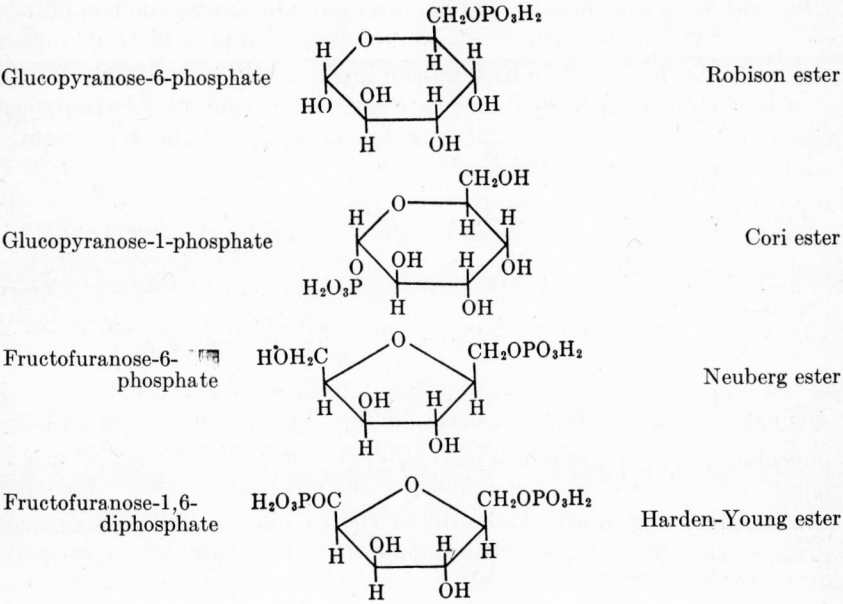

Glucopyranose-6-phosphate — Robison ester

Glucopyranose-1-phosphate — Cori ester

Fructofuranose-6-phosphate — Neuberg ester

Fructofuranose-1,6-diphosphate — Harden-Young ester

Fig. 4-1. Phosphate esters of the hexoses.

phosphate at the expense of phosphate contained in the compound adenosine triphosphate (ATP).[3] In the reaction ATP becomes adenosine diphosphate (ADP) and phosphate is transferred to glucose as follows:

$$\text{Glucose} + \text{ATP} \xrightleftharpoons{\text{Hexokinase}} \text{Glucose-6-phosphate} + \text{ADP}$$

The equilibrium position of this reaction lies far to the right, as can be sensed from the fact that whereas the pyrophosphate bond of ATP yields about 12,000 cal. per mol on hydrolysis, the phosphoric acid ester bond as in glucose-6-phosphate yields but approximately 3000 cal. per mol. ATP, whose mode of formation will be considered in Chapter 15, is hence a carrier of energy-rich phosphate, a source of phosphate available for the synthesis of ester phosphates. The enzyme hexokinase which is responsible for the transphosphorylation of glucose has been isolated in crystalline form from yeast[4,5] and study of the pure enzyme shows that in addition to ATP, the presence of Mg^{++} ions is required for activity. Hexokinase has not been studied extensively in higher plants but evidence of its activity has been found in the potato tuber, in spinach leaves, in seedlings, and in other tissues, and it is highly probable that the enzyme is of universal occurrence in higher plants.

Evidence to be presented in Chapter 30 indicates that hexose phosphates are formed as intermediates in the photosynthetic reduction of CO_2 and that the hexoses themselves are not intermediates in this process. The photosynthetic formation of hexose phosphate may then be a major source of these materials in green tissue.

Conversion of glucose-6-phosphate to glucose-1-phosphate takes place through the mediation of the enzyme phosphoglucomutase. In the equilibrium mixture approximately 95% of the total hexose phosphate is present as the 6-phosphate and approximately 5% as the 1-phosphate. Phosphoglucomutase activity is known to occur in extracts of pea seeds and the potato tuber and the enzyme is doubtless of wide occurrence in higher plants. Formation of fructose-6-phosphate from glucose-6-phosphate occurs in the presence of phosphohexoisomerase; this reaction is so far as known today the sole manner in which glucose and fructose may be interconverted. In an equilibrium mixture approximately 70% of the hexose is present as glucose-6-phosphate and 30% as fructose-6-phosphate. The enzyme has been studied in muscle, in pea meal and in potato tubers. No detailed analysis of the mechanism of the conversion can be given but it should be noted that a change from pyranose to furanose ring structure as well as an alteration in position of the carbonyl group is involved.

Fructose-1,6-diphosphate is formed from fructose-6-phosphate by reaction with ATP in the presence of the enzyme phosphohexokinase. Thus the conversion of glucose to fructose-1,6-diphosphate involves the utilization of two mols of ATP and the intervention of three enzymes. These reactions, which are summarized in Fig. 4-2, are the initial steps in the oxidation of hexose in the organism, whether by respiration or by fermentation. A discussion of their role in respiration will be found in Chapter 15.

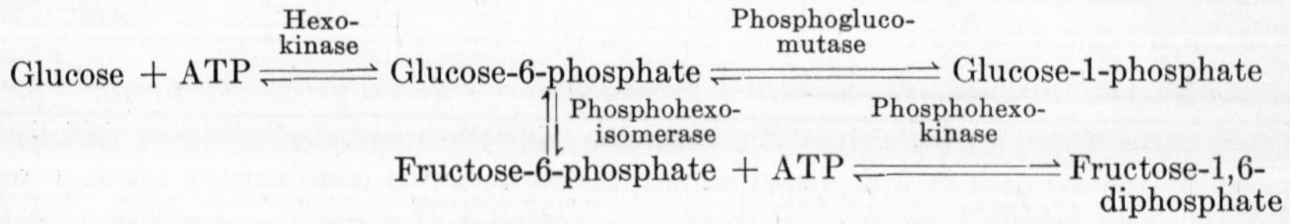

Fig. 4-2. Formation and interconversion of hexosephosphates.

In addition to the hexose phosphate esters, ribose and desoxyribose-phosphates are of general occurrence in combined forms, including in particular the nucleotides in which ribose or desoxyribose phosphate is combined to a purine or pyrimidine base through formation of an N-glycosidic linkage.

Enolization of Hexoses. It has sometimes been suggested that conversion of glucose to fructose might occur through the enol form common to glucose, fructose, and mannose. In mildly alkaline solution, D-glucose is spontaneously although slowly converted to an equilibrium mixture of D-glucose, D-fructose, and D-mannose, the reason being that these compounds arise through the formation of a common enol form, as shown below. At equilibrium the mixture contains approximately 63% glucose, 31% fructose and 2.4% mannose. The conditions for this spontaneous conversion are not, however, fulfilled in the plant organism where the pH is in general less than 7. An additional argument against the interconversion of these sugars by enolization *in vivo* is the fact that rarely if ever are the expected equilibrium amounts found in plant tissues, mannose in particular not occurring in the free form.

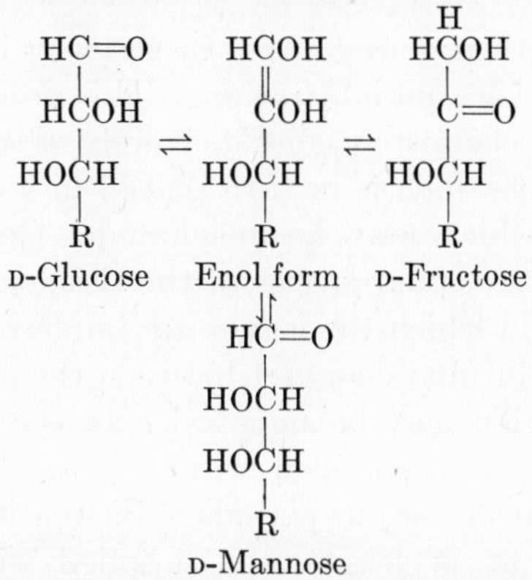

$$
\begin{array}{ccccc}
 & & & & \text{H} \\
\text{HC}{=}\text{O} & & \text{HCOH} & & \text{HCOH} \\
| & & \| & & | \\
\text{HCOH} & & \text{COH} & & \text{C}{=}\text{O} \\
| & \rightleftharpoons & | & \rightleftharpoons & | \\
\text{HOCH} & & \text{HOCH} & & \text{HOCH} \\
| & & | & & | \\
\text{R} & & \text{R} & & \text{R} \\
\text{D-Glucose} & & \text{Enol form} & & \text{D-Fructose}
\end{array}
$$

$$
\begin{array}{c}
\text{HC}{=}\text{O} \\
| \\
\text{HOCH} \\
| \\
\text{HOCH} \\
| \\
\text{R} \\
\text{D-Mannose}
\end{array}
$$

Interconversion of glucose, fructose, and mannose through the common enol form

Formation of Galactose. The conversion of glucose to galactose should involve inversion of the groups on carbon 4. It has been suggested that this might be accomplished by esterification and de-esterification at carbon atom 4 with a molecule such as phosphoric acid. This reaction, which would be of the type known as a Walden inversion, has not, however, been shown to occur either enzymatically or otherwise, and in addition no 4-phosphorylated sugars are known. Nevertheless the possibility that inversion at particular points within the sugar molecule may occur by such a mechanism involving either phosphoric or some other acid is an attractive one.

Formation of Sucrose. Glucose, fructose, and sucrose form the principal soluble sugars of the higher plant. We have seen that rapid interconversion of glucose and fructose in the plant is achieved through the transformation of the hexose-6-phosphates in the presence of the enzyme phosphohexoisomerase. The hexoses are in turn readily converted to sucrose. Direct experimental evidence for this conversion is provided by the fact that sections of leaves or other tissues floated on solutions of glucose or fructose accumulate not only hexose itself but also form sucrose. Experiments of this kind have been done with leaves of clover and wheat, sugar cane, barley, and with slices of potato tuber.[7] In the work of McCready and Hassid,[8] barley leaves were cut from the plant and allowed to respire in the dark for 24 hours to deplete them of

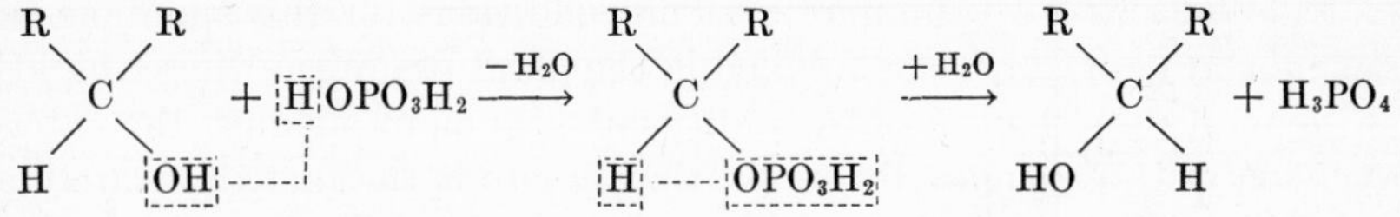

$$\underset{H}{\overset{R}{\diagdown}}\underset{OH}{\overset{R}{\diagup}}C \quad + \quad H|OPO_3H_2 \quad \xrightarrow{-H_2O} \quad \underset{H}{\overset{R}{\diagdown}}\underset{OPO_3H_2}{\overset{R}{\diagup}}C \quad \xrightarrow{+H_2O} \quad \underset{HO}{\overset{R}{\diagdown}}\underset{H}{\overset{R}{\diagup}}C \quad + \quad H_3PO_4$$

Fig. 4-3.   Possible mechanism of Walden inversion in conversion of one stereoisomer into another through intermediary of phosphorolysis.

sucrose. They were then placed either in water or in 10% glucose solution and subjected to vacuum infiltration. This procedure, an extremely useful one in plant biochemistry, consists merely of submerging the tissue in the desired solution which is in turn placed in a container, such as a vacuum desiccator, which may be evacuated. On evacuation, gas is removed from the intercellular spaces of the tissue and on release of the vacuum the solution in which the leaves are immersed is drawn into the tissue. In this way, intimate contact between the cells of a plant tissue and any desired solution may be achieved. In the present experiment, the leaves were placed in distilled water after infiltration and were analyzed for sucrose after various periods of incubation. The data given in Fig. 4-4 show that in leaves not supplied with glucose, sucrose disappears completely within about 20 hours. In leaves supplied with glucose, however, the sucrose content reaches a level of approximately 6% of the dry weight in the same period. Similar results were achieved after infiltration with fructose. Smaller accumulations of sucrose were obtained at the expense of galactose and mannose. These results demonstrate qualitatively then that any of the four sugars may be transformed in the leaf into the glucose and fructose necessary for sucrose synthesis. They also show that an effective mechanism for sucrose production from hexose is present in the leaf. The nature of this mechanism has been the object of intensive investigation. The enzyme invertase or β-fructosidase which hydrolyzes sucrose to its component monosaccharides is widely distributed in plant material and the reaction catalyzed by this enzyme has been thoroughly characterized. In this reaction the equilibrium lies far toward the side of hydrolysis and since reversal of the reaction to a detectable extent has not been achieved, it does not seem possible that invertase participates in sucrose synthesis in the plant. Although intact leaves or other tissues readily form sucrose from monosaccharides, the same reaction will not take place *in vitro* with preparations made from higher plants. Grinding or other disruption of the tissue thus appears to inactivate the sucrose synthesizing mechanism.

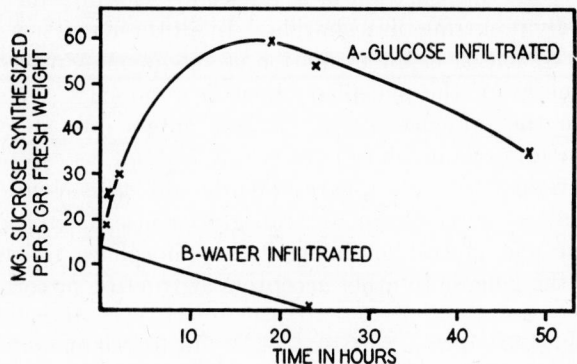

FIG. 4-4. Sucrose synthesis in isolated barley leaves floated on 10% glucose solution. (After McCready and Hassid[8])

That the synthesis of sucrose actually takes place through the intermediary of a phosphorylative mechanism has been shown by the work of Hassid, Doudoroff, and Barker[9] with a microorganism, *Pseudomonas saccharophila*. From the dried cells of this organism an enzyme sucrose phosphorylase can be extracted and purified by fractional ammonium sulfate precipitation. In the presence of sucrose phosphorylase, glucose-1-phosphate and fructose react to form sucrose and inorganic phosphate. Both glucose-1-phosphate and inorganic phosphate enter into the equilibrium as the divalent anions. The equilibrium constant for the reaction

$$\text{Glucose-1-phosphate} + \text{fructose} \rightleftharpoons \text{Sucrose} + H_3PO_4$$

varies with pH owing to the fact that the dissociation constants of glucose-1-phosphate and inorganic phosphate are not identical. Sucrose phosphorylase is specific for glucose-1-phosphate and is unable to utilize

$$K = \frac{(\text{Sucrose})(\text{HPO}_4^=)}{(\text{Glucose-1-phosphate})(\text{Fructose})} = \begin{cases} 0.05 \text{ at } p\text{H } 6.6 \\ 0.09 \text{ at } p\text{H } 5.8 \end{cases}$$

other hexose phosphates. Thus fructose phosphate and glucose cannot replace glucose phosphate and fructose in the synthesis. The enzyme is also unable to combine glucose-1-phosphate with fructose-6-phosphate or fructose-1,6-diphosphate. The presence of free glucose inhibits the reaction between glucose-1-phosphate and fructose, presumably owing to a competition between glucose-1-phosphate and glucose for the enzyme. The specificity is not complete however, since fructose may be replaced by either sorbose or ketoxylose, either of which is able to react with glucose-1-phosphate in the presence of sucrose phosphorylase. In the first case a glucosido-sorboside analog of sucrose is formed, a sugar which is unknown in nature. Similarly, reaction of glucose-1-phosphate with ketoxylose results in formation of a glucosido-xyloside which is likewise a non-naturally occurring disaccharide. In all these reactions, the initial step appears to consist of the formation of an enzyme-glucose complex in which the energy of the phosphate bond is retained. Thus glucose-1-phosphate in the presence of the enzyme enters into an exchange of phosphate with inorganic phosphate in the medium, as can be shown by the use of radioactive phosphate. We might thus consider sucrose phosphorylase as a sort of a glucose transglycosidase, capable of removing glucose from one glycosidic linkage (the phosphate linkage) and of transferring the glucose to other acceptors as fructose or sorbose.

$$\text{Glucose-1-phosphate} + \underset{\text{phosphorylase}}{\text{sucrose}} \rightleftharpoons \underset{\text{glucose complex}}{\text{Sucrose phosphorylase-}} + H_3PO_4$$

Pseudomonas, the organism in which sucrose phosphorylase was first discovered, contains invertase in addition to the phosphorolytic system. Thus there is present in this organism at least one mechanism for the synthesis of sucrose and at least two mechanisms for its degradation. In the hydrolytic system the equilibrium is too far toward the side of hydrolysis to permit of appreciable sucrose synthesis. In the phos-

$$\text{Sucrose} + H_3PO_4 \underset{\phantom{\text{Sucrose}}}{\overset{\text{Sucrose}}{\underset{\text{phosphorylase}}{\rightleftharpoons}}} \text{Glucose-1-phosphate} + \text{fructose}$$
$$\begin{array}{l} + \\ H_2O \\ \downarrow \text{Invertase} \\ \text{Glucose} + \text{fructose} \end{array}$$

phorolytic system on the contrary either synthesis or breakdown may occur, depending on the concentrations of the several reactants. From an energetic standpoint we may say that the energy needed for the formation of the glycosidic bond in sucrose is derived from the energy originally expended in the phosphorylation of glucose with the formation of glucose phosphate. This energy requirement of the initial phosphorylation may be in turn derived from the oxidation of carbohydrate through the intermediary of ATP (Chapter 15). Hexose phosphates may, as indicated earlier, also arise directly as early photosynthetic products and the evidence presented in Chapter 30 suggests that sucrose may be formed in green tissues directly from photosynthetically produced hexose phosphate without the intermediary of the free glucose itself. The sucrose phosphorylase of Doudoroff has not yet been found in higher plants although an intensive search has been made for the enzyme in potato tubers and in other tissues. It is of course possible that still other mechanisms of sucrose synthesis remain to be discovered. That sucrose synthesis does have a direct relation to phosphorylation is, however, shown by the work of Kursanov and Kryukova[11] who have found that leaves of phosphate-deficient plants lose the power to synthesize sucrose from hexose. When phosphate is restored the synthetic power is regained. It seems entirely possible that a sucrose phosphorylase generally similar to that of *Pseudomonas* may be found in the higher plant if appropriate conditions for its isolation can be discovered.

Oxidation of Hexoses. In respiration the hexose molecule is dismembered, first to C_3 units and finally to CO_2 and water. Hexose is also subject, however, to oxidation without dismemberment. Thus the oxidation product of glucose at carbon 6, D-glucuronic acid, is a common constituent of gums and mucilages and presumably arises by direct oxidation of glucose.

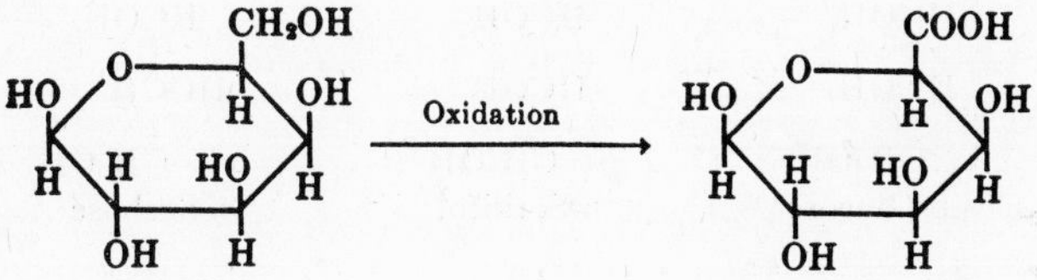

D-Glucose   D-Glucuronic acid

D-Galacturonic acid, which may arise from oxidation of galactose at carbon atom 6, is a component of pectin and of gums and is like D-glucuronic acid a widespread plant product; the corresponding D-mannuronic acid is found as a cell wall constituent of kelp. In none of these cases has any enzyme system for the oxidation been discovered nor has direct proof that the obvious reaction is indeed the actual one been brought forward.

The products of oxidation of sugars at the aldehydic end of the molecule are rarely if ever encountered in higher plants. Thus gluconic acid is not found in plant products, although an enzyme for oxidation of glucose-6-phosphate to phosphogluconic acid has been described in blood. *Aspergillus* and other fungi, as well as many bacteria, contain glucose oxidase or glucose dehydrogenase systems which oxidize glucose itself to gluconic acid. In this reaction H_2O_2 is formed as an additional product. Glucose oxidase is not, however, known in higher plant tissues. Sac-

$$C_6H_{12}O_6 + O_2 + H_2O \xrightarrow[\text{oxidase}]{\text{Glucose}} C_6H_{12}O_7 + H_2O_2$$

Glucose   Gluconic
acid

charic acid, in which both carbon atoms 1 and 6 of D-glucose are oxidized to carboxyl groups can readily be prepared by the action of nitric acid on glucose. Neither saccharic acid nor other six-carbon acids of this type are, however, known as natural plant products.

Reduction of Sugars. There are in nature several sugar alcohols which may be regarded as reduction products of the hexoses and direct evidence is available to show that they are actually produced in this manner by the action of microorganisms. Thus the alcohol D-sorbitol, which is the reduction product of D-glucose, is found in many rosaceous plants, and occurs in particularly high concentrations in fruits including those of the apple, peach, apricot, cherry, and of *Sorbus*. Sorbitol may be reoxidized by various microorganisms of the *Acetobacter* group.

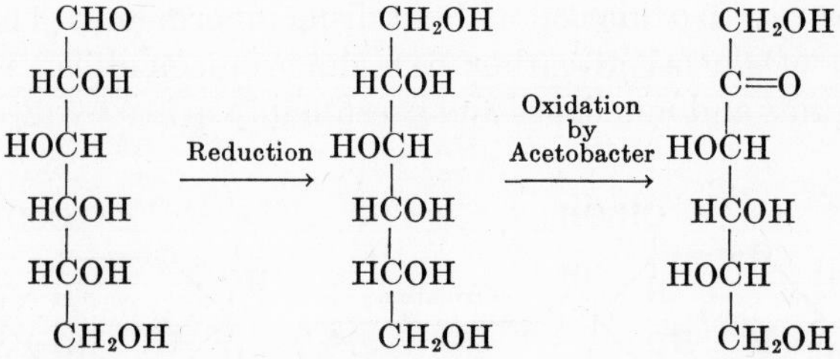

CHO   CH₂OH   CH₂OH

HCOH   HCOH   C=O

HOCH Reduction HOCH Oxidation by Acetobacter HOCH

HCOH   HCOH   HCOH

HCOH   HCOH   HOCH

CH₂OH   CH₂OH   CH₂OH

D-Glucose   D-Sorbitol   L-Sorbose

In the reoxidation however the ketose sugar, L-sorbose, is formed. D-Mannitol, the common reduction product of D-mannose and D-fructose, is of widespread occurrence in higher plants (onion, carrot, parsnip, for example) and occurs occasionally as a crystalline exudate on the bark of olive and ash trees (*Fraxinus ornus*). Mannitol is also formed by microorganisms and during the fermentation of corn leaves in silage, may be formed to the extent of 0.5–3% of the weight of silage. Dulcitol, the reduction product of galactose, is of only rare occurrence but is found in the red algae as well as in the higher plants *Melampyrum nemorosum* and *Euonymus atropurpureus*.

Reduction of a ketose sugar results in the formation of two possible isomers. Thus fructose yields D-mannitol and D-sorbitol which differ only in their configuration at carbon atom 2. It is perhaps noteworthy that the two principal hexose alcohols are thus derivable from the two principal free hexose sugars of the plant.

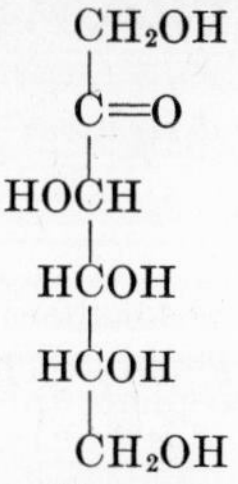

$$\begin{array}{ccc}
CH_2OH & CH_2OH & CH_2OH \\
| & | & | \\
C=O & HCOH & HOCH \\
| & | & | \\
HOCH & HOCH & HOCH \\
| & | & | \\
HCOH & HCOH & HCOH \\
| & | & | \\
HCOH & HCOH & HCOH \\
| & | & | \\
CH_2OH & CH_2OH & CH_2OH \\
\text{D-Fructose} & \text{D-Sorbitol} & \text{D-Mannitol}
\end{array}$$

Sugar alcohols corresponding to sugars other than the hexoses are found as scattered occurrences through the higher plants. Erythritol, a four-carbon sugar alcohol, has been found in grasses, while ribitol, the reduction product of D-ribose, is found in the free form in *Adonis vernalis*. Ribitol is in addition universally found in plant tissues combined as the isoalloxazine ribitol derivative, riboflavin or vitamin B_2. The functions of riboflavin will be discussed in Chapter 15. The reduction product of

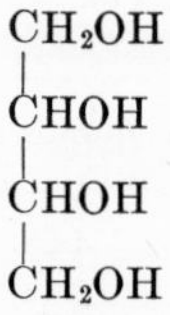

$$\begin{array}{cc}
CH_2OH & CH_2OH \\
| & | \\
CHOH & CHOH \\
| & | \\
CHOH & CHOH \\
| & | \\
CH_2OH & CHOH \\
 & | \\
 & CH_2OH \\
\text{meso-Erythritol} & \text{meso-Ribitol}
\end{array}$$

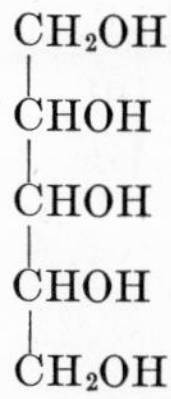

D-mannoketoheptose, D-perseitol, accompanies the seven-carbon sugar in fruits of avocado (*Persea gratissima*).

Inositol. Inositol is a six-carbon cyclic sugar alcohol, and although the term is often used to designate the most widely distributed isomer, there are strictly speaking a number of inositols as well as several inositol derivatives which are found in higher plants. The structures and structural relationships of these compounds are shown in Table 4-2.

Meso-inositol is probably universally distributed in higher plants, being found in concentrations of the order of 1–10 mg. per gram dry weight, and is probably also a universal constituent of other living forms. It is a vitamin of the B complex for higher animals and is required as an exogenous (added) growth factor for several microorganisms. The demonstration by Eastcott in 1928[12] that meso-inositol is the bios I factor of yeast marked, in fact, the first step in elucidation of microbial growth factors of the B group. No specific physiological role can yet be assigned to meso-inositol in higher plants. It is found, however, not only in the free form but also in a variety of derivatives. Thus the hexaphosphoric

TABLE 4-2

Compound	Naturally occurring derivatives

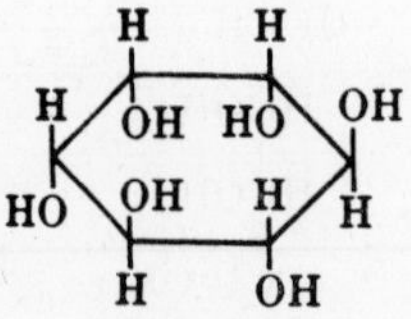

meso-Inositol

Monomethyl meso-inositol (bornesitol)
Monomethyl meso-inositol (sequoyitol)
meso-Inositol hexaphosphate (phytic acid)

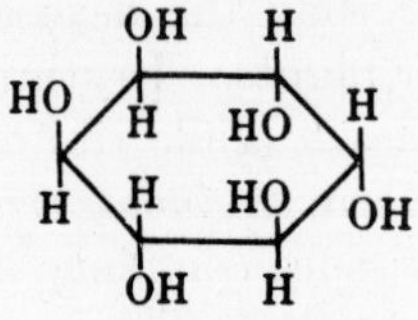

ᴅ'-Inositol

Monomethyl-ᴅ-inositol (pinitol)

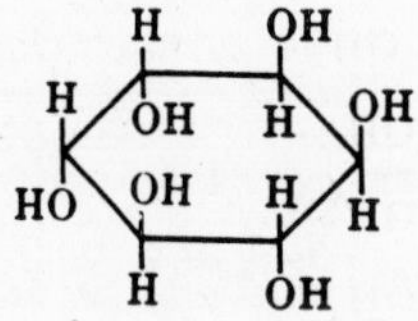

ʟ-Inositol

Monomethyl-ʟ-inositol (quebrachitol)

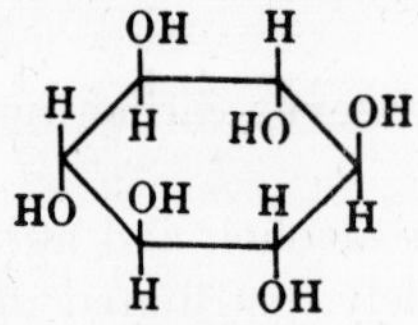

Scyllitol

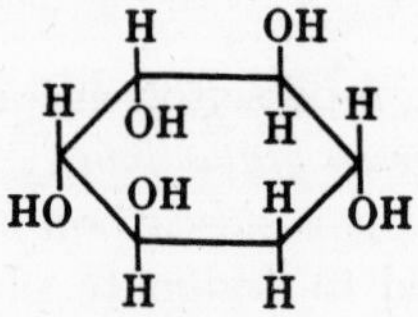

ᴅ-Quercitol

acid ester of meso-inositol, phytic acid, is found in seeds and seedlings and apparently represents a reserve form of phosphate in the seed. The liberation of inorganic phosphate from phytic acid is accomplished by the enzyme phytase which is found in seeds and seedlings as well as in mature plant tissues. Whether phytase is a distinct enzyme or whether phytase activity is merely one attribute of phosphatases in general is not yet wholly clear, although purified leaf phosphatase does appear to possess phytase activity. meso-Inositol is also found as the alcohol of complex lipids, the lipositols, which will be described in Chapter 24. Among the simpler derivatives of meso-inositol are two different monomethyl ethers, bornesitol and sequoyitol. The former is found as an impurity in Borneo rubber, the latter as an exudate on the bark of the redwood *Sequoia sempervirens.*

D-Inositol is very much less widely distributed than meso-inositol and is in fact mainly confined to occurrence as the monomethyl ether pinitol which is found (*a*) as an exudate on the bark of *Pinus lambertiana* ("sugar" pine) and other pines, and (*b*) as a latex constituent of Madagascar rubber. L-Inositol is found in nature as the monomethyl ether, quebrachitol, of which the two best known sources are the latex of the rubber tree, *Hevea brasiliensis,* and the bark of the quebracho tree. In *Hevea,* quebrachitol can apparently be metabolized since the amounts recovered in the latex decrease during starvation (for example after defoliation) of the tree. Scyllitol is a constituent of acorns, dogwood leaves, and of the coconut palm but is primarily known from fishes including the dogfish, sharks and others. Quercitol, a desoxy inositol, is a six-carbon cyclitol containing but five hydroxyl groups, and is found in the oak, especially in the acorn.

Two further inositol derivatives which contain carboxyl groups are quinic acid and shikimic acid.

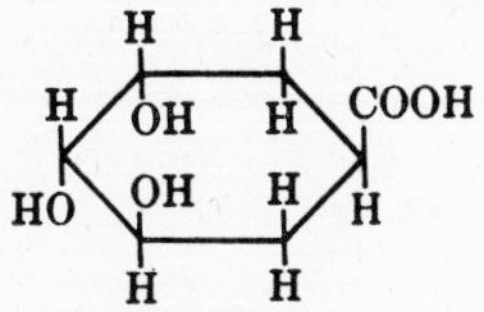

D-Quinic acid 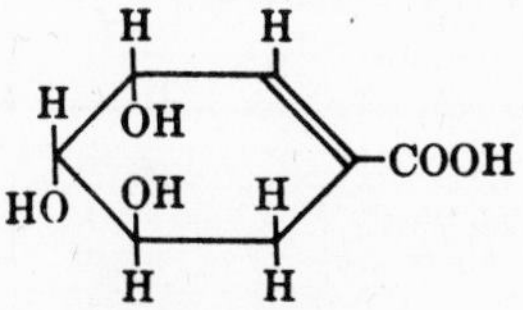   Shikimic acid

Quinic acid appears to be of moderately widespread occurrence in plants, being found in small amounts in many leaves and fruits. Shikimic acid on the other hand is reported only from species of *Illicium.*

It is possible that the inositols are formed by cyclization of the appropriate hexose, and D-glucose would be the hexose appropriate for synthesis of meso-inositol, quantitatively the most important isomer. There is in any case no direct *in vivo* or *in vitro* evidence that such a synthesis does take place in nature, attractive as the possibility might seem.

Ascorbic Acid. Ascorbic acid, vitamin C, is a sugar derivative of nearly universal occurrence in plant tissues but whose function in the plant is still obscure. Ascorbic acid was first isolated by Szent-Györgyi[13] from various higher plant tissues, and identity of the compound isolated with vitamin C was suggested by the same author. Structure determination followed by synthesis[14] led to the conclusion that ascorbic acid is a six-carbon lactone with the lactone ring closed between the first and fourth carbon atoms. Structurally ascorbic acid is a derivative of the ketose L-sorbose and may in fact be synthesized from sorbose.

$$
\begin{array}{cccc}
\text{CH}_2\text{OH} & \text{COOH} & \text{O}=\text{C} & \text{O}=\text{C} \\
\text{C}=\text{O} & \text{C}=\text{O} & \text{HOC} & \text{HOCH} \\
\text{HOCH} & \text{HOCH} & \text{HOC} & \text{O}=\text{C} \\
\text{HCOH} \xrightarrow[+\text{H}_2\text{O}]{-4\text{H}} & \text{HCOH} \xrightarrow{-\text{H}_2\text{O}} & \text{HC} & \rightleftharpoons \quad \text{HC} \\
\text{HOCH} & \text{HOCH} & \text{HOCH} & \text{HOCH} \\
\text{CH}_2\text{OH} & \text{CH}_2\text{OH} & \text{CH}_2\text{OH} & \text{CH}_2\text{OH} \\
\text{L-Sorbose} & \text{2-Keto-L-gulonic} & \text{L-Ascorbic} & \text{L-Ascorbic} \\
 & \text{acid} & \text{acid} & \text{acid,} \\
 & & & \text{keto form}
\end{array}
$$

Structural relationships of L-sorbose and L-ascorbic acid

In aqueous solution an equilibrium appears to be established between the enol forms of ascorbic acid and a 3-keto form, a fact which confers on ascorbic acid some ketonic properties. Ascorbic acid may also be reversibly oxidized by removal of two hydrogen atoms from carbon atoms 3 and 4 with the formation of dehydroascorbic acid, an oxidation carried out by the enzyme ascorbic acid oxidase as described in Chapter 15 and catalyzed also by inorganic copper itself.

$$
\begin{array}{cc}
\text{O}=\text{C} & \text{O}=\text{C} \\
\text{HOC} & \text{O}=\text{C} \\
\text{HOC} \xrightleftharpoons[-2\text{H}]{+2\text{H}} & \text{O}=\text{C} \\
\text{HC} & \text{HC} \\
\text{HOCH} & \text{HOCH} \\
\text{CH}_2\text{OH} & \text{CH}_2\text{OH} \\
\text{Ascorbic acid} & \text{Dehydroascorbic acid}
\end{array}
$$

In plant tissue a variable proportion of the total ascorbic acid is found in this reversibly oxidized state which has lead to the suggestion that ascorbic acid somehow functions as an oxidation-reduction carrier.[15] Oxidation of ascorbic acid beyond the stage of dehydroascorbic acid results in destruction of the compound with loss of vitamin C activity.

General References

Fletcher, Hewitt G., Jr., Advances in Carbohydrate Chemistry. Academic Press, 1947. Vol. 3, p. 45. ·

Guzman Barron, E., Advances in Enzymology. Interscience Publishers, 1943. Vol. 3, p. 149.

Harden, A., Alcoholic Fermentation. Longmans, Green and Company, 1923.

Knight, B. C. J. G., Vitamins and Hormones. Academic Press, 1945. Vol. 3, p. 105.

Onslow, M., Principles of Plant Biochemistry. Cambridge University Press, 1931.

Sumner, J. B., and Somers, G. F., Chemistry and Methods of Enzymes. Academic Press, 2nd ed., 1947.

References

1. Harden, A., and Young, W. J., *J. Chem. Soc.*, **21**, 189 (1905).
2. See review in Kalckar, H. M., *Chem. Revs.*, **28**, 72 (1941).
3. Lohmann, K., *Naturwissenschaften*, **17**, 624 (1929).
4. Berger, L., Slein, M. W., Colowick, S. P., and Cori, C. F., *J. Gen. Physiol.*, **29**, 141 (1946).
5. Kunitz, M., and McDonald, M. R., *J. Gen. Physiol.*, **29**, 143 (1946).
6a. Tankó, B., *Biochem. J.*, **30**, 692 (1936).
  b. Hassid, W. Z., *Plant Physiol.*, **13**, 641 (1938).
  c. Burkhard, J., and Neuberg, C., *Biochem. Z.*, **270**, 229 (1934).
  d. Arreguin, B., and Bonner, J., *Plant Physiol.*, **24**, 720 (1949).
  e. Bonner, J., *Arch. Biochem.*, **17**, 311 (1948).
7. See for example: Virtanen, A. I., and Nordlund, M., *Biochem. J.*, **28**, 1729 (1934).
8. McCready, R. M., and Hassid, W. Z., *Plant Physiol.*, **16**, 599 (1941).
9. Hassid, W. Z., Doudoroff, M., and Barker, H. A., *J. Am. Chem. Soc.*, **66**, 1416 (1944).
10. Doudoroff, M., *Federation Proc.*, **4**, 241 (1945).
  Hassid, W. Z., Doudoroff, M., Barker, H. A., and Dore, W. H., *J. Am. Chem. Soc.*, **68**, 1465 (1946).
11. Kursanov, A. L., and Kryukova, N. N., *Biokhimiya*, **4**, 229 (1939).
  Syssakyan, N. M., *Biokhimiya*, **1**, 301 (1936).
12. Eastcott, Edna, *J. Phys. Chem.*, **32**, 1094 (1928).
13. Szent-Györgyi, A., *Biochem. J.*, **22**, 1387 (1928).
14. Herbert, R. W., Percival, E. J., Reynolds, R. J. W., Smith, F., and Hirst, E. L., *J. Soc. Chem. Ind.*, **52**, 221, 482 (1933).
15. James, W. O., and Cragg, J. M., *New Phytologist*, **42**, 28 (1943).

CHAPTER 5

STARCH

Structure of Starch. Starch is the most abundant reserve carbohydrate of the plant world and is found under favorable conditions and in
higher or lower concentrations in most organs of most higher plants, as
well as in many microorganisms. It is of interest therefore that despite
the quantitative importance of starch and despite the great deal of
chemical attention paid to the substance, it is only in recent years that
any clear understanding of its structure has been attained.

Starch on acid hydrolysis gives a quantitative yield of D-glucose.
Enzymatic hydrolysis with amylase yields the disaccharide maltose and
proof that maltose type linkages are involved in the intact starch molecule
was early obtained by Haworth and others.[1] Haworth methylated
starch, using an intermediary acetylation with acetic anhydride followed
by methylation with dimethyl sulfate. When methylated starch was
subjected to acid hydrolysis, 2,3,6-trimethylglucopyranose was isolated
as the main product. Under very mild conditions of hydrolysis considerable yields of methylated maltose were also obtained. These facts
establish that the starch molecule must consist primarily of chains of
glucopyranose units linked through α-glycosidic linkages as in maltose:

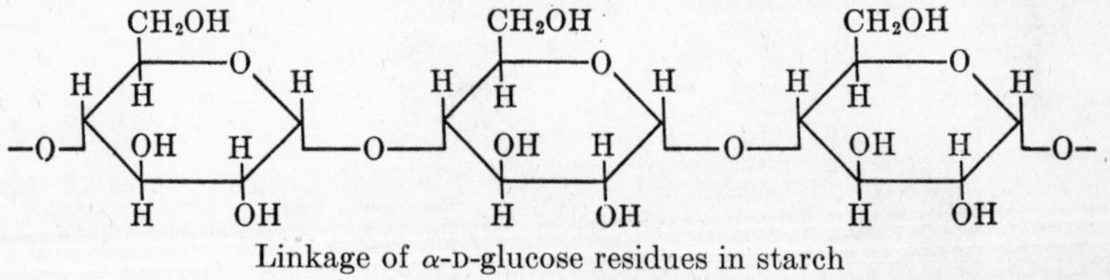

Linkage of α-D-glucose residues in starch

In addition to the trimethylglucose obtained from hydrolysis of
methylated starch, a smaller amount of 2,3,4,6-tetramethylglucose is also
recovered. This must evidently arise from the terminal units of the
starch chains in which carbon atom 4 is free for methylation. The ratio
of tetramethyl to trimethylglucose recovered in the hydrolysis products
provides then a measure of the number of glucose residues linked together
in the starch chain; this means of estimating the chain length is known as
the *end group* assay method.[2] A further general method of end group
assay is periodate oxidation as applied by Hirst to polysaccharides.[3] In

48

the presence of KCl and sodium periodate, oxidation of polysaccharides such as starch or cellulose is confined to the terminal residues which are oxidized with the liberation of formic acid, as illustrated in Fig. 5-1. One molecule of formic acid is produced from oxidation of one nonreducing terminal group, while two molecules of formic acid are produced from each reducing terminal residue. The excess periodate ions remaining after the oxidation may be decomposed by addition of ethylene glycol and the formic acid may be simply determined by titration. Periodate oxidation provides a quick and highly useful method for determination of the number of end groups contained in a polysaccharide and is particularly valuable in cases where only small samples of material are available.[4,5] The chain length of the starch molecule as obtained by these methods

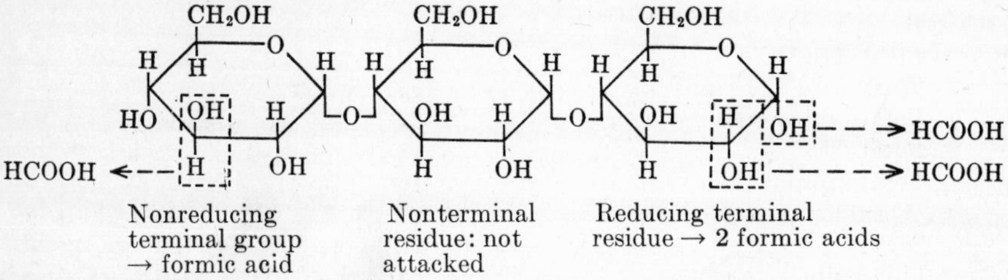

Nonreducing terminal group → formic acid

Nonterminal residue: not attacked

Reducing terminal residue → 2 formic acids

Fig. 5-1. Attack of terminal groups of starch by periodate.

depends on the exact fraction of the starch under consideration, a matter which will now be discussed.

It has long been appreciated that starch consists of two components which differ in physical properties. The two starch constituents are commonly known as amylose, which is more water soluble and less viscous in solution, and amylopectin, less water soluble and more viscous in solution. The two components can be separated by allowing whole starch grains to swell in water at 60–80°C. The amylose then diffuses out and the grains containing the residual amylopectin may be centrifuged off. A second method for separation of the two components consists in precipitation of crystalline amylose from aqueous starch solutions saturated with butanol,[6] or certain other organic solvents. In this case the amylose appears to form insoluble molecular complexes with the precipitant.

Amylose is also primarily responsible for the blue iodine reaction of starch, the iodine color given by a given amount of amylose being approximately six times as intense as the iodine color given by an equal amount of amylopectin. Since amylose, in contrast to amylopectin, takes up iodine quantitatively and stoichiometrically, it is possible to determine the amylose content of whole starch by iodometric titrations. Colorimetric methods for determination of amylose based on the intense color of the amylose-iodine complex have also been proposed.

Amylose makes up 0–35% of whole starch,[7] the proportion varying with the species considered, as is summarized in Table 5-1. It is of special interest that whereas the seed starch of ordinary field corn contains amylose, plants homozygous for the recessive gene *waxy* produce amylose-free seed starch.[7] Similar genes are known in other species and amylose-free seed starch is produced by *waxy* rice, sorghum, and barley. The production of amylose appears then to be controlled by a single gene.

TABLE 5-1. Proportion of amylose in starch of various species[7]

Species	% amylose in starch	Chain length of amylose
Tapioca	17	980
Rice	17	...
Banana	20.5	...
Field corn	22	490
Waxy corn	0	...
Potato tuber	22	980
Potato leaf	18	...
Potato sprouts	46	...
Wheat	24	540
Sago	27	420
Lily, bulb	34	640

Application of the end group assay to amylose shows that the glucose chains of this material contain of the order of 300 to 1000 glucose residues. The molecular weights of amylose as obtained by osmotic pressure measurements further indicate that each molecule must consist of one such chain of glucose residues. Amylose consists therefore of long glucopyranose chains containing as many as 300–1000 glucose residues, with no association of the individual chains into larger complexes.

Application of the end group assay method to amylopectin gives an apparent chain length of 24–30 glucose residues for all starches. The molecular weight of amylopectin as determined by osmotic pressure or ultracentrifuge measurements gives, however, values indicating that a great many more than 30 residues are combined in one unit, the molecules of amylopectin being actually much larger than those in amylose. The nature of the association of the 24–30 residue units into larger aggregates has therefore been a subject for lively discussion. It was early suggested that chains 24–30 glucose residues in length might be associated by secondary valences into small crystalline units or micelles. This idea seems unlikely in view of the fact that no disaggregation of the amylopectin molecules follows acetylation, which would be expected to change radically the binding forces between molecules. Haworth, Hirst, and Isherwood in 1937 brought forth an important new fact in the finding that considerable amounts of 2,3-dimethylglucose are found among the hydrolysis products of methylated starch, together with tri- and tetramethylglucose.[8] The occurrence of this dimethylglucose can be explained on the basis of a branched-chain structure in which carbon atoms 1 and 4 of the glucose residue at the branch point would be involved in the main chain, while carbon atom 6 of the same residue would be involved in formation of the bond to the branch chain. The assumption of a branched-chain structure for amylopectin would also explain the absence of reducing groups in starch, since the aldehydic carbon atom 1 contained in the terminal residue of the branch chain would be involved in formation of the 1,6 branch linkage.

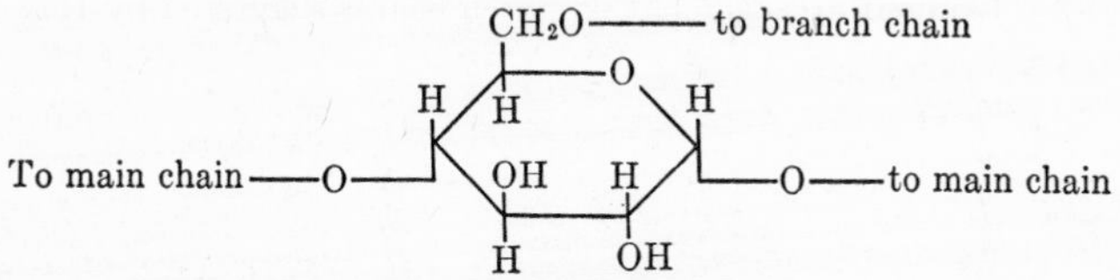

Structure of D-glucopyranose residue of amylopectin involved in branching at carbon atom 6

The early Haworth-Hirst concept of amylopectin structure is then that chains 24–30 glucoses in length are attached to one another through 1,6 linkages.[9] As many as 80 chains might be united in one large molecule, according to molecular weight determinations based on osmotic pressure measurements. Staudinger and Husemann suggest, in contrast to the views of Haworth, that the starch molecule may consist of a central chain of glucose units from which side chains protrude at intervals of several glucose units.[10] The branches would have then to be of the appropriate length so that the structure as a whole would satisfy the overall requirement of one end group per 24–30 glucose residues.

Still another view has been advanced by Meyer.[11] This structure is one in which there is no central backbone chain but in which there is a free and multiple branching of short chains. Meyer's concept is based primarily on work with β-malt amylase, an enzyme which converts amylose quantitatively to maltose, i.e., can apparently attack straight chains consisting of any number of glucose residues linked exclusively through 1,4 linkages. Amylopectin is, however, converted by β-malt amylase to a mixture of maltose (54% of theoretical yield) and a resistant high-molecular-weight residue or residual dextrin. This residue represents, according to Meyer, the central skeleton of amylopectin together with the stubs of the branches, the remainder of the branches having been hydrolyzed off as maltose. A strong support of this interpretation is the fact that the residual dextrin does yield a higher proportion of dimethylglucose on hydrolysis than the original amylopectin. Comparison of data on hydrolysis of methylated amylopectin with that obtained from hydrolysis of the methylated residual dextrin shows that the free portions of the branches must contain 15–18 glucose residues and that these are

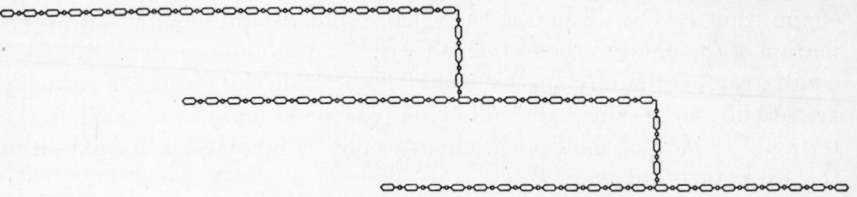

FIG. 5-2. Laminar arrangement of starch chains suggested by Haworth.

FIG. 5-3. Staudinger proposal for structure of starch. The branches are attached to the main chain by 1,6 linkages.

spaced 8–9 residues apart. The concept of amylopectin structure suggested by Meyer is thus in general accord with the presently known facts obtained both from enzymatic and from chemical studies. By similar arguments, Meyer arrives at the conclusion that glycogen, the reserve

carbohydrate of yeast and animals, consists of a multiple-branched structure with the free branch ends 6–7 glucose residues in length.

Starch consists then of chains of glucopyranose residues linked
together through 1,4 β-glucosidic linkages. These chains may be either

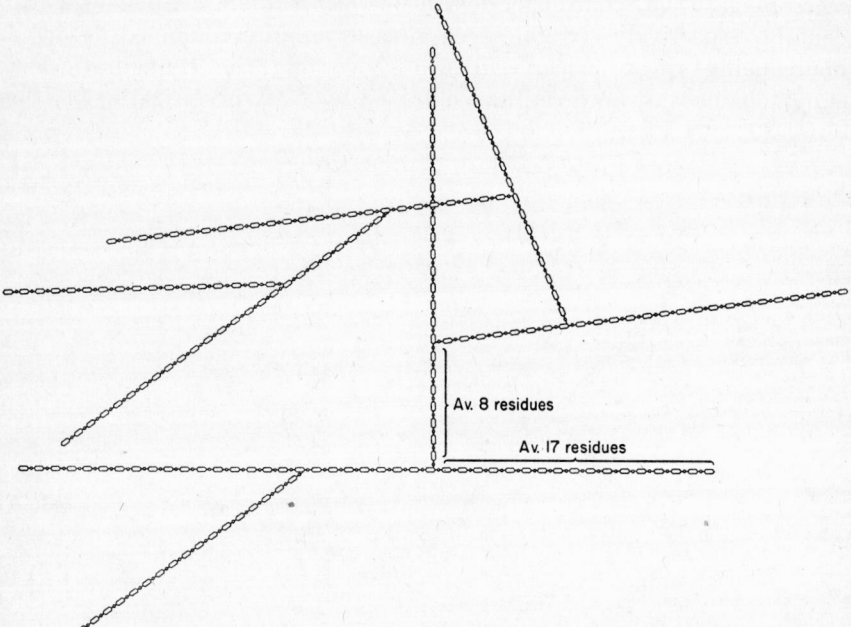

Av. 8 residues

Av. 17 residues

FIG. 5-4. Multiple branched chain of starch according to Meyer.

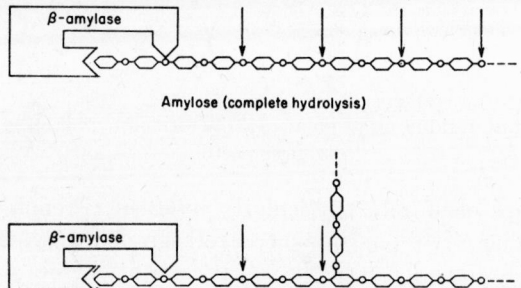

β-amylase

Amylose (complete hydrolysis)

β-amylase

Amylopectin (hydrolysis to branch only)

FIG. 5-5. Mode of attack of amylose and amylopectin by β-amylase. (After Hassid)

unbranched (amylose) or branched (amylopectin). How are these glucose residues arranged in space along the chain axis? Evidence on this
point comes from several sources. The enzyme α-malt amylase splits
starch rapidly into dextrins containing six or twelve glucose units. Hanes

has proposed,[12] principally on this basis, that in starch the chains of glucose residues are looped in wide spirals such that six glucose units constitute one complete coil of the spiral. Some support for this view is the fact that starch is converted by the microorganism *Aerobacillus macerans* into the Schardinger dextrins, two crystalline compounds which contain respectively six and seven glucose residues combined in closed nonreducing rings. Clear evidence concerning the configuration of starch chains was, however, first obtained by x-ray investigations of the

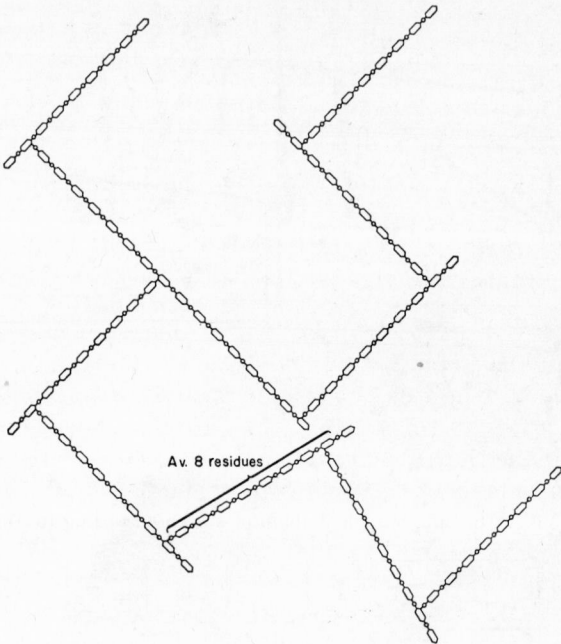

Av. 8 residues

FIG. 5-6. Resistant residue or residual dextrin formed by action of β-malt amylase on amylopectin.

crystal structure of starch, particularly by Bear, French, Rundle, and co-workers. These workers first used retrograded amylose, which had spontaneously become insoluble in and precipitated from water solution. X-ray diffraction examination of films made from such starch leads to the conclusion that retrograded amylose consists of straight or extended chains of amylose arranged in a parallel fashion.[13] Native amylose which has been obtained by butanol precipitation and which is not in the retrograded state gives, on the contrary, an x-ray diffraction pattern consistent with the view that the chains are coiled in the form of helices in which each coil of the helix is made up of six glucose residues.[14] Further confirmation of the helical structure of native amylose as well as information on the arrangement of the helices in crystalline butanol-precipitated amylose can be obtained from the study of the starch-I_2 complex.[14,15]

Iodine Complexes of Starch. It has been noted above that starch possesses the ability of forming colored complexes with iodine. Amylose

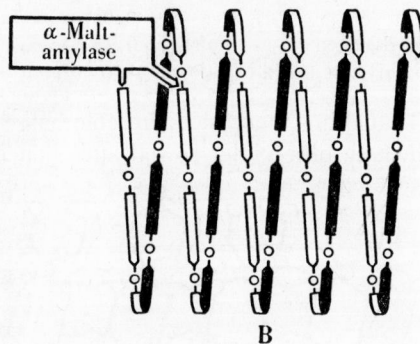

FIG. 5-7. Hypothetical spiral starch model, showing possible manner by which it is attacked by α-malt amylase. (After Hanes[12])

yields an intensely blue iodine complex and amylopectin a much less intensely colored blue-violet complex. Two factors which influence the iodine color given by a particular starch are the length of glucose chain involved and the degree of branching of the chain. Hanes has shown that dextrins obtained as intermediate products in starch hydrolysis by

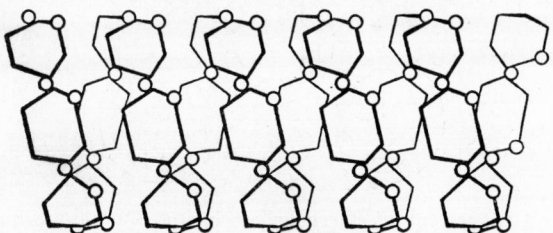

FIG. 5-8. Helical model of starch chain. Heavy lined residues are in front, light residues in back of each turn of the helix. (After Rundle, Foster, and Baldwin[17])

acid or by β-amylose do not give iodine color reaction if they contain six or fewer glucose units as judged by the end group assay method. Furthermore starch dextrins containing eight to twelve glucose units form red rather than blue complexes. Only the longer amylose chains give the typical blue iodine color. The influence of degree of branching of the starch chain on the iodine color reaction is shown in Table 5-2. As the

TABLE 5-2. Influence of extent of branching on color of iodine complex given by various starch derivatives. (After Meyer[11])

Compound	Av. number of branches per glucose unit	Color of I_2 complex
Amylose	0.00	Intense blue
Amylopectin	0.04	Light blue-violet
Residual dextrin* from amylopectin	0.09	Light red
Glycogen	0.09	Brown-red
Residual dextrin* from glycogen	0.18	Light brown

* Dextrin remaining after removal of side chains by β-malt amylase.

average degree of branching increases, the color of the iodine complex passes from blue through blue-violet to red and thence to brown.

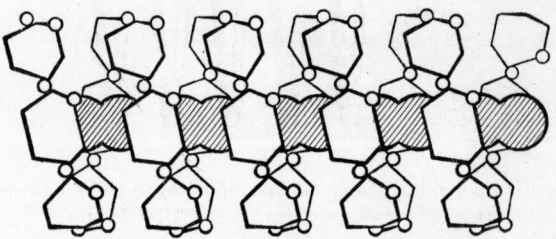

FIG. 5-9. Model of iodine-filled helical starch chain. I_2 molecules arranged along center of helix. (After Rundle, Foster, and Baldwin[17])

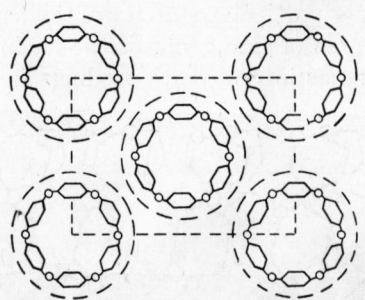

FIG. 5-10. Arrangement of starch helices in wet butanol-precipitated amylose. The centers of the four corner helices are 25.6 × 13.7 Å apart and each turn of the helix is 7.8 Å thick. (After Rundle and Edwards[16])

Amylose takes up very nearly one iodine molecule per six glucose residues. The iodine molecules are held in a regularly oriented fashion so that solutions of the complex show dichroism of flow[16] and so that, additionally, amylose crystals of the complex (as obtained by butanol precipitation) are dichroic. Both the optical and the x-ray data suggest that the amylose-iodine complex may possess a structure similar to that shown in Fig. 5-9. In this structure the iodine molecules occupy the central space or core of the starch helix, one iodine molecule being bound for each complete coil of the chain. The x-ray diffraction pattern given by the iodine-amylose complex also suggests the mode of packing of the helical cylinders into crystals shown in Fig. 5-10. The face dimensions

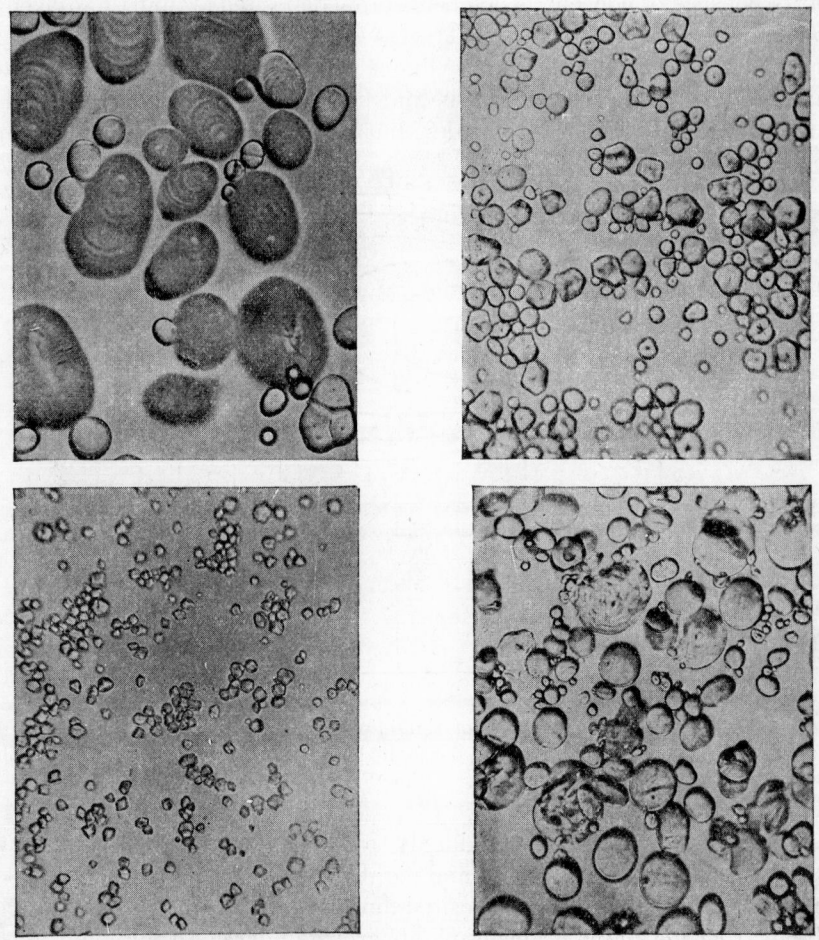

FIG. 5-11. Microphotographs of starch granules ($\times$ 200). (From Chemistry and Industry of Starch, 2nd ed., Academic Press, 1950) upper left, potato; upper right, corn; lower left, rice; lower right, wheat.

of the orthorhombic unit cell containing the helical cylinders are 13.7 and 24.8Å, while the thickness is approximately 7.8Å, a thickness such as to just accommodate one turn of each helix.[15,17]

The Starch Grain. Starch occurs in the plant in the form of starch grains of varying shape and from one to 150 μ in diameter. Starch grains from canna and potato are among the largest, while those from rice and buckwheat are among the smallest known. Each plant species forms grains of a somewhat variable but species-specific form and grains from different plants may be recognized microscopically.[18] The grain ordinarily consists of concentric layers of starch deposited around or adjacent to a hilum or leucoplast. These layers, which may be quite conspicuous, are due to discontinuities in the index of refraction of the deposited material.[19] Within each layer the index of refraction changes continuously from a higher to a lower value, but at the border of each new layer

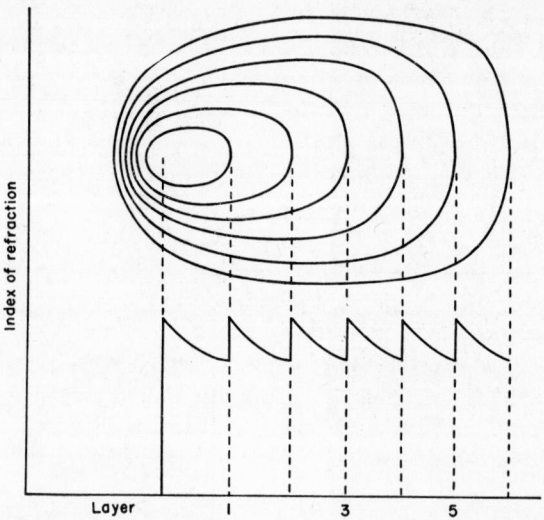

FIG. 5-12. Variations in index of refraction within the individual layers of a starch grain. Each layer starts with dense, highly refracting material and ends with more porous, less refracting material. (After Frey-Wyssling[19])

the index of refraction jumps suddenly to the higher value. The layering is due to the alternation of day and night, with the dense, highly refracting part of each layer being deposited during the day. Layering is hence absent from the starch grains of plants grown under constant conditions. Earlier authors have suggested that a sheath or membrane surrounds the grain but Alsberg[20] in a critical review of this subject has concluded that no such membrane is present and that the membrane-like properties are due to a denser structure of the surface starch. The submicroscopic units of starch which make up the grain are arranged in a radial fashion and radial cracks develop on swelling with agents such as chloral hydrate. Whether the units which are radially arranged are the starch helices or larger units is unknown.

Amylases. The enzymes which hydrolyze starch are known as amylases and have been studied extensively since 1833 when the starch digesting action of wheat amylase was described by Payen and Persoz. Seed amylases have been studied in particular detail although similar enzymes are of very wide occurrence in other plant tissues as well. Dormant starchy seeds such as the cereals contain β-amylase, the enzyme which converts amylose quantitatively to maltose and which is able to hydrolyze amylopectin to the residual dextrin. As the seed germinates α-amylase appears. This enzyme rapidly attacks both amylose and amylopectin with the production of dextrins containing of the order of six glucose units. For this reason the ability of starch to give a blue color with iodine is rapidly lost in presence of α-amylase. In a second slower phase of the reaction maltose, glucose and maltotriose are slowly liberated.

Beta-amylase free of α-amylase can be obtained in extracts of dormant seeds, particularly of barley. In mixtures of β- and α-amylase such as are contained in extracts of germinating barley, the α-amylase may be selectively destroyed by treatment with acid solutions, as for example treatment at pH 3.3.[21] Beta-malt amylase may be selectively destroyed by heating a mixture of the two amylases to a temperature of 70°C for 15 minutes, particularly in the presence of calcium ions and at a pH of 6–7.[22] Selective adsorption has also been used to separate the two amylase components. Beta-amylase may thus be adsorbed on alumina at pH 3.8 while α-amylase may be removed from 40% alcohol solution by adsorption on starch. These and other methods have been used for the preparation of the starch hydrolyzing enzymes in crystalline form.[23] Crystalline β-amylase as prepared from the sweet potato, an unusually rich source of the enzyme, possesses a pH optimum of pH 4–5 and appears to be a protein without any readily detectable prosthetic group. Crystalline α-amylase has been prepared from germinating barley, the principal rich plant source of the enzyme.

The α-amylase which appears during germination of seeds is in part at least present in the dormant seed in a bound state from which the active enzyme may be liberated by proteolysis.[24] Thus when preparations of dormant barley seed are incubated with the proteolytic enzyme papain, a large increase in amylase activity takes place, the increase being predominantly due to appearance of α-amylase. In other cases, as in seeds of oat, β-amylase is also liberated in this manner. The increase in amylase activity which occurs in seeds during germination, as in barley during the malting process, is probably due to release of the bound amylase by native proteases of the seedling.

A third amylase, the liquefying enzyme, is known from the work of Waldschmidt-Leitz and Mayer.[25] This enzyme occurs like β-amylase in dormant seeds. Like β-amylase, also, it is adsorbed on alumina but can be separated from the latter by adsorption on kaolin. The liquefying enzyme converts solutions of native starch to a less viscous state, the reaction product staining blue with iodine and possessing a chain length of about thirty units by the end group assay. It is of interest to note that preparations of the liquefying enzyme also show phosphatase activity, i.e., are able to liberate inorganic phosphate from organic forms such as glycerophosphate. Whether this is truly a property of the enzyme or whether it has any bearing on the mode of action of the liquefying enzyme is not known.

The Mechanism of Starch Synthesis. Since the amylases bring about the degradation of starch and since enzymatic reactions are believed to be reversible, it was only natural for early investigators to examine the possibility of starch synthesis in the presence of amylase. No success has been attained in the reversal of the amylase reaction since the equilibria appear to lie far toward the side of hydrolysis. Since neither maltose nor the dextrins occur in the plant cell, it is clear that reversal of amylase induced starch hydrolysis, if it takes place at all, does so *in vivo* under conditions which do not prevail in the test tube. We now know that both the breakdown and the synthesis of starch actually take place in the organism in the presence of quite a different enzyme, phosphorylase. It was discovered by Cori, Colowick, and Cori in 1937 that animal tissues possess an enzyme system capable of degrading glycogen in the presence of inorganic phosphate, with the production of Cori ester, glucose-1-phosphate.[26] Hanes has shown that a similar reaction takes place with starch under the influence of an enzyme, phosphorylase, which he found to be present both in pea seeds and in potato tubers and which is now known to be widely distributed in plant tissues.[27] In the presence of this enzyme an equilibrium is established in which starch, inorganic phosphate, and glucose-1-phosphate participate. The reaction from left to

$$\text{Starch} + H_3PO_4 \underset{\text{Phosphorylase}}{\rightleftharpoons} \text{Glucose-1-phosphate}$$

right represents breakdown of starch, not in this case by simple hydrolysis such as is brought about by amylase but rather by a phosphorolysis in which the elements of phosphoric acid are added to each severed hexose-hexose bond. The reaction from right to left represents starch synthesis from Cori ester. The energy needed for the synthesis of the hexose-hexose bond comes then from the energy of the phosphate linkage in glucose-1-phosphate and is derived originally from the energy used in phosphorylation of glucose to glucose-1-phosphate. The starch phosphorylase reaction thus resembles that discussed earlier in which sucrose is synthesized from Cori ester and fructose. In both cases the hydrolytic breakdown of the carbohydrate are reactions in which the equilibria are far over toward the side of hydrolysis. Synthesis of the carbohydrate in nature takes place therefore by a different reaction which starts with a different material, i.e., Cori ester, and the reaction carried out is one in which an equilibrium more favorable to synthesis is involved.

The equilibrium constant of the starch phosphorylase reaction is determined by the concentrations of the $HPO_4^=$ and glucose-1-$HPO_4^=$ anions and hence depends on the pH of the solution which in turn deter-

$$\frac{(HPO_4^=)}{(Glucose-1-HPO_4^=)} = K = 2.2$$

mines the proportion of the divalent anions to other anions present in solution. Thus at pH 6.0 the ratio of total inorganic P to glucose-1-phosphate-P in the equilibrium mixture is 5.7. The starch concentration does not appear in the equilibrium expression since the starch is present as a solid phase and hence maintains a constant concentration.

It was noted by Hanes that purified phosphorylase preparations show an induction or lag phase in which the formation of starch from Cori ester starts but slowly. This induction phase may be overcome by adding a trace of starch to the system as a priming agent. Dextrins are less effective as priming agents than starch, while glucose is even inhibitory.[31] Glycogen is a good priming agent for plant phosphorylase and amylopectin for animal phosphorylase. The quality which determines a good priming agent appears to be the presence of many short starch chains or branch points.

The starch produced by potato phosphorylase is similar to amylose in that it consists of long unbranched chains containing 200–300 glucose units.[28] These long chains are formed by addition onto the chains already present in the small amount of starch or other priming agent needed to start the reaction. It is important from this standpoint that the long unbranched chains of synthetic starch cannot serve as priming agents. The usual starch phosphorylase may be termed a 1,4 linkage phosphorylase since it appears to link glucose residues only through these two positions and does not form or attack glucose residues linked through the 1,6 positions. A separate factor is therefore responsible for the formation of the 1,6 or branch point linkages of amylopectin. This factor like phosphorylase is found in potato tubers, where it has been called Q enzyme by Haworth, and is a substance which in conjunction with phosphorylase yields an amylopectin-like substance rather than amylose.[29] Q enzyme or isophosphorylase produces 1,6 linkages from glucose-1-phosphate but does so only on the terminal residues of chains of 1,4 linked α-glucosidic residues.

Muscle phosphorylase has been isolated in crystalline form[30] and found to require adenylic acid (adenine-ribose-5-phosphate) as a coenzyme. Potato phosphorylase which has also been purified does not appear to require adenylic acid as a cofactor,[31] although this may merely mean that the adenylic acid is firmly bound in the protein as a prosthetic group. No other activator appears to be needed. Phosphorylase is widely distributed in plant tissues and is probably universally present in cells which have the ability to form starch. Much of the phosphorylase activity of tissues such as those of seedlings of various kinds is present in the insoluble particles of the cytoplasm. Leaf phosphorylase is similarly associated with the chloroplast.[33]

The glucose-1-phosphate formed as the result of action of phosphorylase on starch is transformed to glucose-6-phosphate by the enzyme phosphoglucomutase, which has been discussed earlier. Glucose-6-phosphate may in turn suffer any one of several fates including conversion to fructose-6-phosphate and ultimate utilization in respiration. It is clear therefore that in the plant the regulation of starch concentration is not a simple matter, even when as in the potato the amount of amylase is negligible in comparison to the phosphorylase present. In general it should be expected however that factors such as respiration which should act to lower the glucose-6-phosphate concentration, should tend to promote starch breakdown. Factors causing a lowering of the inorganic phosphate level should, on the contrary, tend to cause starch synthesis and the same should be true of simple pH changes, increases in acidity favoring increased starch synthesis owing to the effect of pH on the equilibrium between starch and Cori ester. The regulation of the starch level in the higher plant has as yet been experimentally analyzed in terms of the phosphorolytic system only by Arreguin and Bonner in the case of the potato tuber.[32] Potato tubers when stored at temperatures above

TABLE 5-3. Changes in sugars and sugar derivatives in potatoes stored at various temperatures. (After Arreguin and Bonner[32])

Substance	Initial (from common storage)	Amount in % dry wt. of tissue after 2 weeks at temp. indicated			
		0°	9°	16°	25°
Starch	67.0	61.0	65.0	63.0	64.0
Glucose	0.62	0.79	0.73	0.49	0.56
Fructose	0.17	1.50	0.34	0.22	0.15
Sucrose	1.07	6.65	1.25	0.75	0.84
Glucose-1-phosphate	0.00	0.17	0.04	0.00	0.00
Glucose-6-phosphate	3.50	0.70	0.66	4.20	4.50
Fructose-6-phosphate	0.17	2.50	1.05	0.25	0.35
Triose phosphate	0.37	0.94	1.07	0.57	0.26

approximately 10° C retain their starch. At temperatures between 0° and 10° starch is rapidly lost and sucrose accumulates. In addition to these changes in starch and sucrose, changes in the phosphorylated components occur as summarized in Table 5-3. At high temperatures potato tubers contain much glucose-6-phosphate, at low temperatures much fructose-6-phosphate, and it is therefore clear that starch degradation at low temperatures involves changes in the phosphorylated sugars. No changes in amount of phosphorylase of the potato were found, but it was discovered that potatoes stored at high temperature accumulate an

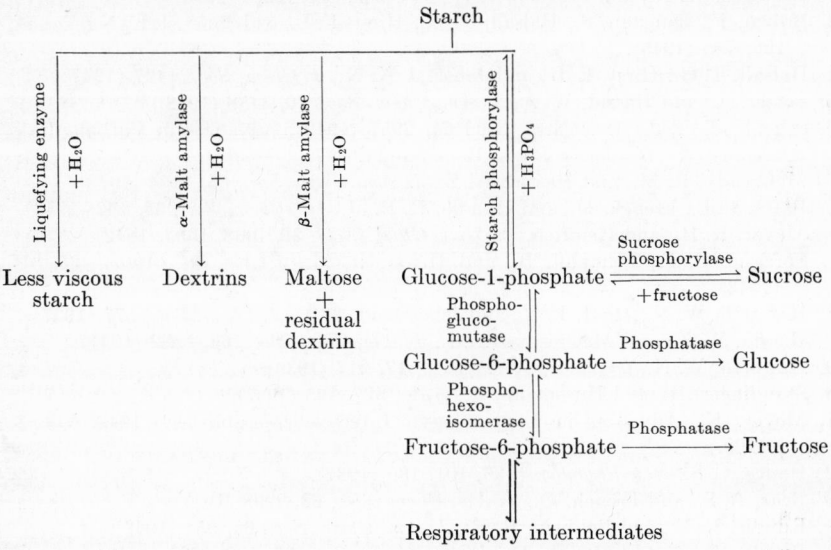

FIG. 5-13. Summary of enzymatic reactions of starch.

inhibitor of this enzyme and that the inhibitor disappears when the tubers are moved to low temperatures. Changes in phosphohexoisomerase activity and in activity of the sucrose synthesizing system were also apparent. It would seem that starch breakdown in potatoes stored at low temperature is induced by (1) disappearance of the phosphorylase inhibitor and (2) increased activity of systems capable of withdrawing glucose-1-phosphate for other syntheses such as that of sucrose.

General References

Pigman, W. W., and Goepp, R. M., Jr., Carbohydrate Chemistry. Academic Press, 1948.

Frey-Wyssling, A., Die Stoffausscheidungen der höheren Pflanzen. Springer, 1935.

Geddes, W. F., Advances in Enzymology. Interscience Publishers, 1946, Vol. 6, p. 415.

Hassid, W. Z., Quart. Rev. Biol., 18, 311 (1943).

Hassid, W. Z., Wallerstein Laboratories. Communications, **9,** 135 (1946).
Haworth, W. N., *Chemistry & Industry,* **17,** 917 (1939).
Hopkins, R. H., Advances in Enzymology. Interscience Publishers, 1946, Vol. 6, p. 389.
Kerr, R. W., Chemistry and Industry of Starch. 2nd ed. Academic Press, 1950.
Meyer, K. H., Advances in Enzymology. Interscience Publishers, 1943, Vol. 3, p. 109.

References

1. Haworth, W. N., and Percival, E., *J. Chem. Soc.,* 1342 (**1931**).
   Irvine, J., and MacDonald, J., *J. Chem. Soc.,* **129,** 1502 (1926).
2. Haworth, W. N., and Machemer, H., *J. Chem. Soc.,* 2270 (**1932**).
3. Brown, F., Dunstan, S., Halsall, T. G., Hirst, E. L., and Jones, J. K. N., *Nature,* **156,** 785 (1945).
4. Halsall, T. G., Hirst, E. L., and Jones, J. K. N., *J. Chem. Soc.,* 1427 (**1947**).
5. Potter, A., and Hassid, W. Z., *J. Am. Chem. Soc.,* **70,** 3488 (1948).
6. Schoch, T. J., *J. Am. Chem. Soc.,* **64,** 2957 (1942); Advances in Carbohydrate Chemistry. Academic Press, 1945, Vol. 1, 247.
7. McCready, R. M., and Hassid, W. Z., *J. Am. Chem. Soc.,* **65,** 1154 (1943).
   Bates, F. L., French, D., and Rundle, R. E., *J. Am. Chem. Soc.,* **65,** 142 (1943).
   Meyer, K. H., and Heinrich, P., *Helv. Chim. Acta,* **25,** 1038, 1639 (1942).
   Sprague, G. F., Brimhall, B., and Hixon, R. M., *J. Am. Soc. Agron.,* **35,** 817 (1943).
8. Haworth, W. N., Hirst, E. L., and Isherwood, F. A., *J. Chem. Soc.,* 577 (**1937**).
   Hassid, W. Z., and McCresdy, R. M., *J. Am. Chem. Soc.,* **63,** 1632 (1941).
9. Haworth, W. N., *Chemistry & Industry,* **17,** 917 (1939).
10. Staudinger, H., and Husemann, E., *Ann.,* **527,** 195 (1937).
11. Meyer, K., Advances in Enzymology. Interscience Publishers, 1943, Vol. 3, p. 109.
12. Hanes, C. S., *New Phytologist,* **36,** 101, 189 (1937).
13. Bear, R. S., and French, D., *J. Am. Chem. Soc.,* **63,** 2298 (1941).
14. Rundle, R. E., and Baldwin, R. R., *J. Am. Chem. Soc.,* **65,** 554 (1943).
15. Rundle, R. E., and French, D., *J. Am. Chem. Soc.,* **65,** 558, 1707 (1943).
16. Rundle, R. E., and Edwards, F. C., *J. Am. Chem. Soc.,* **65,** 2200 (1943).
17. Rundle, R. E., Foster, J. F., and Baldwin, R. R., *J. Am. Chem. Soc.,* **66,** 2116 (1944).
18. Reichert, E., Carnegie Institution. Publication 173, 1913.
19. Frey-Wyssling, A., Submikroskopische Morphologie des Protoplasmas und seiner Derivate. Borntraeger, 1938.
20. Alsberg, C. L., *Plant Physiol.,* **13,** 295 (1938).
21. Ohlsson, Erik, *Z. physiol. Chem.,* **189,** 17 (1930).
22. Kneen, E., Sandstedt, R. M., and Hollenbeck, C. M., *Cereal Chem.,* **20,** 399 (1943).
23. Balls, A. K., Thompson, R. R., and Walden, M. K., *J. Biol. Chem.,* **163,** 571 (1946).
    Balls, A. K., Walden, M. K., and Thompson, R. R., *J. Biol. Chem.,* **173,** 9 (1948).
    Schwimmer, S., and Balls, A. K., *J. Biol. Chem.,* **176,** 465 (1948).
24. Ford, J., and Guthrie, J., *J. Inst. Brew.,* **14,** 61 (1908).
    Davidson, J. N., *J. Agr. Res.,* **70,** 175 (1945).
25. Waldschmidt-Leitz, E., and Mayer, K., *Z. physiol. Chem.,* **236,** 168 (1935).
26. Cori, C. F., Colowick, S. P., and Cori, G. T., *J. Biol. Chem.,* **121,** 465 (1937).
27. Hanes, C. S., *Proc. Roy. Soc. (London),* B **128,** 421; **129,** 174 (1940).

28. Hassid, W. Z., Cori, G. T., and McCready, R. M., *J. Biol. Chem.*, **148,** 89 (1943).

29. Haworth, W. N., Peat, S., and Bourne, E. J., *Nature*, **154,** 236 (1944).

30. Green, A. A., Cori, G. T., and Cori, C. F., *J. Biol. Chem.*, **142,** 447 (1942).

31. Green, D. E., and Stumpf, P. K., *J. Biol. Chem.*, **142,** 355 (1942).

32. Arreguin, B., and Bonner, J., *Plant Physiol.*, **24,** 720 (1949).

33. Yin, H. C., and Sun, C., *Science*, **105,** 650 (1947).
    Yin, H. C., *Nature*, **162,** 928 (1948).

CHAPTER 6

INULIN AND THE WATER-SOLUBLE POLYSACCHARIDES

Inulin. The polyfructoside inulin replaces starch as the reserve carbohydrate in a number of species, particularly in the *Compositae*. Thus inulin constitutes a major reserve material in tubers of *Dahlia* and *Helianthus*, in roots of *Taraxacum* including the rubber-bearing species *kok-saghyz* and in chicory (*Cichorium intybus*) as well as in the roots, stems, and leaves of *Parthenium argentatum*, the guayule. Inulin is also found scattered through a wide variety of other families of dicotyledonous plants as well as in the families *Liliaceae* and *Amaryllidaceae* of the monocotyledons. In some species as in guayule, inulin is found to the exclusion of starch. In others, as in *Iris xiphium*, both starch and inulin are present. In the inulin-bearing species, the material is deposited under

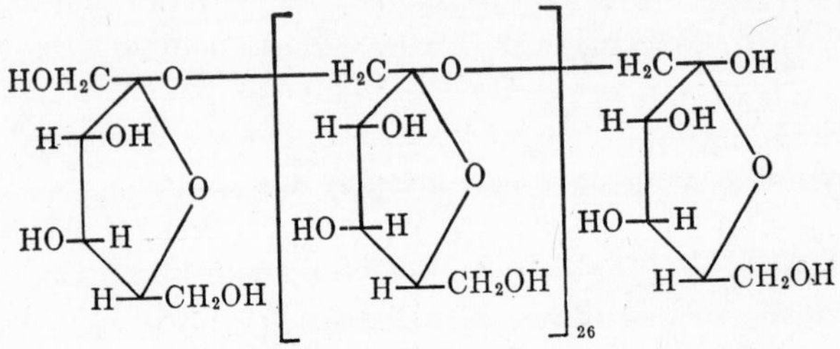

FIG. 6-1. Structure of inulin showing arrangement of fructofuranose residues in chains.

conditions favorable to carbohdyrate storage just as is starch in starch-bearing plants, and inulin is also mobilized when conditions demand as is starch.[1] The role of inulin as a reserve sugar is then well established.

Inulin is soluble in hot water and in fact may be prepared by extracting the tissue with hot water, followed by precipitation of the inulin at low temperatures, as 0°C. Inulin on acid hydrolysis gives a quantitative yield of D-fructose. Methylation followed by hydrolysis yields 3,4,6-trimethyl-D-fructofuranose and a smaller portion of 1,3,4,6-tetramethyl-D-fructofuranose. The D-fructofuranose units are therefore bound together through 1,2 linkages and the ratio of tri- to tetramethylfructose in the hydrolysis products indicates an average chain length of approximately twenty-eight fructose units.[2]

66

Inulin in contrast to starch possesses a definite reducing power, indicating the presence of an appreciable number of free carbonyl groups in the molecule. Quantitative estimation of the chain length of inulin is then possible by the determination of the number of these free reducing groups on the assumption that one such group occurs in the terminal residue of each fructofuranoside chain. The oxidation is best carried out with periodate, which in the presence of potassium chloride oxidizes the terminal residue quantitatively as described earlier, with the production of formic acid. Application of the method to inulin gives a chain length value of twenty-five residues, in good agreement with the value of approximately twenty-eight residues obtained by methylation end group assay.

Inulin is not attacked by amylases. It is, however, hydrolyzed by β-fructosidase (invertase) at a very slow rate, about 0.02% of the rate at which sucrose is attacked.[3] The pH optimum for β-fructosidase hydrolysis of inulin is 3.5–3.7, lower, however, than the pH optimum of 5.0–5.5 found for sucrose hydrolysis by the same enzyme. In addition an active inulase is known from fungi, especially from *Aspergillus niger* and *Penicillium glaucum* and a similar enzyme is also found in yeast, in *Pneumococcus* and in the hepato intestinal juice of the snail, *Helix pomatia*. It is not known however to occur in higher plants and in general it may be said that, with the exception of β-fructosidase, enzymes for which inulin is a substrate have not been found in higher plants.

Our understanding of the mechanism of fructosan formation has been greatly increased by the work of Hestrin and Avineri-Shapiro.[4] These investigators worked with the polyfructoside formed by the bacteria *Aerobacter levanicum*. The structure of this material is not identical with that of inulin since the fructose residues are linked through carbon atoms 2 and 6 rather than through atoms 1 and 2 but the mechanism involved may nevertheless be of general application. Autolyzed cells of *Aerobacter levanicum* contain an enzyme, levan sucrase, in whose presence sucrose is dissimilated essentially to free glucose and polyfructoside. Some addi-

$$n \text{ Sucrose} \rightarrow n(\text{D-glucose}) + (\text{D-fructose})_n$$

Sucrose   Free glucose   Fructosan (levan)

Overall reaction concerned in fructosan formation in *Aerobacter*

tional sucrose appears to be merely split by the invertase of the organism with liberation of equimolecular amounts of glucose and fructose. The enzymatic reaction is reversible although the rate in the direction of sucrose synthesis from levan is slow compared to the reverse rate. The energy for formation of the fructose-fructose bond in this synthesis is derived then from the energy stored in the sucrose bond. No phosphate transfer appears to be involved in the reaction, and production of levan fails to take place from fructose or from dextrose, indicating the absence of any highly active phosphorolytic system in the organism. An enzyme system similar to that described for *Aerobacter* has been found in another bacteria, *Bacillus subtilis*.[5] Whereas with *Bacillus subtilis* the enzyme is adaptive and appears only when the organism is grown on medium containing sucrose, in *Aerobacter* the enzyme is constitutive and is present in the cells under all conditions.

No evidence that inulin synthesis in the higher plant parallels the course of levan synthesis in lower organisms is yet available. It should be stressed, however, that investigations of the metabolism of inulin in higher plants have been far less numerous than those of starch and have in recent years mainly concerned the possibility of phosphorolytic inulin synthesis and breakdown. Enzymes for this type of inulin metabolism do not appear to exist, at least in the guayule.

The Water-Soluble Polysaccharides. A number of polysaccharides which occur in higher plants and more particularly as products of microorganisms, have in common the property of cold water solubility. This property is associated with short chain lengths, not more than six to ten sugar residues being ordinarily associated in a single chain. The water-soluble polysaccharides are in a sense intermediate between the disaccharides on the one hand and the polysaccharides on the other.

Water-Soluble Fructosans. Short chain length fructosans are found widely in the grasses. These substances may be isolated from cold water extracts of tissue after exhaustive extraction of soluble sugars with 80% ethyl alcohol. The fructosan of barley leaves has been methylated, hydrolyzed, and a mixture of 1,3,4-trimethyl- and 1,3,4,6-tetramethyl-D-fructofuranose obtained, indicating that the substance consists of chains of fructofuranose units linked through 2,6 linkages.[6] The chains, which are apparently unbranched, are approximately ten fructose units in length. A similar sugar has been obtained from leaves of rye. Asparagosin, obtained from asparagus roots, is a polyfructoside in which approximately ten D-fructose residues are linked through carbon atoms 1 and 2 as in inulin.[7]

A different type of water soluble fructosan has been isolated from roots of timothy (*Phleum pratense*). This sugar, which makes up over 3% of the weight of roots, on methylation and hydrolysis yields principally 1,3,4-trimethylfructofuranose and is believed to consist of chains of fifteen to sixteen D-fructofuranose residues linked through carbon atoms 2 and 6.[8] Similar fructosans have been obtained from *Poa trivialis* and from rye. A fructosan obtained from roots of *Iris* apparently contains a branched-chain structure since it yields on methylation and hydrolysis equal amounts of 3,6-dimethyl- and 1,3,4,6-tetramethylfructose.[9]

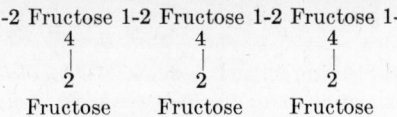

Fig. 6-2. Possible structure of branched-chain fructosan, irisin, of *Iris* rhizomes. The fructose is present in the furanose form.

Water-Soluble Barley Glucosan. Although the leaves of barley contain a fructosan, the roots contain moderate quantities (0.4% on a dry weight basis) of a water-soluble glucosan which may be isolated by methods similar to those used with the fructosans of leaves.[10] This glucosan yields only glucose on acid hydrolysis while hydrolysis following methylation yields primarily 2,3,4-trimethylglucose. The sugar appears to consist of glucopyranose chains linked through the 1,6 positions and containing seven to eleven glucose residues per chain on the basis of molecular weight determinations. In this polysaccharide the glucose residues are therefore linked as they are in the disaccharide gentiobiose. The polysaccharide is not attacked by β-malt amylase. Hehre has elucidated the mechanism of glucosan formation in the microorganism *Leuconostoc mesenteroides*.[11] This glucosan, or dextran as it is known to microbiologists, differs from those of the higher plant by possessing a branched-chain structure in which a main chain with 1,6 linkages may contain 1,4 linked branch residues. Dextran formation proceeds from sucrose as a substrate and, as with levan, according to the equation:

$$n\text{-Sucrose} \rightarrow (\text{D-glucose})_n + n(\text{D-fructose})$$

Sucrose   Glucosan   Free fructose

Overall reaction concerned in glucosan formation in *Leuconostoc*

Water Soluble Mannosan. When *Penicillium charlesii* is grown on glucose media, a mannose-containing water-soluble sugar is produced. The compound consists of mannopyranose units linked through positions 1 and 6 into chains containing approximately nine mannose units.[12] This sugar is then the mannose analog of the barley root glucosan.

General References

Evans, T. H., and Hibbert, H., Advances in Carbohydrate Chemistry. Academic Press, 1946. Vol. 2, p. 203.

McDonald, E. J., Advances in Carbohydrate Chemistry. Academic Press, 1946. Vol. 2, p. 253.

Archbold, H. K., *New Phytologist*, **39,** 185 (1940).

References

1. Archbold, H. K., *New Phytologist*, **39,** 185 (1940).
2. Haworth, W. N., Hirst, E. L., and Percival, E. G. V., *J. Chem. Soc.*, 2384, (**1932**). Irvine, J. C., and Montgomery, T. N., *J. Am. Chem. Soc.*, **55,** 1988 (1933).

3. Weidenhagen, R., in Bamann-Myrbäck, Die Methoden der Fermentforschung, 1940.
4. Hestrin, S., Avineri-Shapiro, S., and Aschner, M., *Biochem. J.*, **37**, 450 (1943). Hestrin, S., and Avineri-Shapiro, S., *Biochem. J.*, **38**, 2 (1944).
5. Doudoroff, M., and O'Neal, R., *J. Biol. Chem.*, **159**, 585 (1945).
6. Challinor, S. W., Haworth, W. N., and Hirst, E. L., *J. Chem. Soc.*, 1560 (**1934**).
7. Schlubach, H. H., and Böe, H., *Ann.*, **532**, 191 (1937).
8. Schlubach, H. H., and Sinh, O. K., *Ann.*, **544**, 101, 105, 111 (1940).
9. Schlubach, H. H., Knoop, H., and Liu, M. Y., *Ann.*, **504**, 30 (1933).
10. Hassid, W. Z., *J. Am. Chem. Soc.*, **61**, 1223 (1939).
11. Hehre, E. J., and Sugg, J. Y., *J. Exptl. Med.*, **75**, 339 (1942). Hehre, E. J., *J. Biol. Chem.*, **163**, 221 (1946).
12. Haworth, W. N., Raistrick, H., and Stacey, M., *Biochem. J.*, **29**, 612 (1935).

PART II. THE CELL WALL AND CELL WALL METABOLISM

CHAPTER 7

CELLULOSE

The Structure of Cellulose. Cellulose is the principal cell wall constituent of the higher plants and is found alike in young growing cell walls and the mature much-thickened walls of fibers. In both cases, cellulose occurs together with varying amounts of associated polysaccharides and other related materials, although secondary walls contain in addition nonpolysaccharide materials such as lignin and silica. It is nevertheless the cellulose to which the physical properties of the cell wall are believed to be due, and the mechanism of plant growth as well as the mechanical stamina of mature plants are both intimately associated with the peculiar properties of this substance. Cellulose may be found in an almost pure form in nature in the bast fibers of ramie (*Boehmeria nivea*) and in the seed hairs of cotton. These fibers are elongated cells with heavily thickened cell walls, and in both cases the mature fibers are dead cells in which the meager remnants of protoplasm are unimportant in comparison with the mass of cell wall. Both ramie and cotton have been much used in the study of the structure of cellulose.

Cellulose on acid hydrolysis gives a quantitative yield of β-D-glucose[1] and, on carefully controlled hydrolysis, cellobiose (4-D-glucopyranose β-D-glucopyranoside) may also be recovered. It was additionally shown by the early studies of Irvine and Hirst and of Haworth that on methylation and hydrolysis a quantitative yield of 2,3,6-trimethylglucose is obtained from cellulose.[2] This fact taken together with the occurrence of cellobiose units in cellulose indicates that cellulose consists of long chains of glucopyranose units linked through carbon atoms 1 and 4 by β-glucosidic bonds. The chains appear to be unbranched since no dimethylglucose is recovered on hydrolysis following methylation. Cellulose methylated in the absence of oxygen yields no detectable tetramethylglucose on hydrolysis, indicating an extremely long apparent chain length for the material. Methylation in the presence of oxygen does not yield accurate information as to chain length since such treatment causes a small amount of degrada-

71

tion of the native cellulose molecule into shorter units and yields as an artifact an erroneously large amount of tetramethylglucose.[3]

Crystallography of Cellulose.[4] The application of purely organic chemical methods establish then that cellulose consists of long chains of glucopyranose units linked through β-glucosidic linkages. The application of physical methods, primarily the techniques of x-ray diffraction and of optics, has given us in addition a clear picture of the orientation and organization of cellulose chains in the cell wall. From a historical point of view also it is of importance to note that the chain structure concept as applied to the structure of biologically important compounds, an

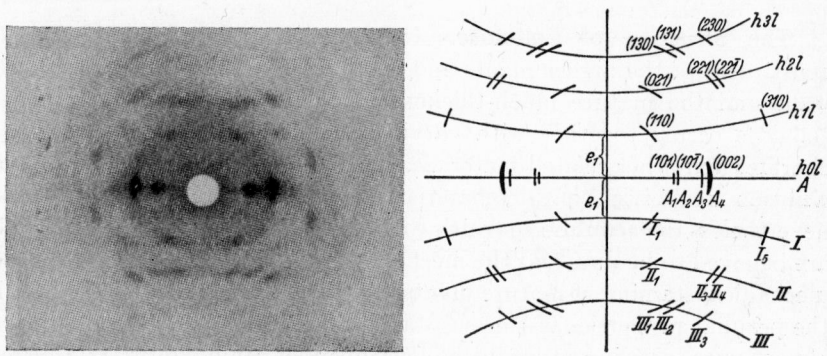

Fig. 7-1. *Left*, fiber diagram produced on plate by diffraction from ramie fiber; *right*, diagrammatic representation of fiber diagram produced by cellulose fiber. *A* is the equator, at right angles to fiber axis. *h1l*, *h2l*, *h3l* are successive hyperbolae or layer lines. The most intense spots, labeled 002, are those due to reflections from the planes which contain the pyranose residues and which are parallel to the fiber axis. (From Frey-Wyssling)

idea which has been exceedingly fruitful in many fields, had its inception in the classical work of Sponsler and Dore[6] with cellulose.

If one prepares a bundle of ramie fibers and permits a beam of monochromatic x-rays to fall upon it (Fig. 7-1) and if one allows the beam after passing through the bundle to fall on a photographic plate, an x-ray diffraction pattern is obtained on the plate. The ability of cellulose fibers to diffract x-rays shows that the fiber is crystalline and contains atoms regularly arranged in a series of intersecting planes to form a crystal lattice. X-rays are refracted by crystals, since the scattering centers in a particular crystallographic plane act as a mirror toward the x-rays incident on them. If the crystal is large enough and contains a sufficient number of refracting planes, then interference patterns result from interaction between the reflections of the successive planes. The general law for a maximum in the reflection pattern is known as Bragg's law and is given by

$$n\lambda = 2d \sin \theta \tag{1}$$

in which lambda is the wave length of the incident x-rays, d the distance between the identical reflecting planes, and θ is the angle which the incident beam makes with the crystal planes. The scattering centers in a crystal plane are the individual atoms of which the crystal is composed. Since these atoms in many crystals recur regularly at intervals which are of the order of the length of x-rays, 1–20 Å, we may use x-ray crystallographic methods to tell us about the distances between identically positioned atoms in a crystal.

In the determination of crystal structure by the help of x-rays, the angle at which a beam of x-rays of a given wave length is reflected from a given set of atomic planes is determined, and the corresponding distance between the planes then calculated. The diffracted beams are recorded as spots on a photographic plate which is placed behind the crystalline object. The spots nearest the center or undeviated beam are those which arise from planes which are farthest apart in the crystal, while the spots which are farther from the center are those due to the closer sets of planes. The diffraction spots fall on hyperbolae (layer lines) which intersect an axis along the plate and which are related to the distance between the atomic planes. The interval between identical planes at right angles to the fiber axis can be calculated by equation 2 for a fiber irradiated with a beam of x-rays at right angles to the fiber axis,

$$\text{Identity period along fiber axis} = \frac{k \cdot \lambda}{\sin n_k} \tag{2}$$

in which k is the number of the hyperbola numbering from the center and n_k is the angle made by the diffracted beam with the undeviated beam.

Demonstration of the crystalline nature of cellulose by means of x-ray diffraction was first made by Nishikawa and Ono in 1913[5] with fibers of hemp although their paper remained unnoticed until after those of Scherrer and of Herzog and Janke in 1920, both of whom used fibers of ramie. Later authors, including Sponsler and Dore[6] and Meyer and Mark[7], have shown that such fibers give strong diffractions from planes at right angles to the axis and 10.2–10.3 Å apart. The distance separating planes parallel to the fiber axis can be similarly calculated and are, according to Meyer and Mark, 8.3 and 7.9 Å. These three dimensions, 10.3 × 8.3 × 7.9 Å, give then the size of the unit cell of cellulose, the smallest portion of crystal which by mere translation in three directions would reproduce the entire crystal. The angle between the 7.9 and 8.3 Å sides has been estimated by Meyer and Mark as 84°, while these planes make an angle of 90° with those perpendicular to the fiber axis.

Arrangement of a Cellobiose in the Unit Cell.[8] We are now in a position to calculate the number of glucose residues contained in the crystallographic unit cell of cellulose. The volume of the unit cell is 670 Å³ (8.3 × 7.9 × 10.3 Å); this multiplied by the density of cellulose (1.5–1.6) gives the weight of the unit cell. This weight divided by the molecular weight of $C_6H_{10}O_5$ and multiplied by Avogadro's number gives the number of glucose residues per unit cell as four. The problem of the arrangement of four glucose residues within the unit cell was first attacked by Sponsler and Dore who made models of cellobiose using the known interatomic distances. The resulting model (Fig. 7-2) shows that the oxygen-oxygen distance along the disaccharide, from the glucosidic bond at the number one carbon atom of one residue to that at the number four carbon atom of the second residue, is 10.3 Å, a distance identical with that found for the period along the fiber axis. In the same way, the width of the cellobiose residue is found to be approximately 7.5 Å, accounting for the width of the unit cell. Further orientation of cellobiose units in the unit cell can be done by consideration of the relative intensities of the several diffraction spots since it is known that the most intense spots are those resulting from the planes most densely populated with atoms. Such planes are assumed to be those in which the pyranose rings themselves lie. In the unit cell of cellulose four cellobiose residues lie on the four edges of a parallelepiped in such a manner as to form two planes of pyranose residues, the planes being 8.3 Å apart. Each of the four cellobiose residues is shared by the four unit cells to which the edge is common. Since the unit cell must contain four glucose residues, it is assumed that each unit cell contains in addition a cellobiose chain passing through it at the intersection of the diagonals of the cross section of the parallelepiped. This center chain is not in the same orientation as those at the corners of the unit cell but runs in the reverse direction and is out of phase with them by being translated along the fiber axis to the extent of one-half of a residue. Cellulose consists then of chains of glucose residues linked together through 1,4 β glucosidic linkages, the chains being arranged in a regular fashion within a rhombohedral unit cell.

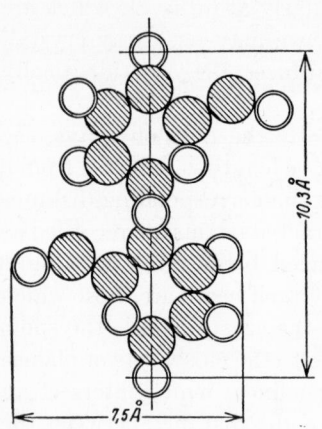

FIG. 7-2. Structure of cellobiose.

Molecular Weight of Cellulose. It is next a matter of importance to consider the length of the individual cellulose chains. The x-ray diffraction method provides a method for the determination of the size of the undisturbed crystal lattice responsible for any particular reflection. The greater the number of reflecting planes in the crystal the sharper the diffraction spot, while small numbers of reflecting planes result in broad, unsharp diffraction spots. From the quantitative relations between the width of each set of diffraction spots and the number of planes in the lattice responsible for the refraction, Hengstenberg and Mark[9] have calculated that the cell wall of the fiber contains cellulose in blocks roughly 55 Å in cross section by 600 Å in length. The chains would thus contain

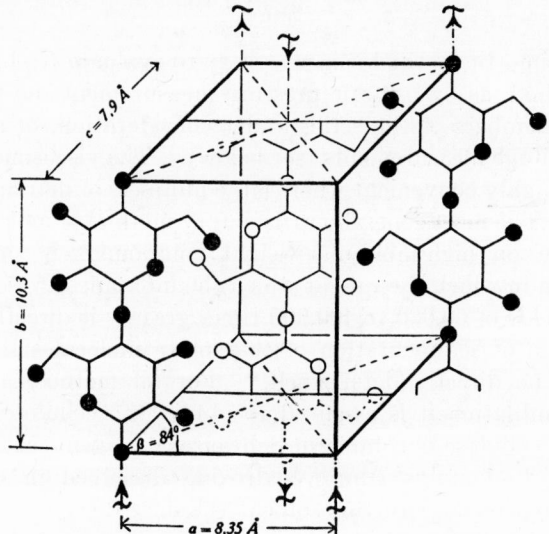

Fig. 7-3. Spatial representation of the unit cell of cellulose. (After Meyer and Misch)

according to this criterion roughly only 120 glucose residues. Molecular weight determinations on cellulose by other methods show however that the chains are in actuality much longer than this. Of the various methods available for this purpose, that of osmotic pressure of the substance is the simplest. The number of molecules in a known weight of material is simply determined by the osmotic pressure of the material in solution. This method is poorly applicable to high-molecular-weight materials, however, since (a) even with high concentrations of the material only low osmotic pressures are obtained, and (b) small amounts of low-molecular-weight impurities contribute unduly to the observed osmotic pressure and cause errors in the direction of lowering the calculated molecular weight.

A second method of molecular weight determination, developed primarily by Staudinger[10], is based on the fact that with substances of a polymeric series, the viscosity of their solutions increases with molecular weight. This relation may be expressed as follows:

$$\eta_{sp} \text{ (specific viscosity)} = K_m \cdot M \cdot c$$

where M is the molecular weight of the material, c the concentration in the solution, and K_m is a constant characteristic of material and solvent used. The specific viscosity is in turn calculated from the measured viscosity from the relation

$$\eta_{sp} = \frac{\eta_{solution}}{\eta_{solvent}} - 1$$

In this viscosimetric method it is necessary to evaluate K_m by some independent method, as by osmotic pressure measurements on lower-molecular-weight members of the series under consideration, or alternatively by ultracentrifuge measurements (see below). The viscosimetric method is, however, highly convenient where large numbers of determinations are to be made. The most generally satisfactory method of molecular weight determination on high-molecular-weight compounds is that of ultra-centrifugation in which the rate at which the molecules are sedimented in centrifugal fields of 50,000 to 400,000 times gravity is directly measured. From the rate of sedimentation of the solute molecules the molecular weight may be directly determined.[11] For all methods of molecular weight determination it is essential to obtain the cellulose in solution. Fortunately, cellulose is soluble in Schweizer's reagent, cuprammonium solution, which contains cupric hydroxide dissolved in concentrated ammonium hydroxide. In this solvent cellulose disperses to individual long-chain molecules and if air is excluded degradation of the chains into shorter chains appears to be slight. Table 7-1 gives a selection of data on the molecular weights of cellulose from different sources and although

TABLE 7-1. Molecular weights of cellulose from various sources and obtained by various methods

Source of cellulose	Method	Molecular wt.	No. of glucose residues per chain	Author (References)
Cotton	Viscosity	330,000	2020	12
Ramie	"	430,000	2660	12
Manila hemp	"	320,000	1990	12
Spruce wood	"	220,000	1360	12
Cotton	Ultracentrifuge	150–500,000	1000–3000	13
Cotton	"	1,500,000	9200	14
Ramie	"	1,840,000	11,300	14

it is clear that there are major differences between methods, sources of cellulose, and investigators, still the molecular weights indicate that cellulose contains 1400–10,000 glucose residues per chain, or a total length of 7000–50,000 Å units per chain. These values are ten times or more the length indicated above for the length of the individual crystalline aggregates in the cell wall. It would appear therefore that in cellulose each chain of glucose residues participates in the formation of more than one crystal lattice; that the chains come together to form a well-ordered array for a distance of 600 Å and then fray out, each chain rejoining with others to form other orderly crystalline units.

Micellar Structure. The individual submicroscopic crystalline aggregates which are responsible for the x-ray diffraction pattern of cellulose are known as micelles, a term first used by the botanist Nägeli in 1858 to denote the submicroscopic building blocks of plant materials. That the plant cell wall is indeed made up of submicroscopic micelles is shown not only by the x-ray data but by a host of other properties including that of swelling in water. When fibers such as those of ramie take up water with consequent increase in diameter, the x-ray diagram remains unchanged, indicating that water molecules have not entered into the crystal lattice but have rather been taken up in spaces between refracting units.[4] Swelling in the direction of the cellulose chains is many times smaller than swelling at right angles to the fiber axis. Thus the primary valence bonds which influence the properties of cellulose along the fiber axis are greater than the lateral forces between chains which operate at right angles to the fiber axis.

A quantitative method for estimation of the total intermicellar space in a cell wall has been suggested by Balls[15] and is based on change of weight of the fiber when the air in the fiber is displaced by a lighter gas. The specific weight of the fiber is determined in air. The air is then displaced by helium and the decrease in weight is determined. This method indicates that about 20% of the cotton fiber consists of intermicellar gas space. The ready penetration of the gas (or of water) through the fiber indicates that the intermicellar spaces are interconnected in one continuous phase.

Many crystalline objects possess different indices of refraction for visible light when the crystal is viewed along different crystallographic axes. Such an object is said to be double refracting and is able to rotate the plane of polarization of a beam of plane-polarized light. When a beam of plane-polarized light falls upon a double refracting object, the beam is resolved into two components which travel through the object with two different velocities corresponding to the two different indices of diffraction. As the two beams emerge from the object they are recombined to a single polarized beam which does not necessarily have the same plane of polarization as the original beam and is in general rotated through a readily determinable angle.   Double refraction may also result from the appropriate assemblage of particles which are not themselves double refracting.[16]   Consider a body consisting of two phases, the one phase an isotropic (non-double refracting) solid, the other an isotropic fluid.   Suppose that the particles of the solid phase are elongated cylinders and that the distance between the cylinders is small compared to the wavelength of light.   It is an empirical fact, also derivable on physical grounds, that a body so constituted is double refracting and that the magnitude of the double refraction depends on the differences between the indices of refraction of the disperse and dispersion phases, as is shown in Fig. 7-4.   If the index of refraction of the dispersion medium just equals that of the solid phase the double refraction of the body as a whole disappears.   If, however, the index of refraction of the dispersion medium is either larger or smaller than that of the solid phase, the body as a whole is double refracting.   This double refraction is known as double refraction of form, or form double refraction, to distinguish it from double refraction due to molecular structure, or intrinsic double refraction.   If the solid phase is itself double refracting owing to its crystallographic nature, the system may still show a double refraction of form, which will then be superimposed on the intrinsic double refraction of the solid phase.   When the body is infiltrated with fluids of varying indices of refraction, a minimum for the double refraction for the body as a whole will occur when the index of refraction of the fluid equals the mean of the indices of refraction of the particulate phase.   In fact, infiltration of the body with fluids of varying indices of refraction may be used to distinguish between intrinsic and form double refraction of an object.   With increasing index of refraction of the immersion fluid, the double refraction of the object approaches a minimum, the index of refraction of the immersion fluid at the minimum being just equal to the mean of the indices of refraction of the crystalline phase.

Cellulose as it is found in cell walls shows double refraction of form[17] and since it does, it must consist of submicroscopic anisodiametric units separated by spaces.   In the cell walls of many cells these spaces are filled with varied substances, such as lignin in the case of lignified tissues, silica in the case of tissues from grasses and Equisetum.   By oxidation of the cellulose of silica-impregnated cell walls with a chromate-sulfuric acid mixture or with potassium chlorate, a silica skeleton may be obtained in which the volume formerly occupied by cellulose is represented by empty spaces.   These silica skeletons exhibit strong double refraction of form as is shown for the cell walls of barley awns in Fig. 7-5.   Similar results have been obtained with lignified cell walls from which all cellulose has been

A

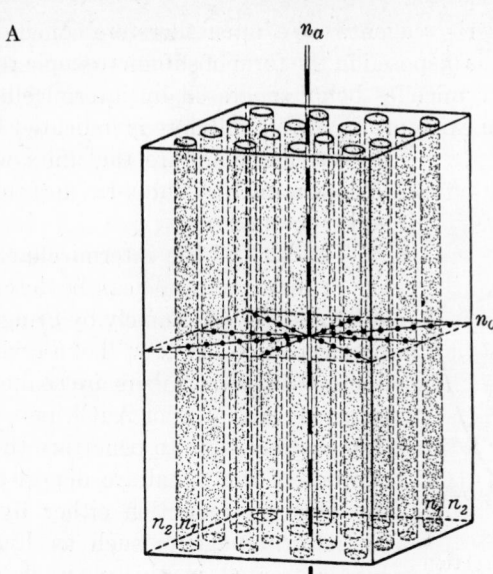

B

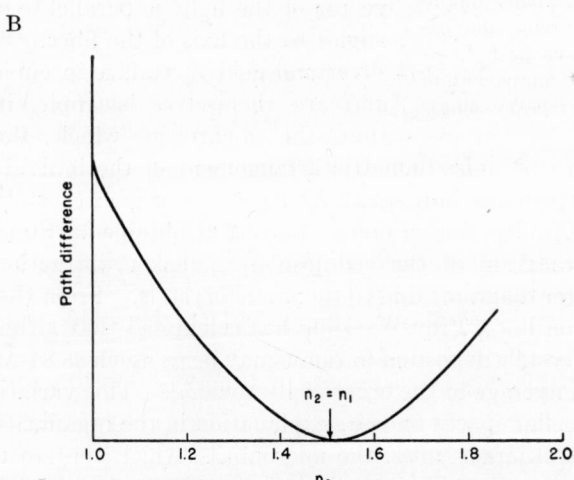

FIG. 7-4.   A. Mixed body constituted of isotropic rods of index of refraction n_1 imbedded in an isotropic matrix of index of refraction n_2. n_a and n_0 are the indices of refraction of the whole assemblage.   (After Frey-Wyssling)   B. Double refraction of the mixed body A, as a function of index of refraction of imbedding medium n_2.   This is constructed on the assumption that $n_1 = 1.50$.

removed by Schweizer's reagent. We must therefore conclude that in the cell wall, cellulose is disposed in the form of submicroscopic rod-shaped units, or micelles, the micelles being separated by intermicellar spaces. These micelles, whose existence in native cellulose is indicated by purely optical methods, are the units whose size is determined by the x-ray diffraction spot width method.

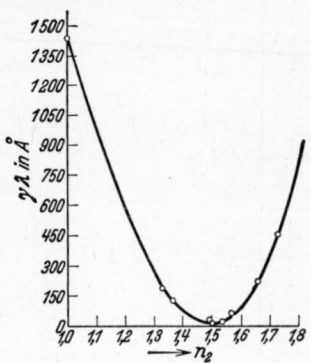

The size of the intermicellar spaces in the cellulose of ramie can be determined in still another way, namely by bringing about the deposition in the wall of foreign crystalline substances. Fibers are soaked in solutions of $AgNO_3$ or of $AuCl$, and, after the salt has had time to penetrate thoroughly, crystals of the metal are deposited within the fiber by reduction either by light or by reducing agents such as hydrazine.[18]

Fig. 7-5. Double refraction of form shown by silica skeletons of the hair of barley awns. *Abscissa:* Index of refraction of the liquid with which skeleton is infiltrated. *Ordinate:* Magnitude of double refraction expressed in terms of path difference between two components of beam. (After Frey-Wyssling)

Fibers treated in this way show strong dichroism, i.e., they are colored when viewed in polarized light and show different colors, depending on whether the electrical vector of the light is parallel to or at right angles to the axis of the fiber. Since both silver and gold crystallize in cubic crystals and are themselves isotropic, it is clear that the dichroism which they cause must be due to an anisodiametric arrangement of the individual metal crystals. When fibers impregnated with silver or gold crystals are subjected to x-ray diffraction, a double pattern is obtained. Superimposed on the fiber diagram of the cellulose is a weaker diffraction pattern (Debye-Scherrer diagram) due to the metal crystals. From the width of these diffraction lines, Frey-Wyssling has calculated that although individual silver crystals deposited in ramie may be as much as 84 Å in width, they are on an average of the order of 10 Å wide.[19] This variation in size of the intermicellar spaces finds its explanation in the organization of the micelles into still larger units, the microfibrils, which contain twenty or more micellar strands. The smaller spaces are those within the microfibril itself while the larger spaces are those between separate microfibrils.

Cellulose does not stain with iodine as does starch, but treatment of cellulose with agents which swell it greatly, as concentrated $ZnCl_2$ or 72% H_2SO_4, cause the material to become stainable with iodine. The stained fibers show a striking dichroism, appearing blue when the electric vector

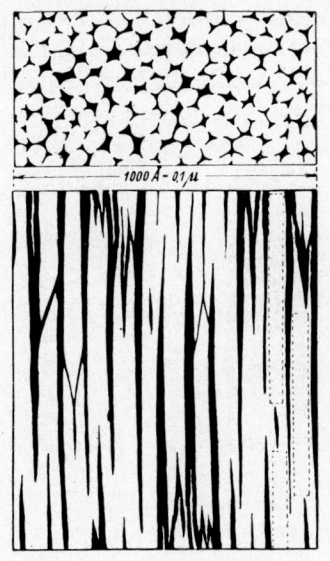

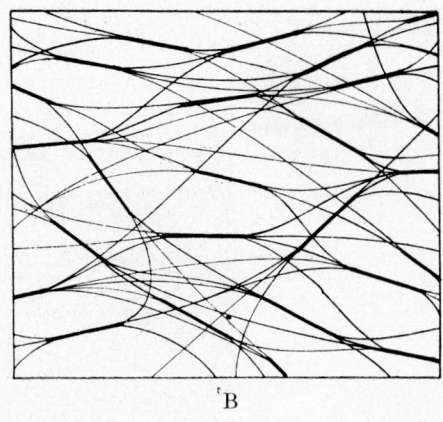

A

Fig. 7-6. A. Micellar structure of secondary cell wall; white spaces, micellar strands. B. Micellar structure of primary cell wall. Solid black lines, micellar strands; white spaces, intermicellar spaces. (After Frey-Wyssling)

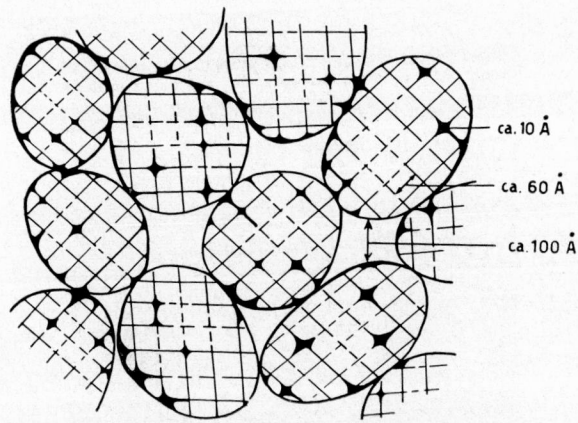

Fig. 7-7. Organization of cell wall into microfibrils composed of micellar strands. Cross section through the cell wall at right angles to cellulose chains. (After Frey-Wyssling[21])

of the incident light is parallel to the axis of the fiber and colorless when the electric vector is at right angles to the fiber. The iodine molecules probably occupy the surfaces of the cellulose micelles and are hence deposited in an oriented fashion in the fiber.[20]

We may picture the cellulose of the cell wall as made up of long chains of glucose residues. Each chain is packed into well-oriented crystalline aggregates with other similar chains. The crystalline aggregates or

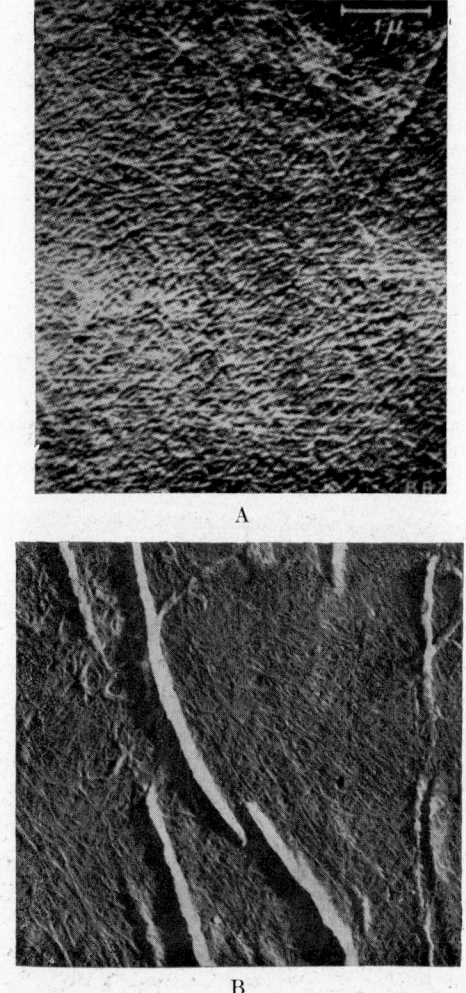

A

B

Fig. 7-8. Structure of the plant cell wall as revealed by electron microscopy. A. Growing primary wall of corn root. B. Primary wall of ramie fiber enlarged 20,000 ×.

micelles are, however, very short as compared to the length of the chains. Each chain must therefore participate in the formation of many micelles and the crystalline micellar units in the wall are separated by regions occupied by cellulose chains which are but poorly packed and are hence amorphous rather than crystalline. Finally the crystalline micelles are

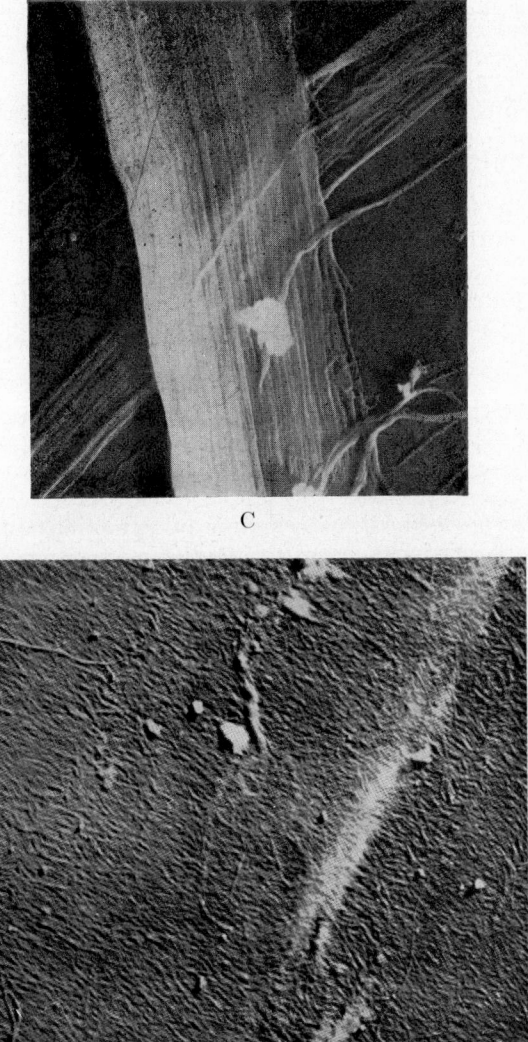

C

D

C. Secondary wall of ramie fiber enlarged 20,000 ×. D. Fine structure of secondary wall of flax fiber enlarged 20,000 ×. (Ref. 21)

separated also by an interconnected series of intermicellar spaces. The nature of the organization of cellulose in cell walls into chains, micelles, and intermicellar spaces was originally deduced by purely physical and chemical methods. Since the elaboration of this picture of the cell wall, it has become possible actually to see these structures with the aid of the electron microscope. The direct observations fully confirm those deductions made on physical and chemical grounds.[21] The electron microscope further shows clearly that the chains of cellulose are not only organized into micellar aggregates with a diameter of roughly 60 Å but that these units are further combined into larger strands, the microfibrils, with a diameter of 200–250 Å. The microfibrils are interwoven in a textile or fabric pattern as is shown in the electron micrographs of Fig. 7-8.

Orientation of Glucopyranose Chains in Cellulose Cell Walls. Study of the x-ray diffraction pattern of the ramie fiber has shown that the glucopyranose chains of cellulose are oriented parallel to the long axis of the fiber. This orientation of the glucopyranose chains is correlated with the indices of refraction of cellulose for light of visible wave lengths which are:

Index of refraction parallel to long axis of fiber = 1.596.
Index of refraction at right angles to long axis of fiber = 1.525.
Difference in two indices of refraction = 0.071.

The difference between the two indices of refraction is fairly great, as great indeed as for many double-refracting inorganic crystals. The greater index of refraction is always parallel to the long axis of the glucopyranose chains. Since the orientation of the two indices of refraction is readily and simply determined by optical means with the polarizing microscope,[16] this instrument may be used to determine the orientation of the cellulose chains in the cell wall. In all cell walls, the chains run in the plane of the wall and are never radial or normal to the cell surface. Orientation of the chains relative to the axes of the cell is, however, varied and dependent on the nature of the cell involved. The cell walls of mature cells consist of numerous superimposed layers which may be divided into the primary and secondary walls.[22] Consider a cross section through a wall which divides two cells from one another. In the center is the middle lamella, flanked on each side by a thin layer, the primary cell wall. It is these layers alone which are present during the active growth in size of the cell. Abutting on the primary wall is the secondary wall, which is deposited after growth in size of the cell has ceased. The secondary wall may be heavily thickened and may consist of several microscopically differentiable layers. The middle lamella consists principally of pectic substances, whereas the adjoining layers of the primary

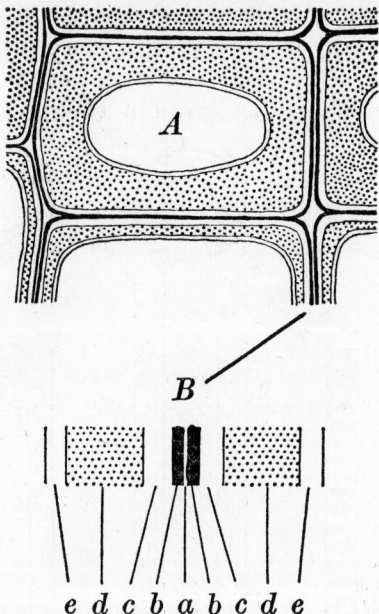

FIG. 7-9. Transverse section through a fiber cell. A. Entire fiber surrounded by fragments of other cells. B. Section through cell wall. *a*, middle lamella intercellular substance; *b*, primary wall; *c*, outer layer of secondary wall; *d*, central layer of secondary wall; *e*, inner layer of secondary wall. (After Bailey[22])

wall as well as the layers of the secondary wall contain cellulose and may be largely composed of this substance, although they may also contain varying amounts of pectic materials, as well as other polysaccharides, polyuronides, lignin, silica, etc. Orientation of the cellulose within each individual layer of a cell wall may be and frequently is different from the orientation in adjoining layers. Figure 7-10 portrays schematically various orientations of the cellulose chains possible in elongated cylindrical cells. In A the chains are in the plane of the wall and are parallel to the long axis of the cell. This arrangement is known as fiber structure, and a close approach to such structure is found in the bast fibers of ramie and nettle. The arrangement in B is similar to that in A except that the chains rather than being exactly parallel to the fiber axis are subject to a certain amount of random scattering in the plane of the wall. The chains are then parallel to the fiber axis only in the over-all statistical sense. It is probable that in all cell walls some scatter of this kind occurs. In arrangement C, the chains pursue a spiral course. The spiral may be either left or right handed and may vary in steepness from a few degrees (computed from the fiber axis) in the secondary wall of hemp, to 30 degrees in cotton and may be even larger in other cases. The secondary walls of the vast majority of fibers possess this type of spiral orientation of the cellulose chains. The angles of ascent of the spirals in adjoining layers may be quite different, as shown in Table 7-2.

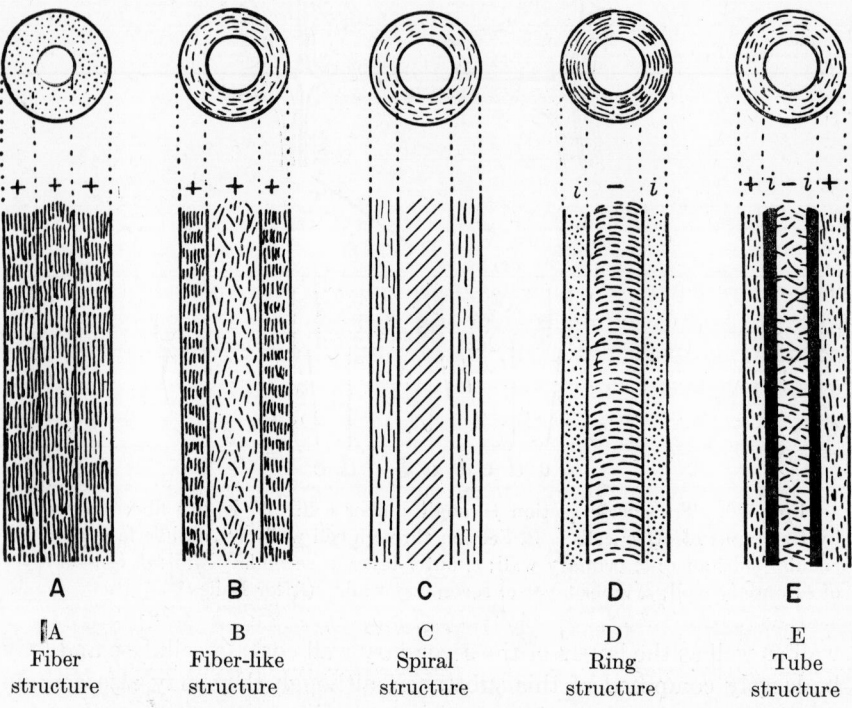

A	B	C	D	E
A Fiber structure	B Fiber-like structure	C Spiral structure	D Ring structure	E Tube structure

FIG. 7-10.   Orientation of cellulose chains in various types of cell walls. *Above:* Transverse sections through cells. *Below:* Longitudinal sections through same series of cells. Chains parallel to section are represented as long dashes. Chains perpendicular to section are represented as dots. Chains at an angle to sections are represented by dashes proportional to their projection on the sections. (After Frey-Wyssling)

In addition to gross layering, the secondary wall may show other microscopically visible fine structure. Thus the cotton fiber reveals in cross section a series of rings of cellulose of alternately greater and less intense stainability with $ZnCl_2$ and I_2. The same kind of lamellated structure is shown by the secondary walls of tracheids of many species.[23] Anderson and Kerr have shown that in cotton the rings in the secondary wall are eliminated when the cotton hair is allowed to develop under conditions of constant temperature and light, with temperature being the more important factor.[24] At temperatures below about 20°, lightly staining cellulose is formed while at higher temperatures the denser cellulose is deposited. Type D of Fig. 7-10 illustrates ring structure in which the

TABLE 7-2. Angle of ascent (angle with fiber axis) of cellulose chains in cell walls of various fibers. (After Frey-Wyssling)

Fiber	Layer of secondary wall	Angle of ascent (optical)	Angle of ascent (x-ray)*
Hemp	Outer	28° left	1.9°
	Inner	2° left	
Ramie	Outer	7.5° right	4°
	Inner	3.2° right	
Flax	Outer	10.1° right	6°
	Inner	5.0° left	
Cotton	Outer	30–35° reversals	32°
	Inner	24° reversals	

* This value is essentially a mean angle of ascent weighted for the relative thickness of the two layers.

chains pursue parallel courses at right angles to the axis of the cylindrical cell. An approach to this ideal ring structure is found only in vessels with annular thickenings, but type E, ringlike or tube structure in which the individual cellulose chains and micelles are subject to considerable scattering and are only statistically directed, is of wide occurrence.

A   B

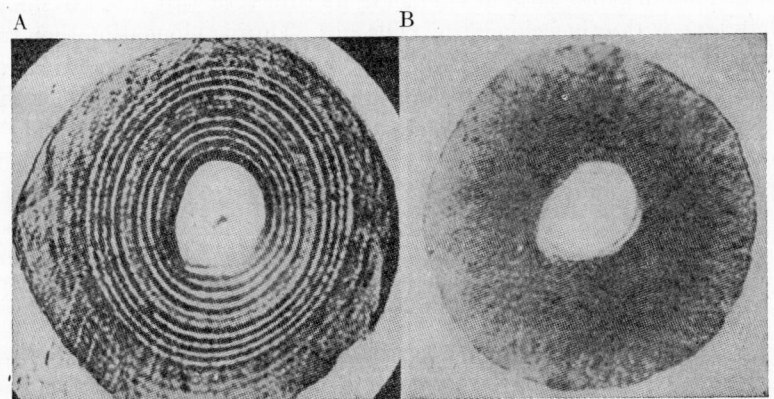

Fig. 7-11. Cross section of cotton fibers. A. Growth rings in swollen cross section of field-grown fiber ($\times$ 550). B. Swollen cross section of fiber grown under continuous illumination and uniform temp. conditions; no growth rings formed ($\times$ 960). (After Anderson and Kerr.[24] Reprinted by permission.)

Thus, the primary walls of many cells possess tube structure or essentially tube structure.[25] The primary walls of the cambium, of the developing and of the mature fiber, of elongated cylindrical parenchymatous cells of the growing stem and root, in fact the primary cell walls of all elongated cells exhibit tube structure. In cells such as the sieve tubes, latex vessels, or cylindrical parenchymatous cells in which no secondary thickening of the wall takes place, the walls of the mature cell retain their tube structure. In fibers, the tube structure of the original primary wall is obscured after the cessation of cell growth by the deposition of secondary walls having fiber or spiral structure.

The relation of the structure of the primary cell wall to the growth of the cell has interested many investigators. It is evident that the tube structure of the primary wall is confined to cells whose growth is particularly marked along one axis. Cells whose growth is equal in all directions, which grow in an isodiametric fashion, such as the fleshy parenchymatous cells of many fruits, possess cell walls in which the cellulose micelles are completely randomly disposed. It is entirely possible that tube structure bears an intimate relation to an isodiametric growth. Thus it is known that when optically anisotropic objects, such as cellophane or cell walls with tube structure, are subjected to shear or stress, the elongated cellulose molecules and micelles tend to become oriented in the direction of stress. This may be shown in model experiments with cellophane or with plant cells themselves. In normal growth, however, tube structure is preserved even though much growth takes place. A parenchymatous cell in the oat coleoptile may grow in length by ten times or more without change in its tube structure and the same is widely true with other elongating cells.[25,26] During elongation then tube structure is not lost. In part this may be due to the intercalation into the wall of new cellulose chains oriented at right angles to the direction of growth. Tube structure is maintained, however, even when growth takes place at low temperatures where wall formation is largely suppressed.[27] In part then it would appear that the growth process involves changes in the primary cell wall which allow the individual chains of cellulose to be pulled past one another without reorientation in the direction of stretch. The elongation of the plant cell is due to plastic stretching of the cell wall by the turgor pressure, or pressure exerted on the wall by the osmotic pressure of the cell contents. Regulation of growth rate is ordinarily achieved in the plant by alterations in cell wall plasticity rather than by alterations in osmotic pressure.[28] In seedlings such as those of oat and sunflower, which have been closely analyzed, application of the plant growth substance, indoleacetic acid, results in increased cell wall plasticity and in increased growth. The mechanism by which such changes in cell wall plasticity are brought about are unknown, except that the mechanism is complex and involves metabolic processes. The end result is, however, that mechanical stretching of walls whose plasticity has been increased by application of growth substance to the tissue causes much less reorientation of cellulose along the direction of stretch than occurs in stretching of walls from tissue not treated with growth substance.[25] The theory has been advanced that increases in plasticity of the primary wall are due to unknown factors which weaken the side by side or lateral association of cellulose and particularly weaken the association between chains. Diminution of the forces holding the chains together would not only account for increased plasticity but also for lessened reorientation of the cellulose chains during stretching.

Metabolism of Cellulose. Cellulose once laid down in the cell wall of the plant does not appear to be reutilized, i.e., is not available as a reserve food substance. A possible exception to this general rule is, however, constituted by the walls of the parenchymatous cells of the endosperm under the scutellum in germinating seeds.[29] Enzymes which attack cellulose are accordingly less widely found in plant tissues than are the starch attacking enzymes. Cellulase or cytase, as cellulose enzymes are called, occurs in seedlings such as those of barley and *Lupinus*.[30] Preparations of cellulase from seedlings are but little active and have been studied only incompletely. Many microorganisms are known to attack cellulose[31] including species of *Penicillium* and *Aspergillus*, pathogens such as *Botrytis*, special wood-destroying organisms as *Polyporus*, and bacteria of many genera, including *Cytophaga*, *Cellfalcicula*, and anaerobic organisms. The properties of cellulase are also known through work on the enzyme of the hepato-pancreatic juice of the snail, *Helix pomatia*. This juice hydrolyzes cellulose to cellobiose, and a separate cellobiase converts cellobiose to glucose.[32] In the case of cellulose-destroying bacteria there is also evidence for the secretion into the culture medium by the organism of two enzymes, cellulase and cellobiase.[33] The two enzymes can be separated by the fact that cellulase is rapidly destroyed by heat at 67°C, whereas the cellobiase is more stable. The pH optimum of the cellulase from *Cellulobacillus* is in the region 5.0–6.0. With the cellulose-destroying *Aspergillus oryzae*, there are also two enzymes concerned, one of which, cellulase, hydrolyzes cellulose to cellobiose with the production also of the higher oligosaccharides, cellotriose, cellotetraose, and even cellohexaose.[34] This enzyme then attacks only the higher cellulose polymers, converting them to lower polymers. The second enzyme, a cellobiase of modified specificity, can be separated from cellulase by preferential adsorption. This enzyme attacks cellohexaose and the shorter sugars including cellobiose with the production of glucose.

Determination of Cellulose. The term cellulose as used in the literature does not always refer to the chemical individual discussed above. It is on the contrary frequently empirically defined as a particular cell wall fraction obtained by a particular treatment. The experimentally obtained fractions may include larger or smaller amounts of other cell wall polysaccharides, particularly xylan and mannan. The classical method for the determination of cellulose is that of Cross and Bevan,[35] which consists in the treatment of the cell wall with boiling NaOH (1%) followed by chlorine and sodium sulfite. This treatment removes lignin and leaves a residue, Cross and Bevan cellulose, which contains not only cellulose but also other cell wall constituents not attacked by the treatment. Many modifications of this method have been suggested, the preliminary alkali extraction often being eliminated.[36] Chlorine may be replaced by hypochlorite as in the methods of Norman and Jenkins[37] and others. The material obtained by these sulfite-chlorination methods contains all the principal cell wall constituents except the polyuronide hemicelluloses which like lignin are removed by the extraction procedure. Results closely similar to those obtained by the Cross and Bevan methods are given by extraction of lignin with chlorine and ethanolamine followed by sulfite treatment of the residue.[38] Ritter and co-workers have developed a procedure for obtaining the whole carbohydrate cell wall material by treatment under conditions which permit the specific removal of lignin.[39] The basis of the determination is chlorination followed by extraction with alcohol or alcoholic ethanolamine. Only the chlorinated lignin derivatives, which are soluble in organic solvents, are removed, and the polyuronide hemicelluloses are left in the tissue. The delignified wall material prepared in this way is termed holocellulose.

The crude cellulose once obtained may be subjected to further analysis. Mannan and galactan may be determined by isolation of mannose and galactose after hydrolysis of the crude cellulose in hot mineral acid. Total pentosan may be determined by distillation in 12% HCl to yield furfural which is determined as the phloroglucide. The crude cellulose corrected for other hexosans, pentosan, residual ash, and lignin, gives an estimate of the pure cellulose.

Crude cellulose may be fractionated into α-, β-, and γ-cellulose according to solubility in 17.5% NaOH as suggested by Cross and Bevan. This treatment removes xylose and variable amounts of the other noncellulosic cell wall constituents. Alpha-cellulose constitutes that portion insoluble in 17.5% NaOH and approximates pure cellulose. Beta-cellulose is the fraction precipitable from the alkaline extract by acidification, while γ-cellulose is the fraction which is alkali soluble but not acid precipitable. The values for α-cellulose are frequently given in the literature and may be used as an approximation of the pure cellulose content of a wood, leaf, or straw, etc. sample. Beta- and gamma-cellulose correspondingly represent noncellulosic polysaccharides removed by the treatment. The relations between the various fractions of the cell wall, whole cellulose and true cellulose are summarized in Fig. 7-12.

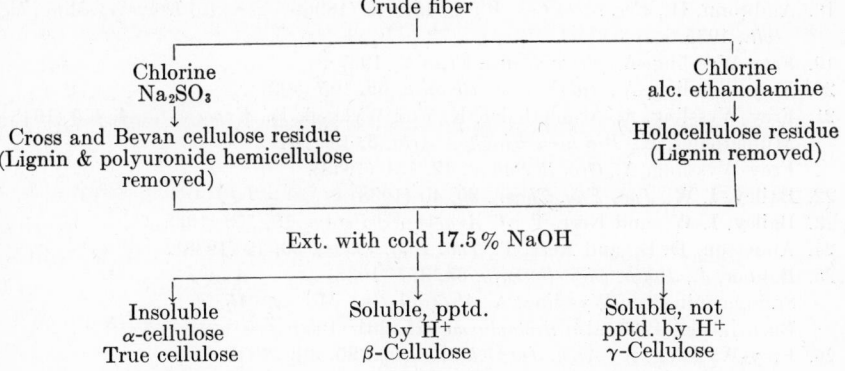

Crude fiber

Chlorine
Na₂SO₃
Cross and Bevan cellulose residue
(Lignin & polyuronide hemicellulose removed)

Chlorine
alc. ethanolamine
Holocellulose residue
(Lignin removed)

Ext. with cold 17.5% NaOH

Insoluble
α-cellulose
True cellulose

Soluble, pptd.
by H⁺
β-Cellulose

Soluble, not
pptd. by H⁺
γ-Cellulose

Fɪɢ. 7-12.   The solubility relations of cellulose and methods for obtaining pure cellulose.

General References

Wise, L. E. (Editor), Wood Chemistry.   Reinhold, 1944.
Ott, E. (Editor), Cellulose and Cellulose Derivatives.   Interscience Publishers, 1943.
Heuser, E., Cellulose Chemistry.   Wiley, 1944.
Pigman, W. W., and Goepp, R. M., Jr., Carbohydrate Chemistry.   Academic Press, 1948.
Frey-Wyssling, A., Die Stoffausscheidungen höherer Pflanzen.   Springer, 1935.

References

1. Willstätter, R., and Zechmeister, L., *Ber.*, **46,** 2401 (1913).
2. Irvine, J., and Hirst, E. L., *J. Chem. Soc.*, **121,** 1213 (1922); **123,** 518 (1923).
   Haworth, W. N., and Leitch, W., *J. Chem. Soc.*, **113,** 188 (1918).
3. Haworth, W. N., Hirst, E. L., Owen, L. N., Peat, S., and Averill, F. J., *J. Chem. Soc.*, 1885 (**1939**).
   Haworth, W. N., Montonna, R. E., and Peat, S., *J. Chem. Soc.*, 1899 (**1939**).
4. For a complete discussion of x-ray crystallography of cellulose see the general references, especially Frey-Wyssling.
5. Nishikawa, S., and Ono, S., *Proc. Phys. Math. Soc. Japan*, **7,** 131 (1913).
6. Sponsler, O., and Dore, W., *Colloid Symp. Monogr.*, **4,** 174 (1926).
7. Meyer, K. H., and Mark, H., *Ber.*, **61,** 593 (1928).
8. Meyer, K. H., and Misch, L. C., *Helv. Chim. Acta*, **20,** 232 (1937).
9. Hengstenberg, J., and Mark, H., *Z. Krist.*, **69,** 271 (1948).
10. Staudinger, H., Die Hochmolekularen organischen Verbindungen.   Springer, 1932.
11. Svedberg, T., and Pedersen, K. O., The Ultracentrifuge.   Oxford University Press, 1940.
12. Staudinger, H., and Feuerstein, K., *Ann.*, **526,** 72 (1936).
13. Kraemer, E. O., *Ind. Eng. Chem.*, **30,** 1200 (1938).
14. Gralén, N., and Svedberg, T., *Nature*, **152,** 625 (1943).

15. Balls, W., Studies on the Quality of Cotton. Macmillan Co., 1928.
16. Ambronn, H., and Frey-Wyssling, A., Das Polarisations Mikroscop. Leipzig, 1926.
17. Frey-Wyssling, A., *Helv. Chim. Acta*, **19**, 900 (1936).
18. Ambronn, H., *Ber. sächs Ges. Wiss.*, **48**, 613 (1896). See also Frey-Wyssling, *loc. cit.*, 1935.
19. Frey-Wyssling, A., *Protoplasma*, **27**, 372 (1937).
20. Frey-Wyssling, A., *Jahrb. wiss. Botanik*, **65**, 195 (1927).
21. Frey-Wyssling, A., Mühlethaler, K., and Wyckoff, R., *Experientia*, **4**, 475 (1948).
    Mühlethaler, K., *Biochim. Biophys. Acta*, **3**, 15 (1949).
    Frey-Wyssling, A., *Growth Symp.*, **12**, 151 (1948).
22. Bailey, I. W., *Ind. Eng. Chem.*, **30**, 40 (1938).
23. Bailey, I. W., and Kerr, T., *J. Arnold Arboretum*, **16**, 273 (1935).
24. Anderson, D. B., and Kerr, T., *Ind. Eng. Chem.*, **30**, 48 (1938).
25. Bonner, J., *Jahrb. wiss. Botanik*, **82**, 377 (1935).
    Summary in Frey-Wyssling, A., Protoplasma Monogr. 15.
    Borntraeger, 1938; also *Protoplasma*, **25**, 261 (1936).
26. Frey-Wyssling, A., *Arch. Jul. Klaus Stiftg.*, **20**, 381 (1945).
27. Bonner, J., *Proc. Natl. Acad. Sci. U.S.*, **20**, 393 (1934).
28. Heyn, A. N. J., *Rec. trav. botan. néerland.*, **28**, 113 (1931); *Botan. Rev.*, **6**, 515 (1940).
29. Brown, H., and Morris, G., *J. Chem. Soc.*, **57**, 458 (1890).
30. Newcombe, F., *Ann. Botany*, **13**, 49 (1899).
31. Waksman, S. A., in Wise, L. E., Wood Chemistry. Reinhold, 1944; also Norman, A. G., and Fuller, W. H., Advances in Enzymology. Interscience Publishers, 1942. Vol. 2, p. 239.
32. Karrer, P., and Schubert, P., *Helv. Chim. Acta*, **9**, 893 (1926); **11**, 229 (1928).
33. Pringsheim, H., Chemistry of the Saccharides. McGraw-Hill, N. Y., 1932. Also Simola, P., *Ann. Acad. Sci. Fennicae*, A **34**, 1 (1931).
34. Grassmann, W., Zechmeister, L., Toth, G., and Stadler, R., *Ann.*, **503**, 167 (1933).
35. Cross, C., and Bevan, E., Cellulose. Longmans Green and Co., 2nd ed., 1911.
36. Ritter, G. J., and Mitchell, R. L., "Mod. Cross & Bevan Method," U. S. Forest Products Lab. Mimeo. R1028, 1934.
37. Norman, A. G., and Jenkins, S. H., *Biochem. J.*, **27**, 818 (1933).
38. Reid, J. D., Nelson, G. H., and Aronovsky, S. I., *Ind. Eng. Chem.*, **25**, 1250 (1933).
    Kurth, E. F., and Ritter, G. J., *J. Am. Chem. Soc.*, **56**, 2720 (1934).
    Van Beckum, W. G., and Ritter, G. J., *Paper Trade J.*, **105**, 127 (1937).
39. Ritter, G. J., and Kurth, E. F., *Ind. Eng. Chem.*, **25**, 1250 (1933).

CHAPTER 8

NONCELLULOSIC POLYSACCHARIDES

Mannans and Galactans. Polysaccharides which yield only mannose on hydrolysis are found in cell walls of the straw of many grasses, of leaves, of wood, of seeds, and of microorganisms. The ivory nut, a palm seed (*Phytelephas macrocarpa*) contains a mannan as a major cell wall constituent.[1] This sugar, which is also found in the date seed, yields on methylation and hydrolysis 2,3,6-trimethylmannopyranose together with a small amount of 2,3,4,6-tetramethylmannopyranose. Thus the mannose residues are linked through 1,4 linkages as in starch or cellulose. The chain length, as indicated by the ratio of tetra- to trimethylmannose among the hydrolysis products, is approximately seventy-five residues. The mannose of ivory nut yields an x-ray diffraction pattern indicating that the mannose chains are arranged in crystalline micelles as in cellulose.

Yeast cells contain a mannan[2] which on methylation and hydrolysis yields equimolecular amounts of di-, tri-, and tetramethylmannopyranose. The presence in the original molecule of 1,2; 1,3; and 1,6 linkages is thus indicated. From the large amount of dimethylmannose obtained it is clear that a much-branched structure is involved, possibly one in which each mannose residue in a principal 1,2 linked chain is in turn linked through a 6,1 linkage to a two-residue branch chain. In each branch, the terminal residue would appear to be attached through carbon atom 1 to either carbon atom 2 or carbon atom 3 of the central residue. At least 200–400 residues are associated in a single molecule.

Mannan, similar to the ivory nut material, occurs as a major cell wall constituent in wood of the coniferous trees, and the quantitative determination of mannan has mainly arisen in connection with wood analysis. The wood mannan is contained in the crude (Cross and Bevan) cellulose fraction of wood but is hydrolyzed by hot 5–10% mineral acid. The mannose may be precipitated from the hydrolysis mixture as the phenylhydrazone under standard conditions and the original mannan thus estimated.[3] Mannan may also be extracted from Cross and Bevan cellulose by 17.5% NaOH at room temperature. The determination of mannan has been discussed by Schorger[3] and more recently by Nowotnowna.[4] The amounts of mannan found in typical woods are shown in Table 8-1.

TABLE 8-1. Amounts of mannan found in various woods. (After Nowotnowna[4])

Species	% mannan
Picea sitchensis	6.0
Pseudotsuga taxifolia	7.1
Taxodium distichum	3.0
Juniperus procera	2.4
Sequoia sempervirens	2.6

A water-soluble mannosan has been reported as a product of species of *Penicillium*.[5] In this sugar approximately nine mannose residues are associated through 1,6 linkages to form a water-soluble mannosan analogous to the water-soluble glucosan of barley roots.

Galactans like mannans are of wide occurrence in secondary cell walls of straw, wood, and seeds. A galactan isolated from lupin seeds yields 2,3,4,6-tetramethyl- and 2,3,6-trimethylgalactopyranose after methylation and hydrolysis.[6] Beta-glycosidic linkages through the 1 and 4

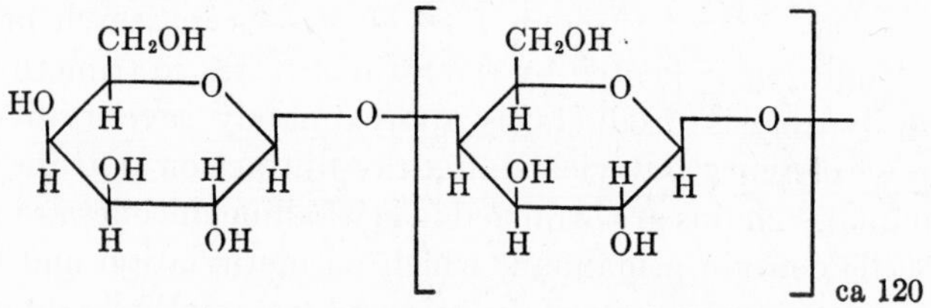

Structure of galactan of lupin seeds[6]

positions are involved. This galactan would appear hence to be an analog of cellulose but consisting of long unbranched chains of galactopyranose rather than of glucopyranose residues.

A galactan isolated from larch wood[15] possesses a branched chain structure rather than the straight-chain structure of the seed galactan. This material contains readily hydrolyzable arabinose and the arabinose-free residue obtained in this way may be methylated and hydrolyzed to yield di-, tri-, and tetramethylgalactopyranose in a 3:1:2 ratio. A possible structure for this polysaccharide is shown in Fig. 8-1.

A water-soluble galactosan has been isolated from the culture filtrates of *Penicillium*. In this compound six to eight galactopyranose units are linked through 1,4 glycosidic linkages, with a glucopyranose unit at the nonaldehydic end of the chain.

The galactan content of woods has been determined by methods based on hydrolysis to galactose followed by oxidation of the galactose to mucic acid. The mucic acid may be readily crystallized out and weighed.[7] Galactan, like mannan, is removed from Cross and Bevan cellulose by cold 17.5% NaOH. Table 8-2 gives the galactan content of various typical woods.

TABLE 8-2. Galactan content of various woods. (After Dore[8])

Wood	% of galactan
Sugar pine	0.35–0.55
Redwood	0.40–0.47
Yellow pine	0.43–0.96

A carbohydrate containing both mannose and galactose is found in the carob bean, *Ceratonia siliqua*.[9] The sugar yields galactose on mild hydrolysis, while more drastic hydrolysis yields mannose in the ratio of three-four mols of mannose to one of galactose. It is probable that all the linkages are between carbon atoms 1 and 2, α-galactosidic bonds and

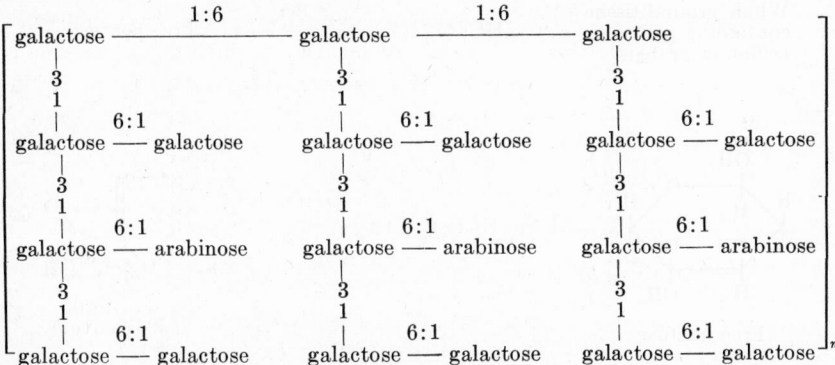

FIG. 8-1. Structure of galactan of larch wood.[15] The terminal arabinose and galactose residues may be interchanged.

β-mannosidic bonds being involved. The carbohydrate is used in the form of crude carob gum as a sizing agent for cloth as well as a base for face creams and other pharmaceuticals.

Xylans and Arabans. Pentosans containing xylose or arabinose are widely distributed as constituents of secondary cell walls. Xylan makes up 25% of the total polysaccharide of some hard woods and constitutes a significant percentage of coniferous woods, while it also is a major constituent of grass straws. When crude cellulose is prepared from plant material by removal of lignin with chlorination and sulfite procedures, xylan, if present initially, will be largely retained in the lignin-free crude cellulose. The xylan may then be removed by prolonged treatment with 5–10% hot mineral acid which hydrolyzes the xylan leaving the cellulose more or less intact. Prolonged treatment with hot alkali results in solution of the xylan, and the degraded polysaccharide may be recovered from the solution by precipitation with alcohol. Xylan from the straw of esparto grass, *Stipa tenacissima*, is made up principally of xylopyranose residues linked in 1,4 glycosidic linkages but contains also arabofuranose in the proportion of one arabinose to approximately 18 xylose residues.[10] The arabofuranose residues can be removed by mild hydrolysis (0.2% oxalic acid), whereas more drastic hydrolysis is needed to hydrolyze the xylose chain itself. The reducing ends of the xylose chains appear to be linked into the central portions of other chains through 1,3 linkages, so that the individual chains are associated into larger molecules.

The xylan of wood is less well understood than that of esparto grass and has not been prepared in a high state of purity.[11] Wood xylan probably also contains 1,4 linked xylopyranose chains since it gives an x-ray diffraction pattern which is generally similar to that of cellulose.

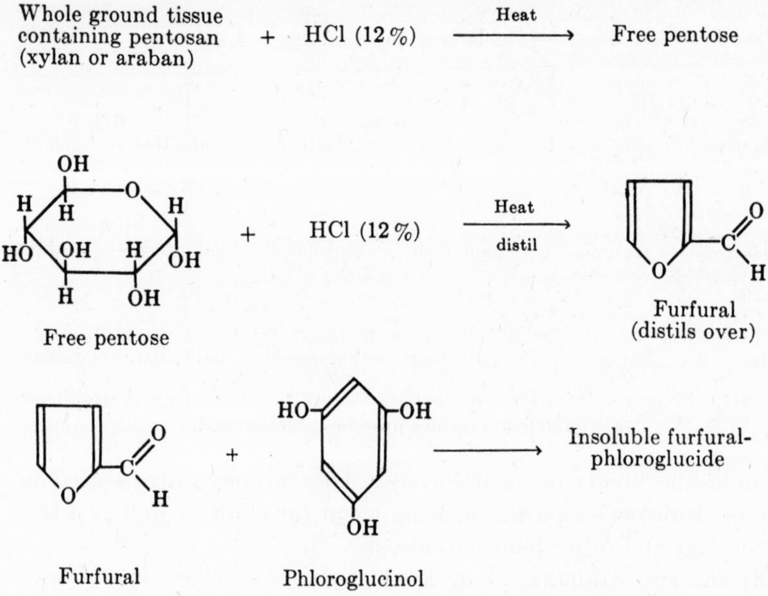

FIG. 8-2. General procedure for determination of pentosan in plant material.

The xylan content of tissues is ordinarily estimated merely by furfural determination on the sample. The dry tissue is heated in 12% HCl, the furfural formed from the pentosan distilled over with steam and the furfural in the distillate allowed to react with phloroglucinol or other reagents which form insoluble furfural derivatives which may then be dried and weighed. Application of this method demands supplementary work to determine whether the pentose involved is xylose or arabinose. This may be done by preparation of the phenylosazone or hydrazone of the pentose liberated by hydrolysis of the sample. The xylan content of the Cross and Bevan cellulose of various tissues is summarized in Table 8-3.

TABLE 8-3. Xylan contents of the Cross and Bevan cellulose of various tissues. (After Norman[12])

Source	Xylan: % of whole cellulose
Straws	
Oat	24.5
Barley	23.5
Wheat	28.6
Woods	
Oak	25.6
Beech	24.0
Ash	20.6
Silver fir	7.3
Douglas fir	5.5
Redwood	11.6
Fibers	
Sisal	24.7
Hemp	18.9
Jute	16.2
Ramie	1.8
Flax	2.5

Arabans accompany pectin in the primary cell wall and are always found in pectin preparations. The araban of peanut and of apple pectin yields, after methylation and hydrolysis, equimolecular amounts of 2,3,4-trimethyl-, 2,3-dimethyl-, and a monomethylarabofuranose. This indicates then that in the polysaccharide, one molecule in three is present as a branch point in the chain, carbon atoms 1, 3 and 5 being probably involved.[13] Hirst has proposed that the structure of this material may be represented as shown below.

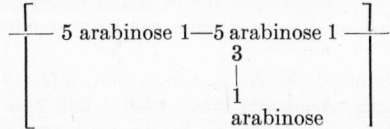

Structure of araban associated with pectin[13]

The main chain consists of arabofuranose residues linked through 1,5 glycosidic linkages. Appended to this are arabofuranose residues linked, probably, through 3,1 linkages to alternate residues in the main chain.

It has sometimes been supposed that the formation of pentosans might proceed from the related hexose polysaccharide by oxidation of each residue at carbon atom 6 to give the intermediate polyuronide, followed by decarboxylation of the polyuronide. That this cannot be the mode of formation of pentosans is shown particularly clearly in the case of the arabans associated with pectin. Whereas pectin itself consists of long unbranched chains of galacturonic acid linked through 1,4 α-glycosidic linkages, the associated araban is much branched, the residues are furanose rather than pyranose, and the linkages involved are 1,5 and 1,2 or 1,3. The arabans, and other pentosans as well, are more probably polymerized from the appropriate phosphorylated pentoses. Neither the phosphorylated pentoses nor the enzymes needed for catalysis of pentosan formation are however known at present.

Enzymes for the degradation of pentosans are known although imperfectly.[14] Xylanase activity occurs together with cellulase in the pancreatic juice of the snail *Helix pomatia*, and similar activity is found in preparations from *Aspergillus oryzae* (takadiastase) and other fungi. Arabanase activity is also found in the fungal preparations.

General References

Pigman, W., and Goepp, R. M., Jr., Chemistry of the Carbohydrates. Academic Press, 1948.

References

1. Lüdtke, M., *Ann.*, **456**, 201 (1927).
   Klages, F., and Niemann, R., *Ann.*, **523**, 224 (1936).
2. Haworth, W. N., Heath, R. L., and Peat, S., *J. Chem. Soc.*, 833 (**1941**).
3. Schorger, A., *J. Ind. Eng. Chem.*, **9**, 748 (1917).
4. Nowotnowna, A., *Biochem. J.*, **30**, 2177 (1936).
5. Haworth, W. N., Raistrick, H., and Stacey, M., *Biochem. J.*, **29**, 612 (1935).
6. Hirst, E. L., *J. Chem. Soc.*, 70 (**1942**).
7. Schorger, A., Chemistry of Cellulose and Wood. McGraw-Hill, 1926.
8. Dore, W., *J. Ind. Eng. Chem.*, **12**, 476 (1920).
9. Lew, B. W., and Gortner, R. A., *Arch. Biochem.*, **1**, 325 (1943).
10. Bywater, R. A. S., Haworth, W. N., Hirst, E. L., and Peat, S., *J. Chem. Soc.*, 1983 (**1937**).
11. Thomas, B., *Paper Ind. and Paper World*, **27**, 374 (1945).
12. Norman, A. G., Biochemistry of Cellulose, the Polyuronides, Lignin, etc. Oxford, 1937.
13. Hirst, E. L., and Jones, J. K. N., *J. Chem. Soc.*, 502 (**1938**); Advances in Carbohydrate Chemistry. Academic Press, 1946. Vol. 2, p. 235.
14. Sumner, J. B., and Somers, G. F., Chemistry and Methods of Enzymes. Academic Press, 2nd ed., 1947.
15. White, E. V., *J. Am. Chem. Soc.*, **63**, 2871 (1941); **64**, 302, 1507, 2838 (1942).

Chapter 9

THE PECTIC SUBSTANCES

The pectic substances are derivatives of polygalacturonic acid and occur particularly in the primary cell wall, although they are also found under certain circumstances in plant juices or extracts. In fact the pectic substances were first described by Payen and by Braconnot[1] as "gelée végétale" and as "the gelatinous principle of plants," since these investigators dealt only with the viscous colloidal solutions of pectin which constitute the form in which the pectic substances are found in plant extracts. Frémy[2] by 1840 recognized that this soluble pectin is a derivative of protopectin, the insoluble pectic substance of cell walls from which soluble pectin may be derived by extraction with boiling water or dilute acid. Still a third type of pectic substance recognized by Frémy was that derived from soluble pectin by treatment of the latter with dilute alkali. This third substance differed from pectin in its property of forming insoluble salts with calcium. Since this derivative yielded acid rather than neutral solutions, it was called by Frémy pectic acid. The pectic compounds may then be divided into three general types:

1. Protopectin: insoluble in water and found only in cell walls.
2. Pectin: neutral, soluble in water and found in plant extracts.
3. Pectic acid: acid, soluble in water, precipitated by calcium ions.

Constitution of Pectic Acid. Pectic acid may be prepared from soluble pectin by hydrolysis of the pectin with dilute alkali followed by repeated precipitation of the acid as the calcium or other salt, coupled with precipitation from water with alcohol. Hydrolysis of the purified product yields galacturonic acid, which accounts for 65–95% of the total, the remainder being made up of arabinose and galactose. It was thought for many years that the arabinose and galactose represented actual integral parts of the pectin molecule, and early workers[3] proposed structures for pectic acid in which a basic unit containing galacturonic acid, arabinose, and galactose in the proportions of 4:1:1 were joined in a ring similar to that of the Schardinger dextrins (Chapter 5). Ehrlich[4] envisaged a basic unit of four galacturonic acid residues linked in a chain or ring with galactose and arabinose as a side chain or chains. We now know, however, that the galactose associated with pectin occurs as a separate galactan, while the arabinose is present in the form of an araban. The two polysaccharides may be extracted from pectin or pectic acid by protracted extraction with 70% alcohol, in which the araban and galactan are ultimately removed.[5] Pectic acid may be further freed of associated polysaccharide by precipitation as the calcium salt. Definitive work on the structure of purified pectic acid was first done by Morell, Baur, and Link in 1934.[6] Their product yielded 95–98% galacturonic acid and was found by a modified end group assay method to possess a chain length of eight to ten residues. Luckett and Smith[7] later subjected pectic acid to exhaustive methylation followed by hydrolysis. Their data indicate that pectic acid consists of pyranose galacturonic acid residues linked in α-glycosidic linkages through the 1,4 positions. A chain length of thirteen residues was obtained for their preparation. Other evidence indicates that in the purified pectic acid used by Morell, Baur, and Link, and by Luckett and Smith, the polygalacturonide chains have been degraded or shortened by the drastic treatments involved in methylation, and that native pectic acid or pectin consists of exceedingly long polygalacturonide chains. Thus, sodium pectate when dried in the form of fibers or films yields an x-ray diffraction pattern[8] showing that the molecules are indeed long enough to be oriented in the fiber under the stress of the drying process. More importantly, application of Staudinger's viscosimetric methods to pectin and pectic acid has been made by Henglein and Schneider.[9] Viscosimetric methods cannot be applied directly to pectic acid or to pectin because the colloidal solutions formed by these compounds are very viscous, almost gel-like in consistency, owing apparently to the high interchain attraction which occurs in the aqueous solvent. Pectic acid or pectin may, however, be converted by treatment with concentrated nitric acid to dinitropectic acid which is soluble in acetone, a solvent in which the interchain association is greatly decreased. Viscosimetric determination of molecular weight is then possible with dinitropectins in acetone or other organic solvents. The measurements of Henglein and Schneider indicate that in pectic acid approximately one hundred galacturonic residues are present per chain while in pectin itself two hundred or more residues per chain are found. Pectic acid possesses a negligible reducing power as would be expected from the long chain length and consequent small proportion of terminal residues. No evidence for the existence of branched chains has been found although such branching is not yet completely excluded.

The carboxyl groups of pectic acid are titratable and the colloidal solution behaves as a monobasic acid having all possible dissociation constants between approximately 2.7 and 4.0.[10] That the galacturonic acid residues should exhibit a range of dissociation constants rather than a particular constant is not readily explainable. Because of its great number of free acid groups, pectic acid is particularly readily precipitated by calcium and other polyvalent cations which act to stabilize the intimate association of the individual chains. Addition of Ca ions to a solution of pectic acid results in immediate formation of a rigid Ca pectate gel, the action of Ca^{++} being not specific but replaceable by Ba^{++}, Sr^{++}, Th^{++++}, and other polyvalent cations.

Structure of Pectin. Pectin is the methyl ester of pectic acid, as was first appreciated by von Fellenberg[11] in 1914. Because of the esterification of its carboxyl groups pectin is neutral, nontitratable, and not precipitable from solution by Ca or other polyvalent cations. Deesterification occurs rapidly in mildly alkaline solution, as for example in 0.1 N

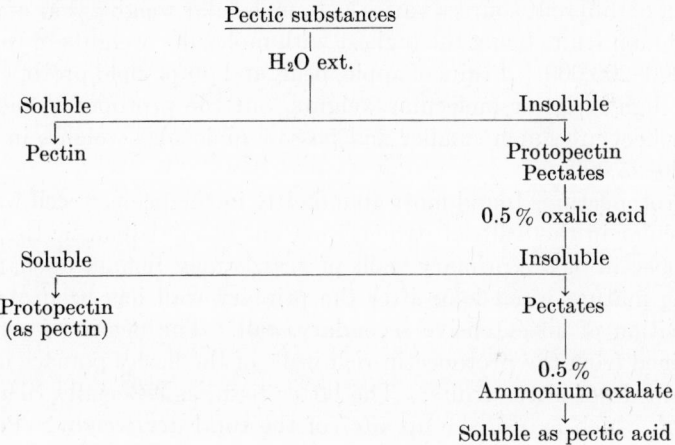

FIG. 9-1. Solubility relationships of the pectic substances.

NaOH and in general at pH values above 7.0. Incomplete hydrolysis of the methyl groups of pectin results in a series of partially esterified materials, the pectinic acids, which range in methoxyl content from none as in pectic acid to approximately 12% in pectin itself. The pectinic acids are intermediate between pectic acid and pectin in salt precipitability and other properties. Pectin is readily precipitated as a gel from its colloidal solution by small concentrations of alcohol, a property not common to other hydrophilic colloids such as the gums. It has been suggested on the basis of model experiments with agar[12] that the methyl polygalacturonide chains of pectin are associated into submicroscopic gel fragments which are coalesced into a macroscopic gel by even slight decrease in hydration.

Protopectin. Protopectin differs from pectin by its insolubility in water, and various proposals have been made to account for this fact.

Thus it was once suggested[13] that pectin may be joined with cellulose to form the insoluble protopectin. The probability of such a compound is, however, remote on the basis of modern information concerning cellulose chemistry. A more plausible possibility is that protopectin consists of methyl polygalacturonide chains still longer than those of pectin. The solubilizing of protopectin would hence consist of cracking of the protopectin molecules to shorter pectin molecules. This interpretation finds substantiation in the work of Henglein and Schneider,[9] who found that while pectic acid or pectin yield dinitro derivatives having one hundred to two hundred residues per chain, dinitropectin prepared by the direct nitration of the protopectin of the fruit yielded materials which appeared to contain as many as a thousand or more residues per chain. Protopectin of different sources varies as to molecular weight, that of grapefruit and lemon fruits being the highest with molecular weights in the range of 100,000–200,000. Fruits of apple, pear, and plum yield protopectins with only slightly lower molecular weights, but the protopectin molecules of sugar beet are much smaller and possess molecular weights in the range 20,000–25,000.

Protopectin is found most abundantly in the primary cell wall, and in particular in the walls of parenchymatous or meristematic tissues. The protopectin of the primary walls of vessels may remain unchanged even in the mature wood long after the primary wall has been obscured by deposition of an extensive secondary wall. The pectin of commerce is obtained from the protopectin-rich walls of the fleshy pomace of apple or the albedo of citrus fruits. The latter tissue is especially rich in protopectin, which may make up 40% of the total dry weight. Protopectin occurs occasionally also in secondary walls, as in those of the collenchyma of tomato stems.

The Middle Lamella. The middle lamella consists largely of a pectic compound which appears to be a mixture of calcium and magnesium pectates. It has been shown by Molisch[14] and others that the middle lamella does contain abundant calcium and magnesium. That the middle lamella pectic compound is not protopectin is shown by the fact that it is not affected by treatment with hot dilute acid, the treatment used for solution of protopectin. It may, however, be dissolved in hot alkali or preferably in solutions of ammonium oxalate, a treatment which dissolves calcium pectate by sequestering the calcium ion. In the aging of the cell wall the middle lamella may undergo changes parallel to those concerned with the protopectin of the primary wall. In the apple the pectate of the middle lamella gradually decreases in amount during the ripening process. This results in the parenchymatous cells of the fleshy tissue being loosened or even completely freed from one another in the overripe fruit. The formation of intercellular spaces is also related to the solution *in vivo* of the middle lamella, and the surfaces of cell walls which face on an intercellular space retain a thin layer of pectic material. In certain tissues such as the leaves of *Dieffenbachia*, this layer is swollen into local warts or knobs of pectate which protrude into the intermicellar spaces.

Pectic Enzymes.[15] Three types of pectic enzymes are known:

1. Protopectinase, which attacks protopectin yielding soluble pectin.
2. Pectinase, or polygalacturonase which attacks pectic acid or pectin, yielding galacturonic acid.
3. Pectase, which attacks pectin, yielding pectinic or pectic acids.

Protopectinase was first described by Bourquelot and Herissey,[16] and Carré has demonstrated the presence of the enzyme in ripening apples and in rutabaga.[17] The enzyme is thought to occur generally in fruits, roots, leaves, and other tissues.[18] Protopectinase is also found in fungi and in bacteria, although in these cases it is accompanied by large amounts of pectinase so that soluble pectin does not accumulate as a product of the enzyme action.[19] No chemical characterization of this enzyme has been carried out.

The enzyme which attacks the middle lamella is apparently not protopectinase but probably rather a pectinase, although the degradation product of the middle lamella enzyme has not yet been determined. The action of the enzyme is particularly evident in the attack of plant tissues by pathogenic fungi. Dissolution of the middle lamella by the fungal enzyme is responsible for the loosening of cell structure which results in softening of the tissue attacked.[20] Sloep has demonstrated the occurrence of an enzymatic degradation of the middle lamella in fruits of *Mespilus* during natural ripening; that the enzyme is of general occurrence in ripening fruits may be inferred.[15] The middle lamella enzyme has also been ascribed a role in leaf and fruit abscission and in penetration of the stylar tissue by the pollen tube.[21] In all these cases, as in the formation of intercellular spaces, the separation of cells at the middle lamella is involved.

Retting of flax, hemp, jute, and other fiber crops depends on the solution of the middle lamella pectic substances by the middle lamella pectinase of microorganisms. This results in the fiber bundles being loosened from the surrounding tissues so that they may be readily removed. Winogradsky[21] in his investigation of the microbiology of retting, isolated an organism, *Granulobacter pectinovorum*, capable of carrying out the process, and Beijerinck and Van Delden[22] have carried out artificial rets with pure cultures of *Bacillus subtilis*. In ordinary practice, however, mixtures of organisms, particularly of *Clostridium* species, are involved. A detailed investigation of the chemical processes involved in flax retting has been made by Eyre and Nodder.[23]

Pectinase, first found in germinating barley,[16] may be determined quantitatively by its action in liberating reducing substances from pectin or pectates. Activity of the enzyme may also be determined by observing the time needed for the enzyme preparation to liquefy a gel of calcium pectate, or by determination of the amount of residual pectin left after action of the enzyme on a pectin solution. Pectinase is present in large amounts in fungi such as *Rhizopus* and *Sclerotinia*, and is present also in takadiastase preparations made from *Aspergillus oryzae*.[24] The enzyme, whose pH optimum is in the region pH 3.0–3.5, is frequently used to remove pectin from fruit juices or other plant products so that they may be filtered, since pectin sols can be filtered only with difficulty. That pectinase must be present in ripe or overripe fruits may be inferred from the loss of total pectin which occurs in such fruit in the terminal stages of ripening. Whole extracts of various fruits (apple, citrus, cucumber) rapidly lose pectin on incubation, as much as 65% of the total pectin disappearing within 24 hours.[25]

Pectase acts on soluble pectin to demethylate the carboxyl groups with the production of pectic acid. In the presence of calcium ions the pectic acid liberated precipitates as a solid gel and qualitative pectase activity determinations may be based on the time needed for calcium pectate gel formation from pectin. The enzyme, whose pH optimum is approximately 6.2, is found universally in leaves, seeds, seedlings, roots, and tubers as well as in fungi and in bacteria. Pectase preparations are also capable of hydrolyzing esters other than pectin and conversely esterase preparations such as *Ricinus* lipase or pancreatic lipase possess pectase activity.[26] The universal or nearly universal occurrence of pectase in higher plant tissues may be a reflection of the general presence in such tissues of esterase activity rather than of the importance of pectase itself.

Pectin in Fruits. The development, maturation, and senescence of fruit involve striking changes in amount and nature of the pectic materials contained in the fruit cell walls, facts first appreciated by Frémy in 1840.[2] During the development of the fruit protopectin is laid down in the primary cell walls and in many cases protopectin accumulates to high concentrations, particularly in the apple, pear, and in citrus species. Middle lamella pectate is likewise accumulated in the fruit, but pectin on the contrary is in general found only in low concentrations up to the time of fruit maturity. With the onset of fruit ripening either on the plant (tomato) or after removal of the fruit (apple, pear, peach, citrus) a characteristic series of processes are set in motion, including conversion of starch to sugar, increase in rate of respiration, evolution by the tissue of ethylene, change in fruit color, and the conversion of protopectin to soluble pectin. All these phenomena are part of the general physiological process of fruit ripening, and the pectic changes are merely to be included as one aspect of this process.

In the apple, which has been most comprehensively investigated, the total pectic substances of the fruit reach a maximum at the time of harvest. When such apples are maintained in storage at a temperature near 1°C, protopectin gradually decreases in concentration, being replaced by soluble pectin, as is shown in Fig. 9-2. Pectate of the middle lamella

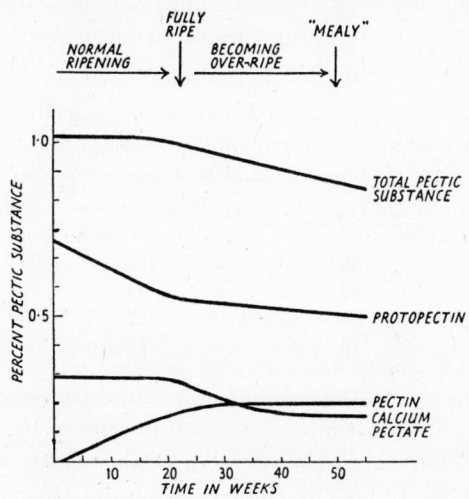

FIG. 9-2. Alterations of pectic substances in apples during ripening at 1°C. (After Norman[36])

decreases only slowly for a considerable period. At the time of commercial edible maturity the apple contains much protopectin, much pectin and appreciable pectate. As storage continues after commercial maturity, the fruit loses total pectin at a progressively increasing rate. Not only does protopectin disappear without any corresponding increase in soluble pectin, but soluble pectin and pectate also decrease in concentration during this phase. With the dissolution of the pectate of the middle lamella, the cells of the overripe fruit tend to come apart readily, and the fruit becomes soft, a typical symptom of the maturation process.

In the ripening apple then we have the picture of transformation of protopectin to pectin and softening of the fruit by removal of middle lamella pectate. In the overripe apple all pectic substances progressively disappear and the fruit becomes oversoft. Entirely similar changes in the pectic constituents take place during ripening of the pear,[30] peach, and tomato.[28] In citrus the changes in pectic constituents are quantitatively smaller but similar in nature.[29]

Since the pectic changes in fruits occur as a part of the ripening process, it is not surprising that factors which influence onset of ripening also influence pectic changes. Thus ripening and pectic changes in the apple and pear occur within a few days during storage at 15°C or higher, but both are greatly delayed at temperatures near 0° (Table 9-1). High concentrations of CO_2 (5–35%) inhibit both ripening and pectic changes, while ethylene in low concentrations greatly hastens the onset of both processes.[31] In species where ripening does not occur as long as the fruit is left on the tree, as in avocado, the initiation of pectic changes is also dependent on picking of the fruit. It is probable that the pectic changes

TABLE 9-1. Correlation of protopectin content with stage of maturity of the apple fruit. Maturity measured by pressure needed to mash fruit. (After Haller[27])

Temp. of storage degrees C	Days in storage needed for:		
	Fruit to mash with 10 lb. load	Protopectin to drop to 0.58%	Pectin to increase to 0.20%
0	100	100	100
4.4	42	44	51
10.0	29	27	31
15.5	16	15	19

during ripening are due to the activity of protopectinase and of the middle lamella pectinase, but we still have no good picture as to the way in which onset of ripening influences the activity of these enzymes or as to the mechanism by which the pectic changes are initiated.

Pectin Jellies. Pectin sols possess the property of forming solid gels in the presence of dehydrating agents; this property is utilized in the manufacture of jams and jellies. Pectin-sugar gels are not to be confused with those formed by calcium pectate since the two types of gels occur under different conditions and depend on different properties of the molecule. The formation of pectin gels occurs best in the pH range 3.1–3.5. At higher alkalinity the gel is soft, while at lower pH levels syneresis follows setting of the gel. The gel is ordinarily formed only in the presence of 65–70% sugar (sucrose or hexose), a concentration which represents an approximately saturated solution of sucrose. That the sugar acts merely as a dehydrating agent in gel formation is indicated by the fact that it may be replaced by other dehydration agents such as alcohol or glycerin. The final gel contains 0.2–1.5% of pectin, the amount depending on the gelling quality of the individual pectin preparation. This is measured commercially by the "jelly grade," which is defined as the number of pounds of sugar which one pound of the pectin will cause to gel under standard conditions of pH 3.2–3.5 and 65–70% sugar.[32] The jelly grade of a pectin may vary from 100 or less to as high as 500, depending on at least two principal factors, degree of esterification and molecular size. Completely demethylated pectin, pectic acid, does not form sugar-acid gels but reesterification of pectic acid with methyl alcohol to regenerate a methylated pectin results also in regeneration of the power of the material to make sugar gels.[33] The partially demethylated pectin, or pectinic acids, are of low quality in jelly formation, and it is important therefore in the preparation of commercial pectin that both enzymatic and alkaline hydrolysis of the pectin be avoided. The salts of pectinic acids do form weak gels at low sugar concentrations, and such gels may find other industrial uses as in paper coatings, thickening agents, and as creaming agents.[34] The relation of chain length of pectin to jelly grade is evident from the fact that greatly degraded short-chain length preparations have poor jelly-making properties, and the same is true of the low-chain length sugar beet pectin. The viscosity of a pectin in solution gives a measure of jelly grade over a considerable range, high viscosity indicating high jelly grade, although, as discussed above, the viscosity of aqueous pectin solutions can give at most only a qualitative measure of chain length.[35]

General References

Bonner, J., *Botan. Rev.*, **2**, 475 (1936); **12**, 535 (1946).
Hinton, C., Fruit Pectins. Chemical Publishing Co., 1940.
Hirst, E. L., and Jones, J., Advances in Carbohydrate Chemistry. Academic Press, 1946. Vol. 2, p. 235.

References

1. Payen, P., *Ann. chim. phys.*, **26**, 329 (1824).
    Braconnot, H. *Ann. chim. phys.*, **28**, 173 (1825).
2. Frémy, E., *J. Pharm.*, **26**, 368 (1840).
3. Nanji, D., Patton, F., and Ling, A., *J. Soc. Chem. Ind.*, **44**, 253 (1925).
4. Ehrlich, F., *Cellulose Chem.*, **11**, 140, 161 (1930).
5. Schneider, G. G., and Bock, H., *Ber.*, **70**, 1617 (1937).
    Hirst, E. L., and Jones, J. K. N., *J. Chem. Soc.*, 452 (**1939**).
6. Morell, S., Baur, L., and Link, K., *J. Biol. Chem.*, **105**, 1 (1934).
7. Luckett, S., and Smith, F., *J. Chem. Soc.*, 1106, 1114, 1506 (**1940**).
8. Palmer, K. J., and Hartzog, M. B., *J. Am. Chem. Soc.*, **67**, 2122 (1945).
9. Henglein, F. A., and Schneider, G. G., *Ber.*, **69**, 309 (1936).
    Schneider, G. G., and Bock, H., *Ber.*, **70**, 1617 (1937).
    Henglein, F. A., *J. makromol. Chem.*, **1**, 121 (1943).
10. Bonner, J., *Kon. Akad. Wet. Amsterdam*, **38**, 3 (1935).
11. von Fellenberg, Th., *Biochem. Z.*, **85**, 118 (1918).
12. Bonner, J., *Botan. Rev.*, **2**, 475 (1936).
13. Sucharipa, R., *J. Am. Chem. Soc.*, **46**, 145 (1924).
14. Molisch, H., Mikrochemie der Pflanze. Jena, 1913.
15. Sloep, A., Thesis, Delft, 1928.
16. Bourquelot, E., and Herissey, H., *J. pharm. chim.* (6) **8**, 145, 1898.

17. Carré, M., and Horne, A., *Ann. Botany*, **41**, 1 (1927).
18. Kertész, Z. I., *Ergebn. Enzymforsch.*, **5**, 233 (1936).
19. Harter, L., and Weiner, J., *Am. J. Botany*, **10**, 245 (1923).
20. Davison, F., and Willaman, J., *Botan. Gaz.*, **83**, 329 (1927).
21. Winogradsky, S., *Compt. rend.*, **121**, 742 (1895).
22. Beijerinck, M., and von Delden, A., *Botan. Centr.*, **96**, 327 (1904).
23. Eyre, J., and Nodder, C., *Trans. Text. Inst.*, **14**, 237 (1924).
24. Branfoot, M., Special Report 33, Dept. Sci. Ind. Res. Lon., 1929.
25. Joslyn, M. A., and Sedky, A., *Plant Physiol.*, **15**, 675 (1940).
26. Kertész, Z. I., *J. Biol. Chem.*, **121**, 589 (1937).
27. Carré, M., *Biochem. J.*, **16**, 704 (1922).
    Haller, M., *J. Agr. Research*, **39**, 739 (1929).
28. Clendenning, K. A., *Can. J. Research*, **19C**, 500 (1941).
29. Goddum, L., *Fla. Agr. Expt. Sta. Bull.* 268, 1934.
30. Hansen, E., *Plant Physiol.*, **14**, 145 (1939).
31. Smock, R. M., *Botan. Rev.*, **10**, 560 (1944).
32. Baker, G. L., and Woodmansee, C. W., *Del. Agr. Expt. Sta. Bull.* 272, Tech. 40, 1948.
33. Baker, G. L., and Goodwin, M. W., *Del. Agr. Expt. Sta. Bull.* 234, Tech. 28, 1941.
34. Baier, W. E., and Wilson, C. W., *Ind. Eng. Chem.*, **33**, 287 (1941).
35. Baker, G. L., and Goodwin, M. W., *Del. Agr. Expt. Sta. Bull.* 216, Tech. 23, 1939.
36. Norman, A. G., Biochemistry of Cellulose, Polyuronides Lignin, etc. Oxford, 1937.

Chapter 10

THE POLYURONIDE HEMICELLULOSES

The polyuronide hemicelluloses are cell wall components widely distributed in the plant world and characterized by (*a*) solubility in dilute alkali and (*b*) ready hydrolysis in hot dilute acid. They are important constituents of wood, straw, seeds, and, in general, of mature heavily thickened walls. For many years even the general nature of the hemicelluloses was poorly understood and it was felt that in a general way these compounds were somehow related to cellulose. We know today, however, that in actuality the polyuronide hemicelluloses are made up of mixed glycosidic chains containing both pentose and uronic acid molecules. The typical polyuronide hemicellulose contains either glucuronic acid and xylose, or galacturonic acid and arabinose, with the former type predominating. Complete hydrolysis to the constituent sugar and uronic acid may be obtained by treatment with hot 1–4% mineral acid. From a qualitative standpoint the hemicelluloses are differentiated from pectin by their solubility in alkali and insolubility in water, a property which also serves to differentiate them from the gums and mucilages which are soluble or dispersable in water.

Purification of Hemicellulose. The separation and purification of the hemicelluloses depend mainly on fractional precipitation of the material by acidification of their alkaline solutions, a method introduced by Clayson, Norris, and Schryver[1] and by O'Dwyer.[2] The pectin-free water-extracted tissue is extracted with cold or hot alkali, the concentration and temperature depending on the tissue. The resulting solution is then acidified stepwise, whereby a series of fractions known in the literature as A1, A2, etc., are precipitated and may be recovered. The most soluble, residual, material in the now acid solution may be precipitated by the further addition of alcohol or acetone. Here again fractions may be precipitated by increasing alcohol concentration. These fractions are known as B1, B2, etc. Another procedure suggested by Norris and Preece[3] for recovery of hemicelluloses from the alkaline extract is precipitation as the copper salt followed by regeneration of the precipitate in dilute acid. Hemicellulose fractions obtained in this manner are not crystalline, are not clearly pure compounds, and this uncertainty as to purity has been a definite stumbling block in the elucidation of hemicellulose structure. In general also, hemicelluloses do appear to occur in the plant as a mixture of related compounds. Occasionally, however, as in the case of the hemicelluloses of the maize cob,[13] the hemicellulose fraction of a given tissue appears to consist mainly or exclusively of one compound. The hemicellulose of maize cobs is sharply precipitated from the alkaline extract by acidification to pH 4.0 and this material yields on hydrolysis 94.8% of xylose and 5.1% of glucuronic acid (both calculated as the anhydrides), a ratio of approximately twenty-five xylose residues to one glucuronic acid residue.

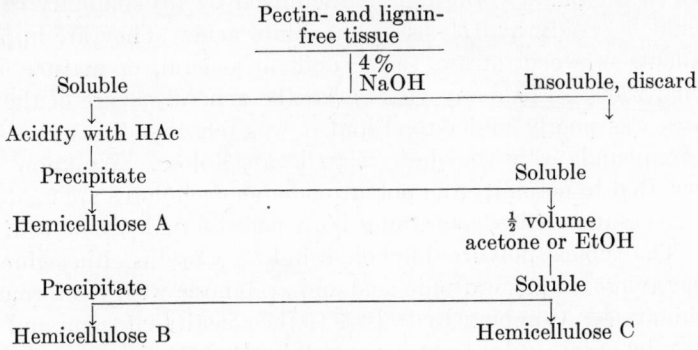

FIG. 10-1. Extraction and separation of the polyuronide hemicelluloses according to O'Dwyer[2] and Norris and Preece.[3]

Composition of Hemicellulose. The hemicelluloses of a great variety of woods have been investigated by O'Dwyer, E. Anderson,[4] and others. Thus oak wood yields a material which gives on analysis approximately 11% uronic anhydride, 85% anhydroxylose, and which contains approximately 2.2% of methoxyl residues.[2] An aldobiuronic acid was also obtained among the hydrolysis products, a material consisting of xylose linked to methylglucuronic acid. The oak wood hemicellulose therefore contains methylglucuronic acid linked to xylose as a structural component. The ratio of pentose to uronic acid in this hemicellulose corresponds to approximately eleven pentose molecules to one methyluronic acid. The hemicellulose of mesquite wood is a typical mixture of more and less soluble constituents.[5] The fraction A precipitated by acid from the alkaline extract contains approximately 10% glucuronic acid, the remainder of the material being mainly xylose. The more soluble fraction B contains almost twice as much glucuronic acid as the less soluble fraction and like the latter is composed of xylose in addition to uronic acid. In general it appears then that the solubility of a given hemicellulose depends greatly on uronic acid content, the fractions of higher acid content being more soluble in acid solution. Woods in general appear to yield a xylose-containing hemicellulose in which a methylglucuronic acid occurs as the acidic group.[4] The wood hemicelluloses have ratios of one uronic acid to seventeen to nineteen xylose units for the less soluble fractions and of one uronic acid to seven to twelve xylose residues for the more soluble fractions. Cocksfoot, a pasture grass (*Dactylis glomerata*), yields a hemicellulose in which arabinose is the principal sugar and in which galacturonic acid constitutes the uronide portion.[6] Considerable quantities of xylose were also found in the preparation although it is possible that this arose from an accompanying xylan rather than as a

TABLE 10-1. Composition of hemicellulose fractions from various sources

| | Per cent of polysaccharide | | | | |
Source	Uronic anhydride	Anhydro-pentose	Methoxyl	Sugars identified	Reference
Maize cobs	5.1	94.8	0.5	Xylose, glucuronic acid	Angell and Norris[13]
Oak heartwood	10.7	84.5	2.2	Xylose, glucuronic acid	O'Dwyer[2]
Mesquite wood (a) Precipitated by acid	10.2	84.5	2.0	Xylose, glucuronic acid	Sands and Nutter[5]
(b) Soluble in acid but precipitated by alcohol	18.0	81.9	3.0	Xylose, glucuronic acid	Sands and Gary[5]
Cocksfoot grass	10.6	53.1	...	Arabinose, xylose, galacturonic acid	Buston[6]

true constituent of the hemicellulose molecule. The hemicelluloses of leaves of *Festuca ovina* and of *Anthoxanthum odoratum* also have been found to contain arabinose combined with uronic acid[7] and the same is true of other leaves and pods.[8] Wheat straw yields a hemicellulose containing uronic acid, arabinose, and xylose in the approximate ratios of $1:1:23$.[9] Hemicelluloses of seeds have also been investigated to some extent. Thus cotton seed hull hemicellulose contains glucuronic acid and xylose in the ratio of one uronic acid to 10–16 xylose units, the ratios depending on the solubility of the fraction.[10]

In summary hemicellulose preparations can be obtained in which xylose and glucuronic acid are intimately combined and make up essentially all the hemicellulose molecule. This is true of the hemicelluloses of corn cobs, oak, mesquite, and other woods as well as of straw of cereals. From other tissues material consisting principally of galacturonic acid and arabinose is obtained and this appears to be true of hemicelluloses of many leaves. The proportion of uronic acid to pentose in the compound determines in part its solubility in acid solution. The less soluble fractions contain one uronic acid to eighteen to twenty pentose residues while the more soluble fractions contain one uronic acid to eight to twelve pentose residues. Many hemicelluloses contain methoxyl groups, and there is reason to believe that this may in general be present in the form of methyluronic acid, as has been indicated by O'Dwyer for oak hemicellulose and as found by Anderson in the case of other woods.

Structure of Hemicellulose. The structure of the hemicelluloses is still not clarified. The acidic groups, although they appear to contribute to the solubility properties of the molecule, can be neither titrated nor

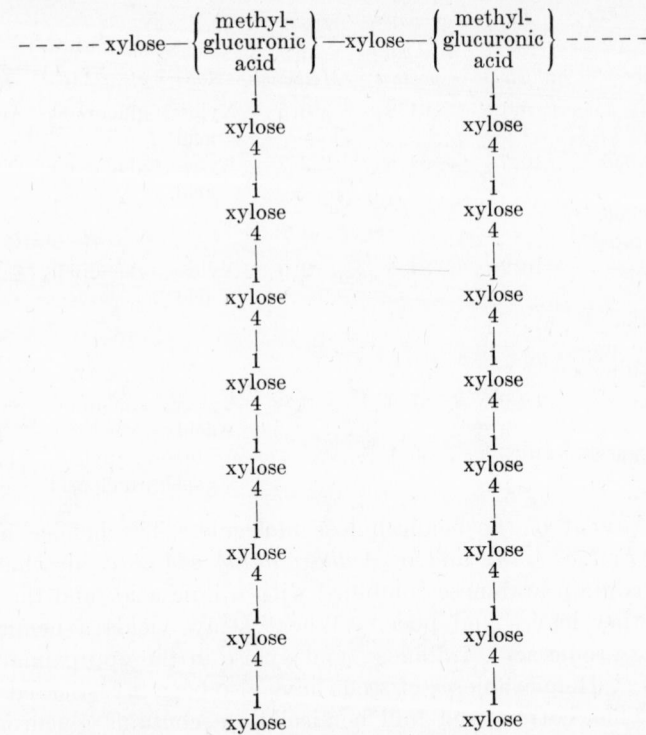

Fig. 10-2. Tentative proposal for the structure of a polyuronide hemicellulose. The exact length of branch chains would be variable.

esterified before extraction. It is probable therefore that in the native cell wall the hemicellulose molecule is bound to other components through the carboxyl group by ester linkages. Since hemicelluloses are nonreducing it is probable that the aldehydic groups are involved in linkages. Methylation and hydrolysis studies[11] on the polyuronide hemicellulose of *Phormium tenax*, New Zealand flax, have yielded 2,3,4-trimethylxylose and 2,3-dimethylxylose in the ratio of one to nine. In addition a methylated hydrolysis-resistant residue of xylose and glucuronic acid is obtained. It is possible then that this particular hemicellulose may consist of a backbone of xylose and glucuronic acid residues to which are linked branches nine xylose residues in length. This picture is in general agreement with that of Anderson, who has suggested[4] that in the wood hemicelluloses, simple short chains of xylose are involved, each chain containing one methyluronic acid residue. The short chains are then connected through additional xylose residues. In the cell wall these units would presumably be further combined, perhaps through esterification at the carboxyl group, as noted above.

Determination of Hemicellulose. Quantitative determination of the polyuronide hemicellulose content of a tissue presents great difficulties, since care must be taken to exclude pectic substances, xylan, and araban from the determination. Preece has suggested the following procedure. The tissue is extracted with ammonium oxalate to remove all pectic material. It is next extracted with 4% NaOH at room temperature (four successive treatments[14]), which may be followed by extraction at 100° with some tissues. The combined extracts are acidified with acetic acid, alcohol is added, and the precipitate dried and weighed as hemicellulose. Appreciable amounts of xylan will if present be included by the hot extraction.

The metabolism of the polyuronide hemicelluloses has been little investigated. Buston has presented data[8] which indicate that hemicelluloses may be partially used up in starving detached leaves while Burkhart[12] has suggested that they may also function as a reserve in alfalfa roots. This observation is surprising in that, in general, cell wall constituents appear to be lost to the metabolism of the plant or at most are only sluggishly available.

General References

Anderson, E., and Sands, L., Advances in Carbohydrate Chemistry. Academic Press, 1945. Vol. 1, p. 329.

Norman, A. G., Biochemistry of Cellulose, the Polyuronides, Lignin, etc. Oxford, 1937.

References

1. Clayson, D., Norris, F., and Schryver, S., *Biochem. J.*, **15**, 643 (1921).
2. O'Dwyer, M. H., *Biochem. J.*, **17**, 501 (1923); **20**, 656 (1926); **22**, 381 (1928); **28**, 2116 (1934).
3. Norris, F. W., and Preece, I., *Biochem. J.*, **24**, 59 (1930).
4. Anderson, E., Kaster, R. B., and Seeley, M. G., *J. Biol. Chem.*, **144**, 767 (1942).
5. Sands, L., and Gary, W., *J. Biol. Chem.*, **101**, 573 (1933).
   Sands, L., and Nutter, P., *J. Biol. Chem.*, **110**, 17 (1935).
6. Buston, H. W., *Biochem. J.*, **28**, 1028 (1934).

7. Bennett, E., *J. Biol. Chem.*, **146,** 407 (1942).
8. Buston, H. W., *Biochem. J.*, **29,** 196 (1935).
9. Weihe, H. D., and Phillips, M., *J. Agr. Research*, **60,** 781 (1940).
10. Anderson, E., Hechtman, J., and Seeley, M. G., *J. Biol. Chem.*, **126,** 175 (1938).
11. McIlroy, R. J., Holmes, G. S., and Mauger, R. P., *J. Chem. Soc.*, 796 (**1945**).
12. Burkhart, B. A., *Plant Physiol.*, **11,** 421 (1936).
13. Angell, S., and Norris, F. W., *Biochem. J.*, **30,** 2155 (1936).
14. Weihe, H. D., and Phillips, M., *J. Agr. Research*, **74,** 77 (1947).

CHAPTER 11

THE GUMS, MUCILAGES, AND GEL-FORMING SUBSTANCES

The gums are a heterogeneous group of acidic substances which have in common the property of swelling in water to form either gels or viscous, sticky solutions. These materials occur as exudates on bark, leaves, or roots of the plant and well-known examples include gum arabic, cherry gum, and damson gum. The related mucilages on the other hand are best known from the seed coats of particular species, as the mucilages of the flax seed. The mucilages merely swell in water with the formation of loose gels. All intermediates in colloidal behavior between gums and mucilages are known, however, and no sharp line of demarcation can be drawn between the two. In contrast to the hemicelluloses, the gums contain constituent sugar residues which may be readily removed by mild hydrolysis, leaving a more difficultly hydrolyzable residue. The residue is characteristically a uronide containing hexose or pentose and uronic acid. In the gums the acidic groups of the uronide are readily titratable and the gum can be prepared as the free acid or as one of various salts.

Gum Arabic. Gum arabic is obtained as an exudate from the bark of several species of *Acacia*, especially *Acacia senegal* of the Sudan, and has been extensively investigated from a chemical point of view.[1,2] The gum, which occurs as a salt, may be purified by solution in water, acidification, and precipitation with alcohol. The purified gum is an acid of equivalent weight 1000–1200 and possesses a dissociation constant of approximately 2×10^{-4}. Hydrolysis with 0.01 N H_2SO_4 removes L-arabinose and L-rhamnose together with D-galactose and leaves a residual polysaccharide which contains galactose and glucuronic acid in the proportions of 3 to 1. Among the products of partial hydrolysis of the resistant residue, the aldobiuronic acid of galactose and glucuronic acid as well as a disaccharide containing two galactose residues have been identified. The residues are all of the pyranose type and 1,3 and 1,6 linkages are involved. On the basis of the products of methylation and hydrolysis it has been proposed that the resistant residue of gum arabic may have the following structure:

115

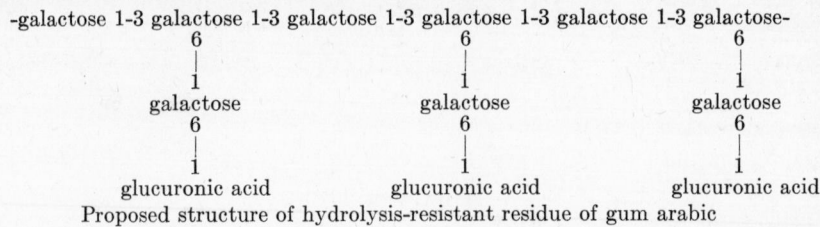

-galactose 1-3 galactose 1-3 galactose 1-3 galactose 1-3 galactose 1-3 galactose-

Proposed structure of hydrolysis-resistant residue of gum arabic

Methylation of the complete arabic acid molecule followed by hydrolysis yields the following seven methylated hydrolysis products:

2,3,4-trimethylrhamnopyranoside
2,3,5-trimethylarabofuranoside
2,5-dimethylarabofuranoside
2,3,4,6-tetramethylgalactopyranoside
2,4-dimethylgalactopyranoside
2,3,4-trimethylglucuronide
2,3-dimethylglucuronide

With the exception of the terminal galactopyranose residues all the galactose appears to be involved in branch linkages. Putting together the probable structure of the resistant residue with the methylation data derived from the intact molecule, Hirst suggests[2] the following type of structure for arabic acid:

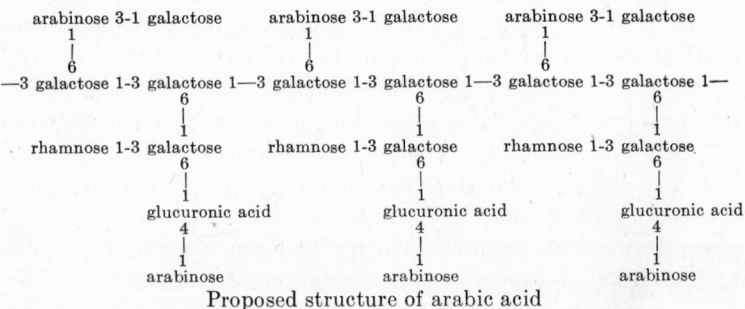

Proposed structure of arabic acid

The basic unit consisting of four galactose units, two molecules of L-arabinose, and one molecule each of L-rhamnose and glucuronic acid, would have an equivalent weight of approximately 1200 or roughly that found by titration. In this scheme the presence of 2,3,5-trimethyl-arabofuranoside among the hydrolysis products is not accounted for, but could be envisaged as due to the occasional absence of a terminal galactopyranose residue.

Mesquite Gum. This gum occurs as an exudate on the bark of the mesquite, *Prosopis juliflora*, which like *Acacia* is a leguminous tree.[3] The gum may be purified by precipitation from acid solution with alcohol, just as in the case of gum arabic, and titration yields an equivalent weight of approximately 1200. Mild hydrolysis (3% H_2SO_4 at 80°C for six hours) results in the removal of arabinose, leaving a resistant residue of galactose and methylglucuronic acid, the three sugars occurring in the proportions of 4 to 2 to 1.[4]   Among the hydrolysis products of the resistant residue are found acidic compounds which are identified as glycosides of methylglucuronic acid and galactose. Methylation and hydrolysis studies show[4] that the methylglucuronic acid is linked through position 1, galactose through positions 1, 3, and 6 while arabinose occurs both as terminal and as nonterminal residues, which are present in the ratio of 1 to 3. These data indicate a possible structure of the nature shown below.

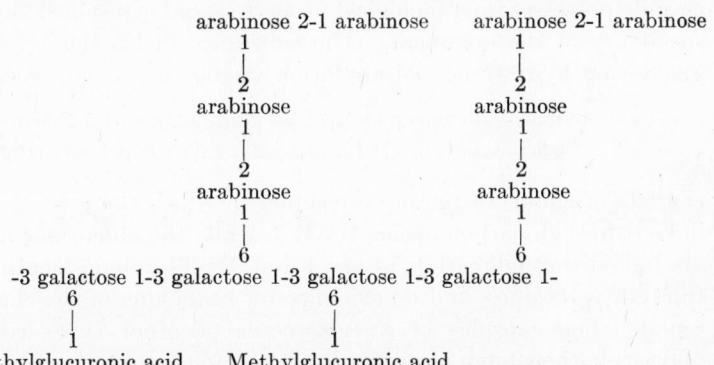

Possible structure of mesquite gum. Arabinose refers to arabofuranose, while the galactose and glucuronic acid are in the pyranose form.

Damson Gum. Damson gum is found as an exudate on the bark of the Damson plum tree. Arabinose is removed by mild hydrolysis and xylose by more drastic treatment, while the hydrolysis-resistant residue contains galactose, mannose, and glucuronic acid.[5] The gum of the cherry tree also contains readily removable arabinose (eight molecules) and xylose (six molecules), while further hydrolysis removes galactose (six molecules). The resistant residue contains mannose and glucuronic acid in the ratios of approximately 3 to 2. Methylation studies show that in both cases highly branched structures are involved.

Gum Tragacanth. Gum tragacanth, obtained as an exudate from *Astragalus gummifer*, a leguminous plant of the middle east, has wide pharmaceutical application as an inert base for pills, as an emulsifying agent, and as a base for pharmaceutical jellies. When the gum is placed in water, a portion passes into colloidal solution and can be filtered from the residue which swells to a bulky jelly. This gum differs from others considered above in that it is a mixture of (a) a galacturonic acid-containing uronide, (b) a galactoaraban, and (c) a glycosidic compound, possibly a sterol.[6] The uronide, tragacanthic acid, contains in addition to galacturonic acid, L-fucose and xylose. The galacturonic acid residues may be linked through 1,4 linkages as in pectic acid, while the fucopyranose and a portion of the xylopyranose are terminal residues and may constitute side chains along the polygalacturonide backbone. The galactoaraban component is likewise highly branched. Hence the gum as a whole is exceedingly complex and is in addition unusual among the gums both as to structure and as to its content of fucose and of galacturonic acid.

Microbial Gums. A microbial gum of great interest is the polysaccharide which makes up the capsule of type III *Pneumococcus* and whose structure has been elucidated by Reeves and Goebel.[7] This polysaccharide possesses immunological properties and is responsible for the type specificity of the organism. The substance yields the 4-β-uronoside of glucose on hydrolysis. Methylation studies have shown that the gum

-4 glucopyranose 1-3 glucuronic acid 1-4 glucopyranose 1-3 glucuronic acid 1-

Structure of type III *Pneumococcus* capsular polysaccharide

consists of chains of pyranose residues in which the glucose residues are linked through carbon atoms 1 and 4, while the glucuronic acid residues are linked through carbon atoms 1 and 3. The chain length is thirty to thirty-five residues, and no evidence for branching of the chain has been found. The capsules of *Pneumococcus* of other types contain other polysaccharides having a wide diversity of structure.[8] Gumlike sheaths are also formed by a variety of other microorganisms and in many cases the sheaths consist of simple fructosans or dextrans. Thus aerobic spore-forming bacilli such as *Bacillus megatherium, mesentericus,* and *subtilis* form loose sheaths of a polysaccharide consisting of fructofuranose linked through 1,6 linkages, as already discussed (Chapter 8). Similar fructosans are formed by some *Pseudomonads* and by *Azotobacter,* while *Leuconostoc* species utilize sucrose with the formation of a dextran capsule. In *Acetobacter xylinum* the sheaths of the individual cells fuse to form a membrane or pellicle over the surface of the culture, the pellicle consisting of cellulose. In still other microorganisms, however, as in the *Rhizobia,* sheaths are produced which are complex gums containing glucose, pentose, and glucuronic acid. In *Rhizobium* as in *Pneumococcus* different strains appear to produce differing polysaccharides.

Mucilages. Mucilages are obtained when seeds of flax, *Plantago,* mustard, or of certain other species are soaked in water. The sticky swollen mass, which arises from the seed coat, may be separated from the seeds by filtration through cloth and the material then precipitated with alcohol. The seed coats in the particular species involved contain an outer layer which consists essentially of the dehydrated components of the mucilage.   The mucilage obtained from flax is a mixture of a gumlike polyuronide, a protein component and microscopically or submicroscopically dispersed cellulose fibrils.   The protein may be removed from the mucilage solution by treatment with protein precipitating agents as phosphotungstate.   The purified gumlike polysaccharide on hydrolysis yields a resistant residue of galacturonic acid and L-rhamnose, to which in turn D-xylose and L-galactose are less firmly attached.[9]   The latter rare sugar is particularly remarkable for its occurrence in that it belongs to the unnatural L-series.   L-Arabinose has also been reported among the products of linseed mucilage hydrolysis.   Mucilages generally similar to that of flax are obtained from seeds of *Brassica alba* and from *Lepidium sativum*, in both cases cellulose fibrils being dispersed in a protein-gum mixture.   The gums yield mixtures of arabinose, galactose, rhamnose, and galacturonic acid as in the flax seed product.   The mucilage of slippery elm (*Ulmus fulva*) bark yields the same hydrolysis-resistant 2-galacturonopyranosylrhamnose as flax seed mucilage but is combined with D-galactose.[10]   Mucilages from seeds of *Plantago psyllium* and related species have been much investigated,[11] particularly since psyllium mucilage is used medicinally as a bulk-forming material.   This substance yields L-arabinose, D-xylose, and D-galacturonic acid, the arabinose and galacturonic acid being bound as an aldobiuronic acid.   The mucilage of *Plantago fastigiata* is similar in general constitution and consists mainly of arabinose and xylose in the proportion of eight to seventeen pentose molecules in combination with one molecule of D-galacturonic acid.   Methylation of mucilage from *Plantago arenaria* has been carried out by Nelson and Percival[12] and yields on hydrolysis trimethyl-, dimethyl-, and monomethylxylopyranose, tetramethylgalactopyranose, and other unidentified derivatives of arabinose.   Galacturonic acid is present as an aldobiuronic acid with xylose.

In summary, the mucilages consist of protein, cellulose fibers (occasionally), and uronides which usually contain galacturonic acid.   Associated with the galacturonic acid are sugars including rhamnose, galactose, arabinose, or xylose.

The Gel-Forming Substances.   The gel-forming substances, typified by agar, are found in algae, particularly marine algae, and are characterized by their ability to swell in water with the formation of rigid gels. The gel-forming substances may be separated according to the organisms producing them, the red algae producing agar and carrageenin, while the brown algae contain alginic acid, lamarin, and fucoidin, as has been summarized by Tseng.[13]

Agar is obtained from various species of the genus *Gelidium*, a red alga harvested commercially in the Pacific ocean on the coasts of California, China, and Japan.   A similar polysaccharide is found in the genera

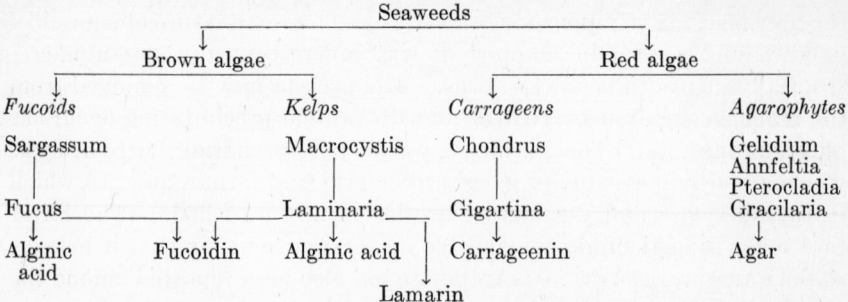

FIG. 11-1. Classification of gel-forming substances according to organisms producing each. (After Tseng[13])

Gracilaria, Pterocladia, and *Ahnfeltia.* The agar, which is a cell wall constituent, is soluble in hot but not in cold water, and the hot solutions when cooled set to a rigid gel, even when the polysaccharide is present in concentrations of 1% or less. The structure of agar has been largely elucidated in recent years and shown to consist mainly of D-galactopyranose residues linked through 1,3 linkages.[14] At the reducing end of each chain of such residues, one L-galactopyranose residue is attached through carbon atom 4. This L-galactopyranose residue is esterified with sulfuric acid at position 6. Thus the basic unit of agar according to Jones and Peat[14] is as follows:

—, -3-D-galactose-1,3-D-galactose-1,3-D-galactose-1,4-L-galactose-6-sulfate

The linear chains contain a hundred or more residues.[15] These chains must in turn be associated into longer units since agar is nonreducing and does not yield any large amount of tetramethylgalactose upon methylation and hydrolysis. Because of the presence of the ethereal sulfate, agar is an acid and can be prepared either as the free acid or as the various salts. The difference in properties between agar and the usual galactans must be largely ascribed to the presence of this one particular feature. Jones and Peat[14] suggest that agar may arise through the condensation of galactose-1-sulfate residues just as starch is condensed from Cori ester, and that the ethereal sulfate on carbon atom 6 arises by an occasional exchange of sulfate between carbon atoms 1 and 6. This attractive idea deserves further investigation from an enzymatic standpoint.

The red alga *Chondrus crispus* or carrageen moss yields a polysaccharide which like agar is a galactose-containing ethereal sulfate. The galactose residues appear to be joined in 1,3 linkages but the sulfuric ester is linked through carbon atom 4 rather than carbon atom 6 as in agar.[16] In addition to galactose, the polysaccharide contains glucose and other unidentified material.

Alginic acid, a cell wall constituent of numerous seaweeds including species of *Macrocystis, Laminaria,* and *Fucus,* appears to be similar to pectic acid but constituted of D-mannuronide rather than of galacturonide residues.[17] The D-mannuronide residues are probably linked through carbon atoms 1 and 4 with β-glycosidic linkages as in pectic acid.[18] In the alga, alginic acid occurs as a mixed salt and may make up 60% of the dry weight of the organism. Alginic acid and its salts have extensive commercial applications, the largest use being as a stabilizer for ice cream. Use of the colloid in ice cream imparts smooth texture and inhibits ice crystal formation. In addition ammonium alginate is used as a creaming agent for latex, and in general alginates find uses as suspending and emulsifying agents.

Fucoidin like agar is an ethereal sulfate but one which gives rise on hydrolysis to a mixture of sugars including the methylpentose, fucose. Lamarin, a cell wall component of *Laminaria,* is a polyglucose derivative in which β-glucopyranose units are linked through carbon atoms 1 and 3[19] rather than through carbon atoms 1 and 4 as in cellulose.

General References

Mantell, C. L., The Water Soluble Gums. Reinhold, 1947.

References

1. Jackson, J., and Smith, F., *J. Chem. Soc.,* 74, 79 (**1940**).
2. Hirst, E. L., *J. Chem. Soc.,* 70 (**1942**).
3. Anderson, E., and Otis, L., *J. Am. Chem. Soc.,* **52,** 4461 (1930).
4. White, E. V., *J. Am. Chem. Soc.,* **68,** 272 (1946).
5. Hirst, E. L., and Jones, J. K. N., *J. Chem. Soc.,* 1174 (**1938**); 506 (**1946**).
6. James, S. P., and Smith, F., *J. Chem. Soc.,* 739, 746, 749 (**1945**).
7. Reeves, R. E., and Goebel, W. F., *J. Biol. Chem.,* **139,** 511 (1941).
8. Stacey, M., Advances in Carbohydrate Chemistry. Academic Press, 1946. Vol. 2, p. 161.
   Haworth, W. N., and Stacey, M., *Ann. Rev. Biochem.,* **17,** 97 (1948).
9. Anderson, E., and Crowder, J., *J. Am. Chem. Soc.,* **52,** 3711 (1930).
   Tipson, R. S., Christman, C. C., and Levene, P. A., *J. Biol. Chem.,* **128,** 609 (1939).
10. Gill, R. E., Hirst, E. L., and Jones, J. K. N., *J. Chem. Soc.,* 1025 (**1946**).
11. Anderson, E., and Fireman, M., *J. Biol. Chem.,* **109,** 437 (1935).
    Anderson, E., Gillette, L. A., and Seeley, M. G., *J. Biol. Chem.,* **140,** 569 (1941).
12. Nelson, W., and Percival, E., *J. Chem. Soc.,* 58 (**1942**).
13. Tseng, C. K., *Sci. Monthly,* **58,** 24 (1944); *Econ. Bot.,* **1,** 69 (1947).
14. Forbes, I. A., and Percival, E. G. V., *J. Chem. Soc.,* 1844 (**1939**).
    Jones, W. G. M., and Peat, S., *J. Chem. Soc.,* 225 (**1942**).
15. Barry, V. C., and Dillon, T., *Chemistry & Industry,* 167, 1944.
16. Buchanan, J., Percival, E. E., and Percival, E. G. V., *J. Chem. Soc.,* 51 (**1943**).
17. Nelson, W., and Cretcher, L., *J. Am. Chem. Soc.,* **54,** 3409 (1932).
18. Hirst, E. L., Jones, J. K. N., and Jones, W. O., *J. Chem. Soc.,* 1880 (**1939**).
19. Barry, V. C., Dillon, T., and McGettrick, W., *J. Chem. Soc.,* 183 (**1942**).
    Barry, V. C., *J. Chem. Soc.,* 578 (**1942**).

CHAPTER 12

LIGNIN

Lignin occurs as a constituent of secondary cell walls, and is found in general in hard mature tissues such as those of wood and straw. It may occur in amounts varying from a few per cent in woods of some deciduous trees to 50% or more on a dry weight basis in the woods of coniferous species. The lignin is deposited in the interstices between the micro-fibrils and micellar strands of cellulose and the other microcrystalline components of the wall. Because of this mode of distribution removal of lignin leaves the cell wall morphologically intact but with an interconnecting system of spaces replacing the lignin. Lignin differs from all other major cell wall constituents thus far discussed in that it is not a carbohydrate or simple carbohydrate derivative, but is rather a condensation product of one or more types of aromatic nuclei into a high-molecular-weight aromatic complex.

Methods for Obtaining Lignin. Lignin cannot apparently be obtained in an unchanged state free of other cell wall constituents. Methods for its separation are based on either (a) the removal of cellulose and other materials to leave lignin as an insoluble residue or (b) the conversion of lignin into soluble derivatives or fragments which may be extracted from the other cell wall constituents. The method of Klason, much used at one time, depends on treatment of the tissue in 72% sulfuric acid, a treatment which hydrolyzes and removes cellulose and other polysaccharides.[1] The residue, diluted with water and filtered off, represents a crude lignin preparation, although because of the high acid concentration employed, it is to be doubted that it represents unaltered lignin. An important modification of this method is that of Willstätter and Zechmeister in which 40–42% (fuming) HCl is used to dissolve out cellulose and other polysaccharides.[2] In this procedure also the lignin is left as an insoluble remainder. Freudenberg and others remove the cell wall polysaccharides by successive treatment with alkali to remove hemicelluloses, hot dilute acids to remove readily hydrolyzable pentosans, etc., and cuprammonium solution to remove the cellulose.[3] In this procedure a certain amount of lignin is lost by solution in the alkali and in the cuprammonium solution. The lignin obtained should, however, represent the native lignin of the cell wall rather more closely than that obtained with the other methods, because of the relatively mild procedures used.

An important method for removal of lignin from wood, a method used in the paper pulp industry, is the treatment of the material with hot aqueous sulfite under pressure. This procedure yields water-soluble sulfonated lignin derivatives. Similarly, treatment of wood with bromine or chlorine yields halogenated lignin derivatives which are soluble in alcohol and which may then be readily removed. Neither the sulfite nor the halogen product has, however, been extensively used in the study of lignin structure.

Treatment of lignin-containing tissue with alkali results in solubilization of the lignin, the ease of solution depending on the tissue involved. Thus the lignin of straw and of corn cobs can be extracted with 1.5–2% alcoholic NaOH in the cold and may be recovered by acidification after removal of the alcohol.[4] Wood lignin is more difficult to extract; treatment with 4–5% NaOH (aqueous or alcoholic) in an autoclave at eight atmospheres pressure for one or more hours is necessary. In this case also the lignin may be recovered by removal or dilution of the alcohol and acidification of the extract. Alkaline delignification constitutes a second important paper pulp process. Lignin may also be extracted in whole or part by heating in HCl dissolved in methyl, ethyl, butyl, or amyl alcohol; in ethanolamine, in various phenols, and in other solvents. In all these cases it is probable that the lignin molecule is altered in the course of solubilization.

General Properties of Lignin. Lignin is amorphous both in the cell wall and in the isolated condition. In the cell wall lignin is isotropic but differs optically from other cell wall constituents in possessing the remarkably high index of refraction of 1.61. Lignin preparations are ordinarily colored brown or yellow but in addition show absorption bands in the ultraviolet between 2600 Å and 2900 Å, a property indicating the presence of aromatic nuclei. Lignin is unstable to oxidizing agents as hypochlorite, permanganate, and H_2O_2, the oxidation yielding ultimately simple fatty and dicarboxylic acids.

The elementary analysis of lignin yields 62–65% carbon and 5–6% hydrogen, and it is hence clear that highly unsaturated aromatic compounds must be involved. Methoxyl groups occur in lignin, the amount varying from 10% to 21%, depending on plant species and on the method of isolation of the material. Free hydroxyl groups which can be methylated with dimethyl sulfate are also present. According to Freudenberg these hydroxyl groups represent secondary and tertiary alcohols to the exclusion of primary alcohol groups. Free phenolic groups are calculated to comprise only 0.6% of lignin.[5] Lastly, lignin contains ether linkages which in the case of spruce lignin amount to approximately 9% of the whole.

Degradation Products of Lignin. An indication that lignin contains aromatic nuclei is given by the results of dry distillation of lignin under reduced pressure. Thus dry distillation of sprucewood lignin yields eugenol and isoeugenol.[5,6]

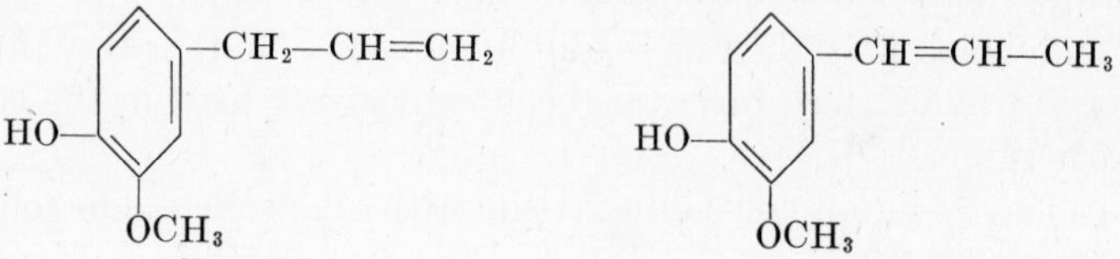

Eugenol   Isoeugenol

Lignin of corn cobs or of corn stalks yields on distillation a more complex mixture including catechol, phenol, guaiacol, o- and p-cresol, n-propyl-guaiacol, and other aromatic compounds.[7]

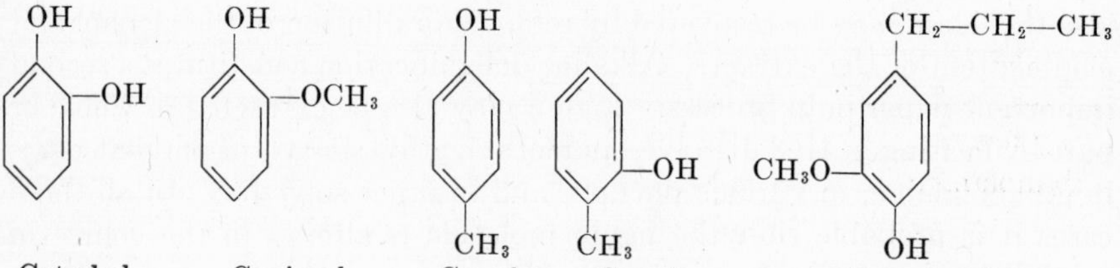

Catechol   Guaiacol   p-Cresol   o-Cresol   n-Propylguaiacol

Distillation with zinc dust in an atmosphere of hydrogen has been carried out with corn cob lignin and here again, catechol, guaiacol, and n-propyl-guaiacol are obtained.[8] Potassium hydroxide fusion has been used in the degradation of lignin and from the fusion products a mixture of compounds including acetic, oxalic, and protocatechuic acids as well as catechol have been isolated in substantial yields.[9,18]

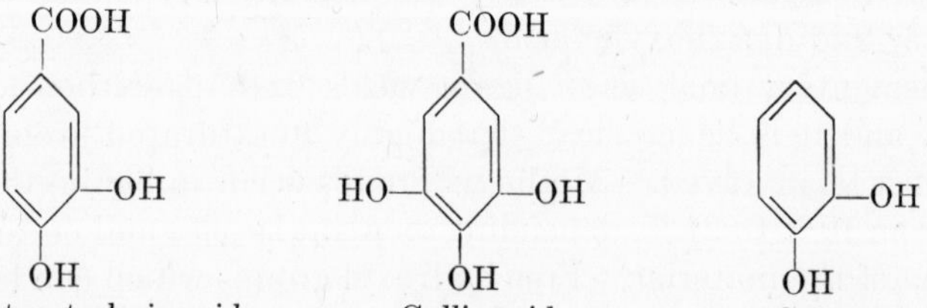

Protocatechuic acid   Gallic acid   Catechol

These results again indicate the probable participation of phenolic nuclei in the constitution of lignin. Freudenberg has used an extension of the alkali fusion technique for identification of the nuclei which actually participate in the lignin molecule.[10] The fusion products were methylated with dimethylsulfate and diazomethane and then oxidized with permanganate. Thus spruce lignin was treated with 70% NaOH at 165°C for ninety minutes. The products were next methylated and then oxidized. Veratric and isohemipinic acids were isolated in this way in yields of 21% and 12% of the original lignin, respectively. If ethylation rather than methylation was applied to the fusion products, 3-methoxy-4-ethoxybenzoic acid was obtained, showing that the methoxyl group at

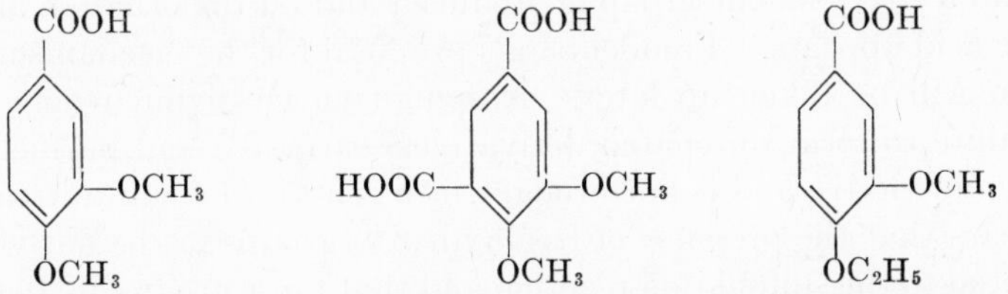

Veratric acid   Isohemipinic acid   3-Methoxy-4-ethoxy-benzoic acid

position 3 in veratric acid is present in the original lignin, but that the second phenolic group is freed or produced during the alkali treatment. Both veratric and isohemipinic acids are themselves partially destroyed by the treatment. If the amounts recovered are corrected for this destruction spruce lignin may be calculated to yield 81% of isohemipinic acid and 33% of veratric acid.[5] Although these figures are of course only semiquantitative, still the important fact is that the two substances can apparently account for a large proportion of the lignin molecule.

Sulfonation of Lignin. When material containing lignin is heated with an aqueous solution of acid sulfite, the lignin is converted to water soluble materials, a procedure which as mentioned above is the basis of one type of delignification of wood used in the preparation of paper pulp. The soluble products formed are salts of lignin-sulfonic acid derivatives. Acidification of the sulfite liquor results in the precipitation of a mixture of the free lignosulfonic acids, although a more soluble residue also remains in solution.

When coniferous sulfite lignin is refluxed with hot alkali the aromatic aldehyde vanillin is produced in yields which may amount to as much as 6% of the weight of lignin.[11] Similar treatment of lignin from hardwoods such as oak and maple yields syringaldehyde in addition to vanillin.

The mechanism of the sulfonation reaction has been studied by Hägglund and Carlsson[12] who found that for each sulfonic acid group introduced into the lignin molecule, one phenolic hydroxyl group is set

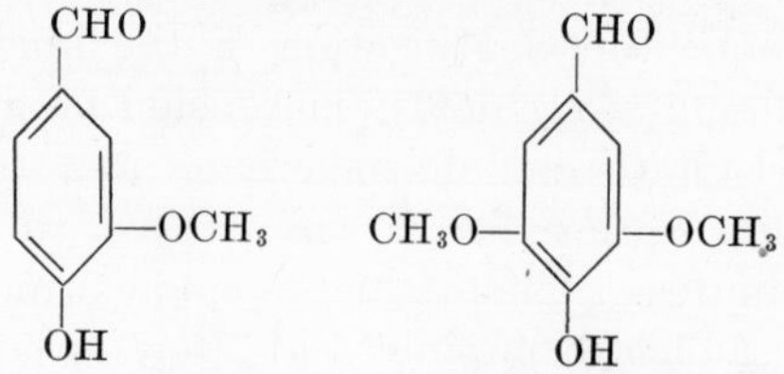

Vanillin   Syringaldehyde

free, an observation confirmed by Freudenberg.[13] In lignin sulfonation
a phenolic linkage must therefore be broken, with the result that on the
one hand a free phenolic group is produced and on the other an aliphatic
sulfonic acid appears. Freudenberg's proposal for the mechanism of this
reaction will be taken up later. Although native lignin reacts readily
with sulfite to form sulfonated derivatives, extracted and purified lignin
is quite unreactive and is not brought into solution by sulfite treatment.
It appears that the presence of free hydroxyl groups in the native lignin
is essential to the sulfonation reaction and that these groups are destroyed
or masked during the isolation of lignin. Thus if wood is methylated
with diazomethane, the lignin can no longer be removed with bisulfite.
If wood is acetylated before isolation of the lignin, the free hydroxyl
groups are protected and the isolated, deacetylated lignin is fully reactive
toward acid sulfite.[14]

Structure of Lignin. No final structure can be assigned to lignin at
present. Nevertheless, on the basis of the considerable amount of
accumulated evidence, very concrete suggestions as to the possible general
form of the lignin molecule may be put forward. It is evident that the
molecule must consist of aromatic nuclei, and Klason early suggested that
lignin might be formed by the condensation of coniferal alcohol. More
recently Freudenberg[5] has developed a detailed proposal as to the structure of lignin. In the first place the degradation experiments show that
lignin of spruce must contain aromatic nuclei of the types I and II below,
while hardwood lignin must also contain nuclei of type III.

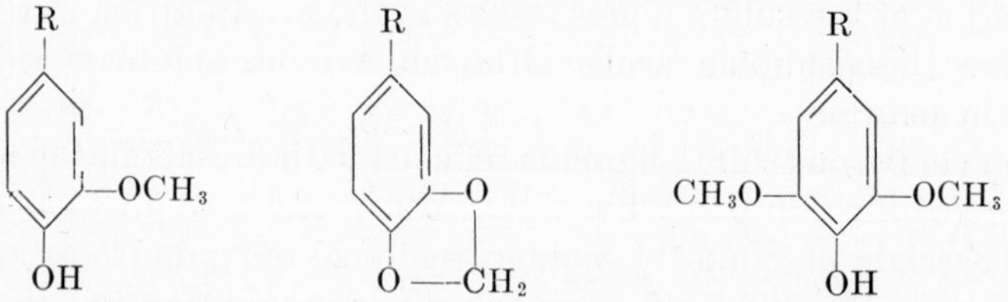

I   II   III

According to Freudenberg, lignin may be built up from molecules of the
type of isoeugenol which are condensed to form products of the dehydro-

diisoeugenol type. This condensation, which does in fact take place under mildly oxidizing conditions, could then be continued to form chains

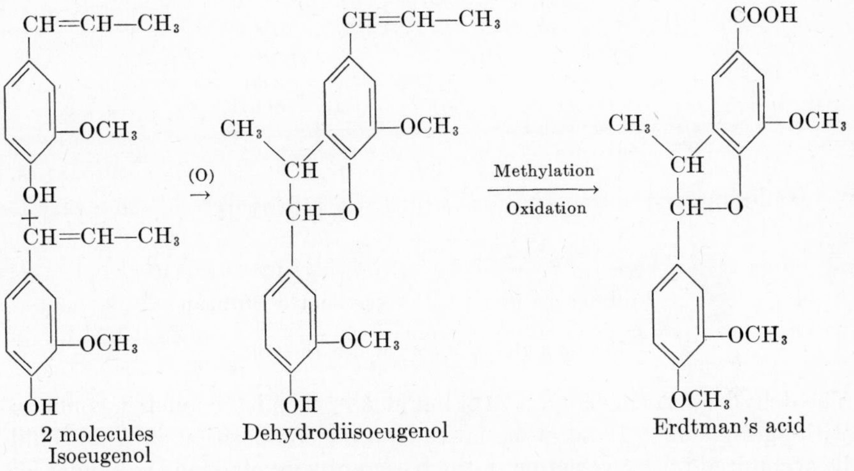

Fig. 12-1. Condensation of isoeugenol to form dehydrodiisoeugenol as a model for formation of lignin-like materials.

of indefinite length. Nuclei other than isoeugenol may also participate in the formation of the polymer, including the propylene derivatives of the two aromatic nucleii II and III shown above. In addition, the propylene side chain might be replaced by other three-carbon side chains of the types shown below.

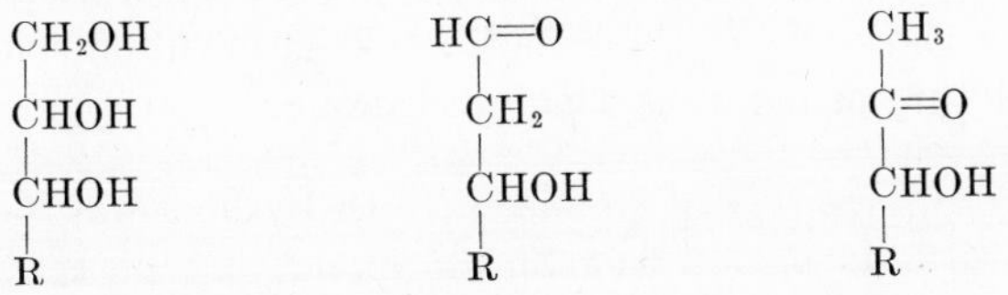

Support of Freudenberg's proposal for the structure of lignin lies in the fact that dehydrodiisoeugenol may be methylated and oxidized, in which case it yields 20% of veratric acid and 5% of isohemipinic acid, amounts not greatly different from those obtained by similar treatment of spruce lignin.

Erdtman's acid, which is obtained by mild oxidation of methylated dehydrodiisoeugenol, reacts with sulfurous acid, a reaction resembling that found in the case of lignin, since in both cases one phenolic group is liberated per mol of sulfite taken up.[15]

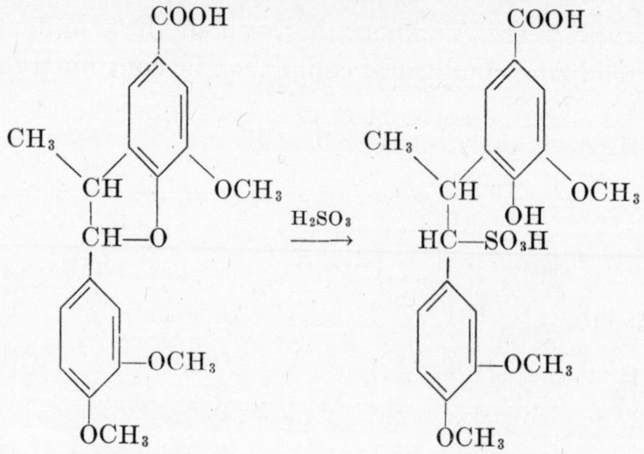

Erdtman's acid   Sulfonated Erdtman's
                                     acid

Reaction of Erdtman's acid with sulfite

The dehydrodiisoeugenol model of lignin does then have much in common with lignin itself. In addition, latitude in the composition of lignin could be accounted for by variation in the basic units involved in the condensation. Even so, Freudenberg's picture of lignin structure, while it accounts for many properties of lignin, does not satisfactorily account for all, and in particular does not suggest any ready interpretation for the presence of dimethylpyrogallol derivatives in hardwood lignins. Various modifications of the Freudenberg structure have therefore been put forward. Chief among these is the suggestion by Russell[16] and by Ritter and others[17] that the basic unit involved is a flavanone rather than a benzpyrane derivative, i.e., the three carbon atoms of each side chain may all be involved in formation of a second ring. This conclusion, arrived at by Ritter from the study of the structure of lignin sulfonic acids, leads to a tentative picture of lignin as a polyflavanone.

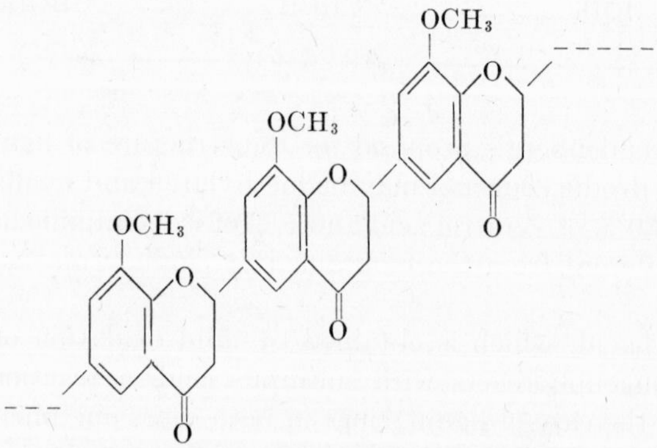

Proposed polyflavanone structure of lignin

Polyflavanone of this structure has been synthesized by Russell and reported to resemble gymnosperm lignin in general characteristics. It is still too early to draw a conclusion as to the detailed structure of lignin, but the main outlines of the structure involved are at least now evident.

Metabolism of Lignin.[18] Various authors have suggested that lignin might be formed from hemicellulose, pentosan, or pectic substances in the cell wall. That the conversion of a polysaccharide or uronide to a complex aromatic polymer should take place in the wall would seem however to be improbable. It is more likely that the basic eugenol-like units may be manufactured in the cytoplasm of the cell and then escape into the wall where they are polymerized. Nothing is known of the biochemistry of this process. Lignin once deposited does not appear to be re-utilized by the plant and is on the contrary lost to metabolism. It is among the most resistant of native plant materials to the attacks of microorganisms, and is decomposed little or not at all over short periods of time. Wood-destroying fungi attack lignin slowly over periods of months or years but nothing approaching complete breakdown of lignin by microorganisms has yet been demonstrated. Lignin is similarly utilized little if at all by the animal organism.

General References

Wise, L. E., Wood Chemistry. Reinhold, 1944.

Brauns, F. E., *Fortschr. Chem. organ. Naturstoffe*, **5**, 175 (1948).

Brauns, F. E., The Chemistry of Lignins. Academic Press. In press.

References

1. Klason, P., *Cellulosechem.*, **4**, 81 (1923).
2. Willstätter, R., and Zechmeister, L., *Ber.*, **46**, 2401 (1913).
3. Freudenberg, K., Zocher, H., and Dürr, W., *Ber.*, **62**, 1814 (1929).
4. Phillips, M., *J. Am. Chem. Soc.*, **49**, 2037 (1927).
5. Freudenberg, K., *Ann. Rev. Biochem.*, **8**, 81 (1939).
6. Pictet, A., and Gaulis, M., *Helv. Chim. Acta*, **6**, 627 (1923).
7. Phillips, M., and Goss, M. J., *Ind. Eng. Chem.*, **24**, 1436 (1932).
   Bridger, G. L., *Ind. Eng. Chem.*, **30**, 1174 (1938).
8. Phillips, M., and Goss, M. J., *J. Am. Chem. Soc.*, **54**, 1518 (1932).
9. Phillips, M., and Goss, M. J., *J. Biol. Chem.*, **114**, 557 (1936).
10. Freudenberg, K., Janson, A., Knopf, E., and Haag, A., *Ber.*, **69**, 1415 (1936).
    Also Freudenberg, K., *et al.*, *Ber.*, **70**, 500 (1937); **71**, 1810 (1938).
11. Kürschner, K., *J. prakt. Chem.* **118**, 238 (1928).
    Tomlinson, G. H., II., and Hibbert, H., *J. Am. Chem. Soc.*, **58**, 345 (1936).
12. Hägglund, E., and Carlsson, G., *Biochem. Z.*, **257**, 467 (1933).
13. Freudenberg, K., Sohns, F., and Janson, A., *Ann.*, **518**, 62 (1935).
14. Hibbert, H., and Steeves, W. H., *J. Am. Chem. Soc.*, **59**, 1768 (1937).
15. Freudenberg, K., Meister, M., and Flickinger, E., *Ber.*, **70**, 500 (1937).
16. Russell, A., *Science*, **106**, 372 (1947).
17. Ritter, D. M., Pennington, D., Olleman, E., Wright, K., and Evans, T., *Science*, **107**, 20 (1948).
18. Wise, L. E., Wood Chemistry. Reinhold, 1944.

CHAPTER 13

COMPOSITION OF THE CELL WALL

It is the purpose of this chapter to bring together the material of the foregoing chapters on the individual cell wall constituents, to summarize the properties and reactions of these constituents, and to present an over-all picture of the relative amounts of the individual components found in the cell walls of various tissues. It is surprising that only rarely have complete balance sheets for the composition of cell walls been prepared. Individual investigators have, rather, concerned themselves with one or another component or have carried out their analyses on whole tissues in such a way that composition of the cell wall alone cannot be calculated. It will be of importance in future biochemical investigations to separate clearly the cell wall from the protoplasmic contents in analyses of tissues.

Table 13-1 gives a summary of the chemical and physical character-

TABLE 13-1. A summary of the reactions of the several principal cell wall constituents. (Modified from Frey-Wyssling[1])

Substance	Behavior in polarized light	Absorption in ultraviolet light
Cellulose	Double refracting, optically positive	Little absorption
Pectic substances	Statistically isotropic	" "
Lignin	Optically isotropic (Lignin residues show double refraction of form)	Strong absorption
Noncellulosic polysaccharides	In general double refracting	Little absorption
Polyuronide hemicelluloses	Isotropic	" "
Cuticular substances	Double refracting, optically negative	Strong absorption

Substance	Solubility and hydrolysis behavior	Solubility in Cross and Bevan treatment
Cellulose	a. Soluble in cuprammonium solution	
	b. Soluble in concentrated (above 72%) sulfuric acid	
	c. Insoluble in hot dilute acid or alkali	Insoluble
Protopectin	a. Insoluble in cuprammonium solution	
	b. Soluble in dilute (3–5%) H_2O_2 at 50°C and in warm dilute ammonium oxalate	

130

TABLE 13-1 *(Continued)*

Substance	Solubility and hydrolysis behavior	Solubility in Cross and Bevan treatment
	c. Soluble in hot dilute acid	Soluble
Calcium pectate	a. Insoluble in cuprammonium solution	
	b. Soluble in dilute H_2O_2 at 50°C and in warm dilute ammonium oxalate	
	c. Soluble in hot alkali	Little soluble
Lignin	a. Insoluble in cuprammonium solution	
	b. Insoluble in concentrated sulfuric acid	
	c. Little soluble in hot dilute acid and alkali	Soluble
Noncellulosic poly-saccharides	a. Hydrolyzed in hot dilute acids	
	b. Soluble in cold concentrated or hot dilute alkali	
	c. Soluble in cuprammonium solution	
	d. Insoluble in hot dilute H_2O_2 (3–5% at 50°C)	Little soluble
Polyuronide hemi-celluloses	a. Soluble in cold 4–5% NaOH	Soluble
Cuticular substances	a. Insoluble in cuprammonium solution	
	b. Insoluble in concentrated cold or hot dilute acid	
	c. Soluble in hot concentrated alkali	
	d. Soluble in hot dilute alcoholic KOH	Partially soluble

Substance	Color reactions
Cellulose	a. Dichroic violet color with chlorzinc iodine
	b. Blue color with I_2-KI in concentrated sulfuric acid
	c. Dichroic color with congo red
	d. Little or no color with ruthenium red
Pectic substances	a. Red color with ruthenium red
Lignin	a. Red color with phloroglucinol in HCl
Noncellulosic polysaccharides and polyuronide hemicelluloses	a. No specific staining reactions
Cuticular substances	a. Red color with scarlet H
	b. Orange-red color with Sudan III

istics of the individual cell wall components and can serve as a general guide to the separation or identification of wall constituents. Particular attention should be paid to the noncellulosic polysaccharides including araban, xylan, mannan, and galactan, as well as to the polyuronide hemicelluloses, since these materials show somewhat similar and partially overlapping solubility properties.[2] The noncellulosic polysaccharides are soluble in hot dilute or cold concentrated alkali and are less readily removed by the concentration of cold alkali (4–5%) needed to remove many of the polyuronide hemicelluloses. Alkaline extracts containing the latter may be expected, however, to contain small amounts of polysac-

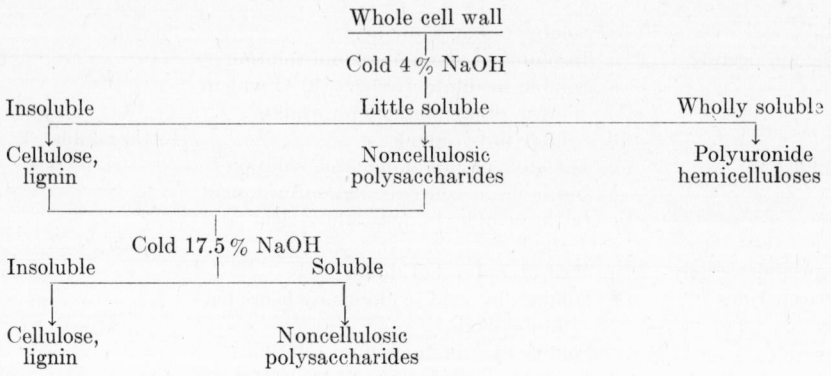

FIG. 13-1. Fractionation of cell wall material with alkali. (General method of O'Dwyer[3])

charide as well. Isolation and determination of noncellulosic polysaccharides may be done on Cross and Bevan cellulose from which the polyuronide hemicelluloses as well as lignin have been removed by chlorination and sulfite treatments.

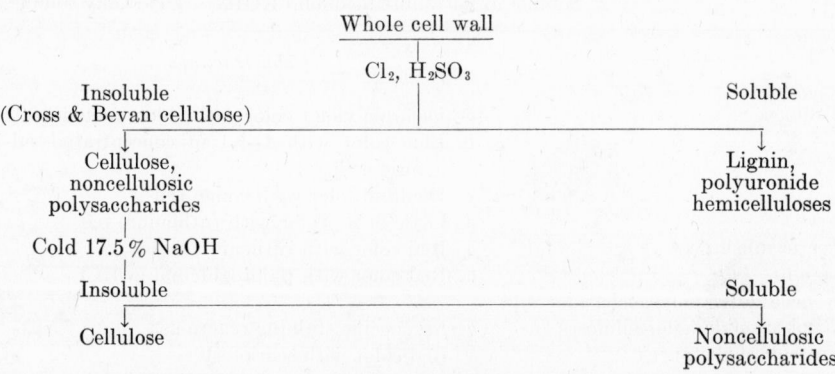

FIG. 13-2. Fractionation of cell wall material after Cross and Bevan[4] treatment for removal of lignin and polyuronide hemicellulose.

Three general methods for fractionation of cell wall material are given in Figs. 13-1, 13-2, and 13-3. In the first procedure, the tissue is initially extracted with cold dilute alkali for removal of polyuronide hemicelluloses and the residue then extracted with more concentrated alkali for removal of noncellulosic polysaccharides. This procedure is primarily useful for preparation of polyuronide hemicelluloses. In the second method, the Cross and Bevan extraction is initially applied, and the insoluble residue is then fractionated by cold concentrated alkali. This procedure, which is the one which has been most commonly used, gives only a partial picture of cell wall composition and is deficient primarily in any accurate estimate of the polyuronide hemicellulose. Probably the most satisfactory procedure for estimation of individual cell wall constituents is that in which an initial extraction with chlorine and alcoholic ethanolamine (or a modification of this procedure[6]) is applied. This extraction

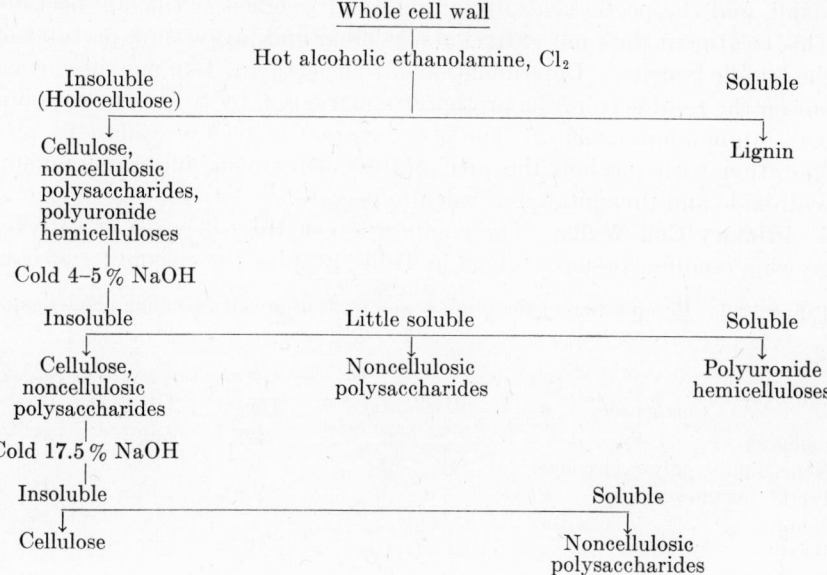

FIG. 13-3. Fractionation of cell wall material after removal of lignin with ethanolamine. (Modified after Ritter and Kurth[5])

removes primarily lignin, and the insoluble residue may now be fractionated with alkali for separation of cellulose, noncellulosic polysaccharides, and polyuronide hemicelluloses.

The separation and estimation of pectic substances is not included in the schemes 13-1, 2, and 3 since pectic substances make up only a very small proportion of woody tissues. The amount of pectic materials in woody tissues can, however, be determined analytically.[19] For this purpose the wood is first extracted with cold 4% NaOH to remove polyuronide hemicelluloses and is then heated for two to four hours at 100°C with 0.05 N HCl. This treatment transforms protopectin to soluble pectin which may then be filtered off, converted to pectic acid by hydrolysis with dilute alkali, and precipitated as calcium pectate for gravimetric determination.   A portion of the pectic content of wood appears to remain in the residue after this treatment and may represent pectates of the middle lamella.   In any case, chlorine-sulfite treatment of the residue liberates further small amounts of alkali extractable pectic material.

The determination of the pectic constituents of pectin-rich primary walls presents a much simpler problem than does the determination with woody tissues.   The whole cell wall material may be heated in hot, dilute (0.05 N) acid to convert protopectin to pectin.   The water soluble pectin thus obtained is then converted to pectic acid with 0.1 N alkali, and the pectic acid precipitated and weighed as calcium pectate. This treatment does not extract the calcium and magnesium pectates of the middle lamella.   Determination of this pectic fraction may be carried out on the residue from the protopectin extraction by treatment with hot 0.5% ammonium oxalate.   The pectic extract is freed of oxalate by precipitation with alcohol, the precipitate taken up in dilute ammonium hydroxide and precipitated as calcium pectate.

Primary Cell Walls.   The composition of the cell walls of actively growing seedling tissues is given in Table 13-2 for the coleoptiles of oats

TABLE 13-2.   Composition of the primary cell walls of growing seedling plant tissues

		Per cent of total cell wall material		
		Corn coleoptile		Sunflower
Component	Oat coleoptile[7]	After[8]	After[10]	hypocotyl[9]
Cellulose	42	40	36	38
Noncellulosic polysaccharides	38	32	30	8
Pectic substances	8	..	13	46
Lignin	...	..	...	8
Wax	...	10	21	...
Protein	12	..	...	...
Total	100	82	100	100

and corn and for the sunflower hypocotyl.   All the analyses are incomplete in the sense that the noncellulosic polysaccharide fraction undoubtedly includes also the hemicellulose.   In the two cereals, cellulose and noncellulosic polysaccharide make up the bulk of the growing wall.   In the case of the oat coleoptile a small amount of protein appears to be firmly bound to the cell wall material.   In the sunflower hypocotyl noncellulosic polysaccharides are also present but are quantitatively of less importance, and cellulose and pectic substances make up the bulk of the cell wall.   All these tissues are marked by low lignin content.   The primary wall once deposited probably does not change greatly in composition and such changes as do occur in wall composition during maturation of the tissue are in the main associated with the deposition of the secondary walls, especially those of the woody elements.   These changes have been studied by Griffioen[9] in the hypocotyl of the sunflower. Sunflower plants were allowed to develop in the field and samples from the hypocotyl region removed periodically. Figure 13-4 shows that the overall composition altered during growth by extensive deposition of noncellulosic polysaccharide ("pentosan") and lignin, while pectic substances, characteristic of the primary wall, diminished in concentration

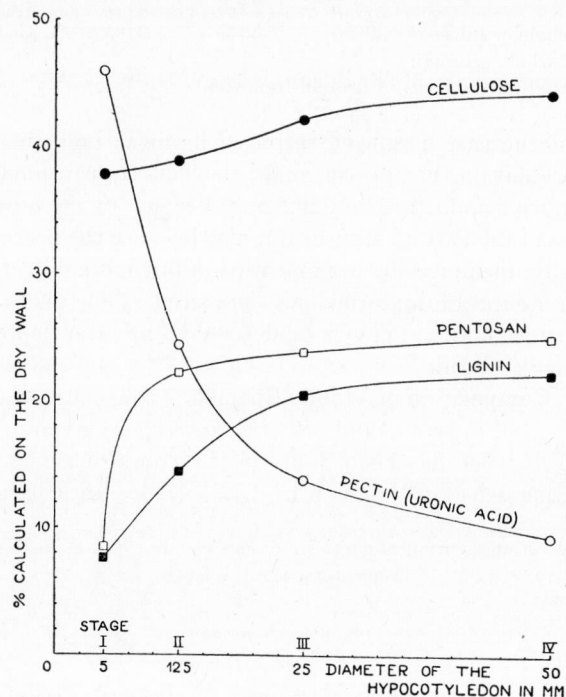

FIG. 13-4. Changing composition of the hypocotyl during development of sunflower plants in the field. (After Griffioen[9])

as they were diluted by nonpectic constituents. Similar observations have been made in the maturation of wood of *Fraxinus* and of *Pinus*, as shown in Table 13-3. In these experiments the wood was separated into three portions: (a) the youngest elements immediately adjacent to the cambium, (b) the recently matured wood, and (c) the mature sapwood. The youngest tissues, those in which the wall material is presumably mainly the primary wall, generally resemble the primary wall tissues cited above in having a moderate cellulose content together with high pectin and noncellulosic polysaccharide contents. As the wood matures the pectic substances decrease markedly in concentration as lignin is deposited. The parallelism between decrease in concentration of pectic

TABLE 13-3. Changing composition of wood during maturation. (Allsop and Misra[11])

Fraction*	Fraxinus elatior			Pinus sylvestris	
	Cambium	New wood	Sapwood	Cambium	Sapwood
Whole (C and B) cellulose	28.6	54.9	59.3	31.8	62.4
Pectin	30.5	4.1	1.6	20.9	1.0
Uronide not in pectin	6.9	5.8	4.2	5.0	3.0
Lignin	6.5	24.6	21.2	10.8	26.4
Xylan, % of whole cellulose	26.9	33.7	24.5	15.0	7.9
Mannan, % of whole cellulose	....	....	....	21.9	6.2

* Recalculated on the basis of protein-free wood.

substance and increase in concentration of lignin in maturing tissues has led to the speculation that pectin might itself be transformed into lignin. Such a transformation is unlikely not only because of the basic difference in the natures of the two substances but also because the pectic substances do not actually disappear during maturation but merely decrease in concentration as nonpectic materials are deposited. The pectic substances of the primary walls can of course be detected long after deposition of the adjacent secondary wall.

Cell Wall Composition of Woody Species. The cell wall composition of mature wood of *Fraxinus* and of *Pinus silvestris* is given in Table 13-3. A great deal of work has been done on the composition of woods,[12] of which a further selection is given in Table 13-4. In general, woods of

TABLE 13-4. Composition of some American woods. (U. S. Forest Products Laboratory, after Wise[12])

Species	% of whole wood			% of Cross & Bevan cellulose	
	Total pentose	Cross & Bevan cellulose	Lignin	Pentose	α-Cellulose
Pinus ponderosa	7.4	57.4	26.6	6.8	62.1
Libocedrus decurrens	10.6	41.6	37.7	9.1	46.9
Sequoia sempervirens	7.8	48.4	34.2	7.4	78.8
Pinus monticola	7.0	59.7	26.4	5.3	64.6
Pseudotsuga taxifolia	6.0	61.5	....	5.3	....
Quercus densiflora	19.6	58.0	24.8	22.8	56.8
Prosopis juliflora	14.0	45.6	30.5	17.8	76.5
Hicoria ovata	18.8	56.2	23.4	21.9	76.3
Eucalyptus globulus	20.1	57.6	25.1	21.0	68.9

temperate tree species are characterized by moderate cellulose content, 25–40%, high lignin content, 25–35%, and variable amounts of the noncellulosic polysaccharides. Barks in general are even higher in lignin content than is wood (Table 13-5) and contain in addition moderate amounts of water- and alcohol-soluble substances such as tannins.

TABLE 13-5. Composition of bark of various trees. (After Richter[13])

Species	Lignin	Total pentosan	Ether-sol.	H_2O-sol.	Acid-alcohol-sol.
Black spruce	45.8	8.8	2.3	10.5	12.1
White birch	49.2	16.5	8.0	5.8	12.2
Beech	37.0	13.7	1.2	4.0	13.2
White maple	35.7	14.4	2.5	2.9	20.6

Cell Wall Composition of Straws and Grasses. The vegetative portions of cereals such as wheat, rye, and oat are characterized by their high content of polyuronide hemicelluloses, which make up 20–35% of the dry weight of the whole straw. The content of noncellulosic polysac-

TABLE 13-6. Composition of the straws of various cereals

	Per cent of whole straw		
Component	Wheat straw[14]	Rye straw[15]	Oat straw[15]
Ash	4.4	3.5	7.1
True cellulose	38.6	46.5	43.8
Xylan	}31.9	8.3	9.3
Polyuronide hemicellulose		33.4	22.8
Lignin	17.2	19.5	18.5
Protein	1.1	1.9	1.8
Pectin	1.0	0.3	1.1
Total	94.2	113.4	104.4

charide, mainly xylans, is relatively low and lignin content is lower than in wood. Straw tends to contain much mineral, detectable as ash, which reflects in part the silica content of the cell walls of the grasses. While the analyses of straw given in Table 13-6 were carried out on material which

TABLE 13-7. Composition of various forage grasses. (After Buston[16])

	Per cent of whole grass			
Component	Poa trivialis	Phleum pratense	Festuca rubra	Festuca pratensis
Ash	9.8	8.2	8.3	7.9
Pectic substances	0.8	1.5	0.9	1.6
Polyuronide hemicellulose	16.2	17.2	20.7	18.5
Cross & Bevan cellulose	22.2	20.8	21.4	24.1
Xylan in Cross & Bevan cellulose	3.2	4.3	3.9	3.7
H_2O-soluble	29.7	23.2	23.0	24.9
Lignin	8.5	....	....	....
Total	87.2	....	....	....

consisted mainly of stem tissue, those on forage grasses given in Table 13-7 were done on material which represents mainly leaves. Again, however, cellulose and polyuronide hemicelluloses are the principal constituents, with xylans making up a minor proportion of the whole. The forage grasses also differ from mature straw in containing strikingly less lignin as is shown in the one example of Table 13-7.

Table 13-8 gives data on the composition of the corn cob which, like other organs of the cereals, contains cellulose and polyuronide hemicelluloses as major constituents, with lignin making up a smaller portion of the total. In the highly woody bamboo stem on the other hand lignin makes up over 30% of the total weight, as in the other woods listed in Table 13-4.

TABLE 13-8.   Cell wall constituents of various plant materials

Component	Corn cob[17] % of cob	Hop flower[18] % of flower	Phyllostachys reticulata (Bamboo) % of stem
Ash	1.4	.....	2
True cellulose	38.3	30.8	41
Polyuronide hemicellulose	42.4	9.7	14
Noncellulosic polysaccharides	0.5	5.4	10
Lignin	16.7	54.1	32
Pectin	0.5	.....	...
Protein	3.2	.....	1
Total	103.0	100.0	100

In summary there are relatively few tissues for which complete balance sheets of cell wall composition can be drawn up. From the data at hand it seems that cellulose tends to make up one-fourth to one-half of the total, while lignin varies from very low levels in young tissues to 35% or more in mature woody tissues. Pectic substances in general present the reverse picture. While pectic substances may make up nearly 50% of the cell wall material of young rapidly growing cell walls (primary walls), they are in general found to the extent of 0.5–1.5% in mature tissues in which secondary cell wall formation has been completed. It is in the nature of their polyuronide hemicellulose and noncellulosic polysaccharide fractions that mature cell walls differ from one another most conspicuously. Of the noncellulosic polysaccharides, xylan is quantitatively important in a wide range of cell walls, including those of straws, leaves, and woods, and may make up from a few to 30% of the Cross and Bevan cellulose fraction. The polyuronide hemicelluloses appear to constitute 5–20% of the whole cell wall in many instances, although in particular cases as in the corn cob they may constitute almost one-half of the whole material.

References

1. Frey-Wyssling, A., Ernährung und Stoffwechsel der Pflanzen. Zürich, 1945.
2. Anderson, E., and Sands, L., Advances in Carbohydrate Chemistry. Academic Press, 1945, Vol. 1, p. 329.
3. O'Dwyer, M. H., *Biochem. J.*, **17**, 501 (1923); **20**, 656 (1926); **22**, 381 (1928).

4. Cross, C., and Bevan, E., Cellulose. Longmans Green & Co., 2nd ed., 1911.

5. Ritter, G., and Kurth, E., *Ind. Eng. Chem.*, **25,** 1250 (1933).

6. Wise, L. E., Murphy, M., and D'Addieco, A. A., *Paper Trade J.*, **122,** 35 (1946).

7. Thimann, K. V., and Bonner, J., *Proc. Roy. Soc. (London)*, **B113,** 126 (1933).

8. Nakamura, Y., and Hess, K., *Ber.*, **71,** 145 (1938).

9. Griffioen, K., *Rec. trav. botan. néerland.*, **35,** 323 (1938).

10. Wirth, P., *Ber.*, **56,** 175 (1938).

11. Allsopp, A., and Misra, P., *Biochem. J.*, **34,** 1078 (1940).

12. See Wise, L. E., Wood Chemistry. Reinhold, 1944.

13. Richter, G., *Ind. Eng. Chem.*, **33,** 75 (1941). See also Kurth, E. F., *Chem. Revs.*, **40,** 33 (1946).

14. Weihe, H. D., and Phillips, M., *J. Agr. Research*, **60,** 781 (1940).

15. Norman, A. G., *Biochem. J.*, **23,** 1353 (1929).

16. Buston, H. W., *Biochem. J.*, **28,** 1028 (1934).

17. Panassjuk, V. G., *Chem. Zentr.*, **III, 2,** 1506 (1940).
Whistler, R. L., Bachrach, J., and Bowman, D. R., *Arch. Biochem.*, **19,** 25 (1948).

18. Angell, S., and Norris, F. W., *Biochem. J.*, **30,** 2159 (1936).

19. Anderson, E., *J. Biol. Chem.*, **112,** 531 (1935).
Anderson, E., Seigle, L. W., Krznarich, P. W., Richards, L., and Marteny, W. W., *J. Biol. Chem.*, **121,** 165 (1937).

PART III. PLANT ACIDS AND PLANT RESPIRATION

Chapter 14

THE ORGANIC ACIDS OF PLANTS

Nature and Distribution of Plant Acids. Of the wide variety of aliphatic and aromatic organic acids contained in higher plants, one group of aliphatic acids, generally known simply as the plant acids, is distinguished by reason of wide distribution and physiological function. The best known of these plant constituents are malic, citric, succinic, tartaric, and oxalic acids. Less well known but probably of equally wide distribution and importance are isocitric, fumaric, *cis*-aconitic, oxaloacetic and α-ketoglutaric acids. This group of compounds includes then the six-

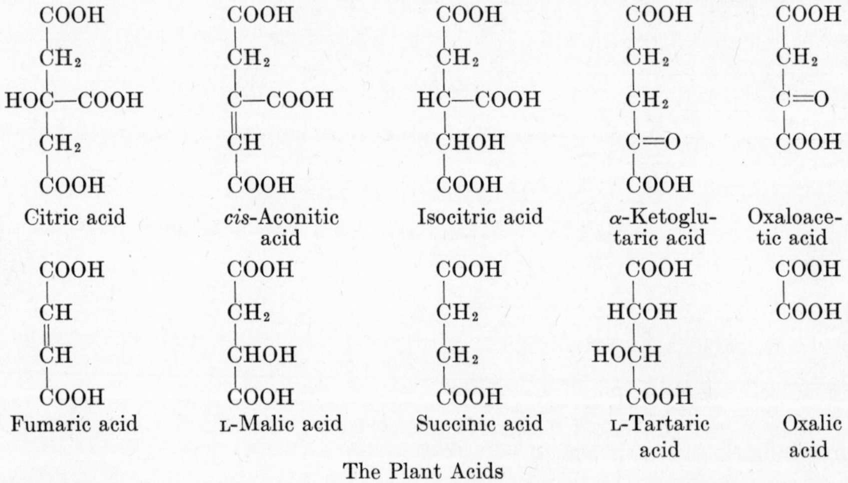

COOH	COOH	COOH	COOH	COOH
CH₂	CH₂	CH₂	CH₂	CH₂
HOC—COOH	C—COOH	HC—COOH	CH₂	C=O
CH₂	CH	CHOH	C=O	COOH
COOH	COOH	COOH	COOH	
Citric acid	*cis*-Aconitic acid	Isocitric acid	α-Ketoglutaric acid	Oxaloacetic acid
COOH	COOH	COOH	COOH	COOH
CH	CH₂	CH₂	HCOH	COOH
CH	CHOH	CH₂	HOCH	
COOH	COOH	COOH	COOH	
Fumaric acid	L-Malic acid	Succinic acid	L-Tartaric acid	Oxalic acid

The Plant Acids

carbon tricarboxylic acids, citric, isocitric, and *cis*-aconitic; the five-carbon dicarboxylic acid, α-ketoglutaric acid; the four-carbon dicarboxylic acids, oxaloacetic, malic, fumaric, succinic, and tartaric; and the two-carbon dicarboxylic acid, oxalic. One or more of these acids have been found in every plant tissue in which they have been carefully sought. In many instances also the acid may occur in large quantities; thus oxalic acid in particular may make up as much as 50 % of the dry weight of some leaves, while the plant acid fraction as a whole makes up in general from a few tenths to 30% of the dry weight of plant tissues.

Citric acid is a general constituent of plant tissues and is probably of universal or nearly universal distribution. It occurs in particularly high concentrations in the fruit of citrus and may make up 60% of the total soluble constituents of the edible portion of the lemon. The concentrations of citric acid reported in a variety of tissues are given in Table 14-1. Isocitric acid, the optically active isomer of citric acid, was discovered in plant tissues long after citric acid itself, and its distribution has been correspondingly less investigated. The D-isomer of isocitric acid is the principal acid of the blackberry[1] and of the leaves of the succulent plants such as *Bryophyllum*.[2] This acid for reasons which will become apparent below is, however, probably widely distributed, although in small concentrations.

Of the four-carbon dicarboxylic acids malic acid is the best known and appears to be of ubiquitous occurrence in plant tissues, making up substantial percentages of the total weight of fruits and leaves. Of the two possible stereoisomers only the L form is found in nature. Succinic and fumaric acids like malic are of wide distribution and are probably of general occurrence, although they occur in general in lower concentrations than does malic acid. Fumaric acid possesses the *trans* configuration at the double bond and the corresponding *cis* isomer, maleic acid, does not occur in plants.

Two keto acids, α-ketoglutaric and oxaloacetic, have not been the subject of as wide inquiry as have the other plant acids. Both, however, have been reported in pea and other seedlings[3] and their close metabolic relation to the other acids of the group would warrant, as indicated below, their inclusion in the group of plant acids.

Tartaric acid is widely known in fruits and leaves and is found in particularly high concentration in the juice of fruits such as the grape. Of the four possible stereoisomers of tartaric acid, only one, L-tartaric acid, occurs in nature.

Oxalic acid, the simplest of the plant acids, is widely distributed and may be present in the amounts of zero to 50% of the dry weight of leaves or fruits. Unlike the other acids which appear to be largely present in the plant in soluble forms, oxalic acid is in general precipitated to a greater or lesser extent as the insoluble calcium oxalate which occurs as characteristic crystalline deposits in the cytoplasm of many species.[4] Calcium oxalate occurs in two forms, i.e., as monoclinic crystals of the calcium monohydrate, and as tetragonal crystals of the calcium trihydrate. The latter is an unstable form which is deposited from solutions of low total osmotic concentration but supersaturated with respect to

TABLE 14-1. Comparative organic acid content of various plant tissues

Fruits

Fruit	Units	Total	Citric	Malic	Iso-citric	Oxalic	Tar-taric	Suc-cinic
Valencia orange[a]	m. equiv. acid							
Immature	per cc.	0.66	0.62	0.021	..	....	..	..
Mature	juice	0.30	0.27	0.026	..	....	..	..
Grapefruit[b]	"							
Immature		0.42	0.35	0.060	..	....	..	..
Mature		0.38	0.35	0.027	..	....	..	..
Lemon[c]	"							
Immature		0.77	0.73	0.044	..	....	..	..
Mature		1.08	1.06	0.029	..	....	..	..
Pineapple[f]	m. equiv. acid per 100 g. concentrate of water-solubles	224	125	73	..	....	34	..
Blackberry[l]	% of total acids	....	0	35	65	....	..	..
Boysenberry[l]	"	....	85	11	4	....	..	..
Pear[m]	"	....	67	33	0	....	..	..
Apple[m]	"	....	Trace	ca. 100	0	....	..	..
Peel of valencia orange[n]	m. equiv. acid per g. fresh weight	0.049	0.004	0.024	..	0.020	..	..
Peel of grapefruit[n]	"	0.045	0.005	0.019	..	0.019	..	..
Peel of lemon[n]	"	0.056	0.004	0.019	..	0.031	..	..

Leaves
(m. equivalents per 100 g. dry leaf)

Leaf	Total	Citric	Malic	Isocitric	Oxalic	Tartaric	Succinic
Rhubarb[d]	438	231	22	...	143	..	0.5
Bryophyllum calycinum[e]	427	42.4	166	213	5.9	..	3.8
Bryophyllum crenatum[f]	252	11	7	235	0.0	..	....
Nicotiana tabacum[g]	191	23	68	...	35	..	0.06
Zea mays[h] (whole plant)	96	3.6	25	...	3.2	..	3.2
Buckwheat[i]	329	12.9	22.6	...	246	..	3.4
Narcissus poeticus[j]	67	2.4	74	...	4.5	..	0.9
Pea[k]	213	80.8	42.6	...	15.1	..	....
Tomato[k]	270	74.4	61.1	...	45.2	..	....
Spinach[k]	380	10.5	8.9	...	309	..	....
Valencia orange[o]	223	27.2	68.0	...	87	..	....

TABLE 14-1 (*Continued*)

Stems
(m. equivalents per 100 g. dry stem)

Stem	Total	Citric	Malic	Isocitric	Oxalic	Tartaric	Succinic
Nicotiana tabacum[g]	138	3.6	27	..	15.3	..	...
Buckwheat[i]	155	7.2	29.2	..	40.0	..	3.2
Pea[k]	181	34.5	46.5	..	7.3	..	...
Tomato[k]	232	12.3	89.3	..	81.3	..	...
Cantaloupe[k]	119	5.9	54.4	..	0.0	..	...

Roots
(m. equivalents per 100 g. dry root)

Root	Total	Citric	Malic	Isocitric	Oxalic	Tartaric	Succinic
Nicotiana tabacum[g]	91	4.7	7.6	..	28.5	..	...
Narcissus poeticus[i]	101	2.5	21.1	..	28.0	..	1.3

[a] Sinclair, W. B., and Ramsey, R. C., *Botan. Gaz.*, **106**, 140 (1944).

[b] Sinclair, W. B., and Eny, D. M., *Plant Physiol.*, **21**, 140 (1946).

[c] Sinclair, W. B., and Eny, D. M., *Botan. Gaz.*, **107**, 231 (1945).

[d] Pucher, G. W., Wakeman, A. J., and Vickery, H. B., *J. Biol. Chem.*, **126**, 43 (1938).

[e] Pucher, G. W., *J. Biol. Chem.*, **145**, 511 (1942).

[f] Bonner, W., Thesis, California Institute of Technology, 1946.
Bonner, W., and Bonner, J., *Am. J. Botany*, **35**, 113 (1948).

[g] Vickery, H. B., Pucher, G. W., Wakeman, A. J., and Leavenworth, C. S., *Conn. Bull.* 442, 1940.

[h] Wadleigh, C. H., and Shive, J. W., *Am. J. Botany*, **26**, 244 (1939).

[i] Pucher, G. W., Wakeman, A. J., and Vickery, H. B., *Plant Physiol.*, **14**, 333 (1939).

[j] Vickery, H. B., Pucher, G. W., Wakeman, A. J., and Leavenworth, C. S., *Conn. Bull.* 496, 1946.

[k] Pierce, E. C., and Appleman, C. O., *Plant Physiol.*, **18**, 224 (1943).

[l] Curl, A. L., and Nelson, E. K., *J. Agr. Research*, **67**, 301 (1943).

[m] Nelson, E. K., *J. Am. Chem. Soc.*, **49**, 1300 (1927).

[n] Sinclair, W. B., and Eny, D. M., *Botan. Gaz.*, **108**, 398 (1947).

[o] Sinclair, W. B., and Eny, D. M., *Plant Physiol.*, **22**, 257 (1947).

calcium oxalate. Such crystalline deposits are found in the cytoplasm of cells of onion scales in particular. The trihydrate can be transformed to the monohydrate in a few hours at 100°C, in a few weeks at 30°, and at an imperceptible rate at 20°C. Calcium oxalate monohydrate, the usual crystal form of calcium oxalate in leaves, fruits, and petioles of many species, deposits when calcium oxalate is crystallized from solutions of either low or high osmotic pressure and is particularly favored over the trihydrate when precipitation occurs at temperatures above 30°C.

Determination of the Plant Acids. Quantitative work on the biochemistry of the plant acids was long hindered by the absence of suitable methods for their determination in small amounts. The older methods of plant acid determination[5] have depended on the quantitative separation of the individual components either by fractional precipitation of the salts or by fractional distillation of the esterified acid fraction.[6] The acids may be extracted from plant material with alcohol or ether provided that the material is first acidified with mineral acid so that the acids are present

Occurrence	Chemical nature	Crystal form
Hedychium, leaf	Ca oxalate monohydrate	
Rhamnus lycioides, leaf	"	
Guaiacum, phloem	"	
Citrus medica, petiole	"	
Begonia, petiole	Ca oxalate trihydrate	
Allium cepa, scale leaf	"	

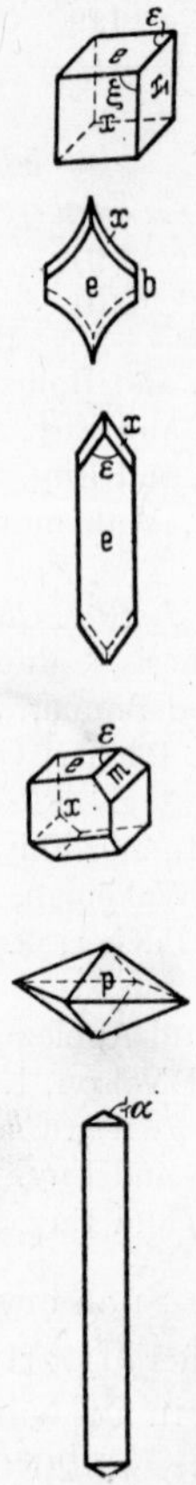

Fig. 14-1. Crystalline forms of calcium oxalate in various tissues. (After Frey-Wyssling[4])

as undissociated molecules. This treatment also removes oxalic acid bound as calcium oxalate. The extract is then concentrated and taken up in water after which the total acids may be determined by titration. The acids may be precipitated from the aqueous extract as the lead salts. The mixed salts are then taken up in dilute ammonium hydroxide in which the salts of citric, malic, and tartaric acids are soluble. The residue contains lead oxalate from which the free acid may be regenerated with H_2S, after which calcium oxalate may be precipitated. The filtrate from the ammonia treatment is also freed of lead with H_2S, after which tartaric acid is precipitated from alcoholic solution as the potassium acid tartrate.

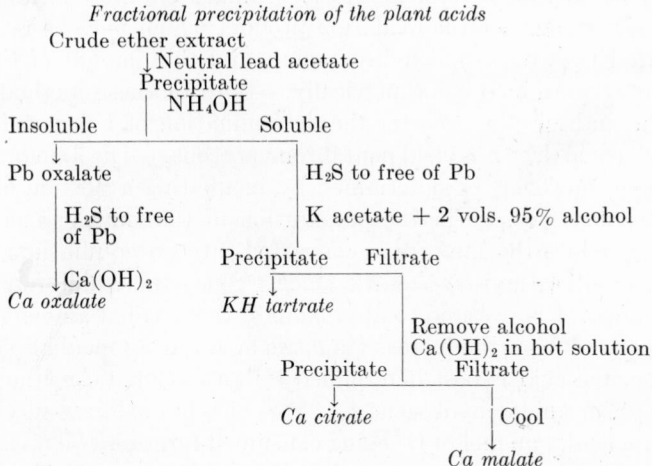

Fractional precipitation of the plant acids

Crude ether extract
↓ Neutral lead acetate
Precipitate

| NH₄OH

Insoluble   Soluble
↓
Pb oxalate H_2S to free of Pb

   H_2S to free   K acetate + 2 vols. 95% alcohol
   of Pb
                 Precipitate   Filtrate
  ↓ $Ca(OH)_2$
Ca oxalate   *KH tartrate*

                            Remove alcohol
                            $Ca(OH)_2$ in hot solution
                Precipitate | Filtrate

                  Ca citrate | Cool
                                Ca malate

The filtrate is used, after removal of the alcohol, for precipitation of calcium citrate and calcium malate, their separation depending on the fact that calcium citrate is more rapidly precipitated in hot than in cold solution, whereas for calcium malate the reverse is true. This general method, which is described in detail in Klein,[5] while feasible, is cumbersome and frequently leads to incomplete separation of the acids considered. In addition, further acids may be present and may confuse the separation.

Determination of plant acids by fractional distillation of the ethyl or methyl esters has also been much used. The acid fraction is esterified with diazomethane or by heating in alcoholic HCl and the resulting mixture of esters distilled under diminished pressure. If sufficient material is available, it is possible to separate esters of the individual acids in this manner. As an analytical method, however, fractional distillation is cumbersome, demands large amounts of material, and is difficult to put on a quantitative basis.

Modern work on the biochemistry of the plant acids has depended on the application of specific micro methods to the determination of individual acids, and Pucher, Vickery, and their colleagues[7] in particular have contributed to this approach. In the methods of Pucher, oxalic acid is precipitated from the acid fractions as calcium oxalate, and the amount of oxalic acid estimated by titration with permanganate in hot solution. The oxalic acid is quantitatively oxidized to CO_2 under these conditions.

A further aliquot of the acid fraction is used for determination of citric and malic acids. The fraction is first oxidized with permanganate in the presence of KBr. Under these conditions citrate yields pentabromo-acetone which may be removed from the oxidation mixture and determined by any of several methods. Malate on the contrary yields an unknown steam volatile oxidation product which may be removed, precipitated with dinitrophenylhydrazine, and the amount of the resulting product determined colorimetrically. Both of these methods are quite specific and may be used for the determination of 1 mg. or less of acid. Isocitric acid does not yield pentabromoacetone. The amount of isocitric acid can, however, be determined by incubating a portion of the crude acid extract with an active preparation of the enzyme aconitase.[8] In this procedure the mixture is converted into an equilibrium mixture of citric, isocitric, and *cis*-aconitic acids. Since the equilibrium constants are known, it is possible to approximate the original amount of isocitric acid present from the increase in citric acid during incubation.

Succinic acid is best determined by an enzymatic method using the enzyme succinic dehydrogenase. This enzyme oxidizes succinic acid to fumaric acid, one mol of O_2 being consumed for each two mols of succinic acid oxidized. The reaction is carried out manometrically in Warburg vessels, and amounts of succinic acid as small as 0.5 mg. may be estimated from the oxygen consumption.[9] Manometric methods are also available for oxaloacetic and α-ketoglutaric acids.[9]

The Krebs Cycle. Although the occurrence of organic acids has been known for many years the function of these compounds in the plant remained entirely obscure until relatively recently. Kostytschev,[10] who considered the question in detail, came to the conclusion as late as 1927 that the organic acids represent breakdown products of amino acid metabolism. This view was originally shared by Ruhland and others[11] who carried out extensive investigations on plant acid metabolism in the 1920's and early 1930's. It is now clear, however, that the plant acids play a central role in cellular respiration both in plants and in animals. The steps leading to this conclusion, and the evidence on which it is based are considered under respiration in Chapter 15. We will here merely summarize the known biochemical interrelations between the various individual acids. These interrelations are based on the work of Krebs and Johnson published in 1937 and the series of reactions which are involved is hence known as the Krebs cycle.[12] The initial reaction of the Krebs cycle is one in which pyruvic acid or a derivative of pyruvic acid condenses with oxaloacetic acid to form *cis*-aconitic or a related six-carbon acid. This reaction involves the removal of hydrogen and water and the evolution of CO_2. The reaction goes readily in living and in ground animal tissues, and both direct and indirect evidence indicates the occurrence of a similar reaction in barley roots, oat seedlings, and in spinach leaves (Chapter 15). *cis*-Aconitic acid is transformed in the presence of the enzyme aconitase into an equilibrium mixture of *cis*-aconitic, citric, and isocitric acids.

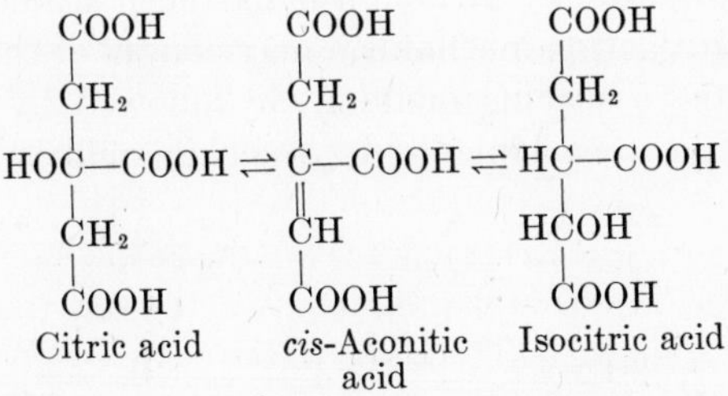

Interconversion of citric, *cis*-aconitic and isocitric acids in the presence of the enzyme aconitase

Isocitric acid is oxidized by the enzyme isocitric dehydrogenase which like aconitase is widely distributed in higher plant tissues. This enzyme removes two hydrogen atoms from isocitric acid with the production of

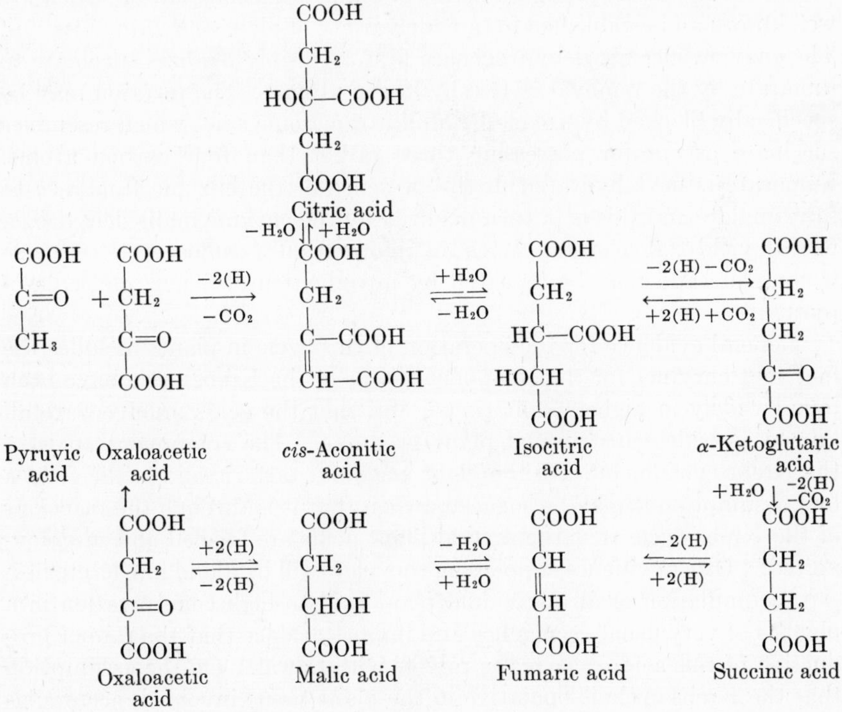

FIG. 14-2. Metabolic interrelations of the plant acids in the Krebs cycle.

oxalosuccinic acid. Decarboxylation of oxalosuccinic acid results in formation of α-ketoglutaric acid. The decarboxylation is enzymatic and the enzyme has been shown to occur and has been studied both in animal and plant tissues.[13] Alpha-ketoglutaric acid is oxidatively decarboxylated to succinic acid by an enzyme system which is as yet but little understood in plant tissues, although it has been studied to some extent in muscle.[14] The reaction mechanism may consist of the decarboxylation of α-ketoglutarate to succinsemialdehyde followed by the oxidation of succinsemialdehyde to succinic acid. Succinic acid is next transformed

$$
\begin{array}{ccc}
\text{COOH} & & \\
| & & \\
\text{C}{=}\text{O} & \text{CHO} & \text{COOH} \\
| & | & | \\
\text{CH}_2 & \text{CH}_2 & \text{CH}_2 \\
| \quad {-\text{CO}_2} & | \quad {-2(\text{H})} & | \\
\text{CH}_2 \xrightarrow{\quad\quad} & \text{CH}_2 \xrightarrow[+\text{H}_2\text{O}]{\quad\quad} & \text{CH}_2 \\
| & | & | \\
\text{COOH} & \text{COOH} & \text{COOH} \\
\alpha\text{-Ketoglutaric} & \text{Succin-} & \text{Succinic} \\
\text{acid} & \text{semialdehyde} & \text{acid}
\end{array}
$$

to oxaloacetic acid through a series of transformations, all of which are well known and established for a wide variety of plant and animal tissues. The enzyme succinic dehydrogenase first reversibly oxidizes succinate to fumarate by the removal of two hydrogen atoms. The reaction may be specifically blocked by use of the inhibitor malonic acid, which resembles succinate except for possessing three rather than four carbon atoms. Fumarate is next hydrated in the presence of the enzyme fumarase to form malate and this is in turn oxidized by the enzyme malic dehydrogenase to yield oxaloacetate. With the reformation of oxaloacetate the cycle is complete and may be repeated by introduction of a new molecule of pyruvate.

General evidence for the operation of this cycle in plants includes the fact that enzymes for the component steps of the Krebs cycle have been found widely in higher plant tissue, and that the acids involved are all known as widely distributed plant products. The component acids of the Krebs cycle include all the plant acids but tartaric and oxalic. As a first assumption it may be suggested then that it is through the reactions of the Krebs cycle that the several plant acids are formed and interconverted. Other evidence supporting this view will be found in Chapter 15.

Accumulation of an individual plant acid to high concentration in a plant is of very usual occurrence and implies at once that the rate of production of this acid exceeds the rate of withdrawal. On the assumption that the Krebs cycle is operative in the plant tissue involved, accumulation of a plant acid in the tissue must be due to a block or partial block of a particular step in the cycle.   Thus the accumulation of malic acid in fruits and leaves must be taken to indicate that the malic dehydrogenase of that tissue is not as active as the other enzymes of the earlier links in the cycle.   Similarly the accumulation of citric or of isocitric acids must be owing to individual blocks in the aconitase or isocitric dehydrogenase systems.   These blocks might be due to low enzyme concentration or to the presence of inhibitors, and it will be of interest to resolve the nature of the defects involved.   We do not at present have the information necessary to explain such preferential accumulations.

As mentioned above neither oxalic nor tartaric acids fit into known organic acid metabolic systems.   It is possible however that oxalic acid may arise through the intermediary of glycolic and glyoxylic acids by some such pathway as that shown below.

$$
\begin{array}{ccccc}
\text{CH}_2\text{OH} & & \text{CHO} & & \text{COOH} \\
| & \rightarrow & | & \rightarrow & | \\
\text{COOH} & & \text{COOH} & & \text{COOH} \\
\text{Glycolic} & & \text{Glyoxylic} & & \text{Oxalic} \\
\text{acid} & & \text{acid} & & \text{acid}
\end{array}
$$

Metabolism of Organic Acids in Leaves.   The changes in organic acids which occur in excised leaves during starvation have been followed in tobacco and in rhubarb by Vickery and others.[15]   In both cases oxalic acid is unaffected by any treatment and is not reutilized even in extreme starvation.   Oxalate would appear then to reenter into metabolism slug-

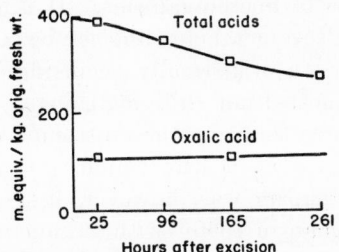 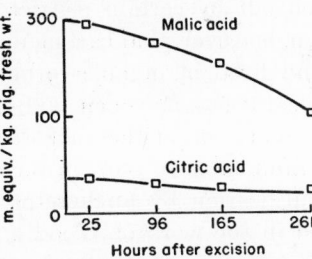

FIG. 14-3.   Changes in organic acids of excised rhubarb leaves maintained in the dark. (After Pucher, Wakeman, and Vickery[15])

gishly if at all.   In tobacco leaves cultured in the dark malic acid rapidly disappears while citric acid appears although in lesser concentration, since approximately one mol of citrate is formed for each two mols of malate utilized.   This transformation might be understandable on the basis of the reactions of the Krebs cycle since it would be possible for malate to be transformed to oxaloacetate and this in turn converted to pyruvate by the oxaloacetate decarboxylase which is known in higher plant tissues.[16] The combination of one mol of pyruvate with one mol of oxaloacetate would then yield one mol of tricarboxylic acid, as found experimentally. In excised leaves of rhubarb both malic and citric acids slowly disappear during incubation in the dark. An unidentified acid accumulates in rhubarb just as does citric acid in tobacco, and it has been suggested that malic acid may be converted in this case to the unknown acid. Excised

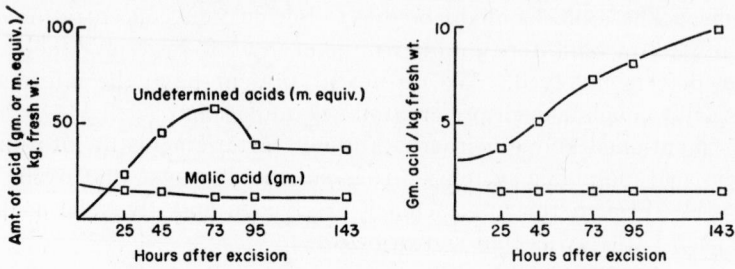

FIG. 14-4. Changes in organic acids of tobacco leaves cultured in the dark.[15] At right: upper curve, citric acid; lower curve, oxalic acid.

leaves of sudan grass and of oat[17] show a still different type of behavior with respect to organic acids during starvation. Here both citrate and malate accumulate at the expense of unidentified precursors.

Carbon Dioxide Fixation. For many years it was considered that carbon dioxide is purely a metabolic product of respiratory and fermentative processes and that fixation of carbon dioxide in organic molecules is restricted to photosynthesis of green plants and to chemosynthesis as carried out by certain restricted groups of microorganisms. It is now known, however, that carbon dioxide is also a metabolite and can be taken up and fixed in organic form in a variety of normally occurring biochemical reactions which take place in most if not all living things. An early indication of this fact was the finding that some heterotrophic bacteria are unable to grow in the absence of CO_2. Clear demonstration of CO_2 utilization by an heterotrophic organism was, however, first presented in the work of Wood and Werkman in 1936[18] with *Propionibacterium*, an organism which occurs commonly in cheese. Wood and Werkman grew *Propionibacterium* anaerobically on a medium containing glycerin as a source of carbon, $CaCO_3$ to neutralize the acids produced, salts, and a small amount of yeast extract as a source of growth factors for the organism. After a period of incubation they determined all the products produced by the organism. It was found in a typical experiment that propionic, acetic, and succinic acids constituted the principal products formed but that in addition CO_2 had actually been utilized during the course of the reaction. Such CO_2 utilization appeared to be correlated with succinic acid production since a strain of the bacteria which utilized but little CO_2 also produced but little succinic acid.

TABLE 14-2. Utilization of CO_2 by two strains of *Propionibacterium*. (After Wood and Werkman[18])

	Strain 1	Strain 2
Amount of glycerol used in m mols	100	100
CO_2 exchange, m mols	−43	−1.1
Propionic acid produced, m mols	+59	+89
Acetic acid produced, m mols	+2	+2.6
Succinic acid produced, m mols	+34	+3.9

Combustion analyses confirmed the increase of total organic carbon of the culture during incubation, the increase agreeing quantitatively with the amount of CO_2 taken up. Conclusive proof of CO_2 utilization by *Propionibacterium* was obtained[19] with the use of CO_2 containing an admixture of the heavy isotope C^{13}. When the organism was allowed to ferment glycerol in the presence of $C^{13}O_2$, the tagged carbon was detected in the carboxyl groups of the succinic acid formed. It was suggested by Wood and Werkman that a mechanism for this reaction might consist in the condensation of pyruvic acid with CO_2 to form oxaloacetate, the oxalo-acetate being in part reduced to succinic acid. The correctness of this suggestion has been abundantly confirmed and the Wood and Werkman reaction is now recognized as a general biological process.

$$CH_3-\underset{O}{\overset{\|}{C}}-COOH + CO_2 \rightarrow HOOC-CH_2-\underset{O}{\overset{\|}{C}}-COOH$$

Pyruvic acid   Oxaloacetic acid
The Wood and Werkman reaction

In higher plants an enzyme catalyzing the reversible breakdown of oxalo-acetic acid to pyruvate and CO_2 has been found in a wide variety of seeds, seedlings, leaves, and roots.[16] The enzyme involved requires the presence of manganese ions in order to carry out the decarboxylation reaction, while the CO_2 fixation reaction requires in addition the presence of adenosine triphosphate. The Wood and Werkman reaction is thus well founded on the basis of *in vitro* as well as *in vivo* experiments.

The use of CO_2 containing isotopic carbon has made it possible to show that in organisms which, unlike *Propionibacterium*, give off large amounts of CO_2 during normal metabolism, CO_2 is also taken up and fixed in organic form, even though the outgo of CO_2 greatly exceeds the uptake. CO_2 fixation in organic acids has been reported for a wide range of hetero-trophic microorganisms, including *E. coli*, *Clostridium* (which produces butyric acid, butyl alcohol, etc.), *Acetobacter*, *Rhizopus*, yeast, and also for animal tissues. Carbon dioxide fixation by roots of barley has also been reported,[20] although not on the large scale in which it occurs in microorganisms. Carbon dioxide fixation is carried on extensively by leaves of a particular group of plants, the succulents, and here the fixation appears to be a major source of organic acids, as is discussed later in this chapter.

The fixation of carbon dioxide by heterotrophic organisms is now known to occur by other reactions in addition to that of Wood and Werkman. Thus, Woods[21] early showed that *E. coli* suspensions can synthesize formic acid from CO_2 according to the reaction:

$$H_2 + CO_2 \rightleftharpoons HCOOH$$

Formic acid formation depends thus on the partial pressure of both CO_2 and H_2. Similarly in the case of the methane-producing bacteria, which oxidize fatty acids with the production of acetic acid and methane, Barker has shown[22] that the reaction proceeds according to the following scheme

$$2CH_3CH_2CH_2COOH + 2H_2O + CO_2 \rightarrow 4CH_3COOH + CH_4$$
$$\text{Butyric acid} \qquad\qquad\qquad \text{Acetic acid} \quad \text{Methane}$$

in which CO_2 is used as a hydrogen acceptor in the oxidation. In both of these reactions, even though CO_2 is fixed in organic form, still the CO_2 functions purely as a hydrogen acceptor and the formation of carbon-carbon bonds is not involved. A further type of reaction in which CO_2 may be fixed into larger molecules has been studied by Lipmann and Tuttle.[23] This involves the reversal of the oxidative decarboxylation of pyruvic acid, a reaction which concerns the oxidation of pyruvic acid in the presence of inorganic phosphate with the production of acetylphosphate. The reaction with *Micrococcus* involves a hydrogen acceptor, in this case cozymase. With extracts of *E. coli* or of *Clostridium* however,

$$\underset{\substack{\| \\ O \\ \text{Pyruvate}}}{CH_3 - C - COOH} + H_3PO_4 \rightleftharpoons \underset{\text{Acetylphosphate}}{CH_3COOPO_3H_2} + CO_2 + 2(H)$$

the reaction takes a somewhat different course involving formic acid which, as explained above, is in equilibrium in *E. coli* with CO_2 and H_2.

$$\underset{\substack{\| \\ O \\ \text{Pyruvate}}}{CH_3 - C - COOH} + H_3PO_4 \rightleftharpoons \underset{\text{Acetylphosphate}}{CH_3COOPO_3H_2} + \underset{\substack{\text{Formic acid} \\ \| \\ H_2 + CO_2}}{HCOOH}$$

That the degradation of pyruvate is reversible to a slight extent was established both by demonstration of pyruvate formation from acetylphosphate and by the introduction of tagged carbon into the carboxyl group of pyruvate from CO_2 in the presence of enzyme and acetylphosphate. CO_2 fixation through the acetylphosphate system is thus well established in microorganisms, and, although the reaction is not known in higher plants, the possibility of its occurrence must be recognized.

A third reaction for the formation of carbon-carbon bonds from CO_2 may also take place as the result of reversal of the oxidative decarboxylation of isocitric acid. In this reaction as explained above isocitrate is first dehydrogenated to oxalosuccinic acid and the latter then decarboxylated to α-ketoglutaric acid. Both reactions are reversible with

$$\text{D-Isocitric acid} + \text{TPN} \rightleftarrows \text{Oxalosuccinic acid} + \text{reduced TPN}$$

measurable equilibrium constants.[24] Thus the oxidation of isocitrate to

$$\text{Oxalosuccinic acid} \rightleftarrows \alpha\text{-Ketoglutaric acid} + CO_2$$

oxalosuccinate in the presence of isocitric dehydrogenase and the hydrogen acceptor, triphosphopyridine nucleotide or TPN, is combined with the decarboxylation of oxalosuccinate to α-ketoglutarate by oxalosuccinic carboxylase. The equilibrium constant for the overall reaction is 7.7×10^3 or far over toward the side of decarboxylation.

$$\text{D-Isocitric acid} + \text{TPN} \rightleftarrows \alpha\text{-Ketoglutaric acid} + CO_2 + \text{reduced TPN}$$

$$K = \frac{(\alpha\text{-Ketoglutarate})(H_2CO_3)(\text{red. TPN})}{(\text{D-Isocitrate})(\text{TPN})} = 7.7 \times 10^3$$

The reaction can be shifted in the direction of CO_2 fixation by coupling with reactions which produce reduced TPN. Thus we may say that the energy required for the fixation of CO_2 in the α-ketoglutarate reaction comes from the oxidative reactions which result in production of reduced TPN.

The entire system for fixation of CO_2 through the α-ketoglutarate-isocitrate reaction has been shown to occur in parsley roots.[13] Reduced TPN drives the reaction as outlined above and fixation of CO_2 can readily be verified from the fact that radioactive carbon supplied to the system as CO_2 can be recovered as carbon present in isocitrate.

The significance of CO_2 fixation as a biological process is twofold. In the first place, CO_2 fixation provides a general means whereby acid groups are formed by organisms, and the acidity of plant tissues in which CO_2 is fixed would hence be expected to depend intimately on the prevalent CO_2 pressure. That this is indeed the case with leaves will be shown below. In the second place, the Wood and Werkman reaction elucidates the nature of the genesis of the four-carbon acids in metabolism. Oxaloacetate and its derivatives malate, furmarate, and succinate may all be derived from pyruvate through the fixation of CO_2, followed by the appropriate further reactions. Oxaloacetate in turn appears to be an essential metabolite in the formation of isocitrate and citrate, both plant acids of quantitative importance. It will be shown in Chapter 15 that the Krebs cycle is an integral part of the respiration of some at least of the tissues of higher plants, and it is of importance, therefore, that the oxaloacetate concentration of the respiring tissue be maintained. Oxaloacetate is, however, not only a metabolite in respiration, but is subject also to decarboxylation and to transamination reactions. It may be expected therefore that oxaloacetate will be withdrawn from the Krebs cycle at a greater or lesser rate. The Wood and Werkman reaction would appear to be the mechanism by which oxaloacetate is restored and maintained in the tissue. Finally, it is highly probable that photosynthesis consists essentially of CO_2 fixation reactions which are basically similar to the fixation reactions discussed above but for which the required energy is derived not from substrate oxidation but from light.

Metabolism of Succulents. It has been shown that many species of plants accumulate acids in high concentrations in their leaves. With the great bulk of plant species this acid, once formed, is relatively stable and

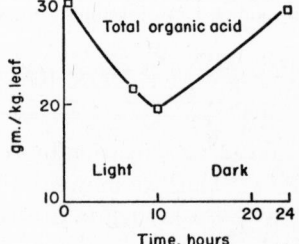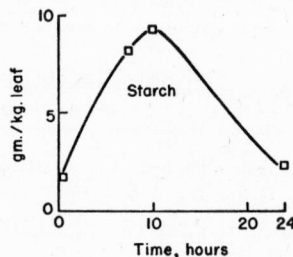

FIG. 14-5. Behavior of organic acids in attached leaves of the succulent *Bryophyllum calycinum* in light and dark. (After Pucher *et al.*[27]) The diurnal fluctuations in acid level are the reverse of those of the photosynthetically produced carbohydrates such as starch. In both cases, the first ten hours are in light, the remainder of the experimental period in dark.

does not disappear except under conditions of stress as in starvation of the leaf. With a few species, on the contrary, acids are formed primarily during the night and disappear again during the day. This diurnal fluctuation in acid content is characteristic of the group of plants known as succulents and the remarkable metabolism which they exhibit is known as succulent or crassulacean metabolism.[25]

The succulent plants are a morphological rather than a taxonomic group. They have in common leaves or photosynthetic stems which consist of thickened spongy tissue. Typical succulents are found in the families *Cactaceae, Euphorbiaceae, Asclepiadaceae, Begoniaceae, Compositae,* and in particular in the *Crassulaceae* of which the genera *Sedum, Crassula,* and *Bryophyllum* are representative. De Saussure first discovered the succulent metabolism in his work on the cactus *Opuntia* when he found that during the night the intake of oxygen by the flattened stems might under certain circumstances greatly exceed the output of CO_2, thus indicating that at night respiration proceeded from sugar to the formation of oxygen-rich compounds.   Heyne (1819) later found that with *Bryophyllum calycinum*, the leaves actually taste more acid at the end of the night than at the end of the day.   A large amount of subsequent work established the principal facts, which are that leaves of succulents either on the plant or excised from it form acids when they are maintained in the dark, and especially at low temperatures, such as 10°C or lower.   When such leaves are removed to the light the acids disappear. Disappearance of acids may also be accomplished in the dark at high tem-

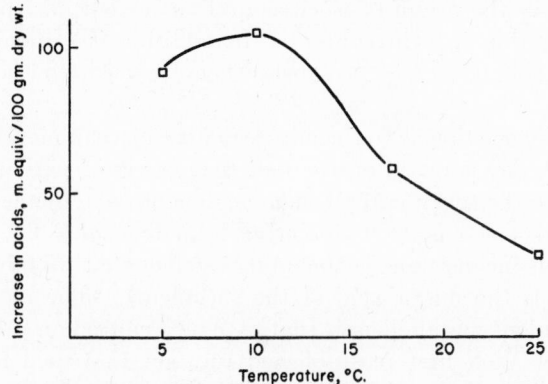

FIG. 14-6.   Effect of temperature on formation of organic acids by leaves of *Bryophyl lum fedtschenkoi*.   Leaves in dark for forty-eight hours.   (After Bonner and Bonner[28])

peratures such as 30°C.   A complication in the study of the metabolic reactions involved has been the fact that the leaves of many succulents contained an unknown acid, referred to in the early work as isomalic or crassulean malic acid.   This acid, which forms a large part of the total, is now known to be isocitric acid, at least in the case of *Bryophyllum, Sedum*, and other typical succulent species.[28]

Early theories as to the mode of acid formation in succulents related these compounds to protein metabolism.   Kostytschev[10] regarded plant acids in the succulents as in other plants as by-products of amino acid degradation, and organic acid production was thought by Kostytschev to parallel protein hydrolysis in the leaf.   Ruhland and Wolf,[11] on the other hand, found that carbohydrates are the source of the plant acids and suggested that in succulents pyruvic acid might be diverted from its normal pathway of oxidation in respiration and transformed to succinic acid through diketoadipic acid.

$$\text{Hexose}$$

$$\text{Hexose} \to \text{HOOCCCH}_2\text{CH}_2\text{CCOOH} \xrightarrow{O_2} \text{HOOCCH}_2\text{CH}_2\text{COOH} + 2\text{CO}_2$$

Diketoadipic acid   Succinic acid

$$\text{CH}_3\text{CCOOH}$$

Pyruvate

$$\frac{1}{2}\text{O}_2$$

$$\text{HOOCCCH}_2\text{COOH} \xleftarrow{-2(\text{H})} \text{HOOCCH}_2\text{CHOHCOOH}$$

Oxaloacetic acid   Malic acid

Normal respiration

Proposal of Wetzel and of Ruhland for formation of organic acids in leaves of succulents[11]

The scheme of Wetzel and Ruhland is, however, unsatisfactory in that it explains neither the origin of isocitric and citric acid nor the reason for diurnal fluctuation in acid content. In addition it supposes reactions involving diketoadipic acid, a substance not known to occur in plant tissue.

A further suggestion as to the nature of crassulacean metabolism made by Bennet-Clark[25] is based on the fact that leaves of various crassulean species contain the seven-carbon sugar, sedoheptose. Bennet-Clark supposed that this sugar is itself converted to malic acid. The idea suffers from numerous inconsistencies including the fact that we now know that isocitric acid is the major acid of the succulents. The most thorough study of succulent metabolism is that of Wolf[26] carried on primarily with *Bryophyllum*. Wolf first reestablished the fact that acid formation in *Bryophyllum* leaves is a function of light and of temperature. With regard to temperature two factors are involved. The maximum amount of acid is accumulated by excised leaves at a temperature of 7°C, but this maximum level is attained only over long periods in the dark (96 hours). At higher temperatures acid formation takes place at a faster rate but never attains as high a concentration level. Over a period of 18 hours, a temperature of 20°C was found to be optimum. At temperatures of 30–35° the leaves actually lost acids. The optimum temperature for acid removal in the light was found to be about 25°. A minimum of 400 meter-candles was needed to effect acid removal and light saturation of the process was attained only with intensities of about 1000 meter-candles. These intensities are of the same order as are needed to effect photosynthesis in the succulent leaf, and it would appear possible then that photosynthesis is involved in acid removal.

Wolf could find no relation of acid formation to either nitrogen metabolism or to breakdown of sedoheptose. On the contrary a direct relation to sugar utilization was found, roughly one mol of acid appearing for each mol of hexose oxidized. The process was inhibited by cyanide as was respiration. Oxygen uptake continued at a high level in leaves forming acids, but CO_2 production was almost completely suppressed. Thus the respiratory quotient, or ratio of CO_2 evolved to oxygen consumed, was below the figure of unity expected for normal respiration of hexose. Pyruvate is possibly involved in the reaction as was shown by experiments with leaf strips floated on buffer solution in vessels suitable for measurement of gas exchange. With leaf strips floated on buffer solutions at 25°C and respiring at the expense of endogenous substrates, the respiratory quotient was found to be moderately high as is shown in Table 14-3. If pyruvate was added oxygen consumption was increased to some extent but CO_2 production was almost completely suppressed, suggesting that the pyruvate utilization involves utilization of the CO_2 normally produced in the cell. Interestingly enough the addition of thiamine completely reversed the influence of pyruvate in decreasing CO_2 production. Wolf

TABLE 14-3. Effect of pyruvate and of thiamine on gas exchange by strips of *Bryophyllum* leaves at 25°C. (After Wolf[26])

	Gas exchange in mm.^{3}/g./hr.			Thiamine added: gas exchange in mm.^{3}/g./hr.		
	O_2	CO_2	R.Q.	O_2	CO_2	R.Q.
No added substrate	−119	+104	0.87	....	....	....
Pyruvate	−160	+6	0.04	−178	+129	0.73

considers that this fact is to be interpreted on the basis of the known pyruvate carboxylase content of *Bryophyllum* leaves, with thiamine (see Chapter 15) acting as a coenzyme to excess apoenzyme present in the tissue. The influence of pyruvate in lowering the respiratory quotient has not been directly related to acid production and it would be of interest to know if the pyruvate supplied actually appears in the leaf in the form of plant acids. It is of interest to note in any case that the effect of pyruvate is inhibited by malonic acid, the specific inhibitor of succinic dehydrogenase, suggesting that pyruvate utilization proceeds through processes involving succinate metabolism. A more recent study of the metabolism of *Bryophyllum* leaves, generally confirming the work of Wolf, has been carried out by Pucher and others.[27]

The production of organic acids by *Bryophyllum* leaves depends intimately on the presence of carbon dioxide as is shown in Fig. 14-7. In the absence of CO_2, the excised leaf makes essentially no acid. With increasing CO_2 pressure, acid formation increases almost linearly up to 0.1% CO_2 and more slowly up to 10% CO_2 in air. The acids formed in excised leaves in the presence of high CO_2 concentration consist of citric, malic, and isocitric acids as they do in normal leaf, but malic acid in particular increases in the leaf in the presence of CO_2.

The fact that acid formation in the leaves of *Bryophyllum* depends on carbon dioxide pressure may be used in an interpretation of the diurnal fluctuation of acid content. In the dark the CO_2 partial pressure inside the fleshy leaves of the succulent will tend to be higher, owing to respiration, whereas in light, photosynthesis will tend to lower the CO_2 content of the internal gas. Thus diurnal fluctuations in acid content may be attributable to diurnal fluctuations in CO_2 partial pressure. This interpretation is strengthened by the observation of Wolf that high CO_2 partial pressures during exposure of *Bryophyllum* leaves to light decrease the rate at which acids are reutilized. The temperature relations of acid forma-

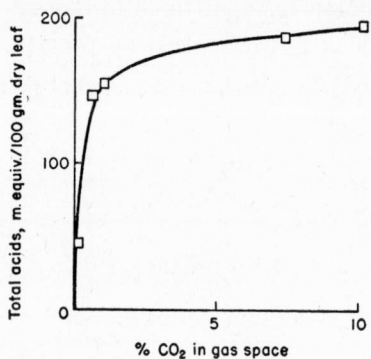

Fig. 14-7. Dependence of organic acid formation in leaves of *Bryophyllum crenatum* on the CO_2 concentration. Leaves incubated in the dark at 3°C for 48 hours. (After Bonner and Bonner[28])

tion tend to reinforce the possible carbon dioxide-induced periodicity, since the night temperatures to which the leaf is normally subject will ordinarily be lower than the day temperatures.

Acid formation in leaves of nonsucculent plants has been investigated only meagerly, but it has been shown that in general nonsucculents do not respond to increased CO_2 partial pressure in the dark by markedly increased acid formation. It would appear probable that the succulents constitute an exceptional case in leaf organic acid metabolism in that they not only synthesize acids extraordinarily freely and extensively but also in that the process is highly responsive to CO_2 pressure. It has been suggested by the work of Wolf that pyruvate may be involved in CO_2 utilization, and further that this process must proceed through the four-carbon acids since it is blocked by malonic acid. These observations may be interpreted on the basis that the Wood and Werkman reaction may be involved. Experiments with isotopically tagged CO_2 show that CO_2 is actually fixed in the malic and other acids formed in the dark by *Bryophyllum* leaves.[28] Oxaloacetic acid cannot be detected in *Bryophyllum* leaves

Pyruvic acid $+ CO_2 \rightleftharpoons$ Oxaloacetic acid $\rightleftharpoons$ Citric, isocitric, and malic acids
Suggested general mechanism of organic acid formation by succulents

in measurable amounts,[27] but the fact that the amount of malate formed in short-term experiments is directly related to the CO_2 partial pressure suggests that a portion at least of the oxaloacetate formed may be reduced directly to malate.   On the other hand a portion of the oxaloacetate must also be oxidatively transformed to isocitric and citric acid.

In summary, the evidence available suggests that acid formation in *Bryophyllum* involves both pyruvate and CO_2.   As a working hypothesis it may be assumed that CO_2 fixation by a reaction similar to that of Wood and Werkman is involved.   In this respect the leaves of succulents would differ from the leaves of other nonsucculent plants in which CO_2 fixation in the dark by nonphotosynthetic reactions takes place at most to a limited extent.

Acid-Base Balance.   The organic acids present in a plant tissue are not ordinarily present exclusively as the undissociated acid molecules but are usually combined to a greater or lesser extent as salts of the mineral cations taken up by the plant.   Thus the organic acid-organic acid salt system of the tissue constitutes a buffer system which is of importance not only in determining the pH of the cell but also in rendering some protection to the cell against external agencies which might otherwise cause drastic alterations in cellular pH.   Looked at in another light, the organic acids serve as anions for a portion of the mineral cations absorbed by the plant.   In some tissues the bulk of the mineral cations are in fact present as the salts of organic acids.   The data of Table 14-4 show that in the

TABLE 14-4.   Distribution of mineral cations between organic acid and inorganic anions in juice of mature citrus fruits.   (After Sinclair and Eny[34])

Fruit	Total cation content	Cations combined with organic acids		Cations combined with inorganic anions (Phosphate, sulfate)
	m. equiv./ 100 g.	m. equiv./ 100 g.	% of total cation	m. equiv./ 100 g.
Valencia orange	5.96	4.26	71.5	1.70
Navel orange	6.06	4.42	72.9	1.64
Grapefruit	5.35	3.05	57.0	2.30

fruits of orange and grapefruit, 57 % to 73 % of the total cations are associated with organic acid anions, the remainder being associated with phosphate, sulfate, and other inorganic anions.   Even though the great bulk of mineral cations is then combined with organic acid anions in the citrus fruit, most of the acid of the fruit is nevertheless present as free acid in this case. In the lemon less than 3 % of the total acid is combined with mineral cation and even in the less acid orange, only 15 % of the total acid was found to be thus combined (Table 14-5).

TABLE 14-5. Proportion of total acid occurring as free acid or as salts of mineral cations in juice of mature citrus fruits. (After Sinclair and Ramsey,[35] Sinclair and Eny[36])

Fruit	Total acids m. equiv./100 g.	Acid combined as salts, m. equiv./ 100 g.	% combined acid
Valencia orange	25.2	4.3	14.6
Grapefruit	41.5	3.1	8.6
Lemon	118.8	3.3	2.9

Sinclair and co-workers[35,36] have shown that in the citrus fruit, the pH of the juice is determined almost solely by (a) total acid concentration and (b) ratio of free acid to salt according to the basic relation between this ratio, the pH and the pK or negative log of the dissociation constant of the acid. This means that small changes in the free acid/salt of acid

$$p\text{H} = p\text{K} + \log \frac{(\text{Acid salt})}{(\text{Free acid})}$$

ratio, such as might be caused by fluctuations in cation uptake during fruit development, will have relatively little effect on fruit pH, and Sinclair has calculated that even were the cations of grapefruit juice to be doubled the resultant change in pH would approximate only 0.3 pH units. Many of the older workers have attempted to discover relations between the titratable acidity of a plant extract and its pH; such relations have in fact been found as in the case of citrus and occasionally of apple fruits. With many tissues including other fruits there is no relation between titratable acidity and pH. This is of course due to the fact that titratable acidity measures only the free acid term in the expression log (acid salt/free acid).

The great importance of the organic acids in the acid-base equilibrium of the fruit has its counterpart in other tissues of the plant. It is well known that during the uptake of inorganic salts by roots or other tissues, the cation uptake may exceed or may be smaller than the anion uptake. In the case of fruits just described, excess cation uptake is accommodated by neutralization of plant acids. In roots, active organic acid production accompanies disproportionate cation uptake while organic acid disappearance accompanies disproportionate anion uptake.[37] The plant acids of the cell in addition to their function as a passive buffer system are thus able to act as a flexible system, changing in quantity in a manner appropriate for accommodation of the mineral uptake problems presented. The operation of the plant acid system is particularly striking in connection with the preservation of ionic balance during uptake and reduction of nitrate salts. The residual cations are then balanced by organic acid anions produced (Table 14-6).

Relation of Nitrogen Nutrition to Organic Acid Content. Of the factors which influence organic acid content of the plant, nitrogen nutrition is of paramount importance. In particular, plants supplied with nitrogen in the form of ammonia contain much lower concentrations of acids than do similar plants grown with nitrate nitrogen. This response to form of nitrogen supply was first found in the tomato. Plants in sand

TABLE 14-6. Influence of form of nitrogen on organic acid composition and distribution of nitrogen in leaves of tobacco plants. (After Vickery et al.[31])

% of N given as NO_3:	100	80	60	40	20	10
Constituent as NH_4:	0	20	40	60	80	90
	(m. equiv./100 gm. dry wt.)					
Total acids	190	163	117	94	77	67
Inorganic cations combined as acid salts	125	130	78	48	29	28
Oxalic acid	35.1	31.9	29.1	21.6	12.1	11.2
Citric acid	23.1	14.9	8.8	1.4	0.9	0.9
Malic acid	68.2	38.3	14.4	2.4	1.6	2.7
	(% of dry wt.)					
Total amide N	0.064	0.066	0.077	0.087	0.151	0.183
Total soluble org. N	1.26	1.26	1.20	1.42	1.64	1.79
Protein N	3.12	3.19	3.29	3.22	3.18	3.06
Soluble sugars	1.49	1.36	1.08	0.97	0.98	1.15

culture which received their nitrogen as ammonia contained less than one-tenth as much of each individual acid as did similar plants supplied with nitrogen as nitrate.[29] A similar response has been found for *Begonia* and also for corn, *Bryophyllum*,[30] and the rubber plant guayule. A particularly intensive study of the response has been made by Vickery and others[31] who grew tobacco in sand culture with complete nutrient media but in which the nitrogen was supplied either as nitrate, as ammonia, or as mixtures of the two forms. Organic acid content of the leaf tissue was found to be a nearly linear function of the proportion of NH_4^+ in the nutrient. Both citric and malic acid were twenty to forty times lower in concentration in the tissue when the nutrient contained 80–90% of the nitrogen as NH_4^+ than when all nitrogen was given as nitrate. Similar changes were found also in root and stem tissues. Table 14-6 includes data on other constituents of the same series of plants. The amides asparagine and glutamine, which make up only a small portion of the total nitrogen, increase in amount as the proportion of ammonia in the medium is increased. Only a slight increase in total soluble organic nitrogen including amino acid, amide, and ammonia nitrogen was found, and protein level was essentially independent of composition of the medium. The content of soluble sugars was markedly decreased in the plants which received high ammonia. In a qualitative way it may be suggested that in plants which are supplied with high ammonium levels in the nutrient, the organic acids are kept at a low level by reactions involving the formation of amino acids and amides from α-ketoglutaric and oxaloacetic acids. The depression in organic acid content cannot, however, be quantitatively accounted for on this basis since only small increases in soluble nitrogen and no increase in protein content occurs. It would appear rather that nutrition with nitrate nitrogen may also actively promote organic acid synthesis, perhaps owing to the accumulation of mineral cations which remain in the cell after reduction of the nitrate anion.

Aromatic Acids of Plants. In addition to the aliphatic plant acids a number of aromatic acids are widely distributed although by no means as nearly universally as the plant acids themselves. The aromatic acids may occur in the free form in certain instances, such as benzoic acid. In general, however, the aromatic acids tend to occur as esters in oils and resins or as glycosides. Thus salicylic acid, o-hydroxybenzoic acid, occurs primarily as the methyl ester and is found as such in oil of spearmint, *Mentha spicata*. Other commonly found aromatic acids are cinnamic, p-coumaric, hydrocoumaric, protocatechuic, gallic, and caffeic acids. The structures of these acids together with notes on their distribution and form of occurrence are given in Table 14-7.

The function and metabolism of the aromatic acids has been but little investigated as compared with the aliphatic acids. It is evident that they may be formed by oxidation of aromatic amino acids such as phenylalanine or they may represent independent metabolic pathways from the unknown precursors of the aromatic nucleus. A function for protocatechuic and caffeic acids is suggested by the fact that they are reversibly oxidizable to the corresponding o-benzoquinones. The oxidation is

Caffeic acid   Benzoquinone of caffeic acid

carried out by the enzyme polyphenoloxidase and the reduction by other as yet not well-defined plant enzyme systems. Caffeic acid and related compounds may be of importance as hydrogen transporting systems in plant respiration, as will be taken up in Chapter 15.

An additional physiological function for protocatechuic acid is sug-

TABLE 14-7.   Distribution of aromatic acids in higher plants

Acid	Structure	Forms in which found	Genera of plants in which found
Benzoic	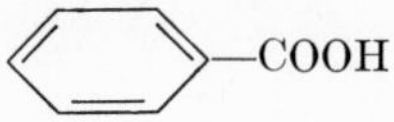 ⬡—COOH	Free	Fruit of *Vaccinium* sp. Leaves of *Gaultheria*
		Esterified with benzyl, ethyl, cinnamyl etc. alcohol. As benzoyl glucose	Resins of *Styrax, Myrrhis, Parthenium, Benzoin.* Fruits of *Vaccinium* sp.
Cinnamic (*trans* isomer)	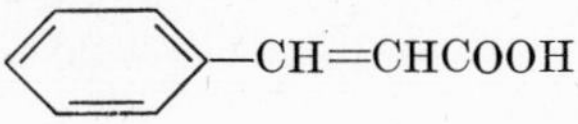 ⬡—CH=CHCOOH	Free	Oil of *Storax*, resin of *Myroxylon* and other legumes, in oil of *Melaleuca*, resin of *Styrax*, etc.
		Esterified with alcohols as methyl, etc.	Resins of *Styrax, Benzoin, Melaleuca, Storax*
		Esterified with terpene alcohols	Amyrin ester in latices of *Moraceae*, ester of partheniol in resin of *Parthenium*
Salicylic	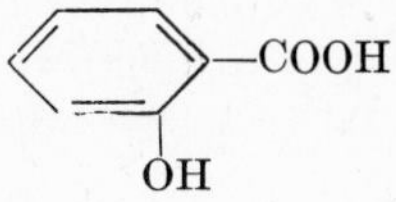 ⬡—COOH, OH	As methyl ester	Leaves of *Chenopodium, Calycanthus, Camellia* Fruits of *Ribes, Fragaria, Rubus, Prunus, Vitis* Flower oil of *Acacia* and other legumes
		As methyl ester glycoside	Flower of *Viola*, leaves of *Gaultheria*
p-Coumaric (*trans* isomer)	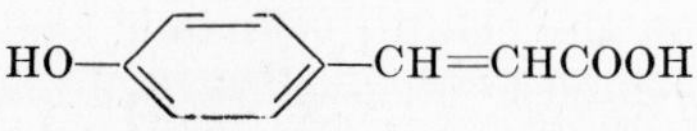 HO—⬡—CH=CHCOOH	Free	Leaves of *Aloe, Daviesia, Trifolium*
		As ester, as ester glycoside Methoxyethyl ester	Leaves of *Aloe, Melilotus* Leaves of *Hedychium*
o-Coumaric (*trans* isomer)	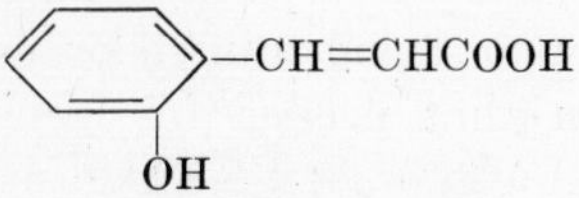 ⬡—CH=CHCOOH, OH	Free and as glycoside	Leaves of *Melilotus*

TABLE 14-7 (*Continued*)

Acid	Structure	Forms in which found	Genera of plants in which found
Melilotic	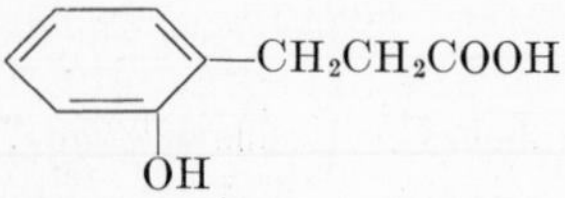 [benzene ring]—CH₂CH₂COOH with OH	Free	Leaves of *Melilotus*
Protocatechuic	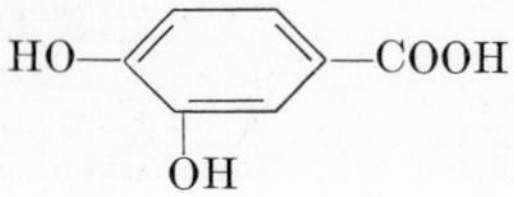 HO—[benzene ring]—COOH with OH	Free	Leaves of *Allium* bulb Flowers of *Hibiscus, Althea*
		As tannin derivative	Bark of *Quercus, Schinopsis*
		As ester with *p*-hydroxybenzoic acid	Leaves of *Grindelia* Fruit of *Bignonia*
Gallic	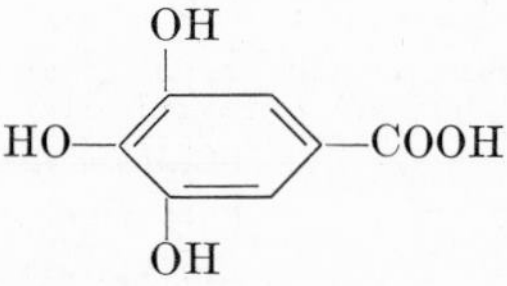 HO—[benzene ring]—COOH with OH and OH	Free	Tannin of galls, leaves of pea
		As ester in glycosides	Tannins of galls on *Quercus, Rhus.* Tannins of many species of leaves, bark, fruits
Caffeic	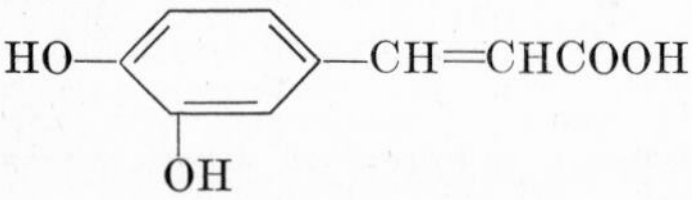 HO—[benzene ring]—CH=CHCOOH with OH	As ester in glycosides	Component of tannins in coffee beans, *Cinchona* bark. Leaves of *Digitalis, Coffea*

gested by the discovery of Walker and others[32] that high concentrations of this acid are present in onion varieties resistant to attacks of a particular fungus: Protocatechuic acid is toxic to the fungus and acts hence as an agent conferring natural immunity on the plant. Cinnamic acid has been shown[33] to be contained in and given off by roots of *Parthenium argentatum*. The cinnamic acid is toxic to the *Parthenium* plant and accumulations of the material in the soil or other substrate reduce growth or actually kill the plant. Other species are less sensitive than *Parthenium* to cinnamic acid, tomato for example being injured only by a hundred times higher concentration than that needed for injury to *Parthenium*.

General References

Klein, G., Handbuch der Pflanzenanalyse. Springer, Vol. 2, 1932.

Wood, H. G., *Physiol. Rev.*, **26**, 198 (1946).

Krebs, H. A., Advances in Enzymology. Interscience Publishers, 1943. Vol. 3, p. 191.

Vickery, H. B., and Pucher, G. W., *Ann. Rev. Biochem.*, **9**, 529 (1940).

Bennet-Clark, T. A., *Ann. Rev. Biochem.*, **6**, 579 (1937).

References

1. Nelson, E. K., *J. Am. Chem. Soc.*, **47**, 1177 (1925).
   Curl, A. L., and Nelson, E. K., *J. Agr. Research*, **67**, 301 (1943).
2. Pucher, G. W., and Vickery, H. B., *J. Biol. Chem.*, **145**, 525 (1942).
3. Virtanen, A. I., Arhimo, A. A., Sundman, J., and Jännes, L., *J. prakt. Chem.*, **162**, 71 (1943).
   Damodaran, M., and Nair, K. R., *Biochem. J.*, **32**, 1064 (1938).
4. Frey-Wyssling, A., Die Stoffausscheidungen der höheren Pflanzen. Springer, 1935
5. Klein, G., Handbuch der Pflanzenanalyse. Springer, Vol. 2, 1932.
6. Franzen, H., and Schucmacher, E., *Z. physiol. Chem.*, **115**, 9 (1921).
7. Pucher, G. W., and Vickery, H. B., *Ind. Eng. Chem., Anal. Ed.*, **13**, 244, 412 (1941).
8. Krebs, H. A., and Eggleston, L. V., *Biochem. J.*, **34**, 1383 (1940).
   Bonner, W., and Bonner, J., *Am. J. Botany*, **35**, 113 (1948).
9. Umbreit, W. W., Burris, R. H., and Stauffer, J. F., Manometric Techniques. Burgess, 1945.
10. Kostytschev, S., Plant Respiration. Blakiston, 1927.
11. Ruhland, W., and Wolf, J., *Ann. Rev. Biochem.*, **3**, 501 (1934).
    Ruhland, W., and Wetzel, K., *Planta*, **3**, 765 (1927); **5**, 503 (1929).
12. Krebs, H. A., and Johnson, W. A., *Enzymologia*, **4**, 148 (1937).
    Krebs, H. A., Advances in Enzymology. 1943. Vol. 3, p. 191.
13. Vennesland, B., Ceithaml, J. J., and Gollub, M., *J. Biol. Chem.*, **171**, 445 (1947).
14. Stumpf, P. K., Zarudnaya, K., and Green, D. E., *J. Biol. Chem.*, **167**, 817 (1947).
    Ochoa, S., *J. Biol. Chem.*, **155**, 87 (1944).
15. Vickery, H. B., Pucher, G. W., Wakeman, A. J., and Leavenworth, C. S., Conn. Agr. Expt. Sta. Bull. 399, 1937; 424, 1939.
    Pucher, G. W., Wakeman, A. J., and Vickery, H. B., *J. Biol. Chem.*, **119**, 523 (1937).
16. Vennesland, B., and Felsher, R. Z., *Arch. Biochem.*, **11**, 279 (1946).
    Gollub, M., and Vennesland, B., *J. Biol. Chem.*, **169**, 233 (1947).
17. Wood, J. G., Cruickshank, D. H., and Kuchel, R. H., *Australian J. Exptl. Biol. Med. Sci.*, **21**, 37 (1943).
    Cruickshank, D. H., and Wood, J. G., *Australian J. Exptl. Biol. Med. Sci.*, **23**, 243 (1945).
18. Wood, H. G., and Werkman, C. H., *Biochem. J.*, **30**, 48 (1936); **32**, 1262 (1938).
    See also Werkman, C. H., and Wood, H. G., Advances in Enzymology. 1942. Vol. 2, p. 135.
19. Wood, H. G., and Werkman, C. H., *J. Biol. Chem.*, **135**, 789 (1940).
20. Laties, G., Thesis, University of California, Berkeley, 1947.
21. Woods, D. D., *Biochem. J.*, **30**, 515 (1936).

22. Barker, H. A., *Arch. Microbiol.*, **7,** 404 (1936); Barker, H. A., Ruben, S., and Kamen, M. D., *Proc. Natl. Acad. Sci. U. S.*, **26,** 426 (1940).

23. Lipmann, F., and Tuttle, L. C., *J. Biol. Chem.*, **158,** 505 (1945).
    Koepsell, H. J., Johnson, M. J., and Meek, J. S., *J. Biol. Chem.*, **154,** 535 (1944).

24. Ochoa, S., *J. Biol. Chem.*, **159,** 243 (1945); **174,** 133 (1948).
    Ochoa, S., and Weisz-Tabori, E., *J. Biol. Chem.*, **159,** 245 (1945).

25. Bennet-Clark, T. A., *New Phytologist,* **32,** 128 (1933).

26. Wolf, J., *Planta,* **15,** 572 (1931); **26,** 516 (1937); **28,** 60 (1938); **29,** 314 (1939); **29,** 450 (1939).

27. Pucher, G. W., Leavenworth, C. S., Ginter, W., and Vickery, H. B., *Plant Physiol*, **22,** 360, 477 (1947); **23,** 123 (1948).

28. Bonner, W., and Bonner, J., *Am. J. Botany,* **35,** 113 (1948).
    Thurlow, J., and Bonner, J., *Arch. Biochem.*, **19,** 509 (1948).

29. Clark, H., *Plant Physiol.*, **11,** 5 (1936).
    Bonner, J., *Botan. Gaz.*, **105,** 352 (1944).

30. Wadleigh, C. H., and Shive, J. W., *Am. J. Botany,* **26,** 244 (1939).
    Pucher, G. W. Leavenworth, C. S., Ginter, W., and Vickery, H. B., *Plant Physiol.*, **22,** 205 (1947).

31. Vickery, H. B., Pucher, G. W., Wakeman, A. J., and Leavenworth, C. S., Conn. Agr. Expt. Sta. Bull. 442, 1940.

32. Walker, J., Link, K., and Angell, H., *Proc. Natl. Acad. Sci., U. S.*, **15,** 845 (1929)

33. Bonner, J., and Galston, A. W., *Botan. Gaz.*, **106,** 185 (1944).

34. Sinclair, W. B., and Eny, D. M., *Proc. Am. Soc. Hort. Sci.*, **47,** 119 (1946)

35. Sinclair, W. B., and Ramsey, R. C., *Botan. Gaz.*, **106,** 140 (1944).

36. Sinclair, W. B., and Eny, D. M., *Plant Physiol.*, **21,** 140 (1946).
    Sinclair, W. B., and Eny, D. M., *Botan. Gaz.*, **107,** 231 (1945).

37. Ulrich, A., *Am. J. Botany,* **29,** 220 (1942).
    Burström, H., *Arkiv Botany,* **32A** (7), 1 (1945).

Chapter 15

PLANT RESPIRATION

Introduction. By respiration is meant the sum total of the biological reactions in which organic material is oxidized to simpler compounds. In aerobic respiration these oxidations are carried out in the presence of oxygen and the substrate acted upon is frequently oxidized completely to CO_2 and H_2O. The simple hexose sugars, glucose and fructose, form the substrate for the respiratory oxidations of many tissues, and in this case the overall reaction involved is the classical one:

$$C_6H_{12}O_6 + 6O_2 \rightarrow 6CO_2 + 6H_2O \quad \frac{CO_2}{O_2} = 1$$

In this complete respiratory oxidation of hexose, six mols of oxygen are taken up and six mols of CO_2 are evolved for each mol of sugar utilized; the respiratory quotient or ratio of CO_2 evolved to oxygen consumed is unity. Although hexose is undoubtedly the most frequent and the major substrate for the reactions of respiration, other substrates may also be utilized, particularly the organic acids and the fatty acids. In the complete oxidation of an organic acid, such as for example malic acid which is richer in oxygen than is carbohydrate, the respiratory quotient is greater than 1.

$$C_4H_6O_5 + 3O_2 \rightarrow 4CO_2 + 3H_2O \quad \frac{CO_2}{O_2} = 1.33$$

Conversely, in the complete oxidation of a fatty acid, poorer in oxygen than carbohydrate, more oxygen must be consumed than CO_2 liberated, and the respiratory quotient is consequently less than 1. In many respiring tissues the respiratory quotient may be found to be slightly greater or smaller than that expected on the basis of oxidation of hexose alone, and this doubtless means then that substrates other than hexose may be oxidized simultaneously with the oxidation of hexose itself.

It is now quite clear that in respiration, either hexose or some other substrate is degraded by a succession of steps involving many different enzymes and involving as intermediates many of the other well-known constituents of plant tissues. This series of graded steps may be conveniently separated into two portions: the initial steps by which carbohydrate is converted to pyruvate, and the terminal phase in which pyruvate is oxidized to CO_2 and water. The initial phase is common to aerobic respiration and to fermentation, the anaerobic degradation of organic material through pyruvate to simpler compounds such as lactic acid or ethyl alcohol. Even though many tissues, including those of higher plants, appear to be equipped with a full complement of enzymes both for fermentation and for aerobic respiration, still in the presence of oxygen respiration largely predominates. Fermentation, which is suppressed in the presence of oxygen, becomes apparent, however, under anaerobic conditions. The suppression of fermentation by oxygen, known as the Pasteur effect, serves then to insure that fermentation will take place only under conditions unsuitable for respiration.

The gaseous exchanges of respiring plant tissues have been the subject of intensive investigation by plant physiologists and much is known concerning such exchange not only as among various tissues and various species of plants, but also as concerns the influence of external factors on the rate and type of gas exchange involved. This discussion will take up a different aspect of respiration, namely the question of the metabolic pathways of respiration, the chemical reactions through which carbohydrates and other substrates are transformed to CO_2 in the respiration of the higher plant. We shall first consider briefly the history of this subject. This is followed by a discussion of the general nature of oxidation and a summary of the best-known respiratory enzymes. We shall discuss in detail the initial reactions of respiration, namely the breakdown of hexose to pyruvate, and then investigate the pathways of pyruvate oxidation. Finally we shall take up the energetics of respiration and the ways in which energy liberated in respiration may be made available to other energy-using cellular processes.

Early Work. The respiration of plants has been the subject of study since the time of Stephen Hales, who showed that plants living in the dark decrease the ability of air to support combustion. That oxygen is taken up in respiration was shown by Ingen Housz (1779), while de Saussure (1804, 1822) measured the CO_2 evolved as well as the oxygen consumed. The study of respiration in relation to growth was initiated by Sachs (1865) who showed that high rates of respiration accompany growth and who introduced the concept of the respiratory quotient or ratio of CO_2 given off to oxygen consumed in respiration. Further work on the chemical mechanism of plant respiration was carried on by Bach (1901). Bach showed that plant extracts possess the ability to oxidize certain phenols in the presence of oxygen; the enzymatic material responsible for this oxidation he termed oxidase. He showed further that plants possess a second enzymatic mechanism for the oxidation of certain phenols by H_2O_2. This enzyme we now know as peroxidase. The concept that these two oxidative enzymes might be a part of the repiratory mechanism constitutes what has been known as the Bach-Chodat theory of respiration, a theory which was developed extensively by Palladin and which is frequently referred to in older textbooks of plant physiology. Further study of the oxidase of Bach by Onslow[1] showed that o-hydroxyphenols are the substrates for this enzyme, while Pugh and Raper in 1927[2] found that the oxidation product is the corresponding o-quinone. The oxidase of Bach and Chodat is therefore what we now know as polyphenoloxidase. This early work, which is summed up in the books of Kostytschev and Onslow showed then that polyphenoloxidase occurs in plant tissues. That this enzyme might play a role in respiration was suggested but not demonstrated in any way.

The early work on the mechanism of fermentation in yeast and/or muscle by Harden and Young, Von Euler, Emden, Meyerhof, and Neuberg found a counterpart in and stimulated work on fermentation in higher plant tissues. Many early workers have shown that higher plant tissues such as meal from seeds, or juice from such tissues as seeds, leaves, and tubers, can ferment hexose, with the production of approximately one molecule of CO_2 per molecule of alcohol formed, a ratio similar to that found in yeast fermentation. Ivanoff[3] and later Bodnar and Hoffner,[4] as well as others, early indicated a role for phosphate in such fermentation by the observation that added phosphate increases the rate of hexose fermentation by pea meal. That pyruvic acid might be an intermediate product of sugar breakdown or glycolysis in higher plant respiration was surmised by analogy with yeast fermentation.

The stages leading from triose or pyruvic acid to CO_2 and water were almost completely neglected in the early work. Zaleski and Marx[5] in 1912 and Bodnar[6] in 1916 demonstrated that such tissues as seeds and potato tubers contain the enzyme carboxylase, an enzyme which decarboxylates pyruvic acid to form CO_2 and acetaldehyde. The presence of acetaldehyde was also demonstrated in fermenting pea meal by methods such as the binding of the aldehyde with bisulfite. With regard to the role of organic acids, Kostytschev in his summary on plant respiration denied these substances any place in respiration, holding with Ruhland and Wetzel (see organic acid metabolism) that organic acids are by-products of nitrogen metabolism rather than intermediates in carbohydrate metabolism. The dehydrogenases which oxidize the plant acids, enzymes discovered by Batelli and Stern and by Thunberg beginning in 1909–10,[7] which we now know to play a central role in respiration, were not considered by the early plant workers as respiratory enzymes.

The Nature of Oxidation.[8] The term oxidation was originally applied to chemical reactions in which a reactant actually gains oxygen; a simple example of this type of reaction is the oxidation of carbon monoxide with oxygen to yield carbon dioxide.

$$CO + \tfrac{1}{2}O_2 \rightarrow CO_2 \tag{1}$$

Oxidation may also consist in the removal of hydrogen from the reactant as in the oxidation of hydroquinone to benzoquinone. Wieland early suggested that even oxidations involving addition of oxygen may be

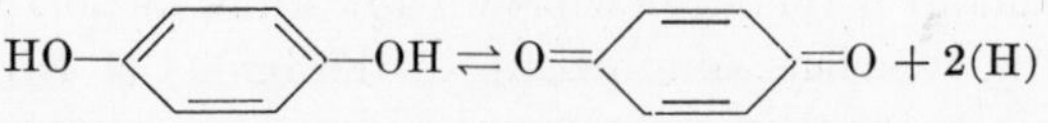

$$HO-\langle\ \rangle-OH \rightleftharpoons O=\langle\ \rangle=O + 2(H) \tag{2}$$

regarded as reactions in which the elements of a molecule of water are added to the reactant, followed by removal of hydrogen, as shown below for the oxidation of acetaldehyde to acetic acid.

$$CH_3CHO + H_2O \rightarrow CH_3C\overset{OH}{\underset{H}{\big<}} OH \rightarrow CH_3COOH + 2(H) \tag{3}$$

Finally, oxidation in its simplest form consists merely in the removal of an electron or electrons from the reactant, as in the case of the oxidation of ferrous to ferric iron.

$$Fe^{++} \rightarrow Fe^{+++} + (e) \tag{4}$$

The oxidative reactions 2, 3, and 4 are partial reactions in that the oxidizing agent is not included in the reactions as written. Neither free electrons nor atomic hydrogen can of course appear as reaction products. On the contrary an oxidizing agent must act as a hydrogen acceptor in reactions 2 and 3, or as an electron acceptor in reaction 4. In all cases the oxidizing agent by accepting either hydrogen or electrons is itself reduced.

Although the reactions typified in 1, 2, 3, and 4 may seem quite disparate at first glance, they do have a common feature, namely the transfer of electrons away from the substance being oxidized. In 1 electrons held by the carbon atom are transferred to the carbon-oxygen bond. In 2 and 3 electrons held by the reactant are removed as hydrogen atoms, while in 4 electrons are removed directly as electrons. In this sense reactions of type 4 represent the simplest type of oxidation. Reduction whether by removal of oxygen or by addition of hydrogen similarly involves basically the addition of electrons to the molecule being reduced.

The types of oxidation listed in reactions 1 to 4 all find their places in the biological oxidations of respiration. The reactions of oxidases with oxygen are examples of type 1, while the dehydrogenation of substrates by dehydrogenases or certain flavoproteins are examples of types 2 and 3. Finally in the iron protein systems such as the cytochromes we have oxidations involving only the transfer of electrons as in type 4.

Oxidative Enzymes. Many of the enzymes involved in biological oxidation consist of proteins or apoenzymes combined with nonprotein prosthetic groups. We may say that while the general nature of the process carried out by the enzyme is determined by the nature of its prosthetic group, the substrate specificity of the enzyme is determined by the nature of the protein apoenzyme. Thus the enzymes which have diphosphopyridine nucleotide (DPN) as their prosthetic group are all dehydrogenases and function in actual removal of hydrogen atoms from substrate. Whether the substrate acted upon is ethyl alcohol, malic acid, or one of several other oxidizable materials, is dependent on the exact protein with which the DPN is combined. With certain of the enzymes of biological oxidation the prosthetic group may be readily removed, for example, by dialysis; in this case the prosthetic group is called a coenzyme. This is true of the dehydrogenases, in which the equilibrium between enzyme and its two components is far to the right.

$$\text{Enzyme} \rightleftharpoons \text{Apoenzyme} + \text{Coenzyme}$$

With other systems as with the flavoproteins, the dissociation constant is small and the coenzyme can be removed by dialysis only when the enzyme is brought to the verge of denaturation as in 0.01 N HCl. In still other cases, as in the heavy metal enzymes, the prosthetic group may not be removed by ordinary dialysis.

The biological oxidations of respiration are carried out by enzyme systems which include several distinct enzymatic components. Oxidation of the substrate is frequently catalyzed by a dehydrogenase which removes two hydrogen atoms from the compound in question and becomes itself reduced. The dehydrogenase is not reoxidized by molecular oxygen, but is rather reoxidized by specific enzymes, such as flavoproteins, which are in turn reduced. In some instances this flavoprotein may be directly reoxidized by oxygen. In general, however, these intermediary enzymes are reoxidized through the intervention of carriers such as cytochrome and in other cases possibly by phenols, compounds which themselves are reoxidized by oxidases. The path of the electrons removed by oxidation of substrate will then be somewhat as shown in Fig. 15-1. In the discussion of the oxidative enzymes of respiration which follows, we shall take up the individual groups of enzymes in the order in which they participate in this transfer of electrons.

The Dehydrogenases. Harden and Young[9] in their investigations on fermentation by cell free yeast juice found that yeast juice could be split by dialysis into two fractions, either of which by itself was incapable of carrying out fermentation, but which together retained the original activity. The dialyzable fraction was thermostable, the nondialyzable fraction thermolabile. To the dialyzable thermostable fraction the name cozymase was applied, and from this work Harden and Young developed the concept of the coenzyme. The problem of the coenzymes of cellular

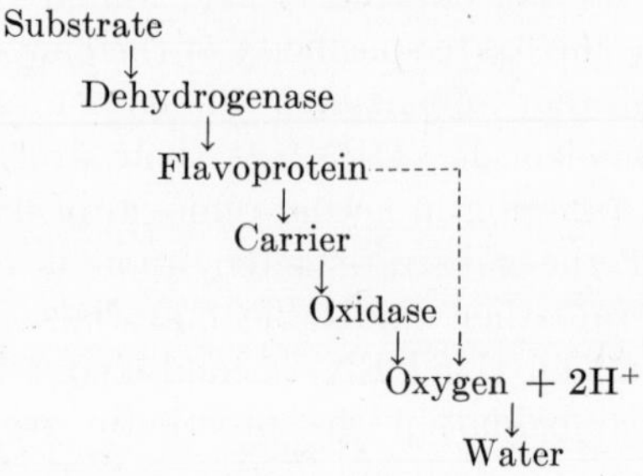

Substrate
↓
Dehydrogenase
↓
Flavoprotein ------┐
↓　　　　　│
Carrier　　　│
↓　　　　　│
Oxidase　│
↓　↓
Oxygen + 2H⁺
↓
Water

FIG. 15-1. General pathway of electrons from substrate to oxygen in biological oxidations.

oxidation was taken up anew by Warburg and Christian in 1932.[10] They worked with the enzyme hexose monophosphate dehydrogenase which is obtained from red blood corpuscles and which catalyzes the reaction:

$$\text{Glucose-6-phosphate} \rightarrow \text{6-Phosphogluconic acid} + 2(H)$$

In this reaction the hydrogen is transferred anaerobically to an acceptor such as the dye methylene blue which can be itself reduced. The enzyme depends for its activity on a dialyzable coenzyme which was isolated in pure form. At the same time similar work was being carried on with cozymase and in 1935, Von Euler, Alberts, and Schlenk showed that this coenzyme is a dinucleotide containing nicotinamide, adenine, ribose, and phosphoric acid.[11] The structure of their material, now known as cozymase, coenzyme I, or diphosphopyridine nucleotide (DPN) is shown below:

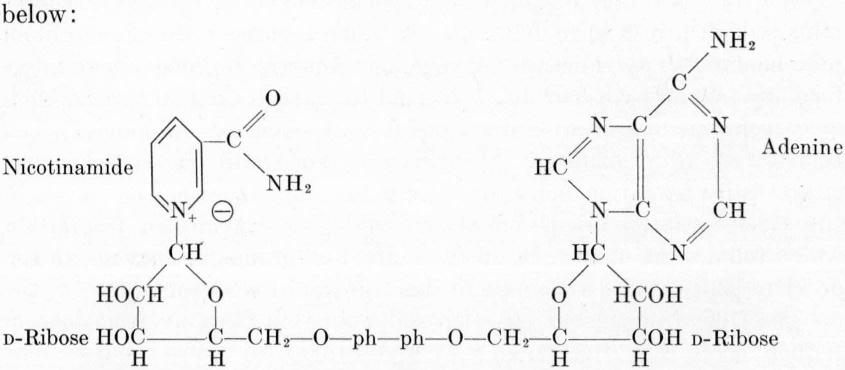

Cozymase: Coenzyme I: Diphosphopyridine nucleotide (DPN)
The phosphate group is represented as ph.

The coenzyme of Warburg and Christian, now known as coenzyme II or triphosphopyridine nucleotide (TPN), is similar to that of DPN except for one additional molecule of phosphoric acid.[12] The structure commonly ascribed to TPN is that shown below; there is still some uncertainty as to the exact location of the third phosphate residue.

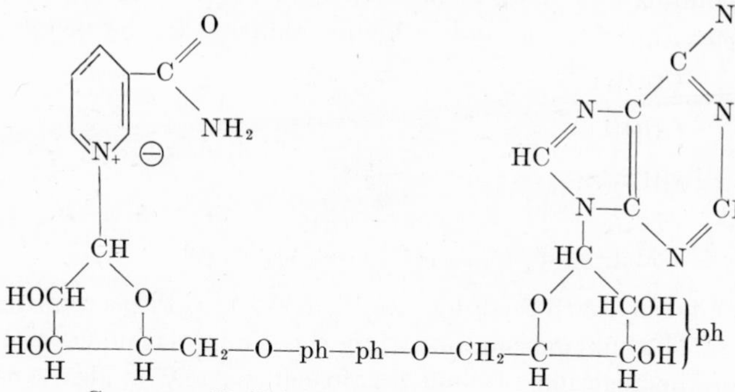

Coenzyme II: Triphosphopyridine nucleotide (TPN)

The electron or hydrogen-carrying group of both nucleotides is nicotinamide which can be reversibly oxidized and reduced and which can therefore act as an acceptor of electrons in oxidation reactions. In the reduction of DPN or of TPN it is the double bond adjacent to the quater-

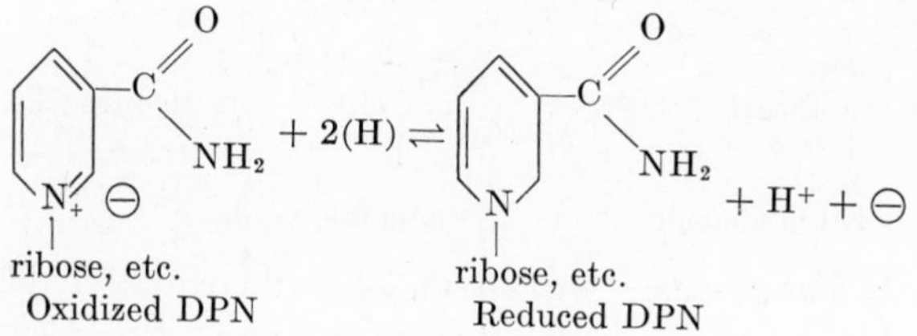

Oxidized DPN   Reduced DPN

nary nitrogen atom which is reduced. The carbon atom takes up one hydrogen atom while the nitrogen atom takes up one electron with the formation of one hydrogen ion. Thus the enzyme hexose monophosphate dehydrogenase consists of a specific protein apoenzyme, which in the presence of TPN is capable of oxidizing glucose phosphate to phosphogluconate, with the coenzyme TPN being itself reduced. The reduced coenzyme can be reoxidized in a variety of ways. It may, for example, be reoxidized by methylene blue although this can occur only if a specific flavoprotein intermediate enzyme is also present. The reduced TPN cannot, however, be directly reoxidized by molecular oxygen. The enzyme is therefore a dehydrogenase; it is capable of activating and accepting electrons of the substrate but cannot transfer these activated electrons ($+H^+$) to oxygen. Although DPN and TPN are very similar, each is nevertheless able to combine only with its own particular apoenzyme and they are not in general interchangeable. Each coenzyme is, however, able to combine with a variety of individual proteins, and each combination yields an enzyme which is specific for a particular substrate.

Malic dehydrogenase. A dehydrogenase for the oxidation of malic to the corresponding keto acid, oxaloacetic acid, occurs widely in nature, in animal tissues, in bacteria, and in higher plants. In the higher plant it

$$\underset{\text{L-Malic acid}}{\begin{array}{c} \text{COOH} \\ | \\ \text{CHOH} \\ | \\ \text{CH}_2 \\ | \\ \text{COOH} \end{array}} + \text{DPN} \underset{\text{Apoenzyme}}{\rightleftharpoons} \underset{\text{Oxaloacetic acid}}{\begin{array}{c} \text{COOH} \\ | \\ \text{C}{=}\text{O} \\ | \\ \text{CH}_2 \\ | \\ \text{COOH} \end{array}} + \text{Reduced DPN}$$

has been found in seeds, seedlings, and leaves.[13] DPN is required, TPN is inactive. The enzyme is inhibited by the accumulation even in minute amounts of the reaction product, oxaloacetate, and is also sensitive to iodoacetate.

Alcohol dehydrogenase. An enzyme for oxidation of ethyl alcohol to acetaldehyde is found in yeast, from which the protein apoenzyme has been obtained in crystalline form by Negelein and Wulff.[14] DPN is required as the coenzyme. This enzyme also is found in higher plant tissues and has been studied in seeds, seedlings, and leaves.

$$\underset{\text{Ethyl alcohol}}{\text{CH}_3\text{CH}_2\text{OH}} + \text{DPN} \underset{\text{Apoenzyme}}{\rightleftharpoons} \underset{\text{Acetaldehyde}}{\text{CH}_3\text{C}\overset{\text{O}}{\underset{\text{H}}{\diagdown}}} + \text{Reduced DPN}$$

Glutamic dehydrogenase. An enzyme for the oxidative deamination of glutamic acid is widely distributed in nature. Presence of such an enzyme in legume seedlings was reported by Damodaran and Nair[15] and it has since been found in oat seedlings, in leaves, and in numerous other species and tissues.[13]

$$\underset{\substack{\text{Glutamic} \\ \text{acid}}}{\begin{array}{c} \text{COOH} \\ | \\ \text{CHNH}_2 \\ | \\ \text{CH}_2 \\ | \\ \text{CH}_2 \\ | \\ \text{COOH} \end{array}} + \text{H}_2\text{O} + \text{DPN} \underset{\text{Apoenzyme}}{\rightleftharpoons} \underset{\substack{\alpha\text{-Ketoglutaric} \\ \text{acid}}}{\begin{array}{c} \text{COOH} \\ | \\ \text{C}{=}\text{O} \\ | \\ \text{CH}_2 \\ | \\ \text{CH}_2 \\ | \\ \text{COOH} \end{array}} + \text{NH}_3 + \text{Reduced DPN}$$

Glutamic dehydrogenase as obtained from liver, which has been particularly studied, can utilize either DPN or TPN. The apoenzyme is further associated with insoluble fragments of the cell and can be obtained in solution only with difficulty. The enzyme as obtained from higher plants is, however, fully soluble and appears to require DPN alone. This dehydrogenase is of first importance to nitrogen metabolism, since in the presence of glutamic dehydrogenase an equilibrium is established between free ammonia and ammonia fixed as the amino group of glutamic acid. In this way, as will be further discussed in Chapter 16, glutamic dehydrogenase participates in nitrogen metabolism as a port of entry for ammonia into organic combination.

Triosephosphate dehydrogenase. Triosephosphate dehydrogenase, which is responsible for the phosphate-coupled oxidation of phosphoglyceraldehyde to phosphoglyceric acid (this oxidation constitutes the only oxidative step in the transformation of hexose to pyruvate), was obtained in pure form from yeast by Warburg and Christian in 1939.[16] The coenzyme of triosephosphate dehydrogenase from higher plants as well as from yeast and mammalian sources is DPN. The enzyme is readily inhibited by iodoacetate or iodoacetamide, reagents which probably in this case as in others cause oxidation of essential sulfhydryl groups of the apoenzyme. The mechanism of the reaction will be discussed under *energy transfer* below.

Isocitric dehydrogenase. An enzyme for the oxidation of isocitric acid occurs in seeds, seedlings, leaves, and probably in all plant as well as animal tissues. The reaction leads to the production of oxalosuccinic acid, an intermediate which is ordinarily immediately decarboxylated by an additional but different enzyme, oxalosuccinic carboxylase, to form α-ketoglutaric acid.

$$
\begin{array}{ccc}
\text{COOH} & & \text{COOH} \\
| & & | \\
\text{CHOH} & & \text{C}=\text{O} \\
| & & | \\
\text{HC}-\text{COOH} + \text{TPN} \rightleftharpoons \text{HC}-\text{COOH} + \text{Reduced TPN} \\
| \quad\quad {\scriptstyle \text{Apoenzyme of iso-}} & & | \\
\text{CH}_2 \quad\quad {\scriptstyle \text{citric dehydrogenase}} & & \text{CH}_2 \\
| & & | \\
\text{COOH} & & \text{COOH} \\
\text{Isocitric acid} & & \text{Oxalosuccinic acid}
\end{array}
$$

TPN is required by the enzyme as found in muscle and is probably required also by the enzyme as found in higher plants, according to results obtained with seedlings[17] and with parsley roots.[18]

The Flavoproteins. In 1932 Warburg and Christian[19] obtained from yeast an enzyme yellowish in color and capable of reversible oxidation by molecular oxygen, the oxidized form of the enzyme being susceptible of reduction by reduced TPN. The enzyme was resolved into an apoenzyme and a prosthetic group, the latter being, however, more tightly bound than in the case of the dehydrogenases. The prosthetic group was shown to be riboflavin phosphate in which D-ribitol phosphate is attached to a substituted isoalloxazine nucleus.[20]

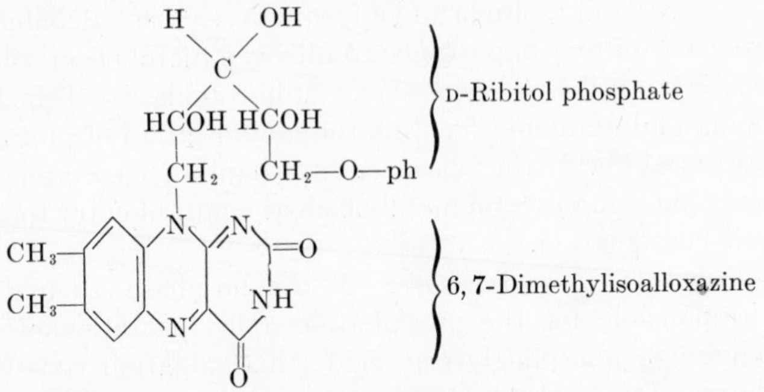

Riboflavin monophosphate
Prosthetic group of original yellow enzyme of Warburg and Christian

Oxidation and reduction of the enzyme is accomplished by the subtraction or addition of two hydrogen atoms from the isoalloxazine nucleus. The enzyme of Warburg and Christian, or "old yellow enzyme" as it is known today, can act *in vitro* as an intermediate hydrogen carrier, coupling the oxidation of hexose phosphate by dehydrogenase to the reduction of oxygen through the following series of steps:

Red. TPN + ox. yellow enzyme → ox. TPN + red. yellow enzyme
Red. yellow enzyme + O_2 → ox. yellow enzyme + H_2O_2

This coupled reaction is probably not a significant feature of respiration because of the relatively low rate of reoxidation of the yellow enzyme by oxygen and in the cell still other yellow enzymes are responsible for reoxidation of reduced codehydrogenases.

Of the yellow enzymes which are known, those which have been carefully studied have with two exceptions been found to consist of apoenzyme combined to a dinucleotide containing riboflavin and adenine.[21] In this

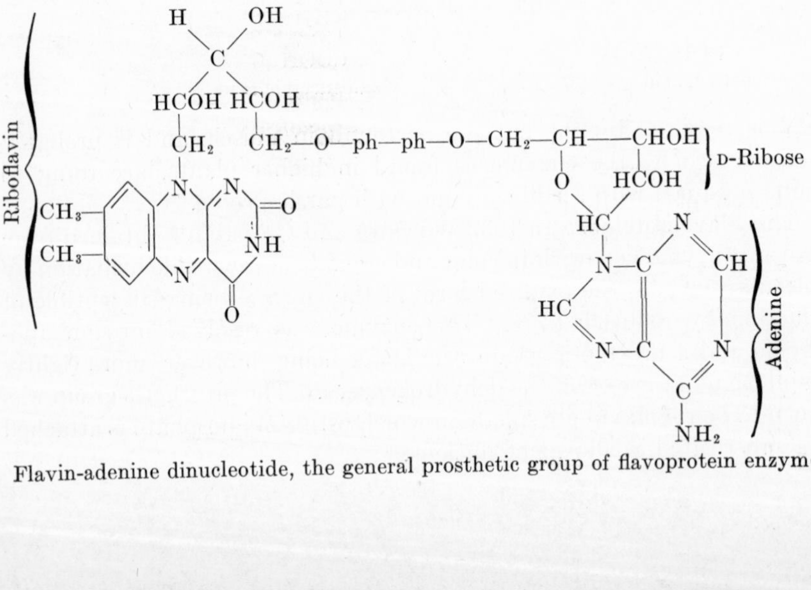

Flavin-adenine dinucleotide, the general prosthetic group of flavoprotein enzymes

dinucleotide, riboflavin phosphate is bound through a pyrophosphate linkage to an adenine-D-ribose phosphate (adenylic acid) identical with that found in the codehydrogenases.

The diaphorases. The diaphorases are characterized by having as their substrates reduced DPN or TPN. Reduced diaphorase is rapidly reoxidized by methylene blue, possibly by other systems, but is not reoxidized by molecular oxygen at an appreciable rate. Reduced diaphorase must therefore in general be reoxidized by still other enzymatic systems. Diaphorase as obtained from tissue is ordinarily associated with cell fragments and does not pass into solution. By treatment of the tissue with ammonium sulfate and dilute warm alcohol (43°), Straub has, however, obtained a soluble diaphorase capable of oxidizing reduced DPN.[22] Lockhart[23] has demonstrated diaphorase in pea and bean seedlings and in potato tubers, while Okunuki[24] has found the enzyme in pollen.

D-*Amino acid oxidase.* Krebs in 1933[25] descr bed the presence in animal tissues of an enzyme capable of oxidatively deaminating the D or unnatural isomers of amino acids with production of the corresponding keto acid and H_2O_2.

$$R—\underset{\underset{\text{D-Amino acid}}{NH_2}}{\underset{|}{CH}}—COOH + O_2 + H_2O \xrightarrow[\text{D-Amino acid oxidase}]{} R—\underset{\underset{\text{Corresponding keto acid}}{O}}{\underset{\|}{C}}—COOH + NH_3 + H_2O_2$$

It is to be noted that the enzyme is auto-oxidizable, i.e., is capable of being reoxidized by molecular oxygen. The enzyme rapidly attacks D-methionine, arginine, phenylalanine, and certain other D-amino acids. The natural or L-isomers on the other hand are not attacked at all. The work of Krebs, Straub,[26] and of Warburg and Christian[27] has shown that the enzyme is made up of a specific protein bound to the flavin-adenine dinucleotide. Although the enzyme appears to be of wide distribution in animal tissues, it has not been described from a higher plant source. It has, however, been found in the fungus *Neurospora.*[28]

Xanthine oxidase. Xanthine oxidase which catalyzes the oxidation of both purines and aldehydes is known from cream, and from the liver of some species of mammals. The enzyme can, for example, catalyze the oxidation of hypoxanthine to xanthine and the further oxidation of xanthine to uric acid.

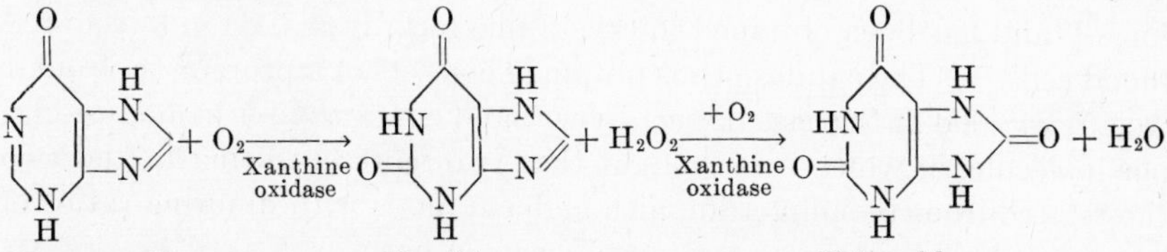

Hypoxanthine   Xanthine   Uric acid

Xanthine oxidase has been purified by Ball[29] and by Warburg and Christian,[27] who have shown that the coenzyme is the flavin-adenine dinucleotide. This enzyme has not been demonstrated in higher plant tissues, although on the basis of the observed products of purine metabolism (Chapter 21) it is possible that it may actually occur.

L-*Amino acid oxidase.* An enzyme for the oxidative deamination of the natural L-isomers of amino acid is, like D-amino acid oxidase, unknown from higher plants, but has been found and studied in liver and kidney[25,30] as well as in bacteria. This enzyme which also appears to be a flavoprotein is further discussed in Chapter 16.

Cytochrome C reductase. Cytochrome C reductase functions as a carrier between cytochrome C and reduced TPN, i.e., it is a diaphorase for TPN and at the same time is rapidly reoxidized by cytochrome C, although it is not reoxidized at an appreciable rate by molecular oxygen. The enzyme as found in yeast and in mammalian tissue is remarkable in that its prosthetic group is identical with that of the old yellow enzyme.[31] A cytochrome C reductase active with reduced DPN is also found in yeast as well as in liver.[32]

Other enzymes. In all, twelve or more flavoprotein enzymes are known today. Despite this fact, not one flavoprotein enzyme has as yet been certainly demonstrated in the higher plant, and but little work has been done with these enzymes in plants. Although diaphorase activity has been demonstrated in higher plant tissues, and although in other cases diaphorase is known to be a flavoprotein, still the identity of plant diaphorase with a flavoprotein is not yet clearly established. That a flavoprotein or proteins may be of general importance in the plant is indicated by the universal occurrence of riboflavin in higher plant tissues.

Porphyrin Proteins. Plants contain a group of enzymes in which the iron-containing prosthetic groups are hematin or related porphyrins. The combined hematin is found in many proteins, which then comprise a variety of iron-containing enzymes.

Catalase. Catalase is found in all higher plants and probably in all tissues of higher plants. It is highly specific, acting essentially only on hydrogen peroxide which is decomposed to water and oxygen.

$$2H_2O_2 \xrightarrow{\text{Catalase}} 2H_2O + O_2$$

The enzyme has been purified from numerous sources, including seedlings,[32] and has been obtained in crystalline form from liver and from red blood cells.[33] The catalase thus obtained consists of a protein, molecular weight 248,000 in the case of beef liver,[34] and contains four heme residues per molecule, of which it is thought that two represent hematin and two the related iron-free biliverdin, although catalases with differing ratios of hematin to biliverdin also appear to occur in nature.[35]

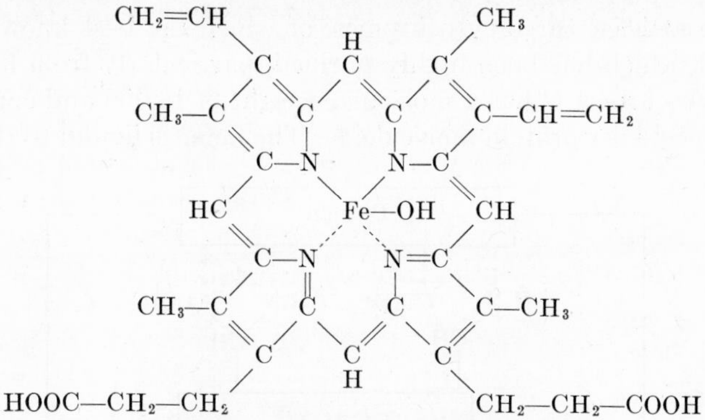

Structure of hematin (H. Fischer)

The iron is in the ferric form and the enzyme is inhibited by HCN, H_2S, and sodium azide, all substances which combine with the ferric atom of the enzyme. The role of catalase in respiration is probably to remove from the tissues H_2O_2 produced in other respiratory reactions, which might otherwise accumulate to toxic levels.

Peroxidase. This enzyme also occurs universally or nearly so in plant tissues, particularly rich sources being horseradish root and the latex of *Ficus* sp. It is active, in the presence of H_2O_2, in oxidizing a wide variety of mono- and diphenols and aromatic amines, as for example *ortho*, *meta*, and *para* cresol, pyrogallol, hydroquinone, and *p*-phenylenediamine. The enzyme exists in horseradish roots in two forms, paraperoxidase or peroxidase I, and a peroxidase II. The latter enzyme has been obtained from horseradish in crystalline form by Theorell.[36] It is a protein of molecular weight 44,000 and contains one molecule of hematin per molecule of protein. Theorell has been able to reversibly separate peroxidase II into the protein apoenzyme and hematin by placing the enzyme in a solution of HCl in acetone at $-15°C$. Recombination of the prosthetic group with the apoenzyme was accomplished in aqueous solution at ordinary temperatures. The iron is present in the ferric form as in catalase and the enzyme is inhibited by HCN, H_2S, sodium azide, and dithionite, all of which combine with the iron atom. The ubiquitous occurrence of peroxidase in plants suggests a function for the enzyme in respiration, but all attempts to find such a function have met with failure up to the present time.

Cytochromes. These heme protein pigments have been known since they were observed in muscle by McMunn in 1886. They function as intermediate carriers between molecular oxygen and sluggishly auto-oxidizable systems such as cytochrome C reductase in the following way:

Reduced flavoprotein + cytochrome → Oxidized flavoprotein + reduced cytochrome

$$2 \text{ Reduced cytochrome} + 2H^+ + \tfrac{1}{2}O_2 \xrightarrow{\text{Cytochrome oxidase}} 2 \text{ Oxidized cytochrome} + H_2O$$

There are at least three cytochromes, of which the best known is cyto-chrome C, which has been highly purified, particularly from heart mus-cle.[37] Cytochrome C has a molecular weight of 13,000 and contains one heme molecule per protein molecule.[38] The heme is bound to the protein

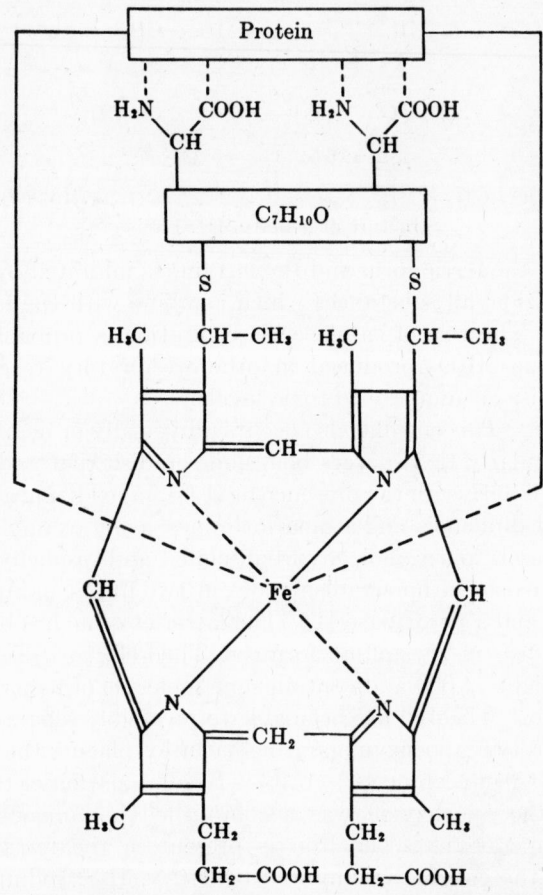

Cytochrome C, showing linkage of iron porphyrin to protein.   (After Theorell)

through two thioether linkages and by two histidine-imidazole nuclei in such a way as to prevent HCN, H_2S, O_2, etc. from approaching the iron atom.   For this reason cytochrome C is neither auto-oxidizable nor inhib-ited by the usual heavy metal poisons.   When cytochrome C acts as an electron carrier the iron is reversibly oxidized and reduced between the ferric and ferrous states.   Cytochrome C, similar if not identical with that of heart muscle, has been isolated from wheat germ.[39]

The enzyme which specifically reoxidizes reduced cytochrome C at the expense of molecular oxygen is known as cytochrome oxidase. The cytochrome C-cytochrome oxidase system as a whole can oxidize other substrates such as various phenols and amines. The system was hence known earlier as "indophenol oxidase," since in the presence of α-naphthol and dimethylphenylenediamine, the blue indophenol is produced.[40] Cytochrome oxidase is associated with the insoluble fragments of the cell. When the active material is heated in water a fraction is obtained in true solution which is necessary for the action of the insoluble oxidase on cytochrome C.[41] The nature of this soluble factor is, however, unknown. The enzyme is inhibitable by HCN, H_2S, and sodium azide and is believed to be a heavy metal protein since the enzyme is also inhibited in the dark by CO, the inhibition being reversed in blue light which causes dissociation of the carbon monoxide-enzyme complex. This behavior is typical of iron proteins and is, so far as known, not true for any other type of heavy metal enzyme.

Copper Proteins. *Polyphenoloxidase.*[42] An enzyme, polyphenoloxidase, has long been known to be in part responsible for darkening of plant tissues after injury. The enzyme has as its substrate monophenols such as p-cresol and tyrosine, oxidizing them to the corresponding o-diphenols and for this reason has also been called tyrosinase. Polyphenoloxidase is,

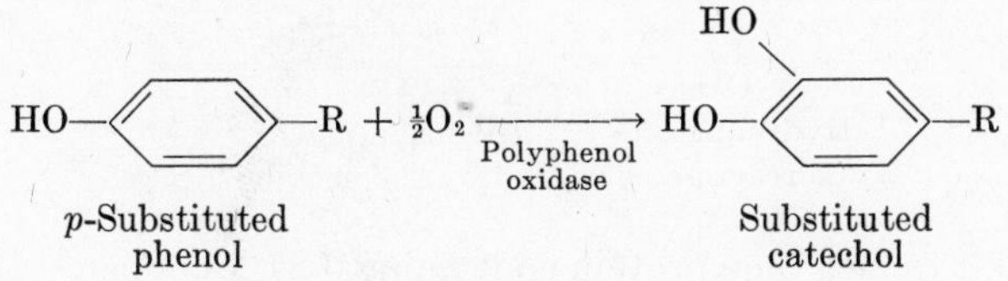

$$HO-\langle\ \rangle-R + \tfrac{1}{2}O_2 \xrightarrow[\text{oxidase}]{\text{Polyphenol}} HO-\langle\ \rangle-R$$

| p-Substituted phenol | Substituted catechol |

however, in addition capable of catalyzing the oxidation of the *ortho*-diphenols to the corresponding *ortho*-quinones. The enzyme is strictly

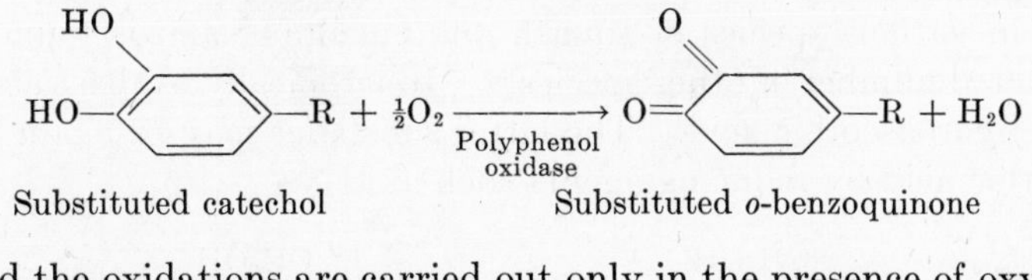

$$HO-\langle\ \rangle-R + \tfrac{1}{2}O_2 \xrightarrow[\text{oxidase}]{\text{Polyphenol}} O=\langle\ \rangle-R + H_2O$$

| Substituted catechol | Substituted o-benzoquinone |

aerobic and the oxidations are carried out only in the presence of oxygen. The quinones produced show a great tendency toward further oxidation and polymerization with the production of colored products, the exact color depending on the phenol used as substrate. Such color development may be used to follow polyphenoloxidase activity although measurements of oxygen uptake are probably preferable in general. The enzyme is inhibited by HCN and by nitro analogs of the normal substrates such as 4-nitrocatechol and p-nitrophenol. Carbon monoxide inhibits the enzyme, but the inhibition is not reversed by light as with cytochrome oxidase.

Polyphenoloxidase has been prepared in greatly enriched form by Graubard and Nelson and by others from fungi,[43] and by Kubowitz from potato,[44] which is a rich source of the enzyme. The protein contains 0.2 to 0.3% copper according to the various authors. Polyphenoloxidase may occur either in the cytoplasm as a constituent of the soluble cytoplasmic protein as in potato and in spinach leaves[45] or as a constituent of the chloroplasts, as in spinach, tea, and chard.[46]

Laccase. This enzyme is found in the latex of species of *Rhus* and is responsible for the oxidation of the phenols urushiol and dehydrourushiol, which are contained in the latex, with production of the black resinous lac. The specificity of laccase differs from that of polyphenoloxidase in that tyrosine and *p*-cresol are not attacked while *p*-phenylenediamine and hydroquinone, which are not attacked by polyphenoloxidase, are attacked by laccase. The enzyme has been purified by Keilin and Mann[47] who

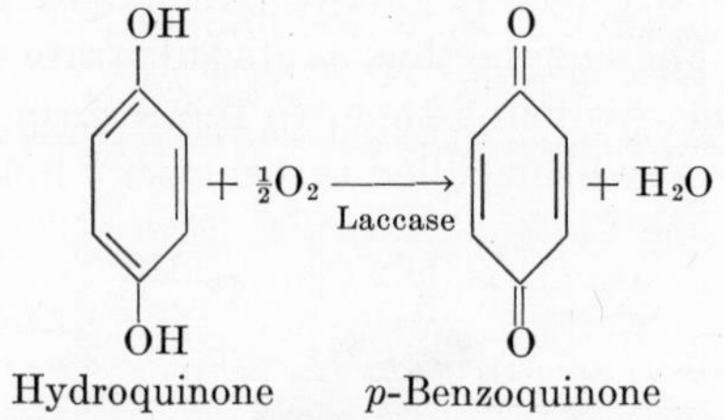

Hydroquinone   *p*-Benzoquinone
Type reaction catalyzed by the enzyme laccase

have found it to be a blue protein containing 0.24% copper. It is moderately widely distributed in higher plants as in cabbage, potato, and sugar beet,[48] but is probably less common than polyphenoloxidase.

Ascorbic acid oxidase. Ascorbic acid oxidase occurs in high concentration in various species of squash and cucumber and is found also in a scattered number of other species.[49] It aerobically oxidizes ascorbic acid to dehydroascorbic acid. The latter substance may be again reduced to ascorbic acid by reducing agents such as H_2S.

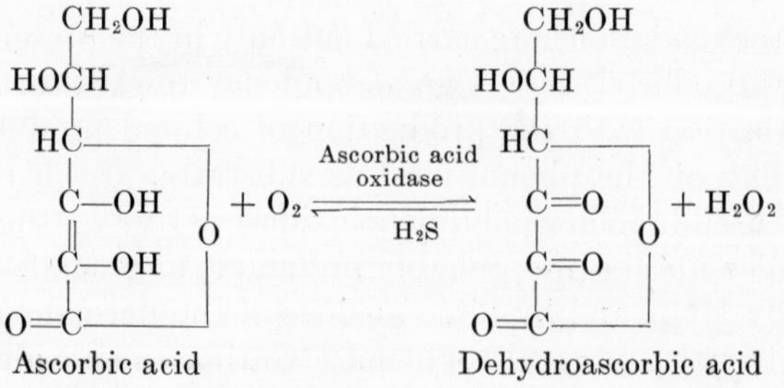

Ascorbic acid   Dehydroascorbic acid

As purified by Powers,[50] Todokoro and Takasugi,[51] and others, it is a blue or greenish blue copper protein containing approximately 0.25% copper.

Enzymes of CO_2 Evolution. The enzymes involved in the liberation of CO_2 in respiration are known as carboxylases and catalyze the liberation of CO_2 primarily from aliphatic alpha keto acids such as pyruvic, oxaloacetic, α-ketoglutaric, and oxalosuccinic.

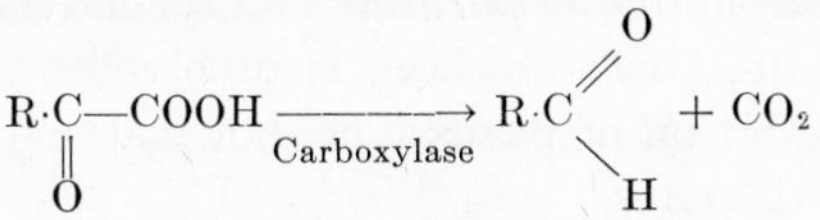

$$
\underset{\underset{O}{\|}}{R \cdot C} - COOH \xrightarrow{\text{Carboxylase}} R \cdot C \underset{H}{\overset{O}{\diagup}} + CO_2
$$

Type reaction catalyzed by carboxylases

Pyruvate carboxylase. The best studied carboxylase is that which acts upon pyruvic acid, converting this substance to acetaldehyde and CO_2. This enzyme, pyruvate carboxylase, was discovered in yeast by Neuberg and Karczog in 1911, and the reaction involved constitutes an essential step in alcoholic fermentation. That the enzyme contains a coenzyme which may be dialyzed off under alkaline conditions was shown by Auhagen in 1932. The coenzyme of yeast pyruvic carboxylase was isolated by Lohmann and Schuster[52] and shown to be the pyrophosphate derivative of the vitamin thiamine.

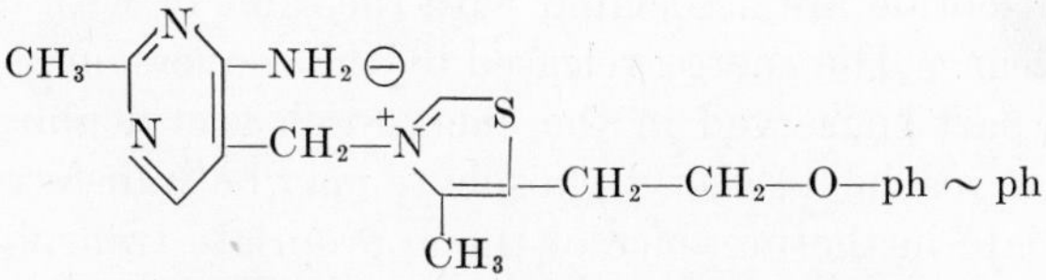

$$
CH_3 - \underset{N}{\overset{N}{\diagdown}} - NH_2 \ominus
$$
$$
- CH_2 - \overset{+}{N} \diagup\overset{S}{} - CH_2 - CH_2 - O - ph \sim ph
$$
$$
\overset{|}{CH_3}
$$

Thiamine pyrophosphate: cocarboxylase

The purified enzyme possesses a molecular weight of approximately 141,000 and has been shown to contain both cocarboxylase and magnesium ions as essential components.[53]

Carboxylase was early found in seeds of pea, bean, and lupin[5] and in potato and beet.[6] More recently, it has been found in leaves of cabbage and other crucifers, barley, *Bryophyllum*, seedlings of *Avena* and in many fleshy roots such as turnip, parsnip, carrot, and parsley.[54] In a number of instances higher plant carboxylase resembles yeast carboxylase and may be readily separated into apoenzyme and cocarboxylase by dialysis at pH 8.5. In the special case of the carboxylase of pea roots, the enzyme differs, however, from yeast carboxylase in that the thiamine pyrophosphate cannot be removed by alkaline washing,[55] since under these conditions pyrophosphate alone is removed with consequent loss of activity of the enzyme. The thiamine in this case remains firmly bound to the apoenzyme. Activity can be restored by addition of pyrophosphoric acid.

Although carboxylase has been found in the tissues of some higher plants, it is not universal. It will be shown below that, in all likelihood, the aerobic respiratory processes in higher plants do not involve the simple decarboxylation of pyruvic acid. It is possible that carboxylase may function only when the plant is subjected to anaerobic conditions with consequent onset of alcoholic fermentation.

Oxidative decarboxylation of pyruvate. A second reaction in which thiamine pyrophosphate, cocarboxylase, is involved as a coenzyme is the oxidative decarboxylation of pyruvic acid, a reaction in which inorganic phosphate also participates.

$$\underset{\text{Pyruvic acid}}{CH_3COCOOH} + H_3PO_4 \rightarrow \underset{\substack{\text{Acetylphosphoric} \\ \text{acid}}}{CH_3COOPO_3H_2} + CO_2 + (2H)$$

The two hydrogen atoms produced in the oxidation are accepted by a flavoprotein which can then be reoxidized by oxygen or by methylene blue. This reaction is carried out by an enzyme system known particularly from bacteria, notably *Lactobacillus delbrückii* and *E. coli*.[56] The complete system requires then the presence of cocarboxylase, flavin-adenine dinucleotide, manganese or magnesium ions, and inorganic phosphate as well as of apoenzyme and pyruvate. Whether the cocarboxylase and the dinucleotide are associated with the same or with different proteins is not clear. The energy released by the conversion of pyruvate to acetate is in part conserved in the energy-rich acetyl phosphate bond. The phosphate residue of acetyl phosphate may be transferred to adenosine diphosphate in the presence of the appropriate transphosphorylase.

$$\underset{\substack{\text{Acetyl phosphoric} \\ \text{acid}}}{\overset{\overset{\displaystyle O}{\displaystyle \|}}{CH_3C}-O \sim ph} + \underset{\text{diphosphate}}{\text{adenosine}} \underset{}{\overset{\text{Transphosphorylase}}{\rightleftharpoons}} \underset{\text{Acetic acid}}{CH_3COOH} + \underset{\text{triphosphate}}{\text{adenosine}}$$

Acetyl phosphate is also able to participate directly in other reactions, which will be further discussed below.

Oxidative decarboxylation of pyruvate may also proceed by a route which does not involve uptake of phosphate and in which acetate rather than acetyl phosphate is the reaction product.[106]

$$\underset{\text{Pyruvic acid}}{CH_3COCOOH} + \tfrac{1}{2}O_2 \rightarrow \underset{\text{Acetic acid}}{CH_3COOH} + CO_2 + H_2O$$

The enzyme, pyruvic oxidase, which catalyzes this reaction has been found in the bacterium *Proteus vulgaris*. Like the carboxylases it contains cocarboxylase and magnesium ions as essential components.

Other carboxylases.   Alpha-ketoglutaric acid is oxidatively decarboxylated to succinic acid by an enzyme requiring cocarboxylase as a coenzyme.[57]   The reaction may proceed through an intermediate such as succinsemialdehyde as shown below:

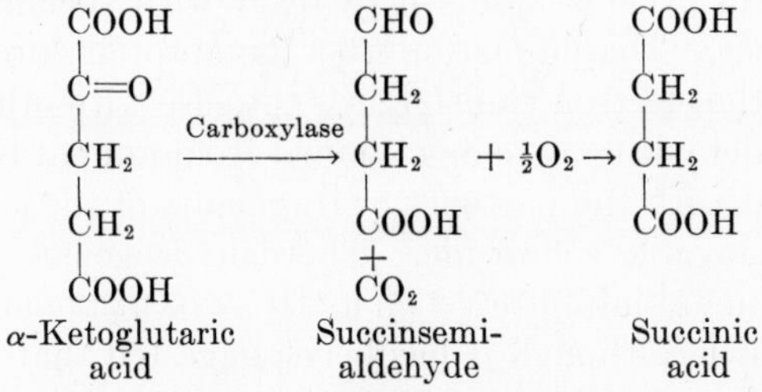

$$
\begin{array}{ccc}
\text{COOH} & \text{CHO} & \text{COOH} \\
| & | & | \\
\text{C=O} & \text{CH}_2 & \text{CH}_2 \\
| \qquad \xrightarrow{\text{Carboxylase}} & | & | \\
\text{CH}_2 & \text{CH}_2 \quad + \tfrac{1}{2}\text{O}_2 \rightarrow & \text{CH}_2 \\
| & | & | \\
\text{CH}_2 & \text{COOH} & \text{COOH} \\
| & + & \\
\text{COOH} & \text{CO}_2 & \\
\alpha\text{-Ketoglutaric} & \text{Succinsemi-} & \text{Succinic} \\
\text{acid} & \text{aldehyde} & \text{acid}
\end{array}
$$

The succinsemialdehyde thus produced would then be oxidized to succinic acid by a separate enzyme.   The α-ketoglutaric acid carboxylase is, however, apparently an insoluble enzyme associated with cellular particles,[81] and these particles have not been resolved into their component enzymes.

The decarboxylation of oxaloacetic acid has been studied by Krampitz and Werkman using an enzyme from the bacterium *Micrococcus lysodeikticus*.[58]   The reaction involved is the reverse of the CO_2 fixation reaction discussed in detail in Chapter 14.   Although the enzyme, like carboxylase, requires magnesium ions for its activity, it does not appear to require cocarboxylase.   Oxaloacetate carboxylase has been found in spinach leaves and in parsley and other roots by Vennesland and Felsher.[54]   These authors have also made the interesting observation that the crystalline seed globulins of the cucurbits (see Chapter 17) possess oxaloacetate carboxylase activity.

Succinic Dehydrogenase.   Succinic dehydrogenase, which is widely distributed in lower and higher plants as well as in animals, catalyzes the establishment of equilibrium between succinic and fumaric acids.   Since

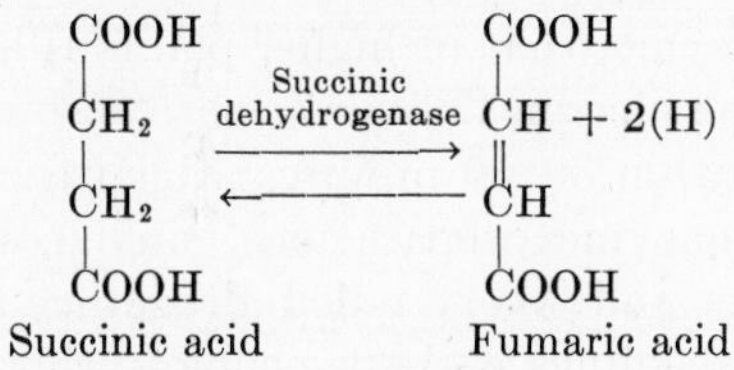

$$
\begin{array}{cc}
\text{COOH} & \text{COOH} \\
| \quad \text{Succinic} & | \\
\text{CH}_2 \quad \xrightarrow{\text{dehydrogenase}} & \text{CH} + 2(\text{H}) \\
| \quad \xleftarrow{} & \| \\
\text{CH}_2 & \text{CH} \\
| & | \\
\text{COOH} & \text{COOH} \\
\text{Succinic acid} & \text{Fumaric acid}
\end{array}
$$

this reaction involves the removal of hydrogen atoms from the substrate it could logically be considered together with the other dehydrogenases in the earlier section on the oxidative enzymes.   Succinic dehydrogenase appears, however, to differ from other dehydrogenases and is therefore here considered separately.

Under anaerobic conditions, methylene blue or other dyes can act as the hydrogen acceptor for the oxidation of succinate in the presence of succinic dehydrogenase. Under aerobic conditions, however, the dehydrogenase commonly forms a system with cytochrome C and cytochrome oxidase although still other systems may react with succinic dehydrogenase in plant tissues. Diaphorase or other flavoproteins are not needed as intermediates in the electron transfer as is the case with other dehydrogenases. Succinic dehydrogenase as obtained from animal tissues is not soluble, is associated with the particles or fragments of the cell, and contains no readily removable coenzyme. The dehydrogenase itself is not inhibited by heavy metal inhibitors such as HCN or azide, and its chemical nature is unknown, although it has been suggested that it may be a flavoprotein.[59] It does not in any case appear to involve a cozymase type of coenzyme. Succinic dehydrogenase is inhibited by malonic acid, the inhibition being a competitive one in which malonic acid competes with

$$
\begin{array}{cc}
& \text{COOH} \\
& | \\
\text{COOH} & \text{CH}_2 \\
| & | \\
\text{CH}_2 & \text{CH}_2 \\
| & | \\
\text{COOH} & \text{COOH} \\
\text{Malonic acid} & \text{Succinic acid}
\end{array}
$$

succinic acid for combination with the enzyme.[60] For this reason, malonic acid is a highly specific inhibitor and may be used to diagnose the participation of succinic dehydrogenase in respiratory systems. The penetration of malonic acid into tissues of higher plants takes place only from solutions of relatively low pH, owing apparently to the fact that penetration must occur as the monovalent malonate.[45] Attention to this detail is apparently of great importance in the demonstration of malonate inhibition of succinic dehydrogenase in living plant tissues.

Although succinic dehydrogenase as found in animal tissues is an exceedingly sturdy enzyme, that of higher plants is less stable. Thus it has been shown that the succinic dehydrogenase of spinach leaves becomes completely inactive within 30–60 minutes after freezing and thawing of the tissue.[45] Active enzyme preparations could not be obtained from ground leaves. Berger and Avery failed altogether to discover succinic dehydrogenase in oat seedlings perhaps owing to this very lability.[13] In both of these tissues, however, succinic dehydrogenase is undoubtedly present and active as a component of the respiratory system since respiration of the living tissue is inhibited by malonate. Active succinic dehydrogenase has been found without difficulty in other plant tissues as wheat embryos,[39] pollen,[61] and other seeds and seedlings.

Initial Breakdown of Carbohydrates. In the tissues and organisms whose respiration has been examined in detail, the initial reactions of respiration consist in the conversion of hexose to pyruvic acid. The pyruvic acid thus formed may then be metabolized in a great variety of ways, the path depending on the organism and the circumstances. The central feature of the initial breakdown of hexose to pyruvic acid is the fact that the conversions involve phosphorylated derivatives of the sugars rather than the sugars themselves. The storing of energy in the phosphate bond by oxidation of such phosphorylated compounds and the release of this phosphate bond energy to other reactions has come to be recognized as an important method by which the energy released in respiration is made available to other energy-consuming processes in the plant.

There is today general agreement that the conversion of hexose to pyruvic acid follows the overall path shown in Fig. 15-2. The general

Hexose
$\updownarrow$
Hexosediphosphate
$\updownarrow$
Triosephosphate (glyceraldehyde phosphate)
$\updownarrow$
Phosphoglyceric acid
$\updownarrow$
Phosphopyruvic acid
$\updownarrow$
Pyruvic acid

FIG. 15-2. Summary of steps in the conversion of hexose to pyruvate in the initial or glycolytic phase of respiration.

steps in this process from hexose to hexosediphosphate have been considered in Chapter 4. We shall now consider these same steps in further detail.

It was observed by Harden and Young in their study of fermentation by cell-free yeast juice that as hexose is fermented, inorganic phosphate is taken up and an organic phosphate, fructose-1,6-diphosphate, appears. Each step in the conversion of hexose to fructose diphosphate has been worked out, and it is known that the process takes place in yeast and in muscle approximately as shown in Fig. 15-3. In this sequence, glucose is first phosphorylated to glucose-6-phosphate by the enzyme hexokinase in the presence of adenosine triphosphate (ATP). The reaction is as follows:[62]

$$\text{Glucose} + \text{adenosine triphosphate} \xrightarrow{\text{Hexokinase}} \text{Glucose-6-phosphate} + \text{adenosine diphosphate}$$

Adenosine triphosphate gives up its terminal phosphate group to the hexose and is hence used up (converted to ADP) in the reaction. The enzyme hexokinase, in whose presence the phosphate exchange takes place, has been studied primarily in yeast and in muscle, and has been isolated in crystalline form from yeast[63] and purified from muscle.[64] Adenosine triphosphate plays in this reaction as in many others the double role of phosphate donor and energy source. The energy released in removal of the third phosphate from ATP (about 12,000 cal. per mol) is greater than that involved in formation of the phosphate bond in glucose-6-phosphate or other hexosephosphates, which is about 3000 calo-

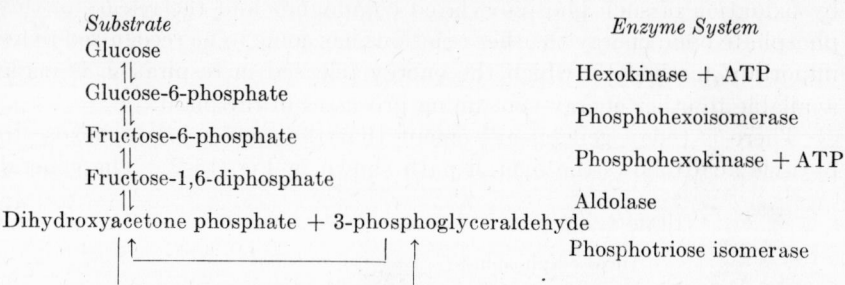

Substrate	Enzyme System
Glucose	
⇅	Hexokinase + ATP
Glucose-6-phosphate	
⇅	Phosphohexoisomerase
Fructose-6-phosphate	
⇅	Phosphohexokinase + ATP
Fructose-1,6-diphosphate	
⇅	Aldolase
Dihydroxyacetone phosphate + 3-phosphoglyceraldehyde	
	Phosphotriose isomerase

FIG. 15-3.　Detailed course of reactions by which glucose is converted to triose-phosphates.

ries per mol.　ATP is then what Lipmann has termed an "energy-rich phosphate" designated as ∼ph in contrast to the "energy-poor phosphate" bonds of ester phosphates such as glucose-6-phosphate which are designated as -ph.[56]　The energy involved in binding of the second phosphate group in adenosine diphosphate (ADP) is equal to that of the third phosphate of ATP, and ADP may also be able to act as a phosphate donor.　Much less energy is involved in formation of the first phosphate bond, that of adenylic acid, which is an ordinary ester phosphate linkage as in the hexose phosphates.　Much good indirect evidence (see below)

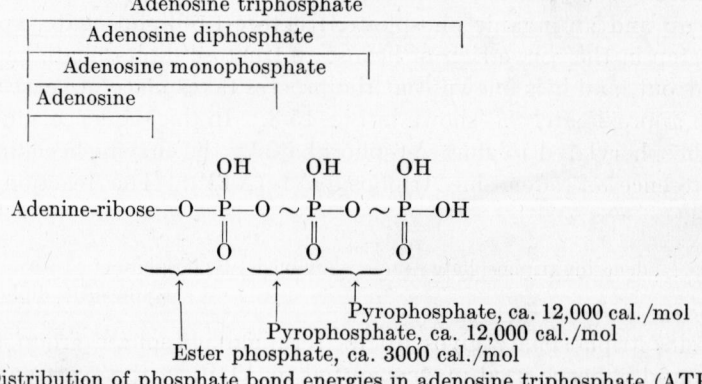

Adenosine triphosphate
Adenosine diphosphate
Adenosine monophosphate
Adenosine

$$\text{Adenine-ribose}-O-\overset{OH}{\underset{O}{P}}-O \sim \overset{OH}{\underset{O}{P}}-O \sim \overset{OH}{\underset{O}{P}}-OH$$

Pyrophosphate, ca. 12,000 cal./mol
Pyrophosphate, ca. 12,000 cal./mol
Ester phosphate, ca. 3000 cal./mol
Distribution of phosphate bond energies in adenosine triphosphate (ATP)

for the existence of hexokinase in higher plants has been obtained, but the enzyme has not been obtained from a plant source free from other enzymes of this series.

The conversion of glucose-6-phosphate to fructose-6-phosphate is catalyzed by phosphohexoisomerase, as discussed in Chapter 4. This enzyme, well known in yeast and muscle, has also been shown to be present in pea meal and in potato tubers.

Fructose-1,6-diphosphate is formed from fructose-6-phosphate by transfer of a second phosphate group from ATP, a transfer catalyzed by the enzyme, phosphohexokinase, which is closely related to the hexokinase discussed above. The hexose diphosphate is now ready for conversion to triosephosphate. In the presence of the enzyme aldolase, one molecule of hexose diphosphate is split into two molecules of triosephosphate, including one molecule each of dihydroxy acetone phosphate and 3-phosphoglyceraldehyde. This important fact was developed by Meyerhof in the years 1934–36. Aldolase has been obtained in crystalline form from muscle[65] and in highly purified form from peas by Stumpf, who has shown furthermore that aldolase activity is widespread in plant tissues.[66] This enzyme appears to have no coenzyme, and, since it is not inhibited by heavy metal inhibitors such as cyanide, is not apparently a heavy metal-containing protein.

The two triosephosphates are interconvertible in the presence of the enzyme phosphotriose isomerase. This enzyme like isomerase catalyzes the establishment of equilibrium between the keto and aldol forms of a phosphorylated sugar, in this case, triosephosphate. At equilibrium about 96% of the total triose is in the form of the ketose, dihydroxyacetone phosphate, and the remaining 4% in the form of glyceraldehyde-3-phosphate.[67]

It should be noted that fructose and glucose-6-phosphates as well as fructose-1,6- diphosphate are subject to dephosphorylation by phosphatases which are widely, probably universally, distributed in plant tissues.[68] The dephosphorylations catalyzed by phosphatases involve the liberation of the phosphate as inorganic phosphate and the production of the free hexoses.

The further degradation of triosephosphate involves the oxidation of glyceraldehyde-3-phosphate to pyruvic acid through a series of steps which were first worked out in detail by Warburg and Christian in 1939,[69] although indicated by earlier work of Needham and Pillai in 1937.[70] The oxidation as noted earlier is accomplished by a dehydrogenase, triosephosphate dehydrogenase, of which DPN is an essential component. Inorganic phosphate is taken up during the reaction and the oxidation product is not 3-phosphoglycerate but is rather 1,3-diphosphoglycerate, in which the phosphate group at position 1 is of the energy-rich carboxyl type and may be transferred to $\sim$ ph acceptors such as ADP in the presence of the appropriate transphosphorylating enzyme. The detailed mechanism of phosphoglyceraldehyde oxidation, which is described later in this chapter, was first worked out only for yeast and for animal tissues, but has been shown by Stumpf[16] to apply also to the system as it occurs in the higher plant.

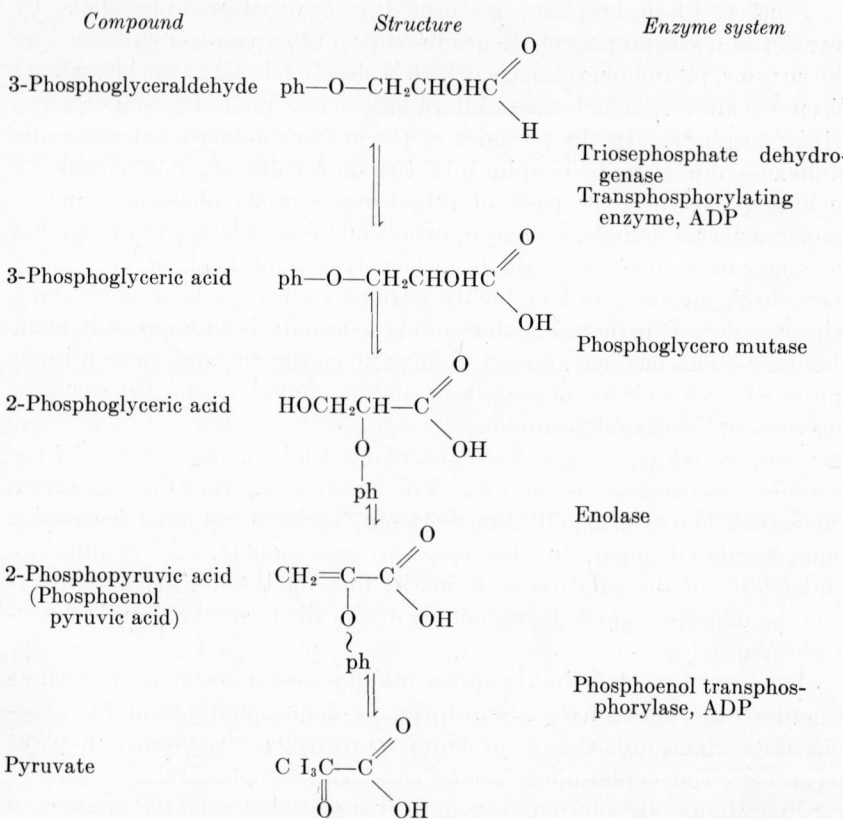

Fig. 15-4. Detail of steps in conversion of triosephosphate to pyruvate.

The 3-phosphoglyceric acid produced through the triose dehydrogenase system is converted to 2-phosphoglycerate by the intervention of the enzyme phosphoglycero mutase, which establishes an equilibrium between the two substances[71] probably through formation of the intermediate 2,3-diphosphoglyceric acid. An equilibrium is in turn established between 2-phosphoglyceric acid and phosphopyruvic acid in the presence of the enzyme enolase, a magnesium protein which was isolated from yeast as the mercury salt by Warburg and Christian in 1941.[72] Enolase is powerfully inhibited by fluoride ions and the commonly observed inhibition of hexose breakdown by fluoride is usually attributed to inhibition of this enzyme. The phosphate bond in 2-phosphopyruvate, like that in the 1 position of 1,3-diphosphoglycerate, is highly labile, and the phosphate can be transferred to $\sim$ ph receptors such as ADP in the presence of the enzyme phosphoenol transphosphorylase.[73]

With the formation of pyruvic acid from phosphopyruvate, the initial breakdown of hexose is complete. In this initial breakdown for each mol of hexose two molecules of energy-rich phosphate from ATP have been used in the formation of hexose diphosphate while four have been reformed, a net gain of two $\sim$ ph per hexose transformed to pyruvate. In addition, during the transformation of hexose to pyruvate, two molecules of DPN have been reduced per molecule of hexose utilized. In aerobic respiration, the reduced DPN is reoxidized by atmospheric oxygen through channels to be discussed below. In anaerobic fermentations, the DPN is reoxidized at the expense of reduction of pyruvic acid or products of pyruvic acid. In the simplest case, that of lactic acid fermentation, pyruvic acid is merely reduced to lactic acid. In alcoholic fermentation, pyruvic acid is decarboxylated to acetaldehyde and CO_2. The resulting acetaldehyde is then reduced to alcohol with concomitant reoxidation of the reduced DPN.

Despite the fact that only a few of the individual enzyme systems of the initial sugar breakdown have been separated and/or even identified in higher plant tissues, there is still an overwhelming body of evidence which indicates that the pathway in higher plants is in general quite similar to that outlined above for yeast and muscle. The evidence is of two principal kinds: (a) the individual compounds typical of yeast and animal carbohydrate breakdown are also found in higher plants and (b) it is possible to prepare complex enzyme systems from higher plant tissues which can carry out the overall glycolytic reaction, with the production of certain of the typical intermediates.

(a) The phosphorylated sugars are well known in higher plants, as discussed in Chapter 4. Fructose diphosphate has been found in a wide variety of tissues including seedlings, leaves, and potato tubers. Fructose- and glucose-6-phosphates have been found in spinach leaves and oat seedlings, while the triosephosphates and phosphoglycerate have been tentatively identified in many tissues. Pyruvic acid has been found widely in higher plants although in low concentrations.

(b) It has been known for a great many years, as pointed out earlier, that plants or plant tissues can ferment hexose anaerobically to CO_2 and alcohol just as in yeast alcoholic fermentation. Press juices of plant tissues such as potato tuber and beet root, as well as seed meal, can carry out the same reaction. More recently it has been shown that enzyme preparations from ground spinach leaves or from oat seedlings are capable of utilizing glucose with the production of fructose diphosphate and phosphoglycerate.[45,74] Addition of ATP to the system results in increased formation of both compounds. As mentioned above, fluoride inhibits sugar breakdown through its inhibition of the enzyme enolase. The respiration of many if not all plant tissues is sensitive to fluoride. That fluoride acts by inhibiting the production of pyruvate from hexose can be demonstrated for spinach leaves and for oat seedlings by the fact that although the tissues treated with fluoride are unable to utilize hexose, they can respire normally if provided with pyruvate.[45,74] James has also shown that fluoride-treated expressed juice of barley leaves converts hexosediphosphate only as far as phosphoglycerate, which then accumulates.[75] Even though these scattered bits of evidence indicate the general pathway of carbohydrate breakdown in the higher plant, it must nevertheless be a matter of great concern to plant biochemistry to place knowledge of the glycolytic enzymes of higher plants on a firmer basis by their actual isolation and characterization.

Oxidation of Pyruvate: The Krebs Cycle. There is today general agreement that pyruvic acid is a direct and primary product of carbohydrate breakdown in many living cells. The fate of the pyruvic acid has been elucidated in the fermentations, and a considerable number of enzymatic reactions of pyruvic acid have been discovered. Until relatively recently, however, the mechanisms of pyruvate oxidation and of CO_2 evolution in aerobic respiration have remained obscure. It is now apparent that this *terminal respiration* is accomplished through the intermediary of the common organic acids, the plant acids. The mechanism by which pyruvic acid is now believed to be oxidized to CO_2 and H_2O was first proposed by Krebs and Johnson in 1937 and is known variously as the Krebs cycle, the citric acid cycle, the tricarboxylic acid cycle, or as the cyclophorase system.

The idea that organic acids function as intermediates in respiration is not new. Thunberg in 1920 and Knoop in 1923 suggested (see reference 76) an ingenious scheme for carbohydrate oxidation in which pyruvate would be first converted to acetate and two molecules of acetate then condensed to succinate. The succinate would then be retransformed to acetate via oxaloacetate and pyruvate. Thunberg knew of the enzymes which oxidize succinic and malic acids as well as of carboxylase, and the scheme offered an explanation for the function of these enzymes in living tissues. The chief drawback of this otherwise attractive scheme was the failure to find any indication for the conversion of acetate to succinate.

The same scheme was revived by Toenissen and Brinkmann (1930) who found that pyruvic acid added to muscle mash resulted in formation of succinic acid, and they suggested that two pyruvate molecules might condense to form an unknown polymerization product which would then be transformed to succinic acid.

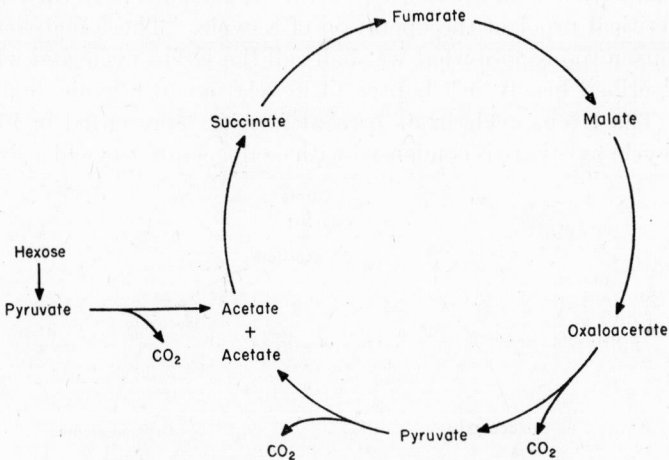

FIG. 15-5. Cyclic oxidation and regeneration of succinic acid suggested by Thunberg and by Knoop to account for pyruvate oxidation.

The first intimation that organic acids do in fact actually function in respiration was contained in the work of Szent-Györgyi who found that small additions of succinic, fumaric, or malic acid served to increase the rate of respiration of muscle tissue preparations or rather to maintain at a high level a rate which would otherwise decrease with time.[77] Since the amount of increased oxygen uptake was far greater than could be accounted for by complete oxidation of the added acid, Szent-Györgyi concluded that the action was a catalytic one and that the added acid served in the mechanism of carbohydrate oxidation not as a substrate but rather as an electron carrier. The Szent-Györgyi suggestion then embodied no cycle, but envisaged the acids as merely acting in the capacity of carriers of electrons between substrate and cytochrome.

In 1937 four new experimental findings of the greatest importance were made by Krebs and Johnson.[76] These were:

1. Citric acid is catalytically active in respiration, just as are succinic, malic, and fumaric acids.

2. Citric, isocitric, *cis*-aconitic and α-ketoglutaric acids are all rapidly oxidized by muscle.

3. Citric and α-ketoglutaric acids are formed in considerable quantities in muscle given oxaloacetate.

4. Succinic acid is formed from fumarate or oxaloacetate *oxidatively*, in muscle in which the direct reduction of these substances to succinate is blocked by presence of malonic acid. In other words, fumarate or oxaloacetate appears to be *oxidized* to succinate.

The demonstration that oxidation of the C_4 acids leads to their reformation is critical proof of the operation of a cycle. These facts led Krebs and Johnson to propose what we shall call the Krebs cycle and which has been described briefly in Chapter 14 in relation to organic acid metabolism. The Krebs cycle in its present form is represented in Fig. 15-6. In this cycle pyruvate is condensed with oxaloacetate to yield a six-carbon

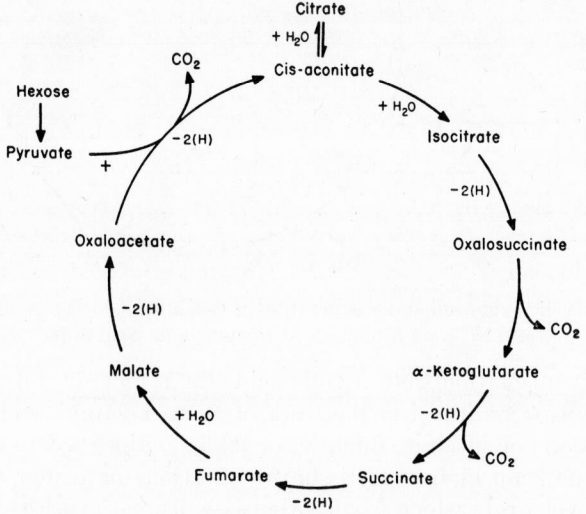

Fig. 15-6. Generalized outline of the Krebs cycle for the oxidation of pyruvate.

acid and CO_2. The six-carbon acid is decarboxylated and oxidized through a series of steps to again yield oxaloacetate. In one complete operation of the cycle, three molecules of CO_2 are liberated and ten electrons are transferred from the substrate. These are just the amounts involved in the complete oxidation of pyruvate to CO_2 and water according to the formulation:

$$\begin{array}{l} CH_3 \\ | \\ C{=}O \quad + \tfrac{5}{2}O_2 \rightarrow 3CO_2 + 2H_2O \\ | \\ COOH \end{array}$$

Pyruvic
acid

We shall now make a stepwise consideration of the individual reactions of the Krebs cycle.

1. The condensation of oxaloacetic and pyruvic acids. When minced muscle is incubated with these two compounds, citric and glutaric acids appear, although in relatively small yields. Martius and Knoop have shown that when the two are incubated in the test tube in the presence of aqueous carbonate and H_2O_2, citric acid is formed in good yield.[78] Krebs and Johnson originally proposed then that citric acid was also the initial condensation product of oxaloacetic and pyruvic acid *in vivo*. Citric, aconitic, and isocitric are, however, held in equilibrium by the enzyme aconitase, the concentrations at equilibrium being approximately as shown below.

	% of total acid at equilibrium[76]
Citric	80%
Aconitic	4%
Isocitric	16%

Thus regardless of which acid is the first formed the other two should also appear. Early experiments with isotopic carbon[79] suggested, however, that aconitic acid might be the first product and citric acid a secondary one. These experiments were done with animal tissue supplied with pyruvic acid and with oxaloacetate containing labeled carbon in a carboxyl group, $HOOCCOCH_2C^*OOH$. If citric acid is the first product of condensation, and if it is actually symmetrical in nature, it must yield only symmetrically labeled products and the labeled carbon, C^*, must be recovered in both of the two COOH groups in the α-ketoglutaric acid formed.

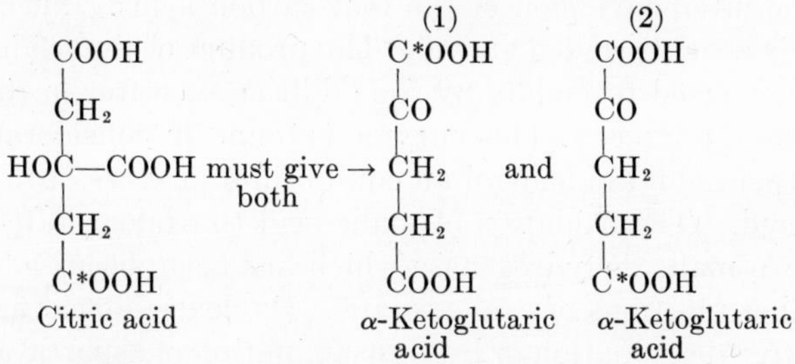

$$
\begin{array}{ccc}
 & (1) & (2) \\
COOH & C^*OOH & COOH \\
| & | & | \\
CH_2 & CO & CO \\
| & | & | \\
HOC{-}COOH \text{ must give} \rightarrow & CH_2 \quad \text{and} & CH_2 \\
| \qquad\qquad \text{both} & | & | \\
CH_2 & CH_2 & CH_2 \\
| & | & | \\
C^*OOH & COOH & C^*OOH \\
\text{Citric acid} & \alpha\text{-Ketoglutaric acid} & \alpha\text{-Ketoglutaric acid}
\end{array}
$$

Actually only one of the two possible α-ketoglutaric acids is formed, (1) above. This has been taken to indicate then that the symmetrical citric acid does not itself participate directly in the cycle. The observed asymmetric distribution of isotopic carbon in α-ketoglutaric acid may on the contrary indicate that citric acid is actually an asymmetric molecule with regard to its enzymatic degradation. The early experiments cannot be taken as conclusive proof that citrate does not directly participate in the Krebs cycle, and it has in fact been shown recently that citrate can be formed directly from oxaloacetate by condensation with acetate.[80]

The exact order of participation of the tricarboxylic acids in the Krebs cycle is therefore still somewhat ambiguous.

The overall reaction between oxaloacetate and pyruvate to yield a six-carbon acid involves the loss of one molecule of CO_2. It is possible that the CO_2 is lost from pyruvate as the first step in the reaction and that a two-carbon derivative of pyruvate rather than pyruvate itself actually participates in the further reaction. It has been suggested that oxidative decarboxylation of pyruvic acid to yield acetylphosphate may be this initial reaction and that acetylphosphate or a related compound may be the two-carbon derivative of pyruvate which is involved.[80]

2. The oxidation of isocitric acid by isocitric acid dehydrogenase. The enzyme for this reaction has been discussed in detail above. In the presence of aconitase, aconitate as well as citrate may be converted to isocitrate and oxidized through isocitric dehydrogenase. Aconitase is the only enzyme known to activate citric acid directly. As pointed out earlier, oxalosuccinic acid is the product of the oxidation and the product actually formed by the dehydrogenase. This acid decomposes reversibly to α-ketoglutaric acid in the presence of oxalosuccinic carboxylase.

3. The oxidative decarboxylation of α-ketoglutarate. Little further can be said concerning this reaction in which both decarboxylation and oxidation are involved. Although the enzyme has not been studied in detail in the plant, still the occurrence of the reaction has been established by *in vivo* experiments with oat seedlings in which oxidation of α-ketoglutarate is known to yield succinate.[74]

4. The interconversion of the four-carbon acids. Succinic dehydrogenase has been discussed above. The product of the oxidation, fumaric acid, is converted to malate by the addition of water in the presence of the enzyme fumarase. This enzyme is found in considerable quantities in most tissues and in general an equilibrium mixture of the two acids is encountered. The oxidation of malic acid to oxaloacetate is due to the well known malic dehydrogenase which has been discussed above.

5. Other reactions of oxaloacetate. Oxaloacetic acid may be formed by oxidative deamination or by transamination of aspartic acid as well as by the addition of CO_2 to the pyruvate molecule. Although CO_2 fixation is well known in lower plants, bacteria, and fungi, it has been less well studied in higher plants other than the succulents. Nevertheless the evidence brought forward in Chapter 14 does show that in nonsucculent higher plants both four-carbon and six-carbon acids may be formed as the result of CO_2 uptake in the dark. Thus Laties has shown that CO_2 fixed by respiring barley roots appears in the acids of the Krebs cycle.[90] Oxaloacetate is also subject to removal from the Krebs cycle by several reactions, of which one is decarboxylation to pyruvate, the reverse of the

Wood-Werkman reaction in which CO_2 and pyruvate react to form oxaloacetate. In addition oxaloacetate is subject to transamination reactions and is used up in the formation of aspartic acid and asparagine. Alpha-ketoglutaric acid is similarly subject to transformation to glutamic acid. Through the two α-keto dicarboxylic acids then respiration is intimately connected with nitrogen metabolism.

There is today general agreement that the biological oxidation of pyruvate involves interconversions among the simple organic acids and that the mechanism proposed by Krebs and Johnson and modified by Krebs and others represents a general mechanism of carbohydrate metabolism. In a following section evidence will be presented which indicates that the general mechanism of the Krebs cycle participates also in the respiration of higher plants.

Pyruvate Oxidation in Higher Plants. Evidence that the respiration of higher plants proceeds through the breakdown of carbohydrate with the production of pyruvic acid as an intermediate has been presented above. A central problem in the study of plant respiration is the mechanism involved in the oxidation of pyruvic acid. Barron[87] has summarized no less than fourteen different pathways for the metabolism of pyruvic acid which have been described in various plant and animal tissues, and it is quite possible that in the higher plant small amounts of pyruvic acid are utilized by a variety of reactions. Evidence that a principal pathway of pyruvate utilization, perhaps the principal pathway in many plant tissues, is a mechanism generally resembling the Krebs cycle comes from three sources: (1) all the acids which are intermediates in the Krebs cycle are found in plants; (2) the enzymes needed by the Krebs cycle for the interconversions of these acids have been found in higher plants in so far as they have been sought; and (3) the small amount of work which has thus far been done on the detailed analysis of pyruvate oxidation in higher plants *in vivo* is in agreement with the view that pyruvate is oxidized through a chain of reactions involving the plant acids in a way generally similar to that postulated by the Krebs cycle.

The plant acids malic, citric, fumaric, and succinic are so widely distributed in plants that they would appear to be universal. Less is known of the distribution of isocitric acid, but it has, however, been identified in fruits and leaves, and in addition, the enzyme for its oxidation is found generally in plant tissues. Aconitic acid is found generally in leaves in small quantities. Oxaloacetic acid and α-ketoglutaric acid are widely distributed in plant materials in low concentration. The principal plant acids of the Krebs cycle are therefore found in plants. It should be pointed out, however, that the occurrence of all these acids together in a single plant has not been demonstrated by modern critical investigation.

The occurrence in higher plants of enzymes for the several steps of the Krebs cycle is summarized in Table 15-1. Malic and isocitric dehydrogenases are commonly found in higher plants, and the same is true for aconitase and fumarase. Succinic dehydrogenase has been found in many seeds and seedlings although, as described above, it is a relatively labile enzyme and is difficult to obtain in cell-free extract from some tis-

TABLE 15-1. Experimental evidence for the reactions of the Krebs cycle in higher plants

Enzyme	Reaction	Evidence	Authors
1.	Oxaloacetate + pyruvate (or derivative) → 6-carbon acid	Presumptive evidence from *in vivo* experiments only. Spinach leaves, oat seedlings, barley roots.	Bonner and Wildman, 1946[45] Laties, 1949,[90] Bonner, 1948[74]
2. Aconitase	Citric ⇌ Aconitic ⇌ Isocitric	General in many species.	Many
3. Isocitric dehydrogenase	Isocitric ⇌ Oxalosuccinic	General in many species.	Many
4. Oxalosuccinate carboxylase	Oxalosuccinic ⇌ α-Ketoglutaric	Found in parsley roots and other leaves, roots.	Vennesland *et al.*, 1947[18]
5. Oxidative decarboxylation of α-ketoglutarate	α-Ketoglutaric ⇌ Succinic	Presumptive evidence from *in vivo* expt. only. α-Ketoglutarate ox. to succinate in oat.	Bonner, 1948[74]
6. Succinic dehydrogenase	Succinic ⇌ Fumaric	Found in many seeds and seedlings. Labile in leaves etc.	Many
7. Fumarase	Fumaric ⇌ Malic	General in many species.	Many
8. Malic dehydrogenase	Malic ⇌ Oxaloacetic	General in many species.	Many

sues, notably leaves. The enzyme system leading to the decarboxylation and oxidation of α-ketoglutaric acid is known in higher plants only from the *in vivo* experiment discussed in relation to Table 15-2.

That the plant acids play a role in plant respiration was first indicated by the observation that the rate of respiration of plant tissue is in some cases increased when plant acids are supplied. This has been shown for potato tuber slices,[88] oat seedlings,[89] barley roots,[90] tomato roots,[91] and spinach leaves.[45] A detailed analysis of the way in which plant acids increase plant respiration has been made for spinach leaves, barley roots, and oat seedlings. In spinach leaves,[45,90] as in the animal tissues with which Krebs originally worked, respiration is markedly inhibited in the presence of malonic acid. This shows that the respiration of spinach leaves passes through succinic dehydrogenase, and that the oxidation of pyruvic acid is hence linked to the oxidation of succinic acid. Fumaric, malic, isocitric, and α-ketoglutaric acids are also oxidized *in vivo* through succinic acid, since in all cases the increases in respiration are inhibited by malonate. In tissues whose respiration is blocked by malonate, succinate accumulates above the level found in the normal tissue. This is because the normal succinate concentration in the tissue is determined both by the rate of production of succinate and by the rate of removal of succinate by succinic dehydrogenase. When removal of succinate is

TABLE 15-2. Oxidative formation of succinate from fumarate and from α-ketoglutarate in malonate-inhibited oat coleoptile tissue. Intact tissue, incubated for 8 hours at 25°C. (After Bonner[74])

Treatment given during incubation period	Final conc. of succinate in tissue, mg./g. dry wt.
None	0.93
Malonate (5 mg./cc.)	1.63
Malonate + fumarate (2.5 mg./cc.)	6.40
None	1.05
Malonate (5 mg./cc.)	1.30
Malonate + α-ketoglutarate (5 mg./cc.)	4.96

blocked by malonate, succinate continues to be formed and merely accumulates. The accumulation of succinate has been demonstrated in malonate-inhibited tissues of spinach,[90] oat seedlings,[74] and barley roots.[90] With these same tissues also, fumarate is rapidly oxidized to succinate just as in the fundamental experiments of Krebs and Johnson with animal tissues (Table 15-2). This result shows then that in plant tissue also, oxidation of the C_4 acids leads to their reformation. Similar experiments have shown that the C_5 and C_6 acids are also oxidized to succinate in malonate-inhibited plant tissue. The data of Table 15-2 show specifically that when α-ketoglutarate is oxidized by oat seedling tissue in the presence of malonate, the product of the oxidation is succinate. This result suggests the presence in the oat seedling of an α-ketoglutarate oxidase generally similar in its action to that of other organisms.

The reaction between oxaloacetate and pyruvate (or a derivative of pyruvate) may also be studied in higher plant tissue by blocking the normal production of pyruvate (from sugar contained in the tissue) with NaF.[45,90] If pyruvate is added to such fluoride-inhibited tissue, it is respired normally. The oxidation of pyruvate depends on the presence of oxaloacetate or a source of oxaloacetate, however, since when the production of oxaloacetate and its precursors is inhibited by malonate the tissue is no longer able to utilize and oxidize pyruvate. In tissue which is inhibited both by NaF and by malonate, added pyruvate is oxidized only when fumarate or a similar source of oxaloacetate is also added. This evidence does not tell us anything concerning the nature of the enzymes involved but does tell us that a reaction of the general nature of the oxaloacetate-pyruvate condensation is involved in pyruvate oxidation by plant tissue.

In general summary, there is good evidence for several higher plant tissues that the course of the reactions involved in respiration comprises first the conversion of hexose to pyruvate. This pyruvate is then oxidized by the plant through further reactions in which the plant acids take part as intermediates. Experimental evidence available at present is in agreement with the view that the steps involved in pyruvate oxidation by higher plant tissues are related to one another in a manner generally similar to the Krebs cycle.

The Terminal Oxidase. The terminal oxidase of respiration is the final stage in the chain of respiratory oxidations through which electrons removed in substrate oxidation are finally combined with oxygen and H^+ ions to yield water. Thus the terminal oxidase is the step in respiration at which oxygen is actually consumed. In animal tissues it appears that cytochrome oxidase is the enzyme responsible for this step. In the tissues of higher plants, on the contrary, at least three different oxidase systems may be involved, the enzyme system varying with the tissue concerned. These are:

Terminal oxidase	Example of tissue
Cytochrome oxidase	Wheat embryo, pollen, oat seedlings
Polyphenoloxidase	Potato tuber, spinach leaves
Nonmetal oxidase	Carrot leaves, avocado fruit

A cytochrome-cytochrome oxidase system has been found in pollen of various species by Okunuki who demonstrated the presence of cytochrome C in this material.[61] The respiration of pollen is inhibited by cyanide and by carbon monoxide. It has been pointed out above that cytochrome oxidase is inhibited by carbon monoxide but that the enzyme-carbon monoxide complex thus formed is stable only in the dark and is dissociated in visible light. Thus the carbon monoxide inhibition of the respiration of cytochrome oxidase-containing tissues prevails only in the dark and is reversed by light. This is true of the carbon monoxide inhibition of the respiration of pollen. The same technique may be used to distinguish between respiration involving cytochrome oxidase and respiration involving polyphenoloxidase which is also inhibited by carbon monoxide but in which the inhibition is not reversed by light. Use of this method indicates that cytochrome oxidase may be involved in the respiration of carrot roots, barley seeds, young and mature barley roots, and young barley leaves.[82] A different method of approach has been used with oat seedlings where it has been shown that the seedling tissues yield an enzyme preparation capable of oxidizing cytochrome C.[74] The first thorough study of the cytochrome-cytochrome oxidase system in plant material is, however, that of Goddard in 1944 in which cytochrome C was isolated in pure form from wheat embryos.[39] An insoluble cytochrome oxidase similar in properties to that of animal tissues was also found in wheat embryos. Spectroscopic investigations have qualitatively revealed the presence of cytochrome in embryos of other cereals, of legumes, in

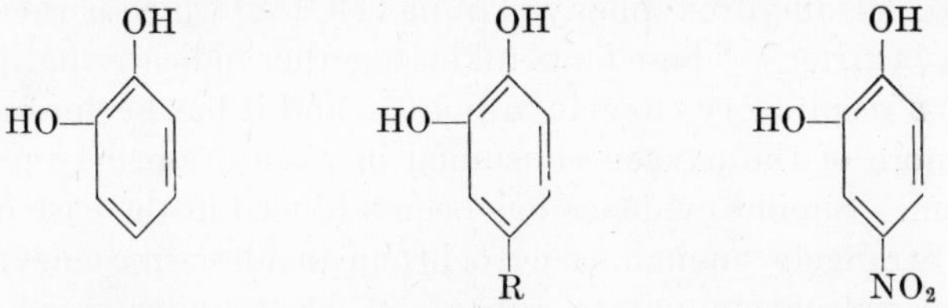

OH   OH   OH

Catechol   p-Substituted   p-Nitrocatechol
(substrate)   Catechols   (inhibitor)
     (substrate)

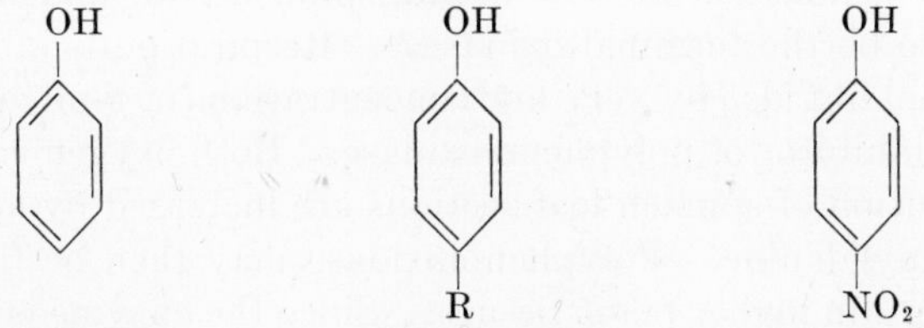

OH   OH   OH

Phenol   p-Substituted   p-Nitrophenol
(substrate)   Phenols   (inhibitor)
     (substrate)

Relationship between normal substrates and inhibitors of polyphenoloxidase

roots of dandelions, turnips, and parsnips, in onion bulbs, and in scattered other tissues.[75] In general, we may suspect that the cytochrome C-cytochrome oxidase system, where it occurs in higher plants, is very similar to the system as found in animal tissues.

The first direct demonstration that polyphenoloxidase is involved in the respiration of a higher plant tissue was that of Boswell and Whiting with the potato tuber.[87] These workers showed that the respiration of tuber slices is largely inhibited by the addition of catechol, a substance which is oxidized by the enzyme, but whose oxidation products are known to slowly inactivate polyphenoloxidase. Still better and more specific inhibitors of polyphenoloxidase are p-nitrocatechol and p-nitrophenol which appear to inhibit the enzyme through their resemblance to the normal substrates. Both oxygen consumption and CO_2 production of potato slices are strongly inhibited in the presence of low concentrations of p-nitrocatechol.[84] Not only can respiration of potato be inhibited by inhibitors of polyphenoloxidase, but the gas exchange can also be increased by addition of certain suitable substrates of the enzyme. Thus addition of protocatechuic acid, a slowly oxidizable substrate of polyphenoloxidase, to intact potato slices increases both oxygen consumption and CO_2 production, the inference being that protocatechuic acid functions as a carrier between polyphenoloxidase and the enzymes responsible for substrate dehydrogenations. Potato juice, additions of which are known to increase the rate of respiration of potato slices, has been shown to contain tyrosine as the active principle.[85] The tyrosine is presumably oxidized to 3,4-dihydroxyphenylalanine (DOPA) which is then the actual respiratory carrier. These facts taken together indicate that polyphenoloxidase is a respiratory enzyme in potato, and it has been estimated that 85% or more of the oxygen consumed in respiration may pass through this system. Similar evidence has been adduced in the case of the sweet potato. Strangely enough, a cytochrome oxidase-like enzyme appears to be also present in potato tubers, at least under some conditions, although cytochrome itself has not been identified in the same material.[107] Polyphenoloxidase is present in the spinach leaf, and in this case also appears to be the terminal oxidase.[45] Respiration is inhibited not only by catechol but also by very low concentrations of p-nitrophenol, another selective inhibitor of polyphenoloxidase. Both oxygen consumption and CO_2 evolution of spinach leaf sections are increased by addition of dihydroxyphenylalanine. Polyphenoloxidase may then be the terminal oxidase in certain higher plant tissues. Since the enzyme is not found in all species or in all tissues, it cannot be the universal oxidase by any means. It seems possible also that polyphenoloxidase may occur in tissues in which the enzyme is not the terminal oxidase. Thus in leaves of tea, polyphenoloxidase does not appear to function in respiration although the enzyme is found in the chloroplasts and is active in injured leaves in the oxidation of the tea tannins, an oxidation associated with the conversion of green to black tea.[46]

It has been stressed above that a flavoprotein enzyme, cytochrome C reductase, intervenes between cytochrome C and the codehydrogenases (DPN and TPN) as an intermediate in hydrogen transfer between substrate and the cytochrome system. No similar carrier or enzyme has as yet been recognized for the step from codehydrogenase to polyphenoloxidase. The polyphenoloxidase of spinach leaves and its substrate dihydroxyphenylalanine do not appear to be able in vitro to directly reoxidize reduced DPN and it is probable that here also an intermediate enzyme is involved.

Ascorbic acid oxidase is not widely distributed in plant tissues, as is ascorbic acid, but is known particularly in seeds and fruits of cucurbits and in various leaves. On the basis of the fact that oxygen uptake by barley leaf extracts in the presence of hexose diphosphate is promoted by additions of ascorbic acid and DPN, James suggests that ascorbic acid may act as a carrier between reduced DPN and the oxidase.[86] Other evidence for the participation of ascorbic acid oxidase in respiration is, however, meager, and even with barley leaves the respiration is not cyanide sensitive, as should be expected if ascorbic acid oxidase were actually involved.[83]

The respiration of mature carrot leaves is not inhibitable by cyanide, and the same is true of the respiration of avocado fruits.[82] Since the three oxidase systems discussed above are all cyanide sensitive, it is probable that carrot leaves and avocado fruits possess still other terminal oxidase systems, possibly nonmetal ones and perhaps, for example, auto-oxidizable flavoproteins.

The Energetics of Respiration.[92] The tendency of any reduced substance to become oxidized may be expressed quantitatively in terms of the oxidation-reduction potential.[93] This quantity may be derived in various ways, but the treatment which follows is based on the general concept of oxidation as the removal of electrons. Suppose we consider a reaction in which a reduced compound is oxidized to its oxidized form plus n electrons.

$$\text{Reduced form [red]} \rightleftharpoons \text{Oxidized form [ox]} + n \text{ electrons [e]} \qquad (1)$$

In many reactions of biological interest two electrons are removed at each oxidative step so that n is equal to 2. The equilibrium constant for such a reaction is given by:

$$\frac{[\text{ox}] \cdot [e]^2}{[\text{red}]} = K \qquad (2)$$

For a second system one may write a similar reaction and equilibrium constant:

$$\frac{[\text{ox}'] \cdot [e']^2}{[\text{red}']} = K' \qquad (3)$$

If the two systems are in equilibrium, $[e]$ must be common to the two and hence:

$$\frac{[\text{ox}]}{[\text{red}]} = \frac{K}{K'} \frac{[\text{ox}']}{[\text{red}']} \qquad (4)$$

If the equilibrium constants K and K' are known, or even if their ratio is known, it is possible to predict what ratio of a given oxidant and reductant can exist in equilibrium with a second oxidant and reductant or, in short, to predict what systems will be able to oxidize which other systems.

It is found that if an electrode of a noble metal such as platinum be placed in a solution containing an oxidation-reduction system (a solution containing a mixture of oxidized and reduced forms of a substance), this electrode will acquire an electron charge which increases with increase in the ratio of reduced to oxidized form of the substance involved, i.e., increases with increase in the reducing tendency of the solution and hence with the electron "concentration" or fugacity of the system, a quantity which we will call $[e_s]$. The electrode itself contains free electrons at some concentration $[e_m]$. The concentration work, w, needed to transfer n faradays of electrons from the electrode to the solution is given in equation 5,

$$w = -nRT \ln \frac{e_m}{e_s} \tag{5}$$

where R is the gas constant and T the absolute temperature. The electrical work involved in the same process is $E_h nF$, where E_h is the potential set up between electrode and solution and F the value of the faraday. Equating the electrical to the concentration work we obtain E_h in terms of the electron fugacity of the solution (equation 6). The quantity e_m is a constant for any particular metal and the whole term $(RT/F) \ln e_m$ is hence a constant.

$$E_h = \frac{RT}{F} \ln e_m - \frac{RT}{F} \ln e_s \tag{6}$$

Rewriting the equation with this change and substituting for $[e_s]$ from equation 2 we obtain equation 7 for the potential of an oxidation-reduction system involving transfer of two electrons:

$$E_h = E_0 - \frac{RT}{2F} \ln \frac{[\text{red}]}{[\text{ox}]} \tag{7}$$

The term E_0 now includes not only $(RT/F) \ln e_m$ but also the term involving K, the equilibrium constant of the oxidation-reduction system. This constant may be determined experimentally for each of several systems. The differences between the constant terms E_0 for various systems are of the form $(RT/F) \ln (K_1/K_2)$ and permit the calculation of the ratio of K_1 to K_2. This ratio as we have already seen permits the arrangement of different systems on a scale with respect to their reducing intensity.

From equation 7 it may be seen that when $[\text{red}]/[\text{ox}] = 1$, then $E_h = E_0$, and it is generally in terms of this E_0 that various systems are compared. It should be emphasized that both E_0, the oxidation-reduction potential of a system at half reduction, and E_h measure the intensity of the oxidizing or reducing tendency of the system and not the amount of oxidation or reduction of which the system is capable. Both E_0 and E_h are quantities concerned with intensity rather than quantity, just as pH is concerned with concentration or intensity of hydrogen ion rather than with titratable acidity.

The reference electrode for the scale of oxidation-reduction potentials is set by convention at the potential between a noble metal electrode and a solution in a normal hydrogen half cell in which the equilibrium between pressure of hydrogen, $p(H_2)$, and hydrogen ions is measured.

$$H_2 \rightleftharpoons 2H^+ + 2(e)$$

The electrode potential of such a cell is given by equation 8 (since $E_0 = 0$, by definition):

$$E_h = -\frac{RT}{F} \ln \frac{\sqrt{p(H_2)}}{[H^+]} = -\frac{RT}{F} pH - \frac{RT}{F} \ln \sqrt{p(H_2)} \qquad (8)$$

The term $-\ln \sqrt{p(H_2)}$ is often known as rH by analogy with pH. It may be seen that by sufficiently changing the hydrogen pressure in equilibrium with a hydrogen half cell at the same pH as a given second oxidation-reduction system under investigation, the E_h of the hydrogen half cell could be made to balance the E_h of the second oxidation-reduction system. The rH is then often used to express this hypothetical pressure of hydrogen, particularly in measurements of the overall oxidation-reduction potential of tissues or cells which contain numerous oxidation-reduction systems.

Equation 7 has been derived for systems which involve the transfer of electrons but in which no transfers of hydrogen are considered. In many biological oxidations, however, hydrogen transfers accompany electron transfers, as described earlier. These reactions are of the general type:

$$A \cdot 2H \rightleftharpoons A + 2H^+ + 2(e) \qquad (9)$$

$$
\begin{array}{ll}
\text{Reduced} & \text{Oxidized} \\
\text{form of} & \text{form of} \\
\text{reactant} & \text{product}
\end{array}
$$

If we rederive equation 7 but include the factor of the hydrogen ions formed during the oxidation, we obtain equation 10, which then includes a third term dependent on the pH of the system.

$$E_h' = E_0' - \frac{RT}{2F} \ln \frac{[\text{red}]}{[\text{ox}]} - \frac{RT}{F} \ln [H^+] \qquad (10)$$

Thus in systems which involve transfer of hydrogen, E_0 is dependent on the pH of the system.

Table 15-3 gives the oxidation-reduction potentials, E_0', all calculated for pH 7, for a selection of the biological oxidation-reduction systems discussed in this chapter. The systems are arranged in order of decreasing potential, that is, each system is capable of oxidizing any system below it in the table. Thus the reduction of oxygen to H_2O possesses the highest oxidation-reduction potential and is capable of oxidizing all other systems in Table 15-3. Conversely, DPN possesses one of the lowest potentials and can be reoxidized by almost all the other systems. It is apparent that the transfer of electrons from reduced DPN to flavoprotein, cytochrome C, cytochrome oxidase, and oxygen, the sequence which actually occurs in biological systems, is the sequence to be expected on purely thermodynamic grounds. On the other hand the oxidation-reduction potential of the malate-oxaloacetate system is higher than that of DPN, and appreciable oxidation of malate by DPN can occur only when the level of reduced DPN is held very low by continued reoxidation of this substance.

TABLE 15-3. The oxidation-reduction potentials of some biological systems. All calculated to pH 7.0

Reduced form	Oxidation products	E_0' volts (Red. potentials)	$T°C$
H_2O	$\frac{1}{2}O_2 + 2H^+ + 2e$	0.815	25
Catechol	Benzoquinone $+ 2H^+ + 2e$	0.792	25
Ferrous cytochrome oxidase	Ferric cytochrome oxidase $+ e$	Unknown	..
Ferrous cytochrome C	Ferric cytochrome C $+ e$	0.29	25
Succinate	Fumarate $+ 2H^+ + 2e$	−0.026	25
Glutamate	α-Ketoglutarate $+ NH_4^+ + 2H^+ + 2e$	−0.030	25
Reduced flavin enzyme	Oxidized flavin enzyme $+ 2e$	−0.063	38
Malate	Oxaloacetate $+ 2H^+ + 2e$	−0.102	37
Lactate	Pyruvate $+ 2H^+ + 2e$	−0.180	35
Alcohol	Acetaldehyde $+ 2H^+ + 2e$	−0.190	30
Reduced DPN	DPN $+ 2H^+ + 2e$	−0.29	30
H_2	$2H^+ + 2e$	−0.414	25

The reduction potentials as given in Table 15-3 are convertible into the free energy change of the same reaction, ΔF_0, by the relation

$$\Delta F_0 = -nFE_0$$

where n is the number of electrons transferred in the reaction and F is the value of the faraday in calories or 23,066 cal. Thus the oxidation of malate to oxaloacetate involves a free energy change of:

$$\Delta F_0 = -2 \times 23,066(+ 0.102) = -4705 \text{ cal.}$$

At the standard concentrations of 1 molal malate and oxaloacetate, the oxidation, carried out reversibly at pH 7.0, would then be spontaneous since ΔF, the change in free energy, is negative. We may make the same calculation for the reduction of DPN. For this reduction, ΔF is positive,

$$\Delta F_0 = -2 \times 23,066(-0.29) = 13,381 \text{ cal.}$$

so energy must be applied in order to make the reaction go. Coupling the oxidation of malate to the reduction of DPN we obtain for the overall reaction

$$\Delta F = 13,381 - 4705 = 8676 \text{ cal.}$$

This reaction would not then proceed spontaneously if all concentrations were maintained at 1 molal but would proceed, as pointed out above, only if reduced DPN were removed as produced. In other cases where the oxidation-reduction potentials are closer, the oxidation of one substrate may be linked to the reduction of another through the intermediary of a carrier such as DPN. This would theoretically be true, for example, of the reduction of pyruvate to lactate by the oxidation of malate, and Green and others[94] have experimentally coupled the oxidation of β-hydroxybutyrate to the reduction of pyruvate using DPN as the intermediate carrier. In reactions of this type the energy liberated by the oxidation of substrate is partially taken up then in the reduction of carriers such as DPN, which may in turn be utilized in the synthesis of other reduced cellular constituents. This is not true of all cellular oxidations and in some cases at least it must be suspected that the energy liberated in oxidation may be dissipated as heat without the production of useful chemical work.[95] This is true particularly of the step from ferrous cytochrome C through cytochrome oxidase to oxygen. The oxidation of cytochrome oxidase by oxygen is apparently not reversible and is not as yet known to be coupled to any other energy storing reaction. The latter may also be true for the oxidation of ferrous cytochrome C by ferric cytochrome oxidase. In this case the energy liberated in the oxidation of ferrous cytochrome C by cytochrome oxidase and the oxidation of this enzyme by oxygen would appear as heat. The standard ΔF_0 for this entire reaction is $-25,880$ calories and Goddard[96] has calculated that under physiological conditions of say 0.01 atmosphere pressure of oxygen in the cell and a ratio of ferrous to ferric cytochrome of 0.1 the ΔF_0 would still be $-23,160$ calories. In the oxidation of hexose to CO_2 and H_2O, the process must be repeated six times with a loss of $6(-23,160)$ calories or $-138,960$ calories, an amount which constitutes almost one-fourth of the roughly 680,000 total calories available from the complete oxidation of hexose.

Transfer of Energy in Biological Systems. It has been shown above that energy liberated in oxidative processes may be transported to other systems in at least two ways, by the intermediary of reduced coenzymes and by the intermediary of energy-rich phosphates. The transfer of energy by reduced DPN or by reduced TPN is possible since the same coenzyme is active with a variety of apoenzymes. A familiar instance of this sort of energy transfer is that of yeast fermentation. In this case, reduced DPN is produced at the step of the oxidation of phosphoglyceraldehyde to phosphoglycerate. The reduced DPN is reoxidized in the reduction of acetaldehyde to alcohol in the presence of alcohol dehydrogenase.

(1) Phosphoglyceraldehyde + DPN → Phosphoglycerate
+ reduced DPN

(2) Acetaldehyde + reduced DPN → Alcohol + DPN

In lactic acid fermentation, as carried on by *Lactobacillus*, step 1 is coupled with the reduction of pyruvate to lactate. In aerobic respiration reduced DPN is apparently in general reoxidized by systems which lead ultimately to oxygen uptake. There is no doubt, however, but that some coupled oxidation-reductions of the above type must also occur.

The transfer of energy through energy-rich phosphate intermediates appears to be the principal manner in which the energy of cellular oxidation is made available to other cellular processes and reactions. Both the formation and the utilization of energy-rich phosphates have been referred to in earlier sections. This section will give an overall summary of phosphorylative energy transfer.

The phosphorylated compounds which occur in nature include those in which phosphoric acid is esterified with hydroxyl groups as in the sugar phosphates, those in which phosphoric acid is bound to nitrogen as in creatine phosphate, those in which phosphoric acid is bound to carboxyl groups as in the acylphosphates, those in which phosphate is esterified with enolic hydroxyl groups, and finally the pyrophosphates in which phosphate is bound to phosphate. The structures and known examples of these several kinds of organic phosphates are given in Table 15-4. It is to be noted that while much is known concerning the occurrence and distribution of organic phosphates in animal tissues, relatively little is known concerning the same compounds in the tissues of higher plants. Thus creatine phosphate, arginine phosphate, and acetyl phosphate have not certainly been identified in higher plant tissues. Adenosine triphosphate while it occurs in microorganisms and in animal tissues has not been isolated in pure form from higher plants, and it has been suggested by Albaum[97] that in the oat plant at least the function of ATP may be filled by a more complex compound structurally different but biochemically interchangeable with ATP and perhaps similar to a diadenine nucleotide isolated from yeast. Seedlings of the mung bean, however,

TABLE 15-4.   Organic phosphates of living organisms

Type of linkage	Compound	Structure	Approximate $-\Delta F$ of hydrolysis of ph bond
Ester	Glucose-6-phosphate	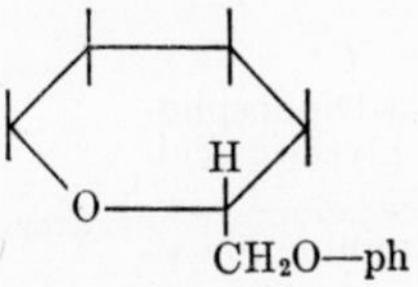	2000–3000
	Fructose-6-phosphate	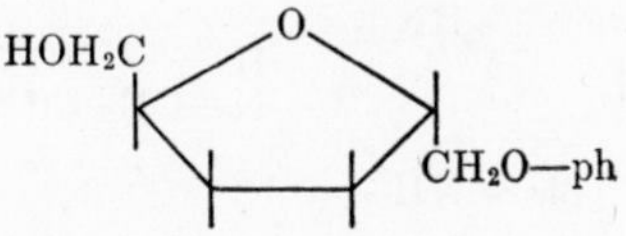	2000–3000
	Glucose-1-phosphate	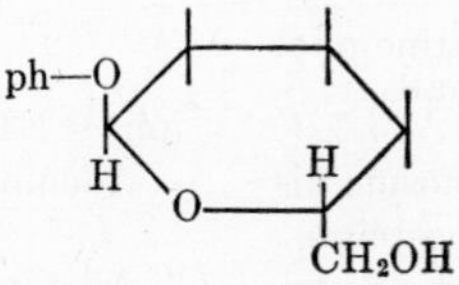	4800
	Fructose-1,6-diphosphate	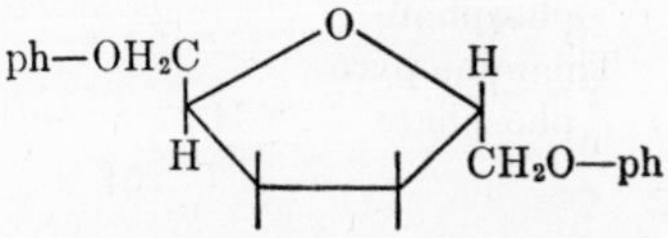	2000–3000
	3-Phosphoglyceraldehyde	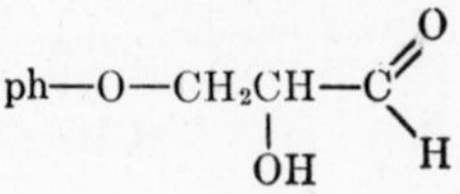	2000–3000
	3-Phosphoglyceric acid	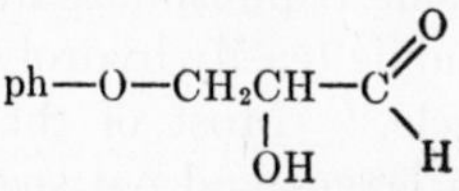	2000–3000
	2-Phosphoglyceric acid	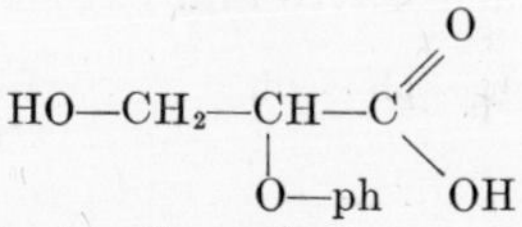	2000–3000
	Adenylic acid (adenosine-5-phosphate)	Adenine-ribose-ph	2000–3000
Enol phosphate	Phosphoenolpyruvic acid	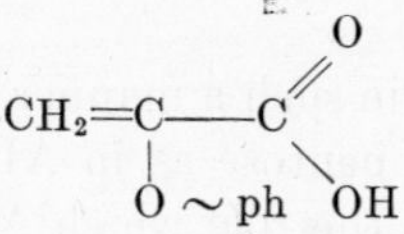	16,000
Carboxyl phosphate	Acetyl phosphate	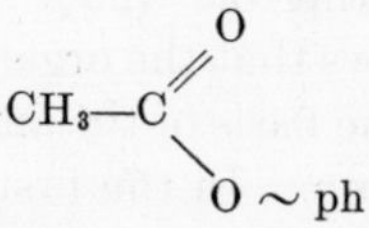	16,000

Structures (by compound):

Glucose-6-phosphate: pyranose ring with H, O, CH$_2$O—ph

Fructose-6-phosphate: HOH$_2$C, furanose ring with O, CH$_2$O—ph

Glucose-1-phosphate: ph—O, pyranose ring, H, O, H, CH$_2$OH

Fructose-1,6-diphosphate: ph—OH$_2$C, furanose ring with O, H, H, CH$_2$O—ph

3-Phosphoglyceraldehyde:
$$ph—O—CH_2CH—C{\overset{O}{\underset{H}{}}}$$
OH

3-Phosphoglyceric acid:
$$ph—O—CH_2CH—C{\overset{O}{\underset{H}{}}}$$
OH

2-Phosphoglyceric acid:
$$HO—CH_2—CH—C{\overset{O}{\underset{OH}{}}}$$
O—ph

Phosphoenolpyruvic acid:
$$CH_2{=}C—C{\overset{O}{\underset{OH}{}}}$$
O ~ ph

Acetyl phosphate:
$$CH_3—C{\overset{O}{\underset{O\sim ph}{}}}$$

TABLE 15-4. (Continued)

Type of linkage	Compound	Structure	Approximate $-\Delta F$ of hydrolysis of ph bond
	1,3-Diphospho-glyceric acid	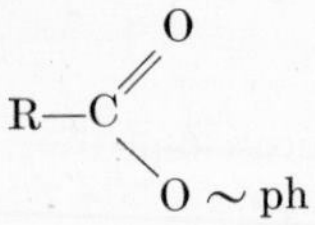 R—C(=O)(O ~ ph)	16,000
Guanidophosphate	Arginine phosphate	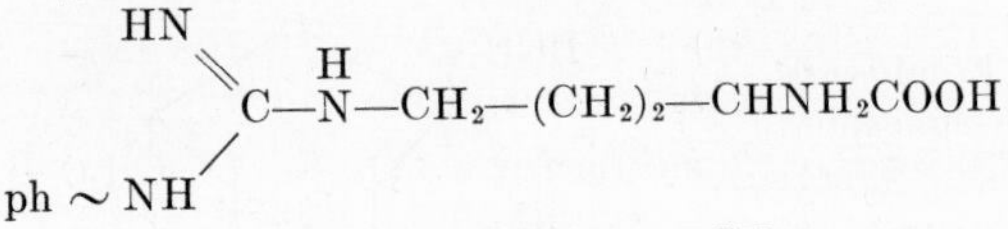 HN=C(ph ~ NH)—N(H)—CH$_2$—(CH$_2$)$_2$—CHNH$_2$COOH	10,000–12,000
	Creatine phosphate	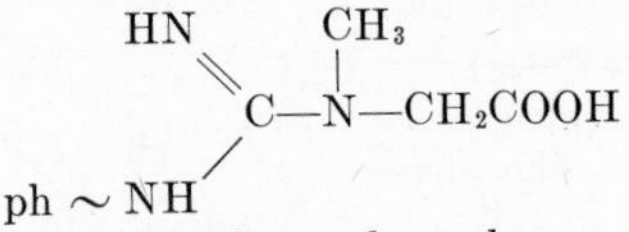 HN=C(ph ~ NH)—N(CH$_3$)—CH$_2$COOH	10,000–12,000
Pyrophosphate	Adenosine diphosphate	Adenine-ribose-ph ~ ph	12,000
	Adenosine triphosphate	Adenine-ribose-ph ~ ph ~ ph	12,000
	Thiamine pyrophosphate	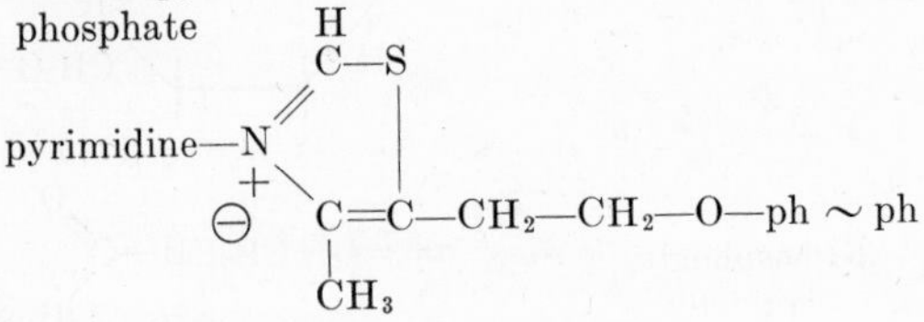 pyrimidine—N⁺(⊖)=C(—C(H)=S)=C(CH$_3$)—CH$_2$—CH$_2$—O—ph ~ ph	10,000–12,000

contain an adenosine triphosphate in which the ratio of adenine : ribose : total phosphate : labile (easily hydrolyzed) phosphate is 1 : 1 : 3 : 2 as it is in the ATP of muscle.[97] Most of the labile phosphate in certain plant tissues as spinach leaves and oat seedlings is not free in solution but is

Adenine-ribose-ph ~ ph ~ ph

Adenine-ribose-ph ~ ph ~ ph
|
ph
|
Adenine-ribose

ATP of animal tissues

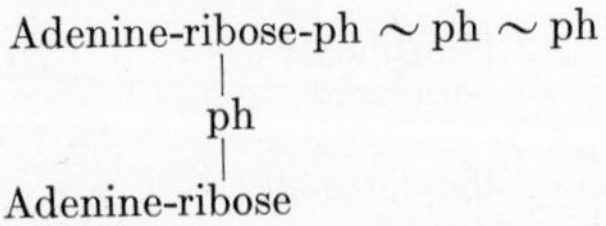

Diadenine nucleotide of yeast, possibly a plant "ATP"

bound to protein in such a manner that the labile phosphate is associated with purine and pentose as in ADP.[98] This protein bound phosphate may play in part the role which ATP plays in other tissues.

Table 15-4 shows that the organic phosphates may be divided into two broad groups on the basis of the amount of energy liberated on removal of the phosphate group. In the first group, represented by the ester phosphates (—ph) the amount of energy thus liberated is of the order of

2000–3000 cal. per mol. In the second group, represented by $N \sim ph$,

$$\overset{O}{\underset{\parallel}{C}} - O \sim ph, P - O \sim ph,$$ and certain other groupings, the energy liberated on hydrolysis is of the order of 10,000–16,000 cal. per mol. Phosphate groupings of all the latter types are referred to as energy-rich phosphates ($\sim ph$).

The complete oxidation of one mol of hexose to CO_2 and H_2O leads to the generation of 12–24 energy-rich phosphate bonds as measured experimentally in animal tissue systems.[99] Thus of the total of 680,000 cal. liberated in the oxidation of glucose to CO_2 and water as much as perhaps 290,000 cal. may be stored in the form of phosphate bond energy. The formation of the energy-rich phosphate bonds in oxidative metabolism takes place in a variety of ways. Of these, a clear example is constituted by the phosphate-coupled oxidation of phosphoglyceraldehyde.[100]

$$\text{Phosphoglyceraldehyde} + \text{DPN} \rightleftharpoons \text{Phosphoglycerate} + \text{reduced DPN}$$

This reaction is, however, coupled with the uptake of inorganic phosphate and the production of ATP. The first step may involve the formation of 1,3-diphosphoglyceraldehyde.

$$\text{3-Phosphoglyceraldehyde} + H_3PO_4 \rightleftharpoons \text{1,3-Diphosphoglyceraldehyde} \qquad (1)$$

Since the new phosphate bond must involve the uptake of at least 1000–2000 calories it is to be expected that only low concentrations of the diphosphate can exist in equilibrium with 3-phosphoglyceraldehyde at any one time. The glyceraldehyde diphosphate is next oxidized by triosephosphate dehydrogenase for which DPN is the coenzyme.

$$\text{1,3-Diphosphoglyceraldehyde} + \text{DPN}$$
$$\rightleftharpoons \text{1,3-Diphosphoglycerate} + \text{reduced DPN} \qquad (2)$$

The new compound, 1,3-diphosphoglycerate, now contains as a result of the oxidation a carboxyl phosphate in which the phosphate bond is of the energy-rich type. This compound is able to transfer its phosphate to ADP in the presence of a specific transphosphorylase with the formation of ATP and in fact this final transfer is essential to the continuation of reaction 2.

$$\text{1,3-Diphosphoglycerate} + \text{ADP} \rightleftharpoons \text{3-Phosphoglycerate} + \text{ATP} \qquad (3)$$

The energy liberated in the oxidation of 3-phosphoglyceraldehyde is thus transformed into the energy of formation of the energy-rich phosphate bond in ATP. The energy stored in ATP may be in turn used for chemical work in other reactions. It should be noted that a phosphate acceptor, ADP, is an essential part of this enzymatic oxidation. In the absence of a phosphate acceptor, the oxidation product, 1,3-diphosphoglycerate, merely accumulates until equilibrium with the reduced form is achieved and further reaction takes place only as the diphosphoglycerate slowly loses its carboxyl phosphate by spontaneous hydrolysis. This reaction presents then the interesting and probably frequent case of an enzymatic oxidation whose progress is conditioned by the presence of an acceptor for the energy liberated in the reaction.

Arsenate can substitute for inorganic phosphate in the oxidation of triosephosphate with the formation, presumably, of a 3-phospho-1-arseno-glyceraldehyde complex.[101] This substance is then oxidized as is the normal diphospho compound. The 3-phospho-1-arseno glyceric acid produced is, however, very unstable and immediately hydrolyzes to form again inorganic arsenate and 3-phosphoglyceric acid. These reactions are shown below.

3-Phosphoglyceraldehyde + H_2AsO_3 → 3-Phospho-1-arsenoglyceraldehyde
3-Phospho-1-arsenoglyceraldehyde + DPN
$$\rightarrow \text{3-Phospho-1-arsenoglycerate} + \text{red. DPN}$$
3-Phospho-1-arsenoglycerate + H_2O → 3-Phosphoglycerate + H_3AsO_3

The overall result of arsenate poisoning of the triosephosphate dehydrogenase system is then that the oxidation of triosephosphate is enabled to proceed without the simultaneous production of energy-rich phosphate. To put it in another way, the oxidation is uncoupled from its associated energy storing mechanism. Similar mechanisms appear to function in many other oxidative processes and arsenate poisoning may accordingly have far-reaching effects on ATP-requiring synthetic processes in the cell without any great disturbance of the normal respiratory process. Loomis and Lipmann have shown that 2,4-dinitrophenol is also able to bring about a similar uncoupling of ATP formation from phosphate coupled oxidation.[102] The exact mechanism of this uncoupling is unknown.

A second reaction in which energy-rich phosphate is formed is the transformation of 2-phosphoglycerate to phosphoenolpyruvate.[103] The product in this case possesses approximately the same total energy as the original reactant, and differs from it only by removal of a molecule of water.

$$CH_2OHCH{-}COOH \underset{+H_2O}{\overset{-H_2O}{\rightleftharpoons}} CH_2{=}C{-}COOH \qquad (4)$$

$$\begin{array}{cc} \mid & \mid \\ O{-}ph & O \sim ph \\ \text{2-Phosphoglyceric} & \text{Phosphoenolpyruvic} \\ \text{acid} & \text{acid} \end{array}$$

In phosphoenolpyruvate, however, much of the total energy of the molecule is concentrated in the phosphate bond, i.e., the energy distribution in phosphoenolpyruvate differs from that in 2-phosphoglycerate in which only a small portion of the energy of the molecule resides in the phosphate bond.   Phosphoenolpyruvate can reversibly exchange its phosphate with ADP to form ATP in the presence of a specific transphosphorylase.

A third reaction in which ATP is generated is the oxidative decarboxylation of pyruvic acid which has been mentioned earlier in this chapter. In the reaction pyruvate is oxidized and decarboxylated simultaneously or approximately simultaneously.   The reaction is coupled to the uptake of inorganic phosphate, and acetyl phosphate is produced.[104]   Acetyl phosphate is capable of exchanging its phosphate group with ADP to form ATP, again in the presence of a specific transphosphorylase.

$$CH_3CCOOH + H_3PO_4 \rightleftharpoons CH_3C\overset{O}{\overset{\parallel}{}}\!\!-O \sim ph + CO_2 + (2H) \qquad (5)$$

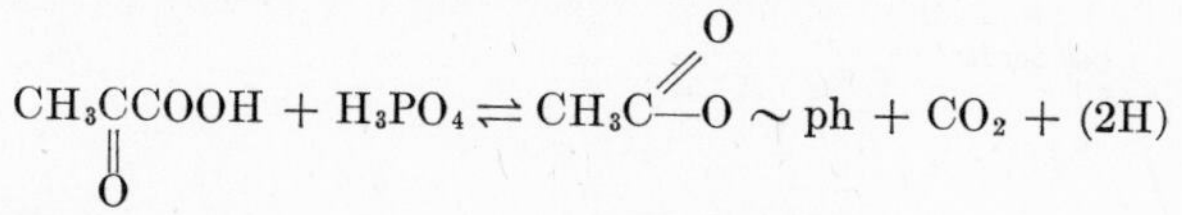

Pyruvic acid   Acetyl phosphate

Other less well understood oxidative reactions which lead to the generation of energy-rich phosphate bonds are found in the reactions of the Krebs cycle.   In particular the oxidations of succinate, fumarate, and of α-ketoglutarate generate energy-rich phosphate bonds.[105]   A possible mechanism in the oxidation of fumarate may be the addition of phosphate at the double bond of fumarate followed by dehydrogenation to phosphoenoloxaloacetate.

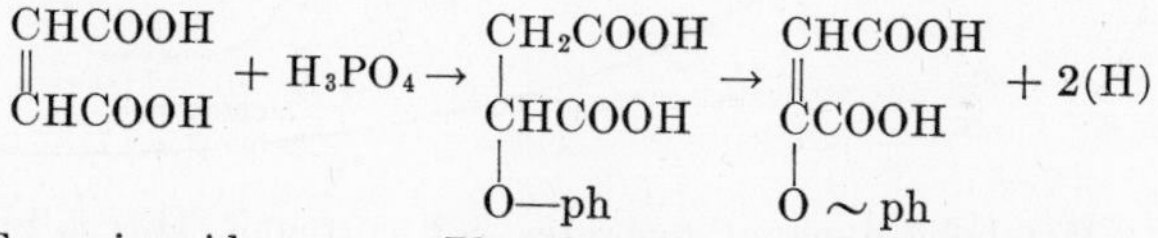

$$\begin{array}{ccc}
\text{CHCOOH} & \text{CH}_2\text{COOH} & \text{CHCOOH} \\
\| \quad\quad + H_3PO_4 \rightarrow & | & \| \quad\quad + 2(H) \\
\text{CHCOOH} & \text{CHCOOH} \rightarrow & \text{CCOOH} \\
& | & | \\
& \text{O—ph} & \text{O} \sim \text{ph}
\end{array} \qquad (6)$$

Fumaric acid   Phosphomalic   Phosphoenoloxalo-
                              acid   acetic acid

The analogy of α-ketoglutarate oxidation to the oxidative decarboxylation of pyruvate suggests that the mechanism of this phosphorylation may be represented as below.

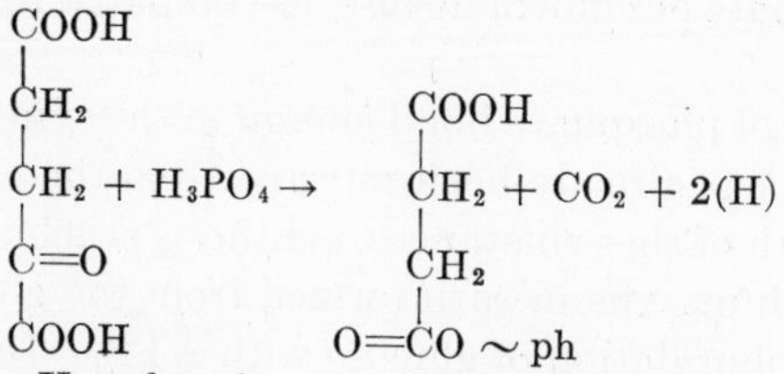

$$\begin{array}{cc}
\text{COOH} & \\
| & \\
\text{CH}_2 & \text{COOH} \\
| & | \\
\text{CH}_2 + H_3PO_4 \rightarrow & \text{CH}_2 + CO_2 + 2(H) \\
| & | \\
\text{C=O} & \text{CH}_2 \\
| & | \\
\text{COOH} & \text{O=CO} \sim \text{ph}
\end{array} \qquad (7)$$

α-Ketoglutaric   Phosphosuccinic
    acid   acid

In both cases the phosphorylated products of the oxidation would be expected to be energy-rich phosphates and hence capable of exchange with ADP to form ATP.

The reactions which have been discussed account for the formation of only ten mols of energy-rich phosphate per mol of hexose oxidized, and of these, two mols must be used in the initial phosphorylation of hexose to fructose diphosphate. There must then be still other reactions in which oxidation is coupled with formation of energy-rich phosphate. It is possible that the transfer of electrons through the pyridine nucleotides to flavoprotein, cytochrome, and oxygen may be coupled with phosphate bond formation in some as yet unknown manner. If each substrate oxidation were so coupled at each of the three levels, pyridine nucleotide-

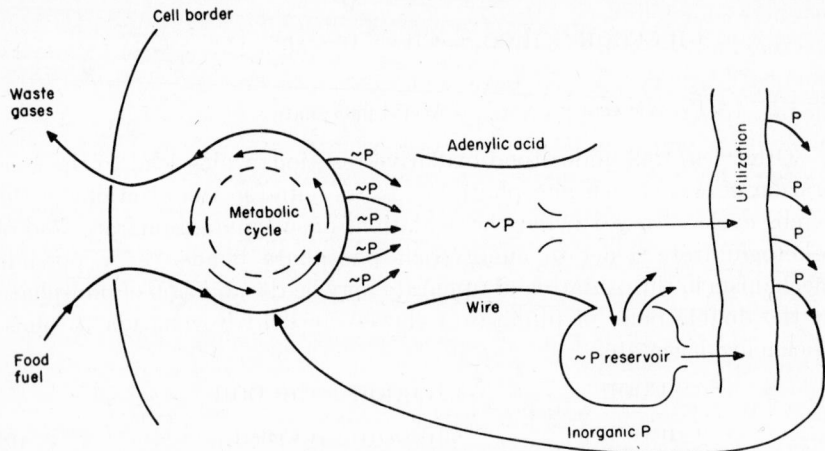

FIG. 15-7. Metabolic dynamo generates $\sim$P current. This is brushed off by adenylic acid, which likewise functions as the wiring system, distributing the current. (After Lipmann[56])

flavoprotein, flavoprotein-cytochrome, and cytochrome-oxygen, then complete oxidation of glucose would yield a maximum of 36 mols of energy-rich phosphate per mol of hexose, as compared to the ten accounted for above.

The utilization of phosphate bond energy in the synthesis of important plant constituents has already been referred to in the cases of starch and of sucrose. In both of these instances, synthesis is achieved from glucose-1-phosphate, which may be in turn formed from the glucose-6-phosphate produced by phosphorylation of glucose with ATP. In other words each glucose residue laid down as starch requires the expenditure of one energy-rich phosphate bond for the formation of the glucosidic linkage. This process is somewhat wasteful since of the 12,000 or more calories of the energy-rich phosphate linkage, at most 2000–3000 calories are conserved in the glucosidic bond. It is entirely possible that the synthetic mechanisms by which cellulose and other cell wall polysaccharides are formed may similarly involve the formation of intermediate hexose phosphate. This would also constitute considerable drain on the energy-rich phosphate pool of the plant. Other reactions in which ATP appears to be an essential reactant but in which the mechanism of the reaction is not wholly clear include the synthesis of peptide bonds (Chapter 17), the fixation of CO_2 to form pyruvate and oxaloacetate (Chapter 14), the preliminary activation of fatty acid molecules prior to oxidation, and the synthesis of fatty acids (Chapter 24). Undoubtedly there are other important reactions in which energy-rich phosphate constitutes the source of energy for cellular synthesis. Energy-rich phosphate is then the connecting link between the energy-liberating reactions of respiration and the energy-using reactions of the cell. Lipmann has portrayed this relation in Fig. 15-7.[56] Here the reactions of respiration are shown as a dynamo generating a current of ATP. The ATP current flows to the site of utilization where it is used as the energy source to run the cellular motors. Poisons such as arsenate and 2,4-dinitrophenol short-circuit the dynamo and permit it to operate without the production of ATP.

General References

Green, D. E., Mechanisms of Biological Oxidations. Cambridge, 1940.
Goddard, D. R., in Höber, Physical Chemistry of Cells and Tissues. Blakiston, 1945.
Sumner, J. B., and Somers, G. F., Enzymes. Academic Press, 2nd ed., 1947.
James, W. O., *Ann. Rev. Biochem.*, **15**, 417 (1946).
Kostytschev, S., Plant Respiration. Blakiston, 1927.
Onslow, M., Principles of Plant Biochemistry. Cambridge, 1931.

References

1. Onslow, M., *Biochem. J.*, **13**, 1 (1919); **14**, 535 (1920).
2. Pugh, C. E. M., and Raper, H. S., *Biochem. J.*, **21**, 1370 (1927).
3. Ivanoff, L., *Biochem. Z.*, **25**, 171 (1910).
4. Bodnar, J., and Hoffner, P., *Biochem. Z.*, **165**, 145 (1925).
5. Zaleski, W., and Marx, E., *Biochem. Z.*, **47**, 184 (1912).
6. Bodnar, J., *Biochem. Z.*, **73**, 193 (1916).
7. See for example Batelli and Stern, *Biochem. Z.*, **28**, 145 (1910); and Thunberg, T., *Skand. Arch. Physiol.*, **24**, 23 (1910).
8. Michaelis, L., and Schubert, M. P., *Chem. Rev.*, **22**, 437 (1938).
9. Harden, A., and Young, W. J., *J. Physiol.*, **32**, Proc. of November, 1904 (1905).
10. Warburg, O., and Christian, W., *Biochem. Z.*, **254**, 438 (1932).
11. Von Euler, H., Albers, H., and Schlenk, F., *Z. physiol. Chem.*, **237**, 1 (1935); **246**, 64 (1937).
12. Warburg, O., Christian, W., and Griese, A., *Biochem. Z.*, **282**, 157 (1935).
13. Information on the occurrence of malic and other dehydrogenases in higher plants can be found in:
   Thunberg, T., *Biochem. Z.*, **206**, 109 (1929);
   Anderson, B., *Z. physiol. Chem.*, **217**, 186 (1933);

Berger, J., and Avery, G. S., *Am. J. Botany*, **30**, 290 (1943); **31**, 11 (1944); Bonner, J., and Wildman, S. G., *Arch. Biochem.*, **10**, 497 (1946).

14. Negelein, E., and Wulff, H. J., *Biochem. Z.*, **289**, 436 (1936); **293**, 351 (1937).
15. Damodaran, M., and Nair, K. R., *Biochem. J.*, **32**, 1064 (1938).
16. Warburg, O., and Christian, W., *Biochem. Z.*, **303**, 40 (1939).
    Stumpf, P. K., *J. Biol. Chem.*, **176**, 233 (1948); *Am. J. Botany*, in press.
17. Berger, J., and Avery, G. S., *Am. J. Botany*, **31**, 11 (1944).
18. Vennesland, B., Ceithaml, J., and Gollub, M., *J. Biol. Chem.*, **171**, 445 (1947).
19. Warburg, O., and Christian, W., *Biochem. Z.*, **254**, 438 (1932); **257**, 492 (1933); **266**, 377 (1933).
20. Kuhn, R., Reinemund, K., Kaltschmitt, H., Ströbele, R., and Trischmann, H., *Naturwiss.*, **23**, 260 (1935).
    Karrer, P., Schöpp, K., and Benz, F., *Helv. Chim. Acta*, **18**, 426 (1935).
21. Warburg, O., and Christian, W., *Biochem. Z.*, **298**, 150 (1938).
22. Straub, F. B., *Biochem. J.*, **33**, 787 (1939).
23. Lockhart, E. E., *Biochem. J.*, **33**, 613 (1939).
24. Okunuki, K., *Acta Phytochim.*, **11**, 249 (1940).
25. Krebs, H. A., *Z. physiol. Chem.*, **217**, 191 (1933); *Biol. J.*, **29**, 162ᴿ (1935).
26. Straub, F. B., *Nature*, **141**, 603 (1938).
27. Warburg, O., and Christian, W., *Biochem. Z.*, **289**, 150 (1938).
28. Horowitz, N. H., *J. Biol. Chem.*, **154**, 141 (1944).
29. Ball, E. G., *J. Biol. Chem.*, **128**, 51 (1939).
30. Stumpf, P. K., and Green, D. E., *J. Biol. Chem.*, **153**, 387 (1944).
    Edlbacher, S., and Grauer, H., *Helv. Chim. Acta*, **27**, 151, 928 (1944).
    Moore, D. H., *J. Biol. Chem.*, **161**, 597 (1945).
31. Haas, E., Horecker, B. L., and Hogness, T. R., *J. Biol. Chem.*, **136**, 747 (1940).
    Haas, E., Hauer, C., and Hogness, T. R., *J. Biol. Chem.*, **143**, 344 (1942).
32. Altschul A. M., Persky, H., and Hogness, T. R., *Science*, **94**, 349 (1941).
    Heppel, L. A., *Fed. Proc.*, **8**, 205 (1949).
33. Sumner, J. B., and Dounce, A. L., *J. Biol. Chem.*, **121**, 417 (1937).
    Laskowski, M., and Sumner, J. B., *Science*, **94**, 615 (1941).
34. Sumner, J. B., and Gralén, N., *J. Biol. Chem.*, **125**, 33 (1938).
35. Lunberg, R., Naerie, M., and Legge, J. W., *Nature*, **144**, 551 (1939).
    Sumner, J. B., Dounce, A. L., and Frampton, V. L., *J. Biol. Chem.*, **136**, 343 (1941).
36. Theorell, H., *Arkiv Kemi, Mineral., Geol.*, **14B** (20), 1, 1940; **15B** (24), 1, 1942; **16A** (2), 1943.
37. Keilin, D., and Hartree, E. F., *Proc. Roy. Soc. (London)*, **B122**, 298 (1937).
38. Theorell, H., and Akesson, A., *J. Am. Chem. Soc.*, **63**, 1804 (1941).
39. Goddard, D. R., *Am. J. Botany*, **31**, 270 (1944).
40. Keilin, D., *Erg. Enzymforsch.*, **2**, 239 (1933).
41. Haas, E., *J. Biol. Chem.*, **152**, 695 (1944).
42. Nelson, J. M., and Dawson, C. R., Advances in Enzymology. Interscience Publishers, 1944. Vol. 4, p. 99.
43. Graubard, M., and Nelson, J. M., *J. Biol. Chem.*, **112**, 135 (1936).
    Keilin, D., and Mann, T., *Proc. Roy. Soc. (London)*, **B125**, 187 (1938).
44. Kubowitz, F., *Biochem. Z.*, **292**, 221 (1937); **296**, 443 (1938); **299**, 32 (1938).
45. Bonner, J., and Wildman, S. G., *Arch. Biochem.*, **10**, 497 (1946).
46. Arnon, D., *Plant Physiol.*, **24**, 1 (1949).
    Laties, G., *Arch. Biochem.*, **20**, 284 (1949).
    Li, L., and Bonner, J., *Biochem. J.*, **41**, 105 (1947).

47. Keilin, D., and Mann, T., *Nature*, **143**, 23 (1939); **145**, 304 (1940).
48. Bertrand, G., *Compt. rend.*, **118**, 1215 (1894); **122**, 1132 (1896).
49. Tauber, H., *Erg. Enzymforsch.*, **7**, 301 (1938).
50. Powers, W. H., Lewis, S., and Dawson, C. R., *J. Gen. Physiol.*, **27**, 167 (1944).
51. Todokoro, T., and Takasugi, H., *J. Chem. Soc., Japan*, **60**, 188 (1939).
52. Lohmann, K., and Schuster, P., *Biochem. Z.*, **294**, 188 (1937).
53. Green, D. E., Herbert, D., and Subrahmanyan, V., *J. Biol. Chem.*, **135**, 795 (1940); **138**, 327 (1941).
    Melnick, J. L., and Stern, K. G., *Enzymologia*, **8**, 129 (1940).
54. Bunting, A. H., and James, W. O., *New Phytologist*, **40**, 262 (1941).
    Vennesland, B., and Felsher, R., *Arch. Biochem.*, **11**, 279 (1946).
    Wetzel, K., *Planta*, **15**, 697 (1932).
55. Horowitz, N. H., and Heegaard, E. V., *J. Biol. Chem.*, **137**, 475 (1941).
56. Lipmann, F., Advances in Enzymology. Interscience Publishers, 1941. Vol. 1, p. 99.
57. Stumpf, P. K., Zarudnaya, K., and Green, D. E., *J. Biol. Chem.*, **167**, 817 (1947).
58. Krampitz, L. O., and Werkman, C. H., *Biochem. J.*, **35**, 595 (1941).
59. Axelrod, A. E., Potter, V. R., and Elvehjem, C. A., *J. Biol. Chem.*, **122**, 85 (1942).
60. Quastel, J. H., *Biochem. J.*, **20**, 166 (1926).
61. Okunuki, K., *Acta Phytochim.*, **11**, 28 (1939).
62. Lutwak-Mann, C., and Mann, T., *Biochem. Z.*, **281**, 140 (1935).
63. Berger, L., Slein, M. W., Colowick, S. P., and Cori, C. F., *J. Gen. Physiol.*, **29**, 141 (1946).
    Kunitz, M., and McDonald, M. R., *J. Gen. Physiol.*, **29**, 143, 313 (1946).
64. Colowick, S. P., and Price, W. H., *J. Biol. Chem.*, **157**, 415; **159**, 563 (1945).
65. Warburg, O., and Christian, W., *Biochem. Z.*, **314**, 149 (1943).
66. Stumpf, P. K., *J. Biol. Chem.*, **176**, 233 (1948).
67. Meyerhof, O., and Kiessling, W., *Biochem. Z.*, **276**, 239 (1935).
68. Moog, F., *Biol. Rev.*, **21**, 41 (1946).
    Yin, H. C., *New Phytologist*, **44**, 191 (1945).
69. Warburg, O., and Christian, W., *Biochem. Z.*, **303**, 40 (1939).
70. Needham, D. M., and Pillai, R. K., *Biochem. J.*, **31**, 1837 (1937).
71. Utter, M. F., and Werkman, C. H., *J. Biol. Chem.*, **146**, 289 (1942).
    Meyerhof, O., and Beck, L. V., *J. Biol. Chem.*, **156**, 109 (1944).
    Sutherland, E. W., Posternak, T., and Cori, C. F., *J. Biol. Chem.*, **179**, 501 (1949).
72. Warburg, O., and Christian, W., *Biochem. Z.*, **310**, 384 (1942).
73. Parnas, J. K., Ostern, P., and Mann, T., *Biochem. Z.*, **272**, 64 (1934).
    Lardy, H. A., and Zeigler, J. A., *J. Biol. Chem.*, **159**, 343 (1945).
74. Bonner, J., *Arch. Biochem.*, **17**, 311 (1948).
75. James, W. O., *Ann. Rev. Biochem.*, **15**, 417 (1946).
76. Krebs, H. A., and Johnson, W. A., *Enzymologia*, **4**, 148 (1937).
    Krebs, H. A., Advances in Enzymology. Interscience Publishers, 1943. Vol. 3, p. 191.
77. Gözsy, B., and Szent-Györgyi, A., *Z. physiol. Chem.*, **224**, 1 (1934).
    Annan, E., *et al.*, *Z. physiol. Chem.*, **236**, 1 (1935); **244**, 105 (1936).
78. Martius, C., and Knoop, F., *Z. physiol. Chem.*, **242**, 1 (1936).
79. Wood, H. G., Werkman, C. H., Hemingway, M., and Nier, A. O., *J. Biol. Chem.*, **139**, 483 (1941).
    Evans, H. A., and Slotin, L., *J. Biol. Chem.*, **141**, 439 (1941).
80. Stern, J. R., and Ochoa, S., *J. Biol. Chem.*, **179**, 491 (1949).

81. Ochoa, S., *J. Biol. Chem.*, **155**, 87 (1944).
   Stumpf, P. K., Zarudnaya, K., and Green, D. E., *J. Biol. Chem.*, **167**, 817 (1947).
82. Merry, J., and Goddard, D. R., *Proc. Roch. Acad. Sci.*, **8**, 28 (1941).
   Marsh, P. B., and Goddard, D. R., *Am. J. Botany*, **26**, 767 (1939).
   Biale, J., Abstract, Boston meeting AAAS, December, 1946.
83. Mikhlin, D., and Kolesnikov, P., *Biokhimia*, **12**, 452 (1947).
   Boswell, J. G., and Whiting, G. C., *Ann. Botany*, New series **2**, 847 (1938).
   Boswell, J. G., *Ann. Botany*, New series **9**, 55 (1945).
84. Baker, D., and Nelson, J. M., *J. Gen. Physiol.*, **26**, 269 (1943).
85. Robinson, E. S., and Nelson, J. M., *Arch. Biochem.*, **4**, 111 (1944).
86. James, W. O., Heard, C. R. C., and James, G. M., *New Phytologist*, **43**, 62 (1944).
87. Barron, E. S. G., Advances in Enzymology. Interscience Publishers, 1943. Vol. 3, p. 149.
88. Bennet-Clark, T. E., and Bexon, D., *New Phytologist*, **42**, 65 (1943).
89. Commoner, B., and Thimann, K. V., *J. Gen. Physiol.*, **42**, 279 (1941).
90. Laties, G., *Arch. Biochem.*, **20**, 284 (1949); **22**, 8 (1949). Also thesis, Berkeley, 1947.
91. Henderson, J. H. M., and Stauffer, J. F., *Am. J. Botany*, **31**, 528 (1944).
92. Kalckar, H. M., *Chem. Rev.*, **28**, 71 (1941).
93. Michaelis, L., Oxidation Reduction Potentials. Lippincott, 1930.
94. Green, D. E., Dewan, J. G., and Leloir, L. F., *Biochem. J.*, **31**, 934 (1937).
95. Winzler, R. J., and Baumberger, J. P., *J. Comp. Cell. Physiol.*, **12**, 183 (1938).
96. Goddard, D. R., in Höber, *Physical Chemistry of Cells and Tissues*. Blakiston, 1945.
97. Albaum, H. G., and Ogur, M., *Arch. Biochem.*, **15**, 158 (1947).
   Albaum, H. G., Ogur, M., and Hirshfeld, A., *Fed. Proc.*, **8**, 179 (1949).
98. Wildman, S. G., Campbell, J. M., and Bonner, J., *J. Biol. Chem.*, **180**, 273 (1949).
99. Ochoa, S., *J. Biol. Chem.*, **151**, 493 (1943).
   Potter, V. R., *J. Biol. Chem.*, **169**, 17 (1947).
   Croes, R. J., Taggart, J. V., Corro, G. A., and Green, D. E., *J. Biol. Chem.*, **177**, 655 (1949).
100. Warburg, O., and Christian, W., *Biochem. Z.*, **303**, 40 (1939).
   Negelein, E., and Brömel, H., *Biochem. Z.*, **303**, 132 (1939).
101. Needham, D. M., and Pillai, R. K., *Biochem. J.*, **31**, 1837 (1937).
102. Loomis, W., and Lipmann, F., *J. Biol. Chem.*, **173**, 807 (1948).
   Croes, R. J., Taggart, J. V., Corro, G. A., and Green, D. E., *J. Biol. Chem.*, **177**, 655 (1949).
103. Lohmann, K., and Meyerhof, O., *Biochem. Z.*, **273**, 60 (1934).
   Meyerhof, O., and Kiersling, W., *Biochem. Z.*, **280**, 99 (1935).
104. Lipmann, F., Advances in Enzymology. Interscience Publishers, 1946. Vol. 6, p. 231.
105. Kalckar, H. M., *Biochem. J.*, **33**, 631 (1939).
   Colowick, S. P., Welch, M. S., and Cori, C. F., *J. Biol. Chem.*, **133**, 359 (1940).
   Ochoa, S., *J. Biol. Chem.*, **149**, 577 (1943); **155**, 87 (1944); **151**, 493 (1943).
106. Stumpf, P. K., *J. Biol. Chem.*, **159**, 529 (1945).
107. Schade, A. L., Bergmann, L., and Byer, A. B., *Arch. Biochem.*, **18**, 85 (1948).
   Levy, H., and Schade, A. L., *Arch. Biochem.*, **19**, 273 (1948).

PART IV METABOLISM OF NITROGENOUS COMPOUNDS

CHAPTER 16

SYNTHESIS AND METABOLISM OF AMINO ACIDS

Amino acids constitute the building blocks of proteins which are in turn the characteristic stuff of protoplasm. There are a great many different kinds of proteins in protoplasm, and these proteins differ from one another in many important ways.   Thus the molecules of the various kinds of proteins differ in size, shape, surface properties, and perhaps most significantly, in enzymatic activities and biological functions. Despite their many individual differences proteins do have in common the fact that they are made up by polymerization of the same relatively small number of kinds of amino acids.   These amino acids, approximately twenty-five in number, are liberated when proteins are hydrolyzed by enzymatic or acid hydrolysis and are combined in the intact protein through peptide linkages, bonds in which the amino group of one amino acid molecule is linked as a substituted amide with the carboxyl group of the next.   Since the amino groups of the amino acids are typically in the α position (adjacent to the carboxyl group), the skeleton or backbone of the protein molecule will look somewhat as pictured below.

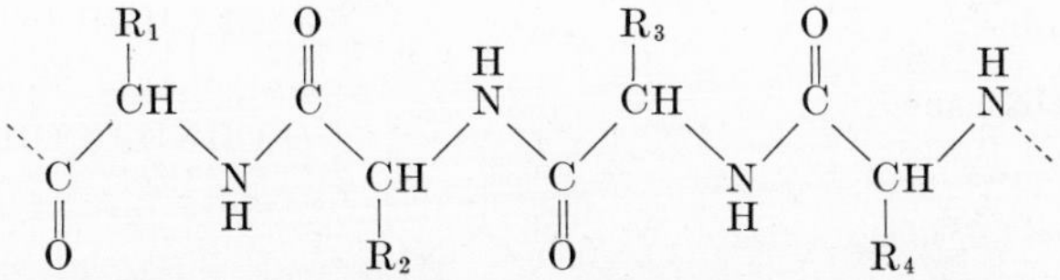

Unit of backbone or skeleton of protein showing α-amino acid residues linked through peptide linkages. R_1, R_2, etc. represent the residues of different amino acids.   Thus for glycine, $R = H$, for alanine, $R = CH_3$, etc.

With few exceptions the amino acids possess the general structure $R \cdot CHNH_2 \cdot COOH$. They differ among themselves, however, in the nature of the group R which is attached to the terminal $-CHNH_2 \cdot COOH$ portion.   In the simplest amino acid, glycine, R is represented merely by a hydrogen atom and glycine therefore has the structure $CH_2NH_2 \cdot COOH$.   In alanine, on the other hand, R is a methyl group and alanine correspondingly possesses the structure CH_3CHNH_2COOH. Several of the amino acids are derivatives of alanine in which the β carbon atom is linked to a further group as in tryptophane (indolealanine), phenylalanine, tyrosine (*p*-hydroxyphenylalanine), etc. The group R may be even more complex and may contain additional amino groups as in arginine or lysine, or may contain an additional carboxyl group as in glutamic acid or aspartic acid. The structures of the principal amino acids are summarized below.

Amino acids commonly considered to be components of proteins and particularly of plant proteins

Name	*Structure*
Glycine	NH_2CH_2COOH
Alanine	CH_3CHNH_2COOH
Valine	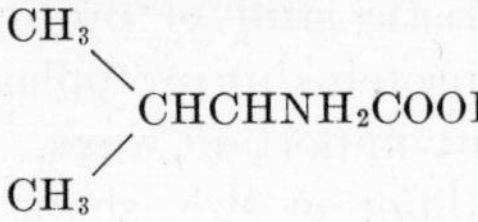 $\begin{array}{c} CH_3 \\ \diagdown \\ CHCHNH_2COOH \\ \diagup \\ CH_3 \end{array}$
Leucine	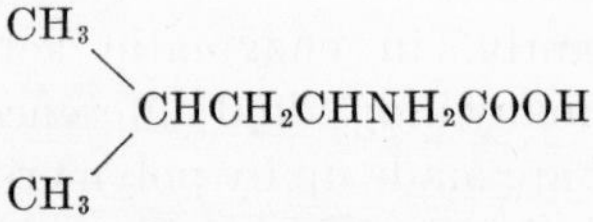 $\begin{array}{c} CH_3 \\ \diagdown \\ CHCH_2CHNH_2COOH \\ \diagup \\ CH_3 \end{array}$
Isoleucine	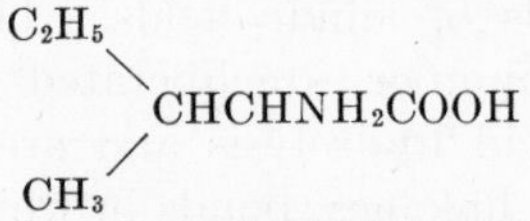 $\begin{array}{c} C_2H_5 \\ \diagdown \\ CHCHNH_2COOH \\ \diagup \\ CH_3 \end{array}$
Norleucine	$CH_3(CH_2)_3CHNH_2COOH$
Serine	$HOCH_2CHNH_2COOH$
Threonine	$CH_3CHOHCHNH_2COOH$
Phenylalanine	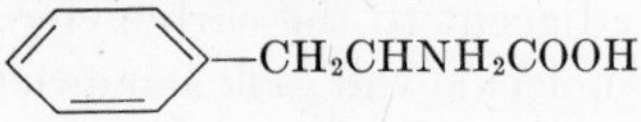 $C_6H_5{-}CH_2CHNH_2COOH$
Tyrosine	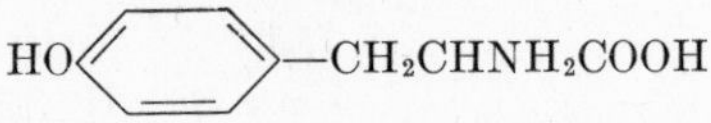 $HO{-}C_6H_4{-}CH_2CHNH_2COOH$
Tryptophane	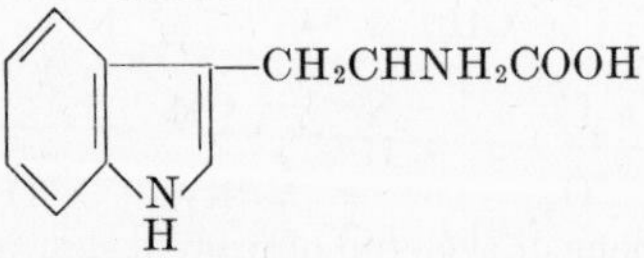 indole${-}CH_2CHNH_2COOH$
Proline	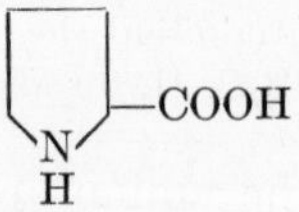 pyrrolidine${-}COOH$
Hydroxyproline	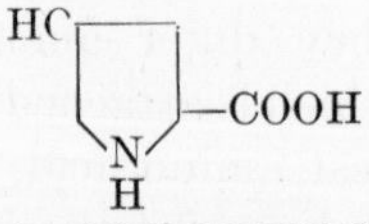 hydroxypyrrolidine${-}COOH$
Cystine	$(-SCH_2CHNH_2COOH)_2$
Cysteine	$HSCH_2CHNH_2COOH$

Methionine	$CH_3SCH_2CH_2CHNH_2COOH$
Aspartic acid	$HOOCCH_2CHNH_2COOH$
Glutamic acid	$HOOCCH_2CH_2CHNH_2COOH$
Hydroxyglutamic acid	$HOOCCH_2CHOHCHNH_2COOH$
Lysine	$NH_2CH_2(CH_2)_3CHNH_2COOH$

$$NH_2$$
$$\diagdown$$

Arginine $\qquad CNHCH_2(CH_2)_2CHNH_2COOH$

$$\diagup\diagup$$
$$NH$$

Histidine

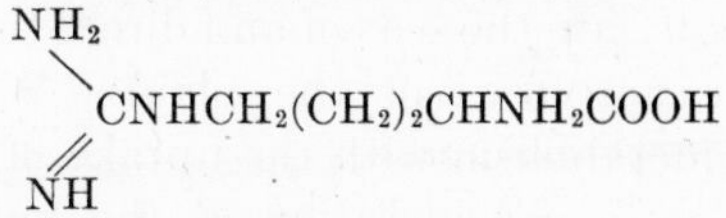

N
‖   ‖—CH_2CHNH_2COOH
N
H

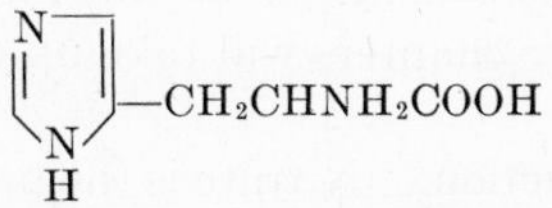

Amino acids found in plants but not commonly as components of plant protein

Name	*Structure*

$$NH_2$$
$$\diagdown$$

Canavanine $\qquad CNHO(CH_2)_2CHNH_2COOH$

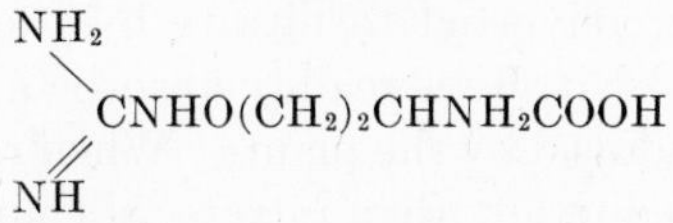

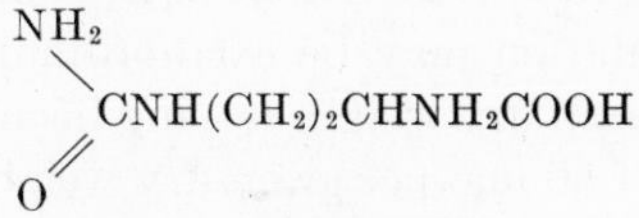

$$\diagup\diagup$$
$$NH$$

$$NH_2$$
$$\diagdown$$

Citrulline $\qquad CNH(CH_2)_2CHNH_2COOH$

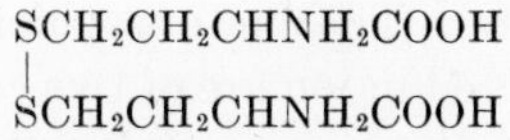

$$\diagup\diagup$$
$$O$$

Homocystine $\qquad SCH_2CH_2CHNH_2COOH$
$\qquad\qquad\qquad\quad$ |
$\qquad\qquad\qquad\; SCH_2CH_2CHNH_2COOH$

Djenkolic acid $\qquad SCH_2CH_2CHNH_2COOH$
$\qquad\qquad\qquad\quad$ |
$\qquad\qquad\qquad\; CH_2$
$\qquad\qquad\qquad\quad$ |
$\qquad\qquad\qquad\; SCH_2CH_2CHNH_2COOH$

$$OH$$
$$|$$

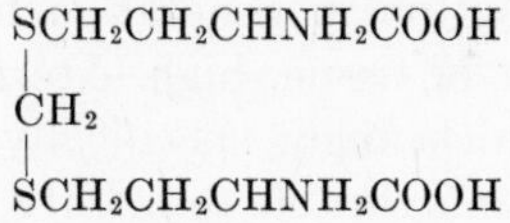

Dihydroxyphenylalanine $\qquad HO-\langle\bigcirc\rangle-CH_2CHNH_2COOH$

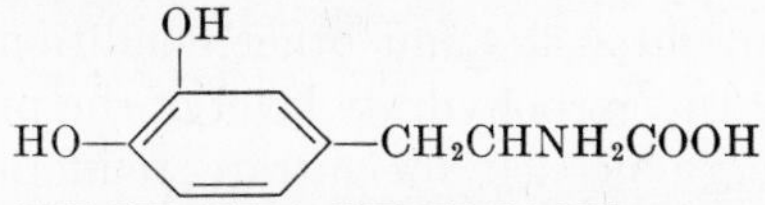

Ornithine $\qquad NH_2CH_2(CH_2)_2CHNH_2COOH$

It is to be noted that the carbon atom to which the α-amino group is attached is in general a center of asymmetry in the amino acid molecule. Thus amino acids (except glycine) are optically active substances. Of the possible isomers of each amino acid, only one is commonly found in nature. The naturally occurring forms of the amino acids are now designated by convention as belonging to the L series. The D-amino acids, while they are known in certain scattered instances as components of biological materials, are frequently referred to as the unnatural isomers.

The following seven chapters of this book will be devoted to the discussion of the formation and metabolism of the nitrogenous compounds of the plant, and of these chapters five concern directly the amino acids and their derivatives, the proteins. It is to be borne in mind throughout that we are primarily interested in the plant proteins and the enzymes which they constitute. We are interested in the synthesis, metabolism, function, and degradation of the plant proteins because these proteins undoubtedly are the seat of and direct the course of all the other manifold metabolic activities of the plant. We shall start the discussion of nitrogen metabolism with the uptake of inorganic nitrogen by the plant, followed by a consideration of the known mechanisms of amino acid synthesis. Later chapters will take up the proteins and protein metabolism proper.

Nitrate Reduction. Nitrate is the principal form of nitrogen taken up from the soil by most species of higher plants. This is in part owing to the fact that other forms of nitrogen such as amino N or ammonium ions are rapidly converted to nitrate by the activities of microorganisms in most soils. Nitrate is readily absorbed and once taken up may be either stored or reduced by the plant. When rate of nitrate uptake exceeds rate of nitrate reduction, then nitrate, of course, accumulates in the plant and such accumulation may be extraordinarily large, more than 100 mg. of nitrate per gram dry weight having been found in peach roots[2] while concentrations of 50 mg. per gram dry weight have frequently been observed in leaves and roots. The factors which favor nitrate accumulation as opposed to nitrate reduction are of two categories: (a) factors which favor rapid ion accumulation in general such as vigorous root aeration, low initial salt content of tissue, high external nitrate level, and absence of competing ions which might specifically depress nitrate uptake, and (b) factors which are unfavorable to nitrate reduction. Such factors include low light intensity and other conditions which act in the direction of limiting the carbohydrate level of the plant, since, as will be seen below, the energy needed for nitrate reduction must come ultimately from carbohydrate oxidation. A few species of plants have further the peculiarity that they appear capable of absorbing nitrate ions but inherently incapable of reducing nitrate. This is said to be true of *Chenopodium album.*[3]

The same factors which influence nitrate accumulation undoubtedly also influence the distribution of nitrate within the plant. In general, in plants which rapidly reduce nitrate, this ion is found in appreciable concentrations only in the root and is at low or zero concentration in the leaves and stems. This is true of many deciduous trees, and Thomas[4] first appreciated the fact that in such plants nitrate reduction takes place exclusively in the roots. In the apple, for example, nitrate reduction is confined to the small roots and amino acids formed from the reduced nitrogen are apparently the form in which nitrogen is translocated to the other portions of the apple tree. In asparagus and in *Narcissus*, as in apple, nitrate reduction is ordinarily confined to the roots.[5] If asparagus roots are cooled below 10°C, however, nitrate reduction is slowed and the material is translocated unchanged from the roots to the tops and there reduced. With the apple or peach, also, conditions of high nitrate supply to the roots result in movements of the ion upward to the aerial parts. In still other species, as tomato, tobacco, cucurbits, and cereals, the roots are not particularly active in nitrate reduction but on the contrary nitrate is translocated through the tissues of the entire plant, with reduction taking place in the leaves as well as in the roots.[1]

The question of the site of nitrate reduction has been studied with excised isolated organs of various plants. In the case of wheat, Burström[6] has shown that both excised roots and excised leaves are capable of reducing nitrate, and he comes to the conclusion that in the intact plant nitrate reduction is probably about equally distributed between roots and leaves. All species of excised roots which have been investigated have proved capable of reducing nitrate[7] and the same is true of the few species of leaves which have been investigated, including pea, tobacco, and wheat.

The reduction of nitrate to ammonia involves the production of nitrite as an intermediate. Thus if tomato plants which have been depleted of nitrogen are suddenly given nitrate, nitrite appears in the tissues within 6–36 hours only to disappear again still later. In normal plants, however, nitrite is not found in appreciable concentrations. Even when nitrite is supplied to plants, it rapidly disappears and is evidently highly unstable, being perhaps subject to further metabolism. With the reduction of nitrate in such nitrogen-depleted tomato plants a rapid parallel depletion of stored carbohydrate is also observed. The energy derived from the respiration of the sugar is presumably used in part as the source of the energy required in NO_3 reduction and the appearance of nitrite is attended by striking (100–300%) increases in respiration rate of the plant.[8] In the case of wheat, photosynthetic activity is found to be similarly increased so that secondary effects of nitrogen additions probably also are at work.[8] In addition to these overall effects on rate of respiration, nitrate reduction is also attended by a change in respiratory quotient; the ratio of CO_2 evolved to O_2 consumed tends to be larger when nitrate is being reduced than when this process is not going on.[9] This is, of course, due to the fact that the oxygen of the nitrate ion must also serve as a hydrogen acceptor in respiration and substitutes for a portion of the oxygen which would otherwise be consumed in respiration.

Nitrate reduction in roots and in many microorganisms proceeds in the absence of light. With green tissues such as leaves, however, there is good evidence that light plays an immediate role in the process. The amounts of light needed to bring about nitrate reduction in excised wheat leaves are somewhat lower than those required to produce significant increases in carbohydrate level by photosynthesis[10] and wave lengths of 3900–5200 Å are primarily effective.[11] The effect of light on nitrate reduction is hence not primarily through effects on photosynthesis but would appear to be on other as yet unknown systems.

An enzyme, nitrate reducase, which was attributed a role in the reduction of nitrate to nitrite, was described by Eckerson in 1924.[12] For the demonstration of nitrate reducase activity according to Eckerson, the water extract of freshly mashed plant tissue is incubated with nitrate and glucose at pH 7.2–7.4. Incubation takes place aerobically for several hours, after which time and in the presence of the enzyme, nitrite is produced. According to Eckerson the activity is largely confined to tissues which do in fact possess *in vivo* the power of nitrate reduction. A variety of workers have further investigated this enzyme, but without concordant results, and in fact the production of nitrite in the Eckerson experimental technique may be largely or wholly due to microorganisms.[13] A variety of enzymes are, however, known which can utilize nitrate as a hydrogen acceptor under anaerobic conditions. An enzyme similar to nitrate reducase, nitritase, is known in the bacterium, *E. coli*.[14] This enzyme enables *E. coli* to utilize nitrate as an hydrogen acceptor under anaerobic conditions. Thus in the absence of nitrate, *E. coli* can utilize succinate as a sole source of carbon under aerobic but not under anaerobic conditions. In the presence of nitrate, succinate can be utilized anaerobically, the nitrate acting as hydrogen acceptor and being reduced to nitrite. Similarly the reduced form of the enzyme xanthine oxidase, which is a flavoprotein normally reoxidized by oxygen, can be reoxidized anaerobi-

$$\text{Xanthine} + \text{xanthine oxidase} \rightarrow \text{Uric acid} + 2\text{H·xanthine oxidase}$$
(Oxidation of substrate by enzyme)

$$2\text{H·xanthine oxidase} + \tfrac{1}{2}O_2 \rightarrow \text{Xanthine oxidase} + H_2O$$
(Aerobic reoxidation of enzyme)

$$2\text{H·xanthine oxidase} + HNO_3 \rightarrow \text{Xanthine oxidase} + HNO_2 + H_2O$$
(Anaerobic reoxidation of enzyme with nitric acid as H acceptor)

tally in the presence of nitrate with the parallel production of nitrite. In the case of xanthine oxidase, oxidation of the substrate (aldehyde or purine) can be directly linked to reduction of nitrate. Similar enzyme systems known in higher plants are the aldehyde dehydrogenase of potato tubers and the lactic dehydrogenase of soybean seedlings, both of which can utilize nitrate as an hydrogen acceptor with the production of nitrite.[15]

Quite evidently, it cannot be stated with certainty what enzyme or enzymes are active in the plant in the reduction of nitrate to nitrite. It would obviously be of great interest to determine further properties of this important enzyme of the nitrogen metabolism of higher plants.

We have seen that there is some reasonable basis for the suspicion that nitrite may be the initial reduction product of nitrate. The final reduction product is, further, undoubtedly ammonia or ammonia in the form of amino groups of amino acids. Two possible intermediates in the reduction of nitrite to ammonia are hyponitrous acid and hydroxylamine.

Compound	Formula	Number of H atoms needed to reduce to NH_3
Nitric acid	NHO_3	8
Nitrous acid	NHO_2	6
Hyponitrous acid	$(NOH)_2$	4
Hydroxylamine	H_2NOH	2
Ammonia	NH_3	0

Hyponitrous acid is insufficiently stable to accumulate in detectable quantities, even if it were an intermediate in nitrite reduction. Hydroxylamine, on the contrary, although it should be detectable if present, has been reported only in a few scattered observations on plant tissues.[16] Hydroxylamine reacts rapidly and spontaneously with α-keto acids to form oximino acids such as shown below for the case of α-ketoglutaric acid. Thus if hydroxylamine is formed as an intermediate in nitrate

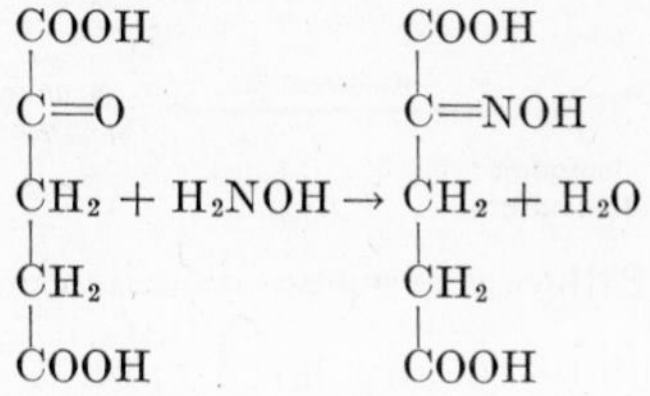

$$
\begin{array}{ccc}
\text{COOH} & & \text{COOH} \\
| & & | \\
\text{C}=\text{O} & & \text{C}=\text{NOH} \\
| & & | \\
\text{CH}_2 + \text{H}_2\text{NOH} & \rightarrow & \text{CH}_2 + \text{H}_2\text{O} \\
| & & | \\
\text{CH}_2 & & \text{CH}_2 \\
| & & | \\
\text{COOH} & & \text{COOH} \\
\alpha\text{-Ketoglutaric} & & \alpha\text{-Oximinoglutaric} \\
\text{acid} & & \text{acid}
\end{array}
$$

reduction, it could conceivably be rapidly removed by reaction with keto acids. Oximinosuccinate and oximinoglutarate are actually both able to act as good sources of nitrogen for oat plants,[17] showing that these compounds may be metabolized and hence could possibly be intermediates in nitrate reduction.

Still other observations indicate that it is entirely possible that in some cases the reduction of nitrate may proceed through other reactions than the series of steps to be expected on the basis of the inorganic chemistry of nitrate. Wheat root brei actively consumes nitrate but without the production of either nitrite or hydroxylamine.[1] In this case nitrate or nitrite could possibly combine with organic compounds to yield organic derivatives which could be the substances actually reduced. This is now known to be the case for *Neurospora* in which one mode of sulfate reduction follows a course through cysteic acid and possibly through sulfinic acid to cysteine.[18] *Neurospora* can, however, also reduce sulfate by an inorganic pathway through thiosulfate.

Although nitrate is the preferred form of nitrogen for most agriculturally important plants, other forms are also available, in particular ammonium ions, nitrite, and certain organic nitrogenous compounds. Ammonia is utilized in preference to nitrate by rice, which is apparently unable to reduce nitrate, at least in the seedling stages.[19] Ammonia is stored in large quantities only in acid plants such as rhubarb and *Begonia*, where it accumulates as ammonium salts. In less acid plants, ammonia appears to be toxic at relatively low concentrations and does

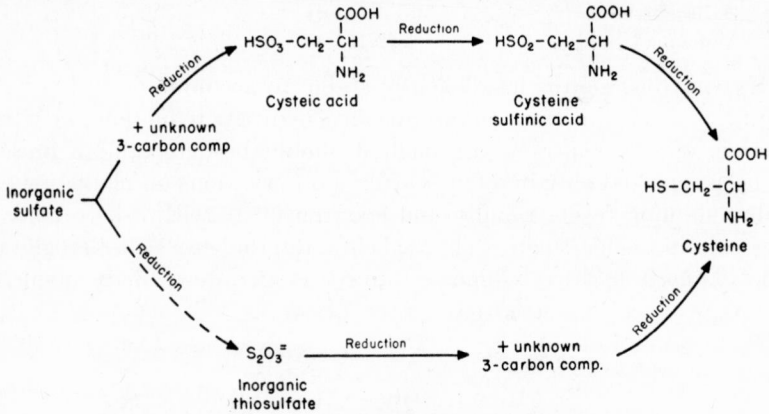

FIG. 16-1. Pathways of sulfate reduction in *Neurospora*.

not ordinarily accumulate to any large extent. Nitrite ions are also toxic to the plant, but may be utilized as a source of nitrogen if applied in relatively low concentrations such as 50 mg. or less of nitrite nitrogen per liter of nutrient.

Organic nitrogen can also be utilized by many species. Experiments designed to show utilization of organic nitrogen must be done under aseptic conditions in order to prevent breakdown of the organic nitrogen to ammonia by microorganisms. Sterile conditions have been used in investigating this problem by a number of the older workers who have shown that wheat and peas can use urea, acetamide, formamide, glycine, and aspartic acid as sole nitrogen sources, while wheat can also utilize the nitrogen of purines and amino acids.[20] Brigham in a comprehensive work[21] found that *Zea mays* can use nitrogen from aspartic and other amino acids, as well as purines and certain proteins. The utilization of protein nitrogen was improved, however, if bacteria were first allowed to grow upon solutions of the material with, presumably, the production of amino acids and ammonia. Other nitrogenous compounds, including caffein and guanidine, were not utilized by corn as nitrogen sources in Brigham's experiments.

The Entrance of Ammonia into Organic Compounds. Ammonia once produced from nitrate is rapidly transformed into organic nitrogen and in particular into amino groups. The nature of this transformation has been particularly studied from the standpoint of the amination and deamination of amino acids. A number of methods by which amino acids are deaminated in nature are known, and these will be taken up individually.

Hydrolytic deamination, in which the amino acid is converted to the corresponding hydroxy acid by addition of the elements of a molecule of H_2O, may be carried out by microorganisms.[22] This simple and chemically straightforward deamination is not, however, found in the higher plant so far as is known.

$$RCHNH_2COOH + H_2O \rightarrow RCHOHCOOH + NH_3 \tag{1}$$
Hydrolytic deamination

Reductive deamination in which the amino acid is converted to the corresponding substituted fatty acid and ammonia by addition of two atoms of hydrogen has been described as an activity of putrefactive bacteria primarily under anaerobic conditions.[22] This reaction also has not been encountered in the metabolism of higher plants.

$$RCHNH_2COOH + 2(H) \rightarrow RCH_2COOH + NH_3 \tag{2}$$
Reductive deamination

A special case of amination and deamination is constituted by aspartic acid. In the presence of aspartase, an enzyme discovered in bacteria by Quastel and Woolf,[23] aspartic acid gives up ammonia to become fumaric acid. The reaction is thus analogous to that which involves fumarase, in which the elements of a molecule of water are added to the double bond of fumaric acid to produce malic acid.

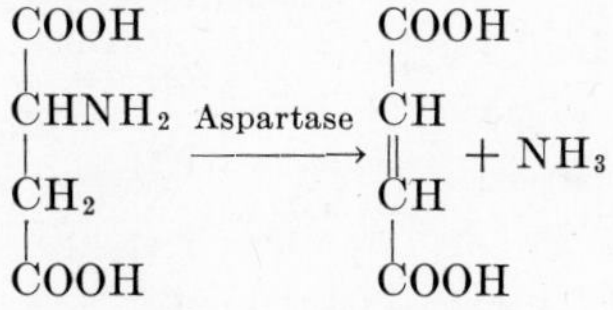

$$
\begin{array}{ccc}
\text{COOH} & & \text{COOH} \\
| & & | \\
\text{CHNH}_2 & \xrightarrow{\text{Aspartase}} & \text{CH} \\
| & & \parallel \quad + \text{NH}_3 \\
\text{CH}_2 & & \text{CH} \\
| & & | \\
\text{COOH} & & \text{COOH} \\
\text{Aspartic acid} & & \text{Fumaric acid}
\end{array} \tag{3}
$$
Deamination of aspartic acid in the presence of aspartase

In the present case the equilibrium is far over toward the side of fumaric acid so that even in the presence of fumaric acid and ammonia, only negligible concentrations of aspartic acid accumulate. Aspartase, a highly specific enzyme which will not attack other amino acids, possesses a pH optimum in the region of pH 7–7.5. The enzyme has been found in leaves of clover and of grasses, as well as in pea seedlings, and is probably widely distributed in the plant kingdom.[24] It is entirely possible that some small amount of transfer of ammonia to organic nitrogen in the plant may take place via the aspartase system. In view of the unfavorable equilibrium it is probable, however, that more ammonia is taken up into organic combination by the following reaction:

$$\underset{\underset{\text{NH}_2}{|}}{\text{RCHCOOH}} + \tfrac{1}{2}\text{O}_2 \rightleftharpoons \underset{\underset{\text{O}}{\|}}{\text{RCCOOH}} + \text{NH}_3 \tag{4}$$

Oxidative deamination

Deamination of amino acids in the plant and in other kinds of organisms takes place primarily through oxidative deaminations of the general type shown in reaction 4 in which the amino acid is oxidized to the corresponding α-keto acid with the consequent production of ammonia. The reaction has been particularly well worked out in the special case of glutamic acid. In the presence of plant or animal tissues, glutamic acid is oxidatively deaminated, and the reaction is reversible.[25] The enzyme involved is glutamic dehydrogenase, an enzyme discussed in Chapter 15, which consists of a protein apoenzyme and a dialyzable coenzyme. DPN is the coenzyme for higher plant glutamic dehydrogenase.[26] The oxidative deamination of glutamic acid proceeds in two steps, the first of which involves the enzyme.[27]

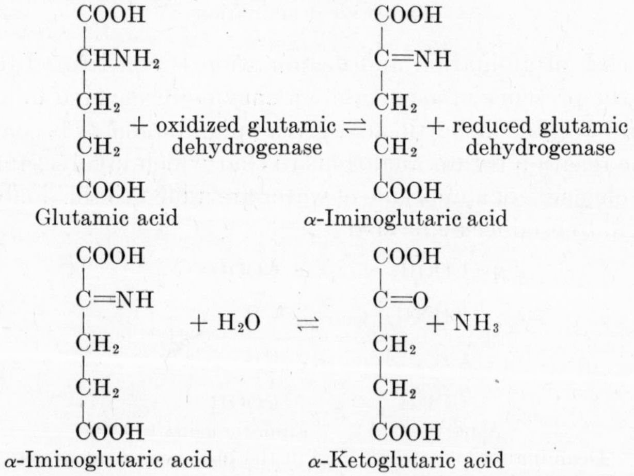

$$
\begin{array}{ccc}
\text{COOH} & & \text{COOH} \\
| & & | \\
\text{CHNH}_2 & & \text{C}=\text{NH} \\
| & & | \\
\text{CH}_2 & +\ \text{oxidized glutamic} \rightleftharpoons & \text{CH}_2 \quad +\ \text{reduced glutamic} \\
| & \text{dehydrogenase} & | \quad\ \text{dehydrogenase} \\
\text{CH}_2 & & \text{CH}_2 \\
| & & | \\
\text{COOH} & & \text{COOH} \\
\text{Glutamic acid} & & \alpha\text{-Iminoglutaric acid}
\end{array}
$$

$$
\begin{array}{ccc}
\text{COOH} & & \text{COOH} \\
| & & | \\
\text{C}=\text{NH} & & \text{C}=\text{O} \\
| & +\ \text{H}_2\text{O} \quad \rightleftharpoons & | \quad +\ \text{NH}_3 \\
\text{CH}_2 & & \text{CH}_2 \\
| & & | \\
\text{CH}_2 & & \text{CH}_2 \\
| & & | \\
\text{COOH} & & \text{COOH} \\
\alpha\text{-Iminoglutaric acid} & & \alpha\text{-Ketoglutaric acid}
\end{array}
$$

In the first step, glutamic acid is dehydrogenated by removal of two hydrogen atoms from the amino group to form α-iminoglutaric acid. In this reaction the DPN moiety of the enzyme is reduced to yield dihydro-DPN. In biological systems, the reduced coenzyme is reoxidized by still other enzyme systems with the ultimate transfer of hydrogen to oxygen of the air. The α-iminoglutaric acid spontaneously decomposes with high velocity to form α-ketoglutaric acid and ammonia. This reaction, like the first, is reversible, so that Von Euler,[27] by starting with ammonia, α-ketoglutaric acid, and reduced DPN, was able to achieve a synthesis of glutamic acid. Through the continued addition of reduced coenzyme and of α-ketoglutarate to the system continued synthesis of glutamic acid may in this way be accomplished.[28] In the cell, then, glutamic dehydrogenase establishes an equilibrium between α-ketoglutaric acid, ammonia, and glutamic acid. Reduced DPN, which is essential to the synthetic reaction, is continuously supplied by the respiratory dehydrogenations in which enzymes requiring DPN participate, as, for example, the oxidation of malic acid to oxaloacetic acid, a reaction which is an integral part of the respiratory mechanism. In the reversible oxidative deamination of glutamic acid, then, we have not only a method for the deamination of glutamic acid but also a means for the entry of ammonia into organic combination. Isotopic evidence suggests that this may be the principal port of entry of ammonia into the amino acids. Thus, when plants are supplied with ammonia containing isotopic nitrogen, more rapid and extensive replacement of nitrogen occurs in glutamic acid than in any other amino acid, suggesting that glutamic acid in the plant is indeed in a readily responsive equilibrium with ammonia.[29]

Glutamic dehydrogenase is found in a wide variety of higher plant tissues, for example in seedlings of legumes,[30] oat coleoptiles,[31] pollen,[32] spinach leaves,[33] and others, and there is reason to suppose that the enzyme may be universally distributed. A corresponding enzyme for deamination of the related aspartic acid is, however, unknown.

Oxidative deamination of a different kind takes place with the non-physiological D-isomers of amino acids as described in Chapter 15. The enzyme for this reaction has been studied in animal tissues[34] as well as in microorganisms[35] and is a flavoprotein which contains as a prosthetic group the riboflavine-phosphate-adenine dinucleotide.[36]

$$\text{RCHNH}_2\text{COOH} + \text{H}_2\text{O} + \tfrac{1}{2}\text{O}_2 \xrightarrow[\substack{\text{acid}\\\text{oxidase}}]{\text{D-Amino}} \text{RCOCOOH} + \text{NH}_3 + \text{H}_2\text{O}_2$$

Oxidative deamination of the D-amino acid to the corresponding keto acid and ammonia

In this reaction the equilibrium is far to the right, and synthesis of appreciable amounts of amino acid by this mechanism could not be achieved under conditions obtaining *in vivo*. Although the oxidase attacks many amino acids the rates at which they are oxidized differ widely, methionine, the most sensitive, being oxidized at least twenty-five times more rapidly than ornithine, one of the poorer substrates.   D-Amino acid oxidase has not been reported in higher plants although it has been specifically sought in some tissues.[33]   It is unlikely, however, that the enzyme is truly absent from higher plants, and further investigation will doubtless reveal its presence in particular species or tissues.

TABLE 16-1.   Relative rates of oxidation of various D-amino acids by D-amino acid oxidase.   (After Horowitz[35])

Rapidly attacked	Slowly attacked	Not attacked
Methionine	Glutamic acid	Serine
Phenylalanine	Isoleucine	Threonine
Norvaline	Alanine	Proline
Citrulline	Aspartic acid	β-Alanine
Arginine	Valine	Glycine
α-Aminobutyric acid	Lysine	
Leucine	Tryptophane	
	Ornithine	

The role of D-amino acid oxidase in metabolism may be in the inversion of any of the unnatural D-amino acids which may happen to be formed in the plant by racemization, by synthesis of racemic mixtures, or otherwise. Thus, a molecule of a D-amino acid could be converted by D-amino acid oxidase to the corresponding keto acid which would then have an opportunity to be transformed to the L-amino acid by transamination (see below).

Oxidases for the L-isomers of the amino acids are known in animal tissues and Edlbacher and Grauer[37] have obtained evidence for individual enzymes which oxidatively deaminate asparagine, L-glutamic acid and certain other amino acids.   A purified flavoprotein enzyme which oxidizes a variety of monoaminomonocarboxylic amino acids is also known.   This enzyme attacks the L forms of leucine, methionine, proline, norleucine, norvaline, phenylalanine, tryptophane, isoleucine, tyrosine, valine, histidine, cystine, and alanine.[38]   Although a similar enzyme is known in the microorganism *Proteus vulgaris*,[39] we have as yet no information concerning L-amino acid oxidases in higher plants.

Transamination.   Ammonia fixed as amino groups in glutamic or aspartic acids is available for the synthesis of other amino acids through the reaction of transamination.   This important biological reaction, which consists of the bodily transfer of an amino group from one compound to a second, was first discovered by the Soviet scientists A. E. Braunstein and M. G. Kritzmann in 1937.[40]   The reaction which they principally studied was that between glutamic acid and pyruvic acid to yield α-ketoglutaric acid and alanine. They showed that in the presence of an enzyme which could be obtained from muscle or from higher plants,

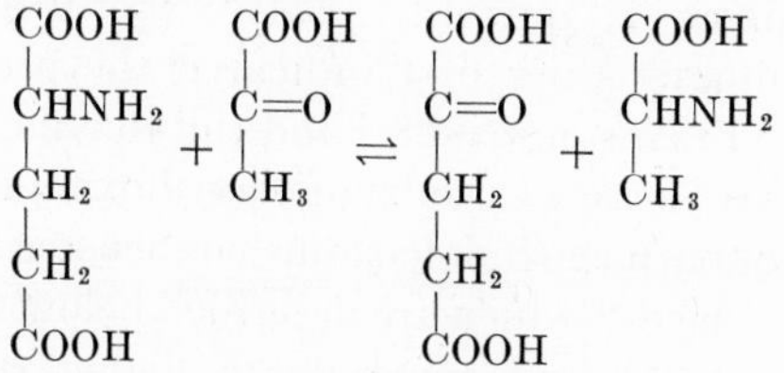

$$
\begin{array}{ccccccc}
\text{COOH} & & \text{COOH} & & \text{COOH} & & \text{COOH} \\
| & & | & & | & & | \\
\text{CHNH}_2 & & \text{C}=\text{O} & & \text{C}=\text{O} & & \text{CHNH}_2 \\
| & + & | & \rightleftharpoons & | & + & | \\
\text{CH}_2 & & \text{CH}_3 & & \text{CH}_2 & & \text{CH}_3 \\
| & & & & | & & \\
\text{CH}_2 & & & & \text{CH}_2 & & \\
| & & & & | & & \\
\text{COOH} & & & & \text{COOH} & & \\
\text{Glutamic} & \text{Pyruvic} & & \alpha\text{-Keto-} & & \text{Alanine} \\
\text{acid} & \text{acid} & & \text{glutaric} & & \\
& & & \text{acid} & & \\
\end{array}
$$

an equilibrium is established between these four substances and that starting with glutamic acid, a synthesis of alanine is possible, while the reverse reaction between alanine and α-ketoglutaric acid yields glutamic and pyruvic acids. Glutamic acid is, however, also able to participate in the transamination reaction with a variety of other keto acids. Oxaloacetate is most rapidly aminated of all the keto acids, with pyruvic acid more slowly reactive. The ratio of the two rates varies greatly in different tissues and two separable enzymes are actually involved.[41] Transfer of amino groups from aspartic acid to pyruvate can be achieved through the mediation of glutamic acid as a carrier as is shown below.

Aspartic acid + α-ketoglutaric acid $\underset{\substack{\text{Glutamic-aspartic}\\\text{transaminase}}}{\rightleftharpoons}$ Oxaloacetic acid + glutamic acid

Glutamic acid + pyruvic acid $\underset{\substack{\text{Glutamic-alanine}\\\text{transaminase}}}{\rightleftharpoons}$ α-Ketoglutaric acid + alanine

Transfer of amino group from aspartic acid to alanine through the intermediary of glutamic acid as a carrier

Glutamic acid is effective in the formation of amino acids other than aspartic acid and alanine. Thus it has been reported that preparations of pea roots can aminate α-ketoisocaproic acid at the expense of glutamic acid with the formation of leucine.[42]

L-Glutamic acid + α-ketoisocaproic acid $\rightleftharpoons$ L-Leucine + α-ketoglutaric acid

The reaction between glutamic acid and α-ketoisovaleric acid to form the amino acid valine is also known in plants.[43] It is possible that phenylalanine and α-aminobutyric acid may be synthesized by transamination reactions although the evidence for these reactions is less complete than for the others.[41] It is clear in any case that the transaminases for reactions other than the glutamic-aspartic and glutamic-alanine systems are present in plant tissues although only in low concentration. It is entirely possible that additional transaminases may ultimately be found to be present in plant tissues in still lower concentrations.

Braunstein and Kritzmann originally described transaminase as possessing a removable coenzyme and this coenzyme is now known to be a derivative of pyridoxine, vitamin B_6. Early experiments of Schlenk and co-workers with animal tissues first indicated the presence of a bound pyridoxine in purified transaminases,[44] and the enzyme has been resolved into a protein apoenzyme and a coenzyme, pyridoxal phosphate, identical with the coenzyme of both the indole-serine condensing enzyme and of the amino acid decarboxylases[45] which are described below. Large quantities of bound pyridoxal are found generally in higher plant tissues[46] and it is probable that pyridoxal phosphate is also the coenzyme of higher plant transaminase.

A mechanism for transamination was first proposed by Braunstein and Kritzmann who suggested that by a series of removals and additions of hydrogen, an intermolecular complex might be formed between the amino and keto acids and this complex then decomposed in the following manner:

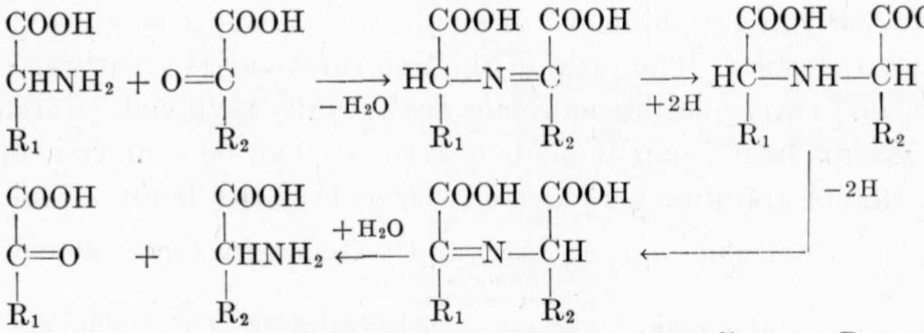

Proposed mechanism of transamination reactions according to Braunstein and Kritzmann

This general type of reaction appears to be in fact that found in transamination with the difference that the enzyme-bound pyridoxal phosphate may act as an intermediate carrier of the amino group. It is of impor-

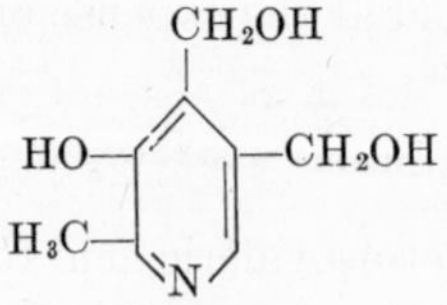

Pyridoxine

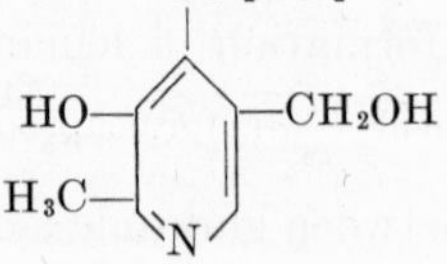

Pyridoxamine

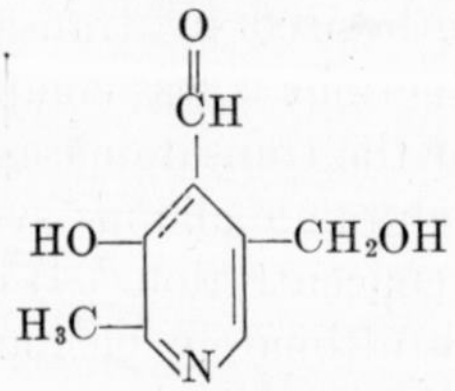

Pyridoxal

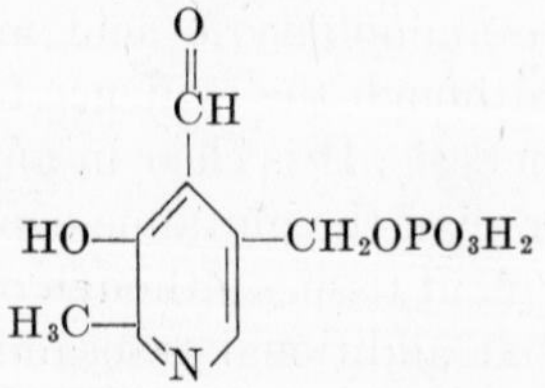

Pyridoxal phosphate

tance in this connection that in addition to pyridoxal a second pyridoxine derivative, pyridoxamine, is found in nature and is biologically effective as a vitamin. Pyridoxamine may be formed *in vitro* by heating pyridoxal with glutamic acid,[47] in which case the amino group of the glutamic acid is transferred to pyridoxal with the formation of pyridoxamine. Although purified transaminase contains a portion of its pyridoxine prosthetic group in the form of pyridoxal phosphate, a further portion is present as pyridoxamine phosphate, suggesting that the interconversion of the two forms may be of importance in the enzymatic function. The mechanism of transamination may then be as follows:

$$\begin{array}{ccc} \text{COOH} & & \text{COOH} \\ | & & | \\ \text{CHNH}_2 + \text{enzyme-pyridoxal} \rightleftharpoons \text{C}=\text{O} + \text{enzyme-pyridoxamine} \\ | & & | \\ \text{R}_1 & & \text{R}_1 \\ \text{Amino acid 1} & & \text{Keto acid 1} \end{array}$$

$$\begin{array}{ccc} & \text{COOH} & & \text{COOH} \\ & | & & | \\ \text{Enzyme-pyridoxamine} + \text{C}=\text{O} \rightleftharpoons \text{Enzyme-pyridoxal} + \text{CHNH}_2 \\ & | & & | \\ & \text{R}_2 & & \text{R}_2 \\ & \text{Keto acid 2} & & \text{Amino acid 2} \end{array}$$

Each amino group exchange may then follow the general course suggested by Braunstein and Kritzmann but with the pyridoxal-pyridoxamine system acting as an intermediate carrier.

The presence of transaminase in higher plants was first described by Braunstein and Kritzmann for seedlings of pea, lupin, and pumpkin.[40] Adler and others reported the occurrence of transaminases in corn, barley, peas, and clover.[48] Virtanen and Laine[42] have studied the transamination reaction in mashes made from pea seedlings and have shown that both glutamic and aspartic acids can react with pyruvate to form alanine, aspartate presumably indirectly and through glutamic acid as a carrier. The corresponding amides glutamine and asparagine do not appear to be able to act as substrates in transamination. Transaminase activity in the developing oat parallels the protein synthetic activity of the growing tissue.[49] A comprehensive survey of transaminase in a wide variety of plant tissues by Leonard and Burris,[50] from which Table 16-2 is taken, shows that in general transaminase activity is highest in roots and in germinating seeds, and is lowest in leaves. Transaminase activity was found to be entirely absent from fruits of apple and tomato. The transaminases of the pea plant have been subjected to study by Rautanen,[51] who found the enzyme in all tissues investigated.

There can be no doubt then but that transaminases are widely distributed in higher plant tissues. The transfer of ammonia from glutamic acid may result in the synthesis of a variety of the other amino acids. The nature of all the amino acids which are formed, directly or indirectly, through the transamination reaction is as yet unknown, but it is evident that transamination does constitute an important link in amino acid synthesis.

TABLE 16-2. Relative rates of glutamic-aspartic transamination reaction in various tissues. (After Leonard and Burris[50]) $Q_T(N)$ = microliters substrate transaminated per hour per milligram N in enzyme preparation

Plant	Part	$Q_T(N)$
Potato	Root	146
	Stem	102
	Leaf	29
Tomato	Root	111
	Leaf	35
Squash	Leaf	45
Barley	Leaf	44
Corn	Germinating seed	116
Pea	Germinating seed	84
Oat	Germinating seed	119

Special Cases of Amino Acid Synthesis. *Arginine.* The synthesis of arginine in animal tissues is known to take place from ornithine according to the steps first proposed by Krebs.[52] Ornithine by addition of CO_2 and two molecules of ammonia becomes successively citrulline and arginine. The arginine in the presence of the enzyme arginase breaks down with the liberation of urea and the reformation of ornithine. This cycle, which is responsible for both urea and arginine synthesis in the animal, has also been studied in detail by Srb and Horowitz[53] in *Neurospora.*

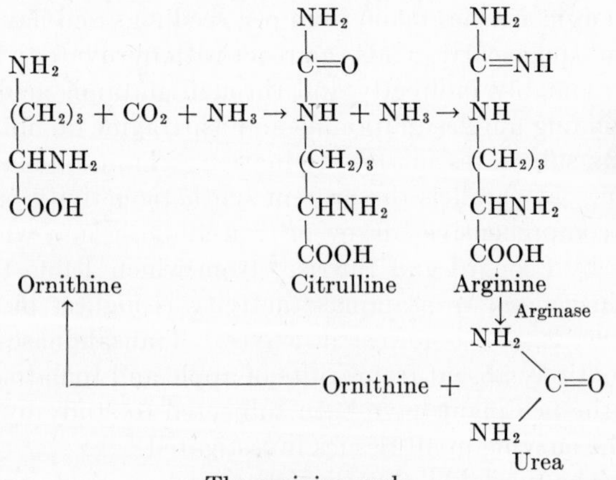

The arginine cycle

Synthesis of arginine from ornithine in *Neurospora* takes place only in intact cells and has not been accomplished in the cell-free tissue extracts which have been prepared from mammalian tissue. Srb and Horowitz were, however, able to separate the reactions involved with mutant strains in which the production of ornithine, of citrulline from ornithine, or of arginine from citrulline was specifically blocked, presumably by the absence of the specific enzyme involved. The enzyme arginase, which splits arginine to ornithine and urea, is present in *Neurospora* as in mammals, and is also found in higher plants as in seeds of the jack bean (*Canavalia ensiformis*).[54] Other evidence for the operation of this mechanism of arginine synthesis in the higher plants is, however, lacking. The enzyme urease, which attacks urea with the liberation of ammonia, is present in all the tissues of the soybean and of the jack bean and probably in other plants as well. Any urea liberated by the action of arginase would hence be expected to be rapidly converted to NH_3.[66] Further relations of ornithine to the amino acids proline and glutamic acid are known from work with animal tissues as well as from investigations with *Neurospora* mutants and similar mutants of *Penicillium*. In *Neurospora* mutant strains have been obtained which can satisfy their arginine requirements with proline, while in *Penicillium*, a particular arginine-requiring mutant can utilize any of the substances arginine, citrulline, ornithine, proline, or glutamic acid.[55] It would seem, thus, that glutamic acid may be a precursor of all these substances. Since a strain is known which requires proline but in which arginine synthesis is not blocked, it is probable that proline is not in the direct chain of reactions leading to arginine, but rather shares a common precursor with ornithine. α-Amino-δ-hydroxy-valeric acid appears to be an intermediate in the conversion of glutamic acid since it is also utilized by the glutamic acid-utilizing strain. These relations are summarized below.

$$
\begin{array}{ccc}
\text{COOH} & \text{CH}_2\text{OH} & \text{COOH} \\
| & | & | \\
\text{CH}_2 & \text{CH}_2 & \text{CHNH}_2 \\
| & | & | \\
\text{CH}_2 \longrightarrow & \text{CH}_2 \longrightarrow & \text{CH}_2 \\
| & | & | \\
\text{CHNH}_2 & \text{CHNH}_2 & \text{CH}_2 \\
| & | & | \\
\text{COOH} & \text{COOH} & \text{CH}_2\text{NH}_2
\end{array}
$$

Glutamic acid α-Amino-δ-hydroxy-valeric acid   Ornithine

$$
\begin{array}{c}
\text{CH}_2\text{---CH}_2 \\
| \quad\quad | \\
\text{CH}_2 \quad \text{CHCOOH} \\
\diagdown \quad \diagup \\
\text{NH}
\end{array}
$$

Proline

Steps involved in the synthesis of proline and ornithine in *Neurospora* and *Penicillium*

Tryptophane. The course of the synthesis of tryptophane in *Neurospora* has been shown to be through condensation of indole and serine,[56] and in fact *Neurospora* mycelium incubated with the two components may yield as much as 45% of the theoretical amount of crystalline tryptophane. This reaction will also proceed *in vitro* in the presence of ground mycelial mat of *Neurospora* provided that pyridoxal phosphate is added as coenzyme.[57] Some information is available also on the course of synthesis of the indole nucleus since a mutant strain of *Neurospora* normally unable to form the indole nucleus was found to be able to form indole when supplied with anthranilic acid. The synthesis of tryptophane may then approximate the course shown below.

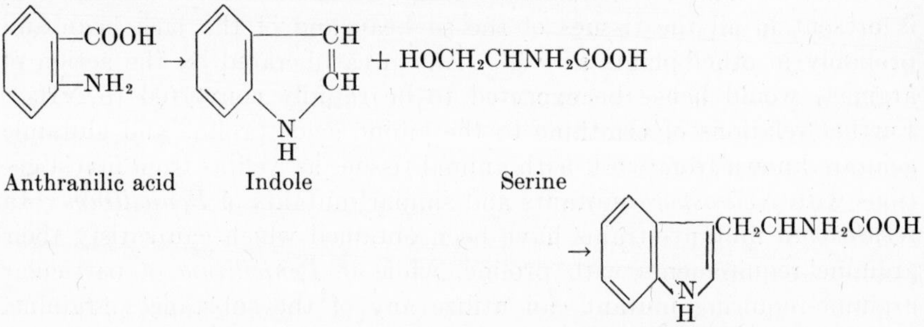

$$\text{Anthranilic acid (—COOH, —NH}_2\text{)} \rightarrow \text{Indole} + \text{HOCH}_2\text{CHNH}_2\text{COOH}$$

Anthranilic acid   Indole   Serine

$$\rightarrow \text{Indole—CH}_2\text{CHNH}_2\text{COOH}$$

Tryptophane

Suggested course of synthesis of tryptophane in *Neurospora*

Wildman, Ferri, and Bonner[58] have obtained evidence that in the spinach leaf tryptophane may be produced from indole and serine, since when intact leaves are infiltrated with a solution of these compounds tryptophane and its metabolic product indoleacetic acid appear (Chapter 29). It is possible then that tryptophane synthesis in the higher plant may follow the pathway determined for *Neurospora*.

Methionine. A special case of amino acid synthesis is constituted by the sulfur-containing amino acids cysteine and methionine. A series

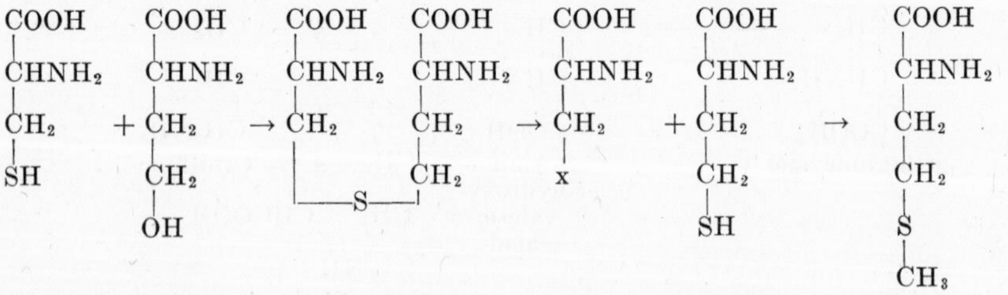

$$
\begin{array}{c}
\text{COOH} \\
\text{CHNH}_2 \\
\text{CH}_2 \\
\text{SH}
\end{array}
+
\begin{array}{c}
\text{COOH} \\
\text{CHNH}_2 \\
\text{CH}_2 \\
\text{CH}_2 \\
\text{OH}
\end{array}
\rightarrow
\begin{array}{c}
\text{COOH} \\
\text{CHNH}_2 \\
\text{CH}_2 \\
\end{array}
\begin{array}{c}
\text{COOH} \\
\text{CHNH}_2 \\
\text{CH}_2 \\
\text{CH}_2
\end{array}
\text{S}
\rightarrow
\begin{array}{c}
\text{COOH} \\
\text{CHNH}_2 \\
\text{CH}_2 \\
\text{x}
\end{array}
+
\begin{array}{c}
\text{COOH} \\
\text{CHNH}_2 \\
\text{CH}_2 \\
\text{CH}_2 \\
\text{SH}
\end{array}
\rightarrow
\begin{array}{c}
\text{COOH} \\
\text{CHNH}_2 \\
\text{CH}_2 \\
\text{CH}_2 \\
\text{S} \\
\text{CH}_3
\end{array}
$$

Cysteine   Homo-   Cystathionine   Unknown   Homo-   Methionine
            serine   cysteine

Biogenesis of methionine from cysteine and homoserine in *Neurospora* of steps in the synthesis of methionine have been worked out by Horowitz and co-workers,[59] again on the basis of a series of mutant strains of *Neurospora* which require methionine or its precursors. Normal *Neurospora* is able to synthesize methionine. The mutant strains in question are ones which cannot carry out the synthesis but which grow only if supplied with methionine. The strains may be subdivided on the basis that a portion of them are able to grow and to produce methionine when supplied with cysteine. Teas, Horowitz, and Fling[60] have shown that cysteine combines with the four-carbon compound homoserine to form the amino acid cystathionine which was isolated and identified from a mutant unable to further convert cystathionine. In the organism, cystathionine is then split to yield homocysteine and an unidentified three-carbon product possibly related to serine. Homocysteine is in turn methylated, presumably by agents such as choline or betaine, to yield methionine. These steps of methionine synthesis are probably reversible since mutants unable to make either cysteine or methionine can synthesize cysteine in the presence of methionine. Methionine is then derived from cysteine and homoserine. Steps involved in the synthesis of cysteine from inorganic sulfate in *Neurospora* have been summarized under *nitrate reduction* above.

Methionine and biological methylation. The methyl group of methionine may be introduced by methylation of homocysteine at the expense of choline, a process known as transmethylation. The reaction which has

$$CH_3—\overset{\overset{\textstyle CH_3}{|}}{\underset{\underset{\textstyle CH_3}{|}}{N^+}}—CH_2CH_2OH$$

Structure of choline ,the trimethyl quaternary nitrogen derivative of aminoethanol

been studied in detail only in *Neurospora*[59] and in mammalian tissues[67] consists in the bodily transfer of the methyl group from a methyl donor, choline in this case, to an acceptor such as homocysteine. Thus a mixture of choline and homocysteine can replace methionine in the diet of the rat for which methionine is an essential amino acid. In the mammal the reaction ordinarily proceeds in the reverse direction, however, synthesis of choline being achieved by the methylation of dimethylaminoethanol at the expense of the methyl group of dietary methionine. Other methyl donors can replace methionine in the synthesis of choline by the rat, the most effective being betaine, the trimethyl quaternary ammonium derivative of glycine, which like choline is also a plant product. The animal then lacks the ability to form labile methyl groups and must depend on an exogenous supply.

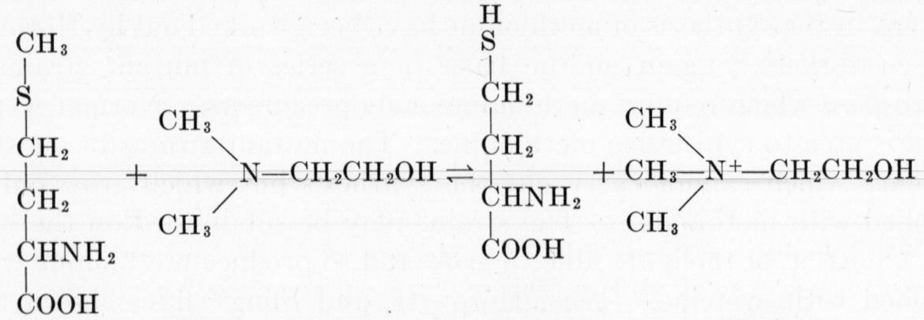

$$CH_3\text{-}S\text{-}CH_2\text{-}CH_2\text{-}CHNH_2\text{-}COOH \;+\; \underset{CH_3}{\overset{CH_3}{N}}\text{-}CH_2CH_2OH \;\rightleftharpoons\; H\text{-}S\text{-}CH_2\text{-}CH_2\text{-}CHNH_2\text{-}COOH \;+\; CH_3\text{-}\underset{CH_3}{\overset{CH_3}{N^+}}\text{-}CH_2CH_2OH$$

Methionine   Dimethylaminoethanol   Homo-cysteine   Choline

Overall course of the reversible transmethylation reaction between methionine and choline

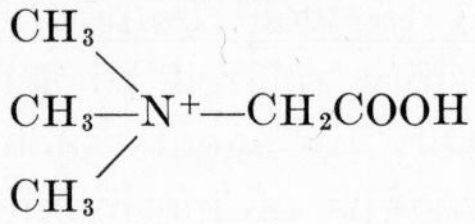

$$CH_3\text{-}\underset{CH_3}{\overset{CH_3}{N^+}}\text{-}CH_2COOH$$

Betaine, the trimethyl quaternary ammonium derivative of the amino acid glycine

Plants including *Neurospora* are able to synthesize labile methyl which is used not only in the synthesis of methionine but probably also in the synthesis of the wide variety of other methylated materials which occur in plants. The basic process by which labile methyl is generated in plants is unknown but several of the steps leading to the synthesis of choline are known from investigations of mutant strains of *Neurospora*.[68] In this organism choline synthesis is carried out by the stepwise methylation of aminoethanol. The source of the methyl groups involved in the process is unknown. It will be evident that a full investigation of the

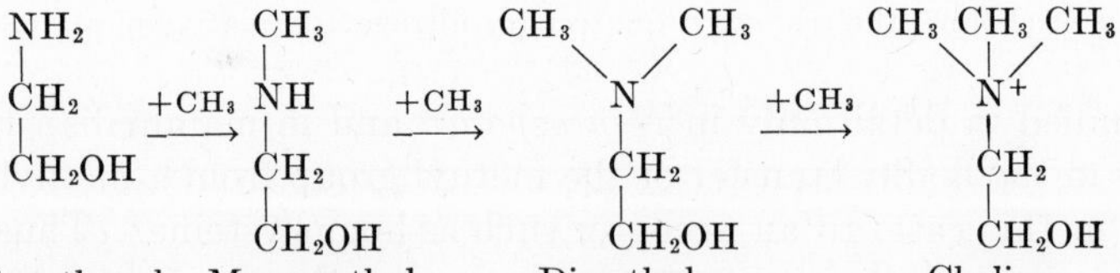

$$NH_2\text{-}CH_2\text{-}CH_2OH \quad \overset{+CH_3}{\longrightarrow} \quad \underset{CH_2OH}{\overset{CH_3}{NH}\text{-}CH_2} \quad \overset{+CH_3}{\longrightarrow} \quad \underset{CH_2OH}{\overset{CH_3\;\;CH_3}{N}\text{-}CH_2} \quad \overset{+CH_3}{\longrightarrow} \quad \underset{CH_2OH}{\overset{CH_3\;CH_3\;CH_3}{N^+}\text{-}CH_2}$$

Aminoethanol   Monomethyl-aminoethanol   Dimethyl-aminoethanol   Choline

Stepwise methylation of aminoethanol in the production of choline in *Neurospora*

mode of formation of labile methyl as well as of transmethylating systems in plants would be of interest not only from the standpoint of general plant biochemistry but also from the standpoint of the labile methyl nutrition of the animal organism.

Threonine. In *Neurospora*, threonine is synthesized from homoserine, as is the related α-aminobutyric acid.[59]

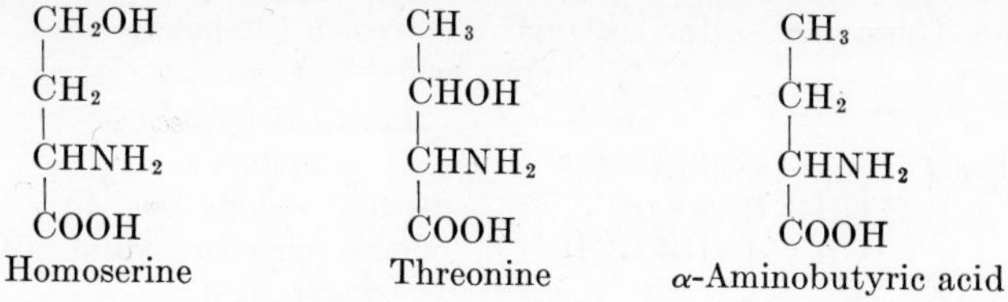

$$CH_2OH \quad CH_3 \quad CH_3$$
$$CH_2 \quad CHOH \quad CH_2$$
$$CHNH_2 \quad CHNH_2 \quad CHNH_2$$
$$COOH \quad COOH \quad COOH$$

Homoserine   Threonine α-Aminobutyric acid

The mechanism of α-aminobutyric acid synthesis may involve oxidation to α-ketobutyric acid and subsequent amination since, as pointed out earlier, this compound is subject to a transamination reaction with glutamic acid.

Phenylalanine and tyrosine.   Early work with transaminase of the pea plant indicated that phenylalanine may arise from phenylpyruvic acid.[42] In the animal, tyrosine is formed from phenylalanine by oxidation of the aromatic nucleus.

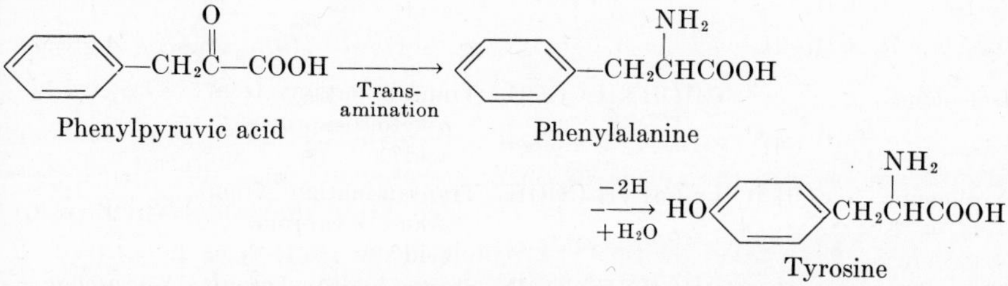

Phenylpyruvic acid   Phenylalanine

Tyrosine

Lysine.   Lysine is apparently formed from α-aminoadipic acid in *Neurospora*,[61] a conversion which requires introduction of an amino group and a reduction.

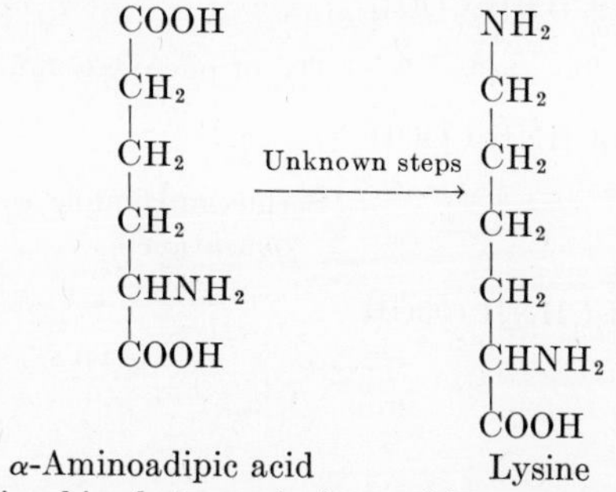

α-Aminoadipic acid   Lysine

Structural relationships between lysine and its precursor, α-aminoadipic acid

Glycine.   Glycine is produced in the animal body by removal of the β-carbon atom from serine by an unknown mechanism.[62]

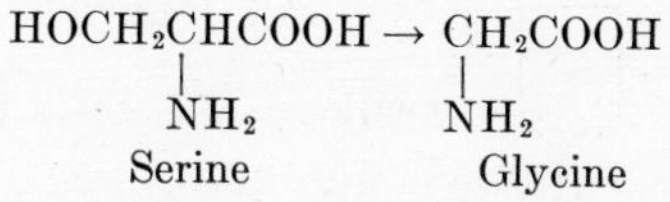

$$HOCH_2CHCOOH \rightarrow CH_2COOH$$
$$NH_2 \qquad NH_2$$

Serine   Glycine

TABLE 16-3. Summary of the reactions involved in the biogenesis of the principal amino acids

Amino acid	Structure	Reactions involved in synthesis	Organisms studied
Glycine	CH_2NH_2COOH	Serine $\longrightarrow$ glycine	Rat
Alanine	CH_3CHNH_2COOH	Transamination from pyruvic acid	General
Serine	$HOCH_2CHNH_2COOH$	Unknown	. . . .
Homoserine	$HOCH_2CH_2CHNH_2COOH$	Unknown	. . . .
Valine	$\begin{matrix} CH_3 \\ \diagdown \\ CHCHNH_2COOH \\ \diagup \\ CH_3 \end{matrix}$	Transamination from α-ketoisovaleric acid	Neurospora, green plants
Leucine	$\begin{matrix} CH_3 \\ \diagdown \\ CHCH_2CHNH_2COOH \\ \diagup \\ CH_3 \end{matrix}$	Transamination from α-ketoisocaproic acid	Rat, Neurospora, pea
Isoleucine	$\begin{matrix} CH_3CH_2 \\ \diagdown \\ CHCHNH_2COOH \\ \diagup \\ CH_3 \end{matrix}$	Transamination from α-ketoisocaproic acid?	. . . .
Norleucine	$CH_3(CH_2)_3CHNH_2COOH$	Transamination from α-keto-n-caproic acid?	. . . .
Threonine	$CH_3CHOHCHNH_2COOH$	Shift of hydroxyl group in homoserine	Neurospora
Phenylalanine	 ⬡—CH_2CHNH_2COOH	Transamination from phenylpyruvic acid	Peas
Tyrosine	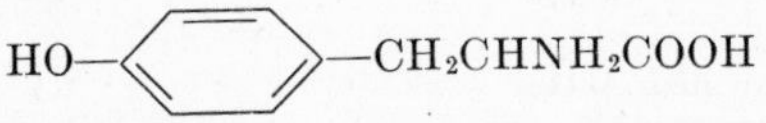 HO—⬡—CH_2CHNH_2COOH	Ox. of phenylalanine	Rat
Tryptophane	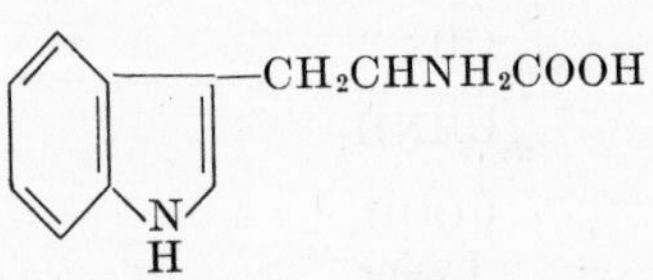 (indole)—CH_2CHNH_2COOH	Serine and indole condensation	Neurospora, spinach leaf
Proline	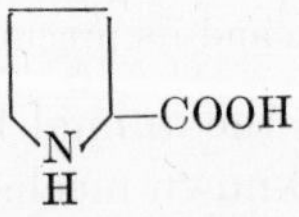 (pyrrolidine)—$COOH$	From glutamic acid	Penicillium
Hydroxyproline	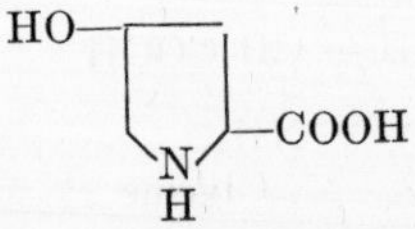 HO—(pyrrolidine)—$COOH$	Ox. of proline?	. . . .
Cystine	$(-SCH_2CHNH_2COOH)_2$	Ox. of cysteine	General

TABLE 16-3 (*Continued*)

Amino acid	Structure	Reactions involved in synthesis	Organisms studied
Cysteine	$HSCH_2CHNH_2COOH$	Reduction of cysteic acid; from a serine-like percursor; reverse of methionine synthesis	*Neurospora*, mammals
Methionine	$CH_3SCH_2CH_2CHNH_2COOH$	Methylation of homocysteine (this from homoserine)	*Neurospora*
Aspartic	$COOHCH_2CHNH_2COOH$	Transamination from oxaloacetic acid	General
Glutamic	$COOHCHNH_2CH_2CH_2COOH$	Introduction of ammonia into α-ketoglutaric acid	General
Lysine	$NH_2CH_2(CH_2)_3CHNH_2COOH$	From α-aminoadipic acid	*Neurospora*
Arginine	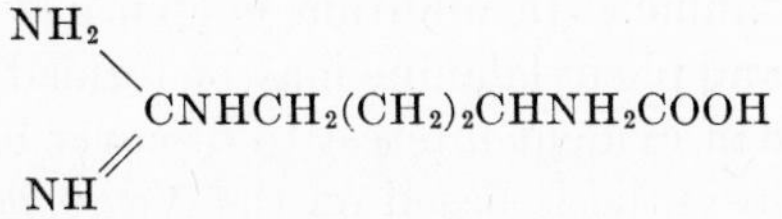	Imidation of citrulline	Mammals, *Neurospora*

$$NH_2$$
$$\backslash$$
$$CNHCH_2(CH_2)_2CHNH_2COOH$$
$$/\!/$$
$$NH$$

Histidine	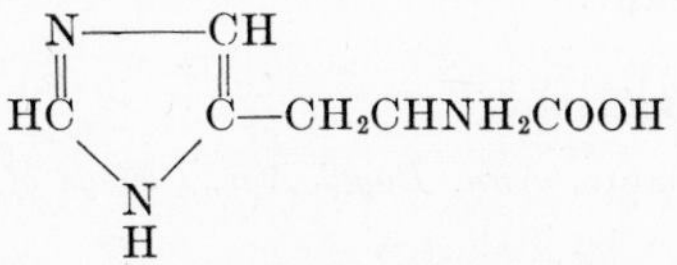	Unknown	. . . .

$$N———CH$$
$$HC\quad\quad C—CH_2CHNH_2COOH$$
$$N$$
$$H$$

Aminobutyric	$CH_3CH_2CHNH_2COOH$	Transamination from α-ketobutyric acid	Rat, pig
Citrulline	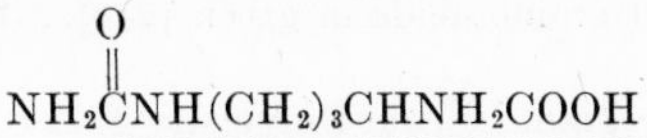 O	Amination of ornithine	Rat, *Neurospora*
	$NH_2CNH(CH_2)_3CHNH_2COOH$		
Ornithine	$NH_2(CH_2)_3CHNH_2COOH$	From glutamic acid via common proline precursor	*Neurospora*, *Penicillium*

Amino Acid Decarboxylation.[63] Many microorganisms, including *Streptococcus* species, *Rhizobium, E. coli*, and others, possess enzymes for the decarboxylation of amino acids to the corresponding amines. The reactions carried out by these enzymes are of the following type:

$$RCHNH_2COOH \xrightarrow{\text{decarboxylase}} RCH_2NH_2 + CO_2$$

Type reaction carried out by amino acid decarboxylase

Amino acid decarboxylase can be separated into a protein apoenzyme and a coenzyme which is identical with and can be replaced by synthetic pyridoxal phosphate. Pyridoxal phosphate appears to be the coenzyme for a whole group of amino acid decarboxylases, including the enzymes for decarboxylation of histidine, ornithine, tyrosine, lysine, arginine, and glutamic acid. The enzyme for decarboxylation of glutamic acid has been shown to be present in bean seedlings,[64] carrot roots, cabbage leaves, and pea seed, and pyridoxal phosphate is the coenzyme for this enzyme just as with the bacterial enzymes.[65] In bacteria, amino acid decarboxylases are produced in response to growth in an acid environment and represent a mechanism for regulation of the environmental acidity by the production of the basic amines. It will be of interest to determine to what extent the enzymes fill the same function in the higher plant.

Conclusions. Elucidation of the mechanisms of amino acid formation has been principally achieved by the study of mutant strains of *Neurospora* or other microorganisms and by the study of amino acid biogenesis in animal tissues. The general course of synthesis of each amino acid suggested by this work is indicated in Table 16-3. Of the twenty-five amino acids listed, something can be surmised about the biogenesis of at least nineteen. With special reference to higher plants, however, we have conclusive information as to the mode of origin of only three amino acids, glutamic acid, aspartic acid, and alanine. In addition evidence is at hand indicating that valine, leucine, and phenylalanine may be formed by transamination reactions. It will be of evident interest to discover how far the mechanisms of amino acid biosynthesis based on the *Neurospora* work actually apply to the higher plant.

General References

Burström, H., The nitrate nutrition of plants, *Ann. Royal Agr. College of Sweden*, **13**, 1 (1946).

Braunstein, A. E., Advances in Protein Chemistry. Academic Press, 1947. Vol. 3, p. 1.

Rautanen, N., On the synthesis of the first amino acids in green plants, *Ann. Acad. Sci. Finland*, Series **A:II**, No. 33, 1948.

Chibnall, A. C., Protein Metabolism in the Plant. Yale University Press, 1939.

Horowitz, N. H., Advances in Genetics. Academic Press, 1950. Vol. 3.

McKee, H. S., *New Phytologist*, **48**, 1 (1949).

References

1. Burström, H., *Ann. Royal Agr. College of Sweden*, **13**, 1 (1946).
2. Davidson, O. W., and Shive, J. W., *Soil Sci.*, **37**, 357 (1934).
3. Marthaler, H., *1937, Jahrb. Wiss. Bot.*, **85**, 76 (1937).
4. Thomas, W., *Science*, **66**, 115 (1927).
5. Nightingale, G. T., and Robbins, W. R., New Jersey Agr. Expt. Sta. Bull. 472, 1928.
   Nightingale, G. T., and Schermerhorn, L. G., New Jersey Agr. Expt. Sta. Bull. 476, 1928.
6. Burström, H., *Ann. Royal Agr. College of Sweden*, **6**, 1 (1937).
7. Bonner, J., and Bonner, H., Vitamins and Hormones. Academic Press, 1948. Vol. VI, p. 225.
8. Hamner, K. C., *Botan. Gaz.*, **97**, 744 (1935).

9. Warburg, O., and Negelein, E., *Biochem. Z.*, **110**, 66 (1920).
10. Burström, H., *Ann. Royal Agr. College of Sweden*, **11**, 1 (1943).
11. Lease, E. J., and Tottingham, W. E., *J. Am. Chem. Soc.*, **57**, 2613 (1935).
12. Eckerson, S., *Botan. Gaz.*, **77**, 377 (1924). See also *Contrib. Boyce Thomp. Inst.*, **3**, 405 (1931); **4**, 119 (1932).
13. Somner, A. L., *Plant Physiol.*, **11**, 429 (1936).
    Arreguin, B., Unpublished. California Institute of Technology, 1943. See however,
    Dittrich, W., *Planta*, **12**, 69 (1930).
14. Stickland, L. H., *Biochem. J.*, **25**, 1543 (1931).
15. Michlin, D., *Biochem. Z.*, **202**, 329 (1928).
    Michlin, D., and Severin, B., *Biochem. Z.*, **237**, 339 (1931).
16. Lemoigne, M., Monguillon, P., and Desveaux, R., *Bull. soc. chim. biol.*, **19**, 671 (1937); *Compt. rend.*, **204**, 1841 (1937).
17. Wood, J. G., and Hone, M. R., *Australian J. Sci.*, **1**, 163 (1938).
18. Horowitz, N. H., Fling, M., Phinney, B. O., and Shen, S., Abstract, Div. Biol. Chem., San Francisco Meeting ACS, p. 45c, 1949.
    Phinney, B. O., *Genetics*, **33**, 624 (1948) (abstract).
19. Bonner, J., *Botan. Gaz.*, **108**, 267 (1946).
20. Schreiner, O., and Skinner, J. J., U. S. Bur. Soils Bull. 87, 1912.
21. Brigham, R. O., *Soil Sci.*, **3**, 155 (1917).
    Hutchinson, H. B., and Miller, N. H. J., *Centr. Bakt. II Abt.*, **30**, 513 (1911).
22. Stephenson, M., Bacterial Metabolism. Longmans, Green and Co., 3rd ed., 1949.
23. Quastel, J. H., and Woolf, B., *Biochem. J.*, **20**, 545 (1926).
24. Virtanen, A. L., and Tarnanen, J., *Biochem. Z.*, **250**, 193 (1932).
25. Thunberg, T., *Biochem. Z.*, **206**, 109 (1929).
    Andersson, B., *Z. physiol. Chem.*, **217**, 186 (1933).
26. Von Euler, H., and Adler, E., *Enzymologia*, **7**, 21 (1939).
27. Von Euler, H., Adler, E., Gunther, G., and Das, N. B., *Z. physiol. Chem.*, **254**, 61 (1938).
28. Adler, E., Von Euler, H., Gunther, G., and Plass, M., *Biochem. J.*, **33**, 1028 (1939).
29. Vickery, H. B., Pucher, G. W., Schoenheimer, R., and Rittenberg, D., *J. Biol. Chem.*, **135**, 531 (1940).
    MacVicar, R., and Burris, R. H., *J. Biol. Chem.*, **176**, 511 (1948).
30. Damodaran, M., and Nair, K. R., *Biochem. J.*, **32**, 1064 (1938).
31. Berger, J., and Avery, G. S., *Am. J. Botany*, **30**, 290 (1943); **31**, 11 (1944).
32. Okunuki, K., *Acta Phytochim.*, **11**, 65 (1939).
33. Bonner, J., and Wildman, S. G., *Arch. Biochem.*, **10**, 497 (1946).
34. Krebs, H. A., *Z. physiol. Chem.*, **217**, 191 (1933), *Biochem. J.*, **29**, 1620 (1935).
35. Horowitz, N. H., *J. Biol. Chem.*, **154**, 141 (1944).
36. Warburg, O., and Christian, W., *Biochem. Z.*, **295**, 261; **296**, 294; **298**, 150 (1938).
37. Edlbacher, S., and Grauer, H., *Helv. Chim. Acta*, **27**, 151, 928 (1944).
38. Blanchard, M., Green, D. E., Nocito, V., and Ratner, S., *J. Biol. Chem.*, **155**, 421 (1944); **161**, 583 (1945).
    Ratner, S., Nocito, V., and Green, D. E., *J. Biol. Chem.*, **152**, 119 (1944).
39. Stumpf, P. K., and Green, D. E., *J. Biol. Chem.*, **153**, 387 (1944).
40. Braunstein, A. E., and Kritzmann, M. G., *Enzymologia*, **2**, 129 (1937).
    See reviews in Kritzmann, *J. Biol. Chem.*, **167**, 77 (1947).
    Braunstein, A. E., Advances in Protein Chemistry. Academic Press, 1947. Vol. 3, p. 1.
41. Green, D. E., Leloir, L. F., and Nocito, V., *J. Biol. Chem.*, **161**, 559 (1945).

42. Virtanen, A. I., and Laine, T., *Biochem. J.*, **33**, 412 (1939); *Biochem. Z.*, **308**, 213 (1941).
43. Rautanen, N., *J. Biol. Chem.*, **163**, 687 (1946).
44. Schlenk, F., and Fisher, A., *Arch. Biochem.*, **8**, 337 (1945).
    Schlenk, F., and Snell, E. E., *J. Biol. Chem.*, **157**, 425 (1945).
45. O'Kane, D. E., and Gunsalus, I. C., *J. Biol. Chem.*, **170**, 425 (1947).
    Schlenk, F., and Fisher, A., *Arch Biochem.*, **12**, 69 (1947).
46. Rabinowitz, J. C., and Snell, E. E., *J. Biol. Chem.*, **176**, 1157 (1948).
47. Snell, E. E., *J. Am. Chem. Soc.*, **67**, 194 (1945).
48. Adler, E., Gunther, G., and Everett, J., *Z. physiol. Chem.*, **255**, 27 (1938).
49. Albaum, H. G., and Cohen, P. P., *J. Biol. Chem.*, **149**, 19 (1943).
50. Leonard, M., and Burris, R. H., *J. Biol. Chem.*, **170**, 701 (1947).
51. Rautanen, N., *Ann. Acad. Sci. Finland*, **A:II**, No. 33, 1948.
52. Krebs, H. A., and Henseleit, K., *Z. physiol. Chem.*, **210**, 33 (1932).
53. Srb, A., and Horowitz, N. H., *J. Biol. Chem.*, **154**, 129 (1944).
54. Damodaran, M., and Narayanan, K., *Biochem. J.*, **34**, 1449 (1940).
55. Bonner, D., *Am. J. Botany*, **33**, 788 (1946).
    Srb, A., Thesis, Stanford University, 1946.
    Fincham, J., *J. Biol. Chem.*, **182**, 61 (1949).
56. Tatum, E. L., and Bonner, D., *Proc. Natl. Acad. Sci. U.S.*, **30**, 30 (1944).
    Tatum, E. L., Bonner, D., and Beadle, G. W., *Arch. Biochem.*, **3**, 477 (1943).
57. Umbreit, W. W., Wood, W. A., and Gunsalus, I. C., *J. Biol. Chem.*, **165**, 731 (1946).
58. Wildman, S. G., Ferri, M., and Bonner, J., *Arch. Biochem.*, **13**, 131 (1947).
59. Horowitz, N. H., Advances in Genetics. Academic Press, 1950.
60. Teas, H., Horowitz, N. H., and Fling, M., *J. Biol. Chem.*, **172**, 651 (1948).
61. Houlahan, M. B., and Mitchell, H. K., *Proc. Natl. Acad. Sci. U.S.*, **34**, 465 (1948).
62. Shemin, D., *J. Biol. Chem.*, **162**, 297 (1946).
63. Gale, E. F., Advances in Enzymology. Interscience Publishers, 1946. Vol. 6, p. 1.
64. Okunuki, K., *Botan. Mag., Tokio*, **51**, 27 (1927).
65. Schales, O., Mims, V., and Schales, S. S., *Arch. Biochem.*, **10**, 455 (1946).
66. Granick, S., *Plant Physiol.*, **12**, 471 (1937); **13**, 29 (1938).
67. du Vigneaud, V., Harvey Lectures Series, **38**, 39 (1942).
    Jukes, T. H., *Ann. Rev. Biochem.*, **16**, 193 (1947).
68. Horowitz, N. H., Bonner, D., and Houlahan, M. B., *J. Biol. Chem.*, **159**, 145 (1945).
    Horowitz, N. H., *J. Biol. Chem.*, **162**, 413 (1946).

Chapter 17

THE PLANT PROTEINS

General. Although the proteins are constituted of amino acids linked in long chains through peptide bonds, they possess as a class new properties both physical and biological which are not characteristic of their amino acid building blocks. The physical characteristics of the proteins are due in part at least to the large size of protein molecules, together with their amphoteric properties. The unique biological properties of the proteins, properties of specificity, are probably due to the great opportunity for unique arrangements of amino acid residues in the molecule and on the surface of the molecule which the protein structure affords.[1]

The most characteristic reaction of the native protein is denaturation, whereby the soluble protein is altered in such a way as to become insoluble.[2] This change, which may be caused by exposure to high temperatures, generally 40–75°C, strongly acid or alkaline solutions, ultraviolet irradiation, precipitation by heavy metal ions, and by many other agents, is ordinarily attended by loss of all biological activity, such as enzymatic properties and immunological specificity. Denaturation consists then of alterations in the shape and arrangement of the protein molecule. In the preparation of proteins from tissues it is necessary to guard against denaturation since this, if it occurs, may lead to loss of enzymatic or other properties of interest. In the preparation of proteins from plant tissues such as leaves, stems, or roots, denaturation is readily brought about but may be kept at a minimum if all manipulations such as grinding of the tissue, centrifugation, filtration, and dialysis are rapidly carried out at low temperatures, 0°C or lower. It is typical of the denaturation process that while the rate of the reaction is very low at low temperatures, and has but a low temperature coefficient, the rate rises rapidly and is attended by an extraordinarily high temperature coefficient at higher temperatures.

Denaturation is associated with (*a*) alterations in the crystallographic arrangement of the peptide chains of the molecule[3] and (*b*) the appearance of free sulfhydryl groups not present before denaturation.[4] These two changes are probably related in that the sulfhydryl groups which appear during denaturation were presumably previously bound in some nonreactive form, possibly as —S-S— groups linking adjacent peptide chains, and are exposed during reorientation or "unfolding" of the chains in denaturation.

Proteins like amino acids are amphoteric, i.e., contain both acidic carboxyl groups and basic groups. Each protein has then its own characteristic isoelectric point, i.e., pH at which the number of ionized negatively charged carboxyl groups just balances the number of positively charged basic groups. At the isoelectric point a protein shows maximum instability, or maximum ease and rapidity of denaturation as well as minimum solubility in aqueous media.

The purification of proteins and their separation from one another make use of techniques which are quite different from those of the organic chemistry of small molecules. The first and most usual method of separation of protein from nonprotein material is that of denaturation. Thus from an extract containing, say, all the soluble components of a plant tissue, the proteins may be removed by heating to 70–100° and filtration or centrifugation of the denatured precipitate from the soluble low-molecular-weight compounds of the supernatant. High concentrations of acid may be used and 5–15% trichloracetic acid is particularly effective in precipitating most proteins even at low temperatures. Similarly, a tissue may be treated with heat, for example by drying in an oven at 70°C, in which case the denatured proteins remain in the residue during a subsequent water extraction. This procedure is widely used in analysis of soluble tissue constituents.

Several classes of proteins (see below) are precipitated from solution by high concentrations of neutral salts such as ammonium sulfate, and this property is very generally utilized in the separation of proteins from one another. In the fractionation of the cytoplasmic proteins of leaves for example, one constituent can be quantitatively precipitated by 0.35 saturated ammonium sulfate while other constituents remain in solution. A portion of these further constituents may then be precipitated by making the solution successively more concentrated in ammonium sulfate. Fractional precipitation may be carried out with acetone, alcohol, or dioxane added to the aqueous protein solution in successively greater concentration. This method has been used for the preparation of many enzymes and in particular in the fractionation of the human blood serum proteins.[5] When these materials are used as protein precipitants it is particularly important to carry out the operation at low temperatures since at ordinary temperatures denaturation may proceed rapidly. It is of importance also that relatively high concentrations of protein be employed since sharp fractionations are much more readily obtained with high than with low protein concentrations. The methods of fractional precipitation have led to the purification and crystallization of a great number of plant proteins, including not only reserve proteins of seeds, discussed below, but also numerous plant enzymes which are summarized in Table 17-1.[6] It is of interest that the first crystalline enzyme, urease, was prepared by Sumner from seeds of the jack bean, *Canavalia ensiformis*.[7]

TABLE 17-1. Crystalline enzymes prepared from higher plants

Enzyme	Source	Date	Author
Urease	Seeds of *Canavalia*	1926	Sumner
Ficin	Latex of *Ficus*	1936	Walti
Papain	Latex of *Carica*	1937	Balls *et al.*
Ascorbic acid oxidase	Fruit of *Cucurbita*	1939	Tadokoro *et al.*
Peroxidase	Root of horseradish	1940	Theorell
Asclepain	Latex of *Asclepias*	1943	Carpenter *et al.*
β-Amylase	Tubers of sweet potato	1946	Balls *et al.*
α-Amylase	Germinating barley	1948	Schwimmer and Balls

A powerful tool for the separation and particularly for the study of the composition of protein mixtures is electrophoresis. As has been pointed out above, protein molecules at pH's other than their isoelectric points are charged and behave in solution as polyvalent ions. Thus when a potential difference is applied to two inert electrodes immersed in a protein solution, the protein molecules carry a portion of the current and migrate toward the appropriate electrode. In its simplest form electrophoresis has consisted of a method for the determination of the isoelectric point of proteins.[8] Very finely divided inert particles, silica or TiO_2, are suspended in the protein solution and are then coated by a layer of protein molecules. A current is passed through the solution, and the rate of movement of the particles determined, ordinarily under a microscope. If this rate of electrophoretic migration is determined for solutions adjusted to different pH values, the pH at which the protein has zero mobility, i.e., does not migrate, can be determined. This pH is then the pH at which the particular protein under consideration possesses zero charge and is hence the isoelectric point. More recently Tiselius[9] has developed electrophoretic methods suitable for analysis of the components of a protein mixture. The protein is dissolved in buffer of known ionic strength and placed in a U-shaped cuvette so designed that the protein occupies the lower portion of the U and is separated by sharp boundaries from the protein-free buffer in the two arms of the cuvette.[10] A current is now passed through the cuvette. The protein migrates in accordance with its electrophoretic mobility, and the buffer-protein boundary in one arm of the cuvette recedes while the similar boundary in the other arm ascends. The moving boundary can be followed optically by a scanning system which determines the rate of change of index of refraction of the solution. Thus as we approach the moving protein boundary from the region of buffer alone, the index of refraction suddenly changes as the protein is encountered. This represents a large rate of change of the index of refraction. As we proceed further into the moving boundary, the index of refraction remains nearly constant since the solution now contains protein in nearly constant concentration. The rate of change of index of refraction therefore again drops to a low level. The scanning pattern of the moving boundary of a pure protein during electrophoresis should ideally then be represented by a single sharp spike. This is never the case, however, since

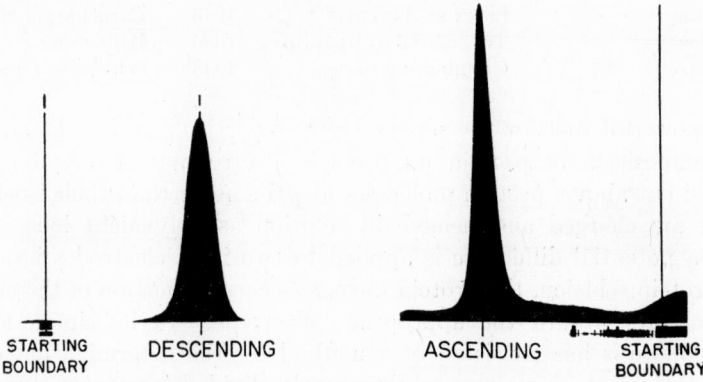

STARTING BOUNDARY   DESCENDING   ASCENDING   STARTING BOUNDARY

FIG. 17-1. Tiselius scanning pattern of an electrophoretically homogenous protein. The diagram shown is for the Fraction I protein of spinach leaf cytoplasm. The two diagrams represent, respectively, the descending and ascending arms of the U-shaped electrophoretic cuvette. (Courtesy of S. G. Wildman)

thermal motions and convection always broaden the boundary somewhat. A typical example of the Tiselius scanning pattern of an electrophoretically homogenous protein is given in Fig. 17-1.

Since individual proteins differ as they do in molecular weight, shape, amino acid constitution, etc., it is most uncommon that two proteins migrate at exactly the same rate in an electric field. Commonly each constituent in a mixture of proteins migrates with its own characteristic rate, and the resulting scanning pattern reveals them as a series of moving boundaries as is shown in Fig. 17-2. Since the indices of refraction of various proteins are not greatly different, comparisons of the total area under each hump in the scanning pattern give us an estimate also of the relative amounts of each component.

Electrophoresis may be used to ascertain the purity of protein preparations or to estimate the number of components present in a protein mixture. More than any other tool it has made possible the study of complex protein mixtures such as the proteins of cytoplasm.[11]

The proteins possess molecules which are sufficiently large to sediment in a centrifuge provided that very high centrifugal forces of the order of 50,000–100,000 times gravity are employed. The use of such centrifuges was first developed by Svedberg[12] for application to molecular weight determination. For this purpose, an optical system essentially similar to that described for electrophoresis is employed, and the moving boundary of the protein as it is centrifuged down can be followed. From the rate of sedimentation and the centrifugal force employed, it is possible to calculate a sedimentation constant for the protein and from this, if the molecular shape is known or can be assumed, the molecular weight can be calculated. In the ultracentrifuge as in the electrophoresis apparatus,

FIG. 17-2. Electrophoretic scanning pattern of a complex mixture of proteins, in this case Fraction II of the cytoplasmic protein of leaves of *Xanthium*. Fraction I protein has been largely removed by salt precipitation. At least six proteins appear to be contained in this mixture. (Courtesy of J. Campbell)

the different components of a mixture ordinarily migrate at different rates, so that ultracentrifuge measurements also yield information relative to the purity of a protein preparation.

Both the electrophoretic method and the ultracentrifuge can be employed for the separation of proteins in a quantitative and preparative way rather than for purely analytical measurements. Thus in electrophoresis the desired component can be caused to migrate away from an undesired impurity or vice versa, while fractional centrifugation can be used to separate heavier from lighter components.

Other methods of protein separation and study include adsorption of proteins on suitable adsorbents,[13] and partition chromatography on starch or filter paper.[14] These methods while apparently of general application have not yet been used as widely as those outlined above. The determination of solubility in the presence of varying amounts of the solid phase has been widely used as a criterion of purity and is a useful check on other methods.[15]

The proteins may be classified in a variety of ways, as for example on the basis of enzymatic activity or on the basis of location in the cell. The following classification, which is a physical one and based primarily on solubility in various solvents, is a useful although not infallible guide for the separation of the proteins into experimentally distinguishable groups.

1. Protamines, which are relatively simple low-molecular-weight proteins with a high content of the basic amino acids.
2. Histones, which like the protamines have a high content of the basic amino acids, but which have higher molecular weights and a greater variety of amino acids than the protamines.
3. Albumins, which are soluble in water, soluble in dilute neutral salt solutions, and precipitable by saturated ammonium sulfate solutions. Albumins are found among the reserve proteins of most seeds, as the leucosin of wheat.
4. Globulins, which are insoluble in water but soluble in dilute neutral salt solutions. These are precipitated by moderate salt concentrations, frequently by half-saturated ammonium sulfate. Globulins are characteristically found among the seed reserve proteins.
5. Glutelins, which constitute an important portion of the reserve protein of cereal seeds (glutenin, oryzenin, etc.). They are insoluble in water or dilute neutral salt solutions but are soluble in dilute alkali.
6. Prolamines which are soluble only in 50–70% alcohol or in dilute acid or alkali but not in water or in neutral salt solutions. The gliadin of wheat and zein of maize are examples of this class.

In addition to these classes of the simple proteins, we also find in nature conjugated proteins or proteins which contain nonprotein molecules attached as prosthetic groups. This group comprises a wide variety of proteins, including many of the enzymes with their various prosthetic groups, the nucleoproteins in which the prosthetic group consists of nucleic acid, the chlorophyll protein of the chloroplast, proteins containing carbohydrate or lipid components, and others. From a practical standpoint it is convenient to divide the plant proteins into the seed proteins on the one hand and the protoplasmic proteins of the active plant tissues on the other. The seed proteins include primarily the substances laid down as reserve food material in the seed and hydrolyzed during germination to yield organic nitrogen for the growing seedling. The seed proteins may again be subdivided into the endosperm proteins of the grains and the reserve proteins of the seeds of dicotyledonous plants contained in the cotyledons or endosperm. The protoplasmic proteins, on the other hand, are characteristic of the actively metabolizing tissues of the plant and include the structural proteins of the cytoplasm as well as the specialized enzyme proteins of the active tissue.

Seed Proteins of the Cereals. The proteins of the wheat seed have been studied very extensively because of the important practical consideration that the seed protein, gluten, contributes the property which

TABLE 17-2. Proteins of the wheat seed. Other cereal seeds contain a similar complement of proteins

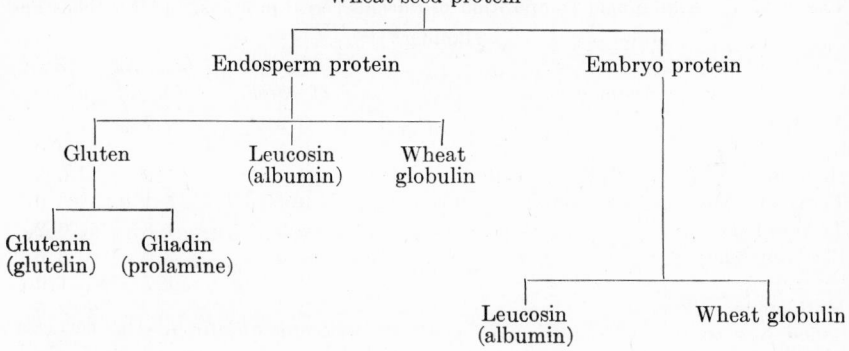

wheat flour possesses of forming in water a spongy, elastic, and cohesive dough. Gluten in turn is made up of two fractions, glutenin, a glutelin soluble in dilute alkali, and gliadin, a prolamine soluble in aqueous alcohol. In addition the wheat seed contains significant amounts of leucosin, an albumin, and of a globulin. The amounts of these proteins present in whole wheat grains are shown in Table 17-3. The gluten

TABLE 17-3. Amounts of proteins present in whole wheat grains. (After Osborne[16])

	Per cent dry weight	
Protein fraction	Spring wheat	Winter wheat
Glutenin	4.68	4.17
Gliadin	3.96	3.90
Globulin	0.62	0.63
Leucosin	0.39	0.36

fraction which makes up the bulk of the seed protein is contained exclusively in the endosperm of the grain whereas the leucosin and globulin are contained in part in the embryo as well as in the endosperm. The molecular weights of these proteins have been determined by osmotic pressure measurements and by ultracentrifugation, the molecular weight of gliadin being approximately 34,000.[17] The glutenin fraction has not proved susceptible to accurate measurements since the molecules associate into loose aggregates to give polydisperse solutions.[18] For the same reason glutenin cannot be subjected to electrophoretic study. The fact that glutenin can be separated into a series of fractions by various means, including precipitation with neutral salt from acid solution, suggests further that it may consist of a mixture of related proteins rather than of an individual pure substance. Gliadin also has been shown to consist of at least two fractions possessing different electrophoretic mobilities.[19] The amino acid compositions of gliadin and glutenin have been investigated by many workers and Table 17-4 gives values taken from the

TABLE 17-4. Amino acid composition of various cereal proteins. (After Block and Bolling[20])

Amino acid	Gliadin of wheat	Glutenin of wheat	Zein of maize
Arginine	3.2%	4.7%	1.6%
Lysine	0.6	1.9	0.0
Histidine	2.1	1.8	0.8
Tyrosine	3.1	5.1	5.9
Tryptophane	0.9	1.8	0.2
Phenylalanine	2.5	2.0	6.6
Cystine	2.3	1.7	1.0
Methionine	2.3	......	2.5
Serine	0.1	0.7	......
Threonine	3.0	......	2.5
Leucine and isoleucine	6.0	6.0	3.0
Valine	3.0	1.0	3.0
Glutamic acid	46.0	27.2	35.6
Aspartic acid	1.4	2.1	3.4
Glycine	1.0	1.0	0.0
Alanine	2.5	4.4	9.9
Proline	13.2	4.4	9.0
Hydroxyproline	......	......	0.0
Ammonia	5.1	4.0	......
Total accounted for	98.3	69.8	85.0

summary of Block and Bolling.[20] A striking characteristic of both proteins is their high content of glutamic acid which reaches 46% in gliadin, while proline, which makes up 13.2% of gliadin, is also an important constituent. The amino acids analyzed account almost quantitatively for the composition of gliadin but an undetermined residue remains in the case of the glutenin fraction.

The proteins of the seeds of other grasses are qualitatively similar to those of wheat. Thus a glutelin is found in rice, and prolamines are found in barley, rye, maize, and rice, although the amount in the latter is small.

Zein, the prolamine of maize, has been subjected to extensive amino acid analyses (Table 17-4) and physical investigation. It is not a homogeneous protein but contains a series of components of which the principal one possesses a molecular weight of 35,000.[22] Small amounts of albumin and of globulin are likewise found in other cereal seeds.

The amounts and composition of the cereal seed proteins are controlled by genetic factors and may be influenced by breeding and selection. These qualities are quantitative characters, governed by many genes, and no simple genetic interpretation of the genetic mechanisms involved has yet been achieved.[23]

TABLE 17-5. Amounts of proteins found in various cereal seeds. (After Onslow[21])

Species	Glutelin %*	Glutelin protein	Prolamine %*	Prolamine protein	Albumin %*	Albumin protein	Globulin %*	Globulin protein
Wheat								
Triticum vulgare	1.0	glutelin	4.2	gliadin	0.3	leucosin	0.6	globulin
Sorghum								
Andropogon sp.	...	....	7.9	kafirin	...	....	...	....
Barley								
Hordeum vulgare	...	....	4.0	hordein	0.3	leucosin	...	....
Rice								
Oryza sativa	1.5	oryzenin	0.1	prolamine	...	....	0.2	globulin
Rye								
Secale cereale	...	....	4.0	gliadin	0.4	leucosin	...	....
Maize								
Zea mays	0.7	glutelin	5.0	zein	...	....	...	....

* Amount of protein in whole dry seed.

Proteins of Seeds Other than Cereals. The seeds of dicotyledonous plants in general contain globulins as their principal protein components. These globulins, of which several have been prepared in crystalline form, appear to be the storage or reserve protein of the cotyledons. Other proteins are present in small amounts but these minor components have been but little studied, and it is not known whether they are also reserve proteins or whether they may not represent the cytoplasmic proteins of the embryonic axis. The seed globulins are obtained by extraction of the seed meal with dilute neutral salt solution such as 5–10% NaCl and may be precipitated from such a solution by 0.5–1.0 saturated ammonium sulfate or by dialysis against distilled water, the latter procedure being particularly suitable for precipitation of the protein in crystalline form. The slow cooling of warm saturated solutions of the protein may also result in crystallization. Edestin, the globulin of hemp seed, as well as the globulins of various cucurbits including squashes and melons may be crystallized by these methods.

The seed globulins possess high molecular weights, values in the range of 200,000 to 400,000 having been found in various cases by both osmotic pressure and ultracentrifuge measurements. A summary of molecular weights of several of the seed globulins is given in Table 17-6.

TABLE 17-6. Molecular weights of some seed globulins. (After Svedberg et al.[24])

Species	Name of protein	Molecular weight
Hemp, *Cannabis sativa*	Edestin	303,000
Brazil nut, *Bertholletia excelsa*	Excelsin	294,000
Almond, *Prunus communis*	Amandin	329,000
Horse chestnut, *Aesculus hippocastanum*	Hippocastanin	430,000

The purified seed globulins of various species show a striking overall similarity in amino acid content, as can be seen in Table 17-7. All are characterized by a high content of arginine as well as by moderately high content of leucine, isoleucine, and valine.

TABLE 17-7. Amino acid composition of various seed globulins. (After Smith and Greene[25])

Amino acid	Content of amino acid in % weight of protein					
	Edestin	Pumpkin	Squash	Watermelon	Cucumber	Tobacco
Arginine	16.7	16.2	16.2	17.9	15.8	16.1
Histidine	2.5	2.2	2.2	2.2	2.3	2.2
Lysine	2.3	2.8	3.0	2.9	2.9	1.6
Threonine	3.1	2.6	2.8	2.9	3.6	4.2
Leucine	7.4	8.0	8.0	7.5	9.1	10.5
Isoleucine	6.2	5.1	5.5	5.7	5.5	6.3
Valine	6.6	6.5	6.5	6.4	7.0	6.7
Tyrosine	4.3	4.4	4.4	4.6	4.6	4.1
Tryptophane	1.2	1.7	1.7	1.9	1.9	1.5
Phenylalanine	5.4	7.2	6.8	7.7	6.5	5.7
Methionine	2.2	2.3	2.3	2.8	2.5	2.2
Cystine	1.3	1.1	1.1	1.1	1.1	1.1

The proteins of the peanut have been subjected to detailed fractionation and found to contain two principal globulin fractions, arachin and conarachin, which differ in isoelectric point and amino acid composition.[26] Both fractions are, however, mixtures, and both in fact contain as a principal component the same globulin, protein A, which makes up about 80% of each. Arachin contains in addition 20% (or more) of a protein B, while conarachin contains some 20% of other minor components. These relations are shown in Table 17-8. Proteins of cotton seed[27] and of soybean[28] have also been extensively investigated.

The Leaf Proteins. Studies on the leaf proteins are relatively recent and meager as compared with those concerned with seed proteins. The leaf proteins are however of special interest, including as they must the full complement of enzymes with which the leaf carries out not only photosynthesis but also the wide range of other synthetic processes of which the leaf is capable.   The proteins of the leaf may be separated into three principal categories based on site within the cell, i.e., the nuclear proteins of the nucleus and nuclear constituents, the chloroplastic proteins of the chloroplast, and the cytoplasmic proteins of the cytoplasm in which nucleus and chloroplast are suspended.   It is doubtful whether the vacuole contains appreciable amounts of proteins, and although cell walls appear to contain small amounts of proteins, it is impossible to be certain that this protein does not merely represent cytoplasmic protein occluded on the wall.

Early studies on the preparation of the leaf proteins were carried out by Chibnall[29] by the following general methods.   Intact leaves are killed and the semipermeability of the cells destroyed by a brief immersion in ether.   The leaves are then pressed.   The clear juice so obtained largely represents the contents of the vacuoles of the leaf cells, and the small

TABLE 17-8.   Composition of peanut proteins according to Irving and Fontaine[26]

Protein	Isoelectric point	% of total protein	Component proteins
Arachin	5.1	63	76% A
			24% B
Conarachin	3.9	33	80% A
			20% minor proteins
Minor proteins	...	3	....

amount of protein contained in it has been termed vacuolar protein by Chibnall, although it may well represent cytoplasmic protein expressed from the small number of cells ruptured by the treatment.   The green residual tissue is then ground and extracted with water.   The green water extract may be separated from the residue of cell walls and unground cells by filtration through a coarse filter.   The extract itself may now be separated by filtration or by centrifugation into two fractions, a green precipitate and a clear filtrate.   The green precipitate consists of chloroplasts or chloroplast fragments, known as chloroplastic protein or chloroplastic material, whereas the filtrate contains the cytoplasmic proteins.   The general method of Chibnall has been used by Menke[30] and others, with various modifications, of which the most important is elimination of the initial step of killing in ether and substitution of grinding in a blendor or other high-speed grinding device for the coarse grinding and pressing steps.   Highly efficient and complete extraction of protein from leaf tissue can be obtained by use of the colloid mill, a high-speed centrifugal grinding machine, used by Wildman and Bonner.[31]   With the colloid mill the protoplasmic contents may be nearly quantitatively separated from cell wall material.   The application of Chibnall's general method shows then that leaf protein may be separated into two main portions, chloroplastic and cytoplasmic protein. The distribution of leaf protein between these two forms is shown in Table 17-9 for the case of spinach leaves. The chloroplasts make up in spinach about one-quarter of the dry weight of the leaf and contain almost 40% of the total nitrogen. The cytoplasmic proteins, on the other hand, make up about 16% of the dry weight and another 40% of the total nitrogen. Roughly similar distributions of nitrogen between chloroplasts and cytoplasm have been obtained for tobacco, tomato, and grasses; Neish has shown in a variety of species that the chloroplasts make up 24–33% of the leaf dry weight.[32] In all these cases the nuclear proteins appear to make up but a negligible proportion of the total.

TABLE 17-9. Approximate distribution of cellular constituents of spinach leaf. (After Wildman and Bonner[31])

Fraction of tissue	% of dry weight	% of fresh weight	% of total N
Cell walls	23.8	2.5	
Protoplasm	76.2	8.2	100*
a. Chloroplasts	26.8	2.7	37.9
b. Cytoplasmic proteins	15.8	1.9	39.3
c. Water-soluble low-molecular-weight substances	33.6	3.6	22.8

* Arbitrarily taken as 100%.

Chloroplastic Protein. The chloroplasts of higher plants are typically disk- or saucer-shaped bodies 3–10 μ in diameter and 1–2 μ in thickness. There are perhaps 50 chloroplasts in a typical cell of the spongy parenchyma of the leaf and 20–100 in a typical palisade cell. Intact chloroplasts may be obtained by the method of Granick[32] which consists in grinding the leaves in a more or less isotonic solution such as 0.5 molal sucrose. The cell wall fragments are then filtered off and the chloroplasts centrifuged free from the residual suspension. The intact chloroplasts thus obtained possess a boundary membrane which behaves as a semipermeable barrier, and the chloroplasts exhibit typical osmotic behavior: swelling, generally with rupture, in hypotonic solution and shrinking in hypertonic solution. It is for this reason that grinding of the tissue must be carried out in isotonic solution. Chloroplasts are frequently also ruptured by the grinding process so that chloroplast fragments rather than intact chloroplasts are often obtained. The drastic grinding methods employed for quantitative release of protein from the tissue results in almost complete rupture of chloroplasts in many species as in spinach. Other species, however, show more resistance to fragmentation of their chloroplasts, and it is a relatively simple matter to obtain intact chloroplasts from tobacco, tea, oat, and other leaves.

The chloroplast is made up of disk-shaped bodies, the grana,[33] which in size are close to the limit of the resolving power of the light microscope and vary in different species from perhaps 0.2 μ in diameter to as large as 2 μ in exceptional cases. The grana of the spinach chloroplast average approximately 0.6 μ in diameter and are somewhat less than 0.1 μ in thickness. A single chloroplast contains of the order of 10 to 100 grana. When spinach leaves are ground in water, the chloroplasts are ruptured as stated above and the grana are released. Secondarily, the grana may floc into irregular masses, but by careful grinding of whole leaves it is often possible to obtain nearly homogeneous suspensions of grana which may then be centrifuged free of the cytoplasmic constituents. All the chlorophyll of the chloroplast appears to be concentrated in the grana, which in turn are imbedded in a colorless matrix, the stroma, the whole surrounded by the limiting membrane. Nothing is known concerning the nature of stroma, and attempts to prepare stroma from intact chloroplasts[31] have shown merely that the solids of stroma make up an extremely small part of the mass of the chloroplast.

The grana possess further structure in that they are composed of stacks of laminae which are readily discernible in the electron microscope. These laminae are of the order of 0.01–0.02 μ in thickness so that a single granum must be composed of numerous layers. The laminar structure of the granum is further evidenced by its properties of laminar double refraction, a property analogous to the form double refraction of regularly oriented anisotropic particles discussed in Chapter 8, but in this case due to the superposition of laminae of different indices of refraction.

In composition the grana are largely protein, lipid, chlorophyll, carotenoid, and ash. Protein makes up 33–50% of the total grana weight, the exact amount depending on the species and state of specimen used. The protein component of chloroplastic material exhibits an amino acid composition roughly similar to that of the whole cytoplasmic protein from the same species. Chlorophyll in the best studied case, spinach, makes up about 8% of the chloroplastic dry weight; ash, including the magnesium of the plastid, 7–8% of the dry weight. Carotenoids constitute in general less than one-fifth of the amount of chlorophyll, i.e., less than 2% of the chloroplastic mass.

The total lipid referred to in Table 17-10 comprises material soluble in ether and in ether-alcohol mixtures and includes chlorophyll and the carotenoids. The latter two components make up, however, but 10% of the total chloroplastic mass, whereas the fraction as a whole may make up as much as 37% of the total. The remaining lipid material includes fats, fatty acids, phospholipids, and possibly sterols.[29] These components still need, however, to be carefully separated and characterized.

A striking feature of chloroplastic proteins is the difficulty which has

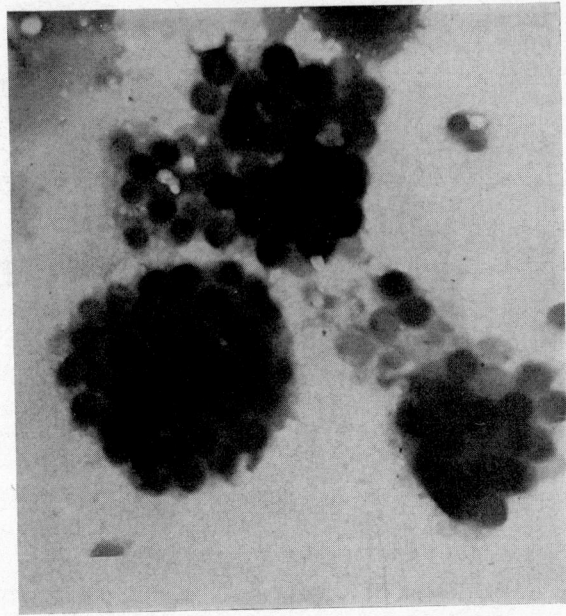

A

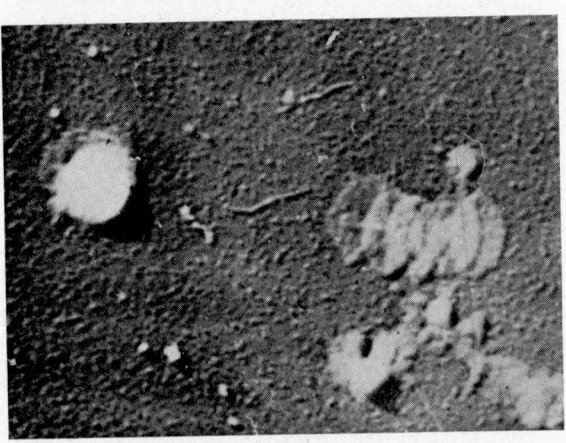

B

Fig. 17-3. A. Structure of chloroplast (spinach) showing grana. Shadowed with gold. The grana have a height of about 800 Å and a diameter of about 6000 Å. Electron microscope photograph. (After Granick[77])

B. Structure of granum (tobacco) as seen in the electron microscope. *Left:* Intact granum. The granum has a considerable height as can be seen by the length of the shadow. *Right:* Granum disintegrated into its lamellae. Enlargement, 1 μ = 2.4 cm. (Courtesy of A. Frey-Wyssling[78]) been experienced in attempts to obtain soluble preparations. The grana material appears to be tightly organized and transformable into soluble form only by rather drastic procedures. Such a procedure has been applied by Smith who has shown that when the highly surface active glycoside digitonin is added to a grana suspension the grana are dispersed in true solution.[35] Digitonin (also bile salts) removes the chlorophyll from the protein-chlorophyll complex, the chlorophyll remaining as a suspension together with the soluble chlorophyll-free proteins. The protein thus obtained has a molecular weight by ultracentrifugation of

TABLE 17-10. Gross composition of chloroplasts or grana of leaves of various species

Species	% Protein	% Total lipid	% Ash	% Chlorophyll	Author
Spinach	40	25	16.9	....	Chibnall[29]
Spinach	48	37	8	7.6–8.3	Menke[34]
Spinach	42–54	26–32	....	6	Bot[34]
Spinach	54	34	7	....	Comar[34]
Sweet pea	33–50	18–30	....	4–6	Bot[34]
Clover	50	22	....	....	Neish[34]

235,000 and is hence much smaller than the original granum. Treatment of grana solutions with 1 N KOH at $-10°$ C for periods of one hour or more also results in solubilization of the particles accompanied by liberation of soluble proteins.[36] In this case the porphyrin residue of chlorophyll remains attached to the protein but in altered form, due to removal of the magnesium atom as well as to perhaps other changes induced by the alkaline treatment. Present methods appear to be inadequate to justify any detailed discussion of the number and kinds of proteins which make up the grana.

Chloroplastic material contains a number of mineral constituents in addition to the magnesium of chlorophyll, as is shown in Table 17-11.

TABLE 17-11. Mineral constituents of chloroplastic material. (After Neish[37])

Plant	Ash % of dry weight	Component in % of ash				
		Mg	P	Fe	Ca	Cu	Na
Clover	8.0	2.0	9.2	0.44	6.2	0.10	0.48
Arctium	5.3	3.0	6.4	1.18	9.6	0.16	0.39
Onoclea	5.2	3.0	4.9	1.08	6.3	0.33	0.74
Elodea	9.7	2.1	4.0	2.56	...	0.16	2.57

Phosphorus and calcium are present in relatively large amounts, and sodium in lesser amounts. Potassium is absent from chloroplasts, according to Neish, but present in about the same amounts as Ca, according to Menke.[38] The presence of iron and copper is particularly interesting and in fact according to Leibich[39] as much as 82% of the iron of spinach leaves is present in the chloroplasts. Of this iron at least four-fifths is found in organic combination. Chloroplasts of some species of plants contain catalase, an iron enzyme, as has been shown for clover, *Arctium* and *Onoclea* by Neish. On the other hand in some species as in spinach the grana seem to have only a low catalase activity.[31] Peroxidase also is generally absent from the chloroplast. The copper of the chloroplast may be in part at least present as polyphenoloxidase since this enzyme is found in chloroplasts of many species including tea, chard, and possibly spinach.[40] Polyphenoloxidase also appears, however, to be present in the cytoplasm of spinach leaves. In any case, these enzymes are present in the grana in insoluble form and cannot be released even by relatively drastic treatment. Thus when grana of tea leaves are extracted with acetone, the polyphenoloxidase activity remains behind in the chlorophyll-free grana residue.

Chloroplasts may contain starch, particularly after periods of intense photosynthesis. It is probable that such starch formation is brought about by activity of starch phosphorylase present in the granum or matrix, acting on the glucose-1-phosphate produced in photosynthesis (Chapter 30).

The Cytoplasmic Proteins. The cytoplasmic proteins, in contrast to the proteins of the grana, are readily soluble and can be investigated with the classical methods of protein chemistry. Preparation of cytoplasmic proteins from leaf tissues involves, as described above, a preliminary grinding of the tissue. The ground tissue is filtered or centrifuged free of cell wall and then centrifuged at a higher speed to free the solution of grana and chloroplasts. The filtrate now contains not only the cytoplasmic proteins but also all the other soluble cellular constituents, including low-molecular-weight compounds, such as carbohydrates, amino acids, and organic acids. The cytoplasmic proteins may be prepared from this complex mixture by precipitation with protein precipitants such as ammonium sulfate, or they may be precipitated by denaturation, for example by heat or by acidification of the solution to a pH of 4 or below. Cytoplasmic proteins may also be obtained from the solution by dialysis of the whole cytoplasm to remove the readily dialyzable low-molecular-weight constituents. A satisfactory method for the preparation of cytoplasmic proteins of leaves and other tissues is summarized in the general scheme reproduced in Fig. 17-4.[31] Here the ground leaves are filtered to free the material of cell wall fragments. The whole protoplasm thus obtained is centrifuged free of grana and the resultant clear supernatant solution used directly for the preparation of the various individual cytoplasmic proteins discussed below. The total cytoplasmic protein obtained in this way amounts to about 8 gm. of protein per kilogram of fresh tissue in the case of spinach, tobacco, and similar leaves.

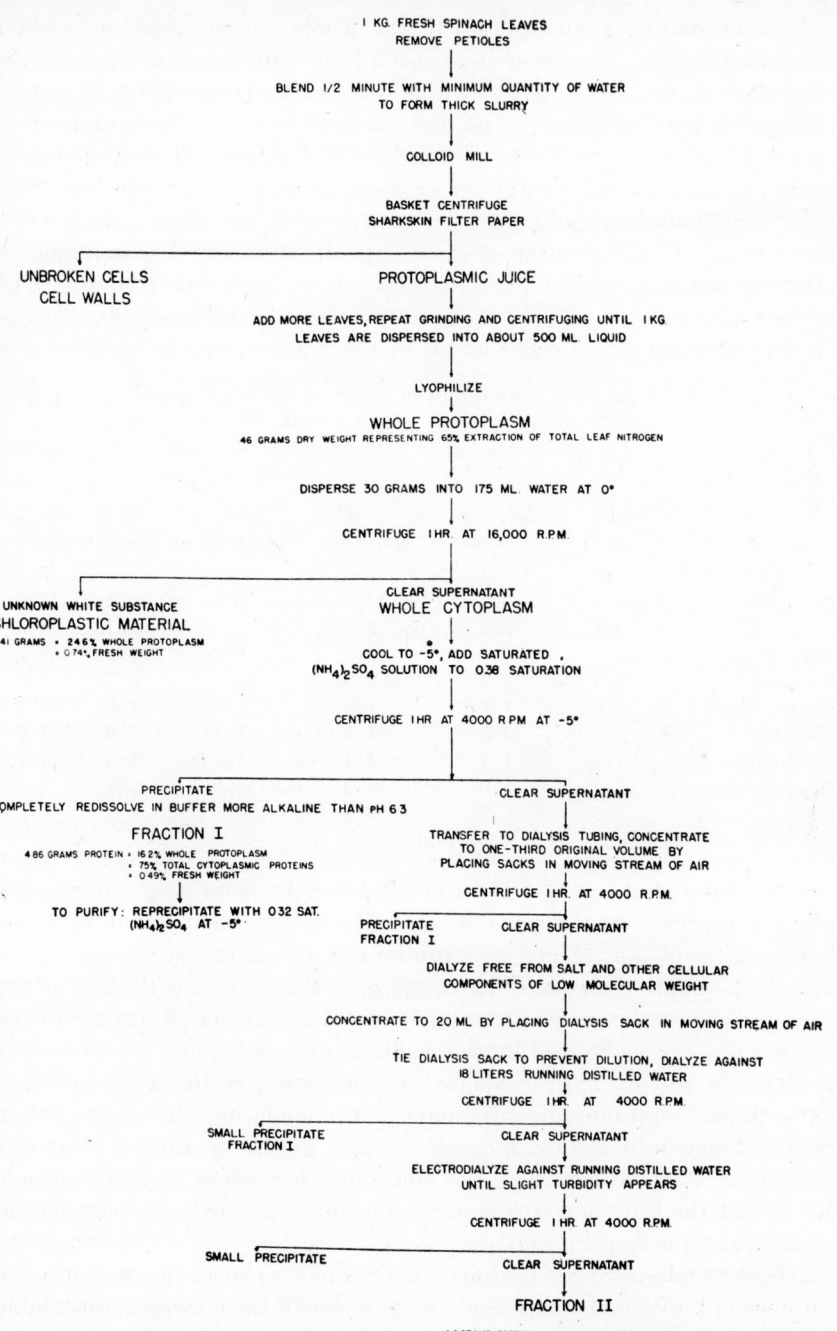

I KG. FRESH SPINACH LEAVES
REMOVE PETIOLES

BLEND 1/2 MINUTE WITH MINIMUM QUANTITY OF WATER
TO FORM THICK SLURRY

COLLOID MILL

BASKET CENTRIFUGE
SHARKSKIN FILTER PAPER

UNBROKEN CELLS   PROTOPLASMIC JUICE
CELL WALLS

ADD MORE LEAVES, REPEAT GRINDING AND CENTRIFUGING UNTIL 1 KG.
LEAVES ARE DISPERSED INTO ABOUT 500 ML. LIQUID

LYOPHILIZE

WHOLE PROTOPLASM
46 GRAMS DRY WEIGHT REPRESENTING 65% EXTRACTION OF TOTAL LEAF NITROGEN

DISPERSE 30 GRAMS INTO 175 ML. WATER AT 0°

CENTRIFUGE 1 HR. AT 16,000 R.P.M.

UNKNOWN WHITE SUBSTANCE   CLEAR SUPERNATANT
CHLOROPLASTIC MATERIAL   WHOLE CYTOPLASM
7.41 GRAMS = 24.6% WHOLE PROTOPLASM
= 0.74% FRESH WEIGHT   COOL TO −5°, ADD SATURATED
                                   $(NH_4)_2SO_4$ SOLUTION TO 0.38 SATURATION

CENTRIFUGE 1 HR AT 4000 R PM AT −5°

PRECIPITATE   CLEAR SUPERNATANT
COMPLETELY REDISSOLVE IN BUFFER MORE ALKALINE THAN pH 6.3

FRACTION I   TRANSFER TO DIALYSIS TUBING, CONCENTRATE
4.86 GRAMS PROTEIN = 16.2% WHOLE PROTOPLASM   TO ONE-THIRD ORIGINAL VOLUME BY
= 75% TOTAL CYTOPLASMIC PROTEINS   PLACING SACKS IN MOVING STREAM OF AIR
= 0.49% FRESH WEIGHT
                                        CENTRIFUGE 1 HR. AT 4000 R.P.M.
TO PURIFY: REPRECIPITATE WITH 0.32 SAT.
$(NH_4)_2SO_4$ AT −5°   PRECIPITATE   CLEAR SUPERNATANT
                                 FRACTION I

                                 DIALYZE FREE FROM SALT AND OTHER CELLULAR
                                   COMPONENTS OF LOW MOLECULAR WEIGHT

                       CONCENTRATE TO 20 ML BY PLACING DIALYSIS SACK IN MOVING STREAM OF AIR

                       TIE DIALYSIS SACK TO PREVENT DILUTION, DIALYZE AGAINST
                                 18 LITERS RUNNING DISTILLED WATER

                                 CENTRIFUGE 1 HR. AT 4000 R.P.M.

SMALL PRECIPITATE   CLEAR SUPERNATANT
FRACTION I

                       ELECTRODIALYZE AGAINST RUNNING DISTILLED WATER
                                 UNTIL SLIGHT TURBIDITY APPEARS

                                 CENTRIFUGE 1 HR. AT 4000 R.P.M.

SMALL PRECIPITATE   CLEAR SUPERNATANT

                                         FRACTION II
                            1.4 GRAMS PROTEIN = 4.6% WHOLE PROTOPLASM
                                       = 25% TOTAL CYTOPLASMIC PROTEINS
                                       = 0.14% FRESH WEIGHT

FIG. 17-4.   Scheme for separating the cellular constituents of spinach leaves.[31]

The amino acid composition of the whole cytoplasmic protein of a variety of plants has been investigated by Chibnall and co-workers[29] and their results, of which a sample is given in Table 17-12, show that the amino acid composition of this material is surprisingly uniform as between widely varied species of plants. Not only do the dicotyledonous plants show uniformity in amino acid composition, but these plants are also very similar to the grasses in amino acid composition of their cytoplasmic proteins. Typical of the composition of the leaf cytoplasmic proteins of higher plants is the high content of arginine which makes up of the order of 14% to 15% of the total protein. Tryptophane is uniformly low, making up of the order of 1.6 to 1.8% whereas lysine makes up in

TABLE 17-12. Amino acid composition of whole cytoplasmic protein of leaves of various species. (After Chibnall[29])

| | Species from which leaf cytoplasmic proteins were obtained* | | | | | |
| | | | | | Spinacia oleracea | |
Amino acid	Zea mays	Ricinus communis	Phaseolus multiflorus	Dactylis glomerata	cyto-plasmic	chloro-plastic
Amide N	5.4	5.1	5.4	5.0	5.6	5.1
Arginine	14.4	12.9	14.9	15.5	14.1	13.9
Histidine	2.1	2.2	2.6	2.3	2.2	3.3
Lysine	6.1	6.5	6.1	6.0	6.2	4.7
Tyrosine	2.3	2.6	2.5	2.3	2.7	2.6
Tryptophane	1.6	1.7	1.6	1.8	1.7	1.7
Cystine	1.1	1.5	1.1	1.4	1.4	1.2
Methionine	1.3	1.4	1.1	1.2	1.3	1.3
Aspartic acid	....	5.6	5.2	4.9	5.5	5.8
Glutamic acid	....	6.7	6.7	7.8	6.5	6.5

* Figures are given as per cent of total protein N.

general somewhat over 6% of the total weight of protein. The cytoplasmic proteins, on the other hand, differ in their composition from the composition of the reserve seed proteins of the same species, as can be seen by comparing the analyses for maize in Table 17-12 with those of the seed protein of the same species in Table 17-4. In the utilization of the reserve protein of the seed by the seedling, it would appear that not only is the seed protein hydrolyzed to its constituent amino acids but that these amino acids must be interconverted to a considerable extent before reassemblage into the cytoplasmic protein of the growing part of the organism. On the contrary, the chloroplastic protein and cytoplasmic protein of the leaf show a striking similarity in amino acid composition, as is shown for spinach in Table 17-12.

The cytoplasmic proteins must be expected to consist of a mixture of a number of individual components since many of the enzymes responsible for the wide variety of activities carried out by the leaf should be contained in this fraction. The work of Wildman and others[41] has shown not only that cytoplasmic protein does indeed consist of several constituents but also that these may be partially separated from one another. Thus spinach leaf cytoplasmic protein can be divided into two principal fractions by fractional ammonium sulfate precipitation. The first fraction is that readily precipitated by relatively low concentrations (0.32–0.38 saturated) of ammonium sulfate, provided only that the precipitation is carried out from solutions relatively concentrated in protein (1–3%). This protein may be purified by repeated precipitation with 0.32 saturated ammonium sulfate and appears to be a homogeneous material with a molecular weight of 500,000 or more. The protein of this fraction, fraction I, makes up 75% to 80% of the total cytoplasmic protein fraction of the spinach leaf. The residual protein, that not precipitated by 0.32–0.38 saturated ammonium sulfate, is constituted of a wide variety of individual components, each present in only small total concentrations. The methods used in separating these two fractions are summarized in Fig. 17-4. The composition of cytoplasmic protein and its fractionation into these two major fractions can also be readily visualized from the electrophoretic diagrams reproduced in Fig. 17-5. The two fractions of spinach leaf cytoplasmic protein differ not only in electrophoretic behavior but also in other respects. Thus the mixture of proteins of fraction II contains a wide variety of enzymes. Many of the enzymes of the cellular respiration appear to be present in this fraction including the dehydrogenases, as described in Chapter 15, as well as aconitase, fumarase, peroxidase, catalase, polyphenoloxidase, and a variety of other enzymes. Fraction I, on the other hand, contains so far as known at the present time only one type of enzyme activity, namely, a phosphatase which is capable of splitting ATP and other energy-rich phosphates to inorganic phosphate. It is also active, although less so, toward glycerophosphate and other ester phosphates. The enzymatic property is closely associated with the fraction I protein and is not separable from fraction I by repeated precipitation or other fractionation procedures. In addition to its other properties the fraction I protein of spinach leaf contains residues of purines, as adenine, bound to the protein moiety.[42] Each purine molecule is further associated with two molecules of phosphoric acid and with one molecule of a pentose, possibly ribose. The purine-pentose-diphosphate moiety may be split from the protein by two-minute hydrolysis with 1 N HCl at 100°C, although the phosphate is released as inorganic phosphate only by further hydrolysis. Of the two molecules of bound phosphate, one is readily hydrolyzed by acid (100% in 16 minutes), while the other is more resistant and behaves as a stable ester phosphate. It is known that many tissues contain phosphate bound to protein or other tissue components[42] and that a portion of this phosphate is rapidly released on acid hydrolysis. In the spinach leaf, at least, the phosphate bound to fraction I protein constitutes the bulk of the tissue-bound phosphate. Tissue-bound phosphate is apparently of

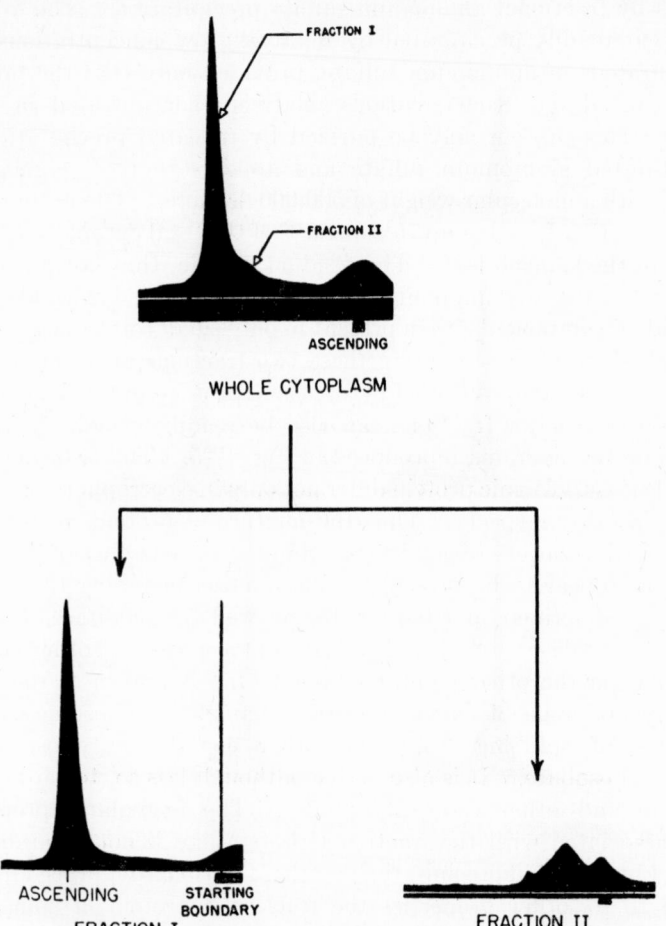

FRACTION I

FRACTION II

ASCENDING

WHOLE CYTOPLASM

ASCENDING   STARTING
BOUNDARY
FRACTION I

FRACTION II

Fig. 17-5. Electrophoretic scanning diagrams of whole spinach leaf cytoplasm and of the two principal components of whole cytoplasm. Only ascending boundaries are shown. (Modified after Wildman and Bonner[31])

metabolic importance in that it is in a state of rapid turnover in metabolically active tissue,[43] the phosphate being continuously renewed from the inorganic phosphate of the cell. Since respiratory inhibitors inhibit this turnover, it is probable that tissue-bound phosphate is in some way related to the energy-rich phosphate formed in respiration. The exact role of tissue-bound phosphate and of fraction I protein is, however, still unknown.

The general relations found for the constitution of the cytoplasmic proteins of spinach leaves apply also to a variety of other species.[41] Thus in tobacco, cocklebur, pea, and other species, a bulk protein makes up 75% to 80% of the total cytoplasmic proteins, and in tobacco this

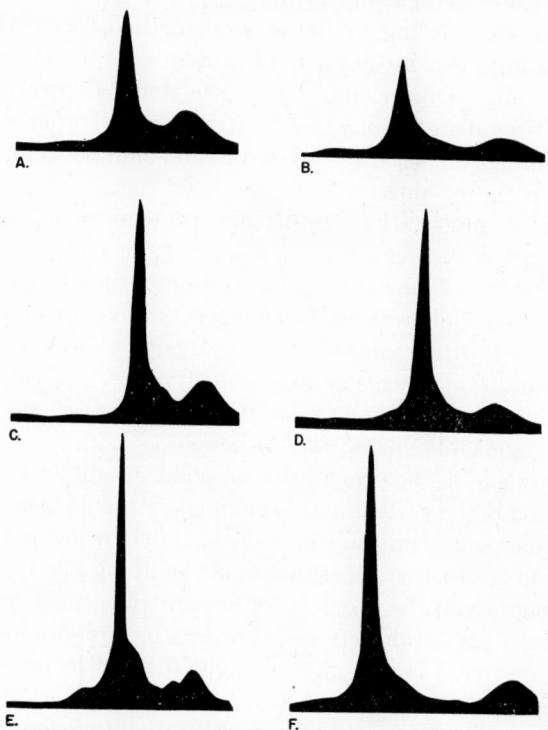

FIG. 17-6. Electrophoretic scanning patterns of whole cytoplasmic protein from various species of leaves. A, *Nicotiana glutinosa;* B, *Nicotiana tabacum*, var. Havana; C, *Brassica pekinensis*, var. Chinese cabbage; D, *Nicotiana tabacum*, var. Turkish; E, *Pisum sativum*, etiolated leaves; F, *Spinacia oleracea.* (Courtesy of S. G. Wildman)

bulk protein possesses the same general properties of high molecular weight and conjugation with purine nucleotide described above for spinach. Similarly, these other species of plants also contain a second protein fraction of greater solubility in ammonium sulfate, a fraction which contains the various activities ordinarily associated with the respiratory enzymes and consists also of a variety of electrophoretically distinguishable components (Fig. 17-2). It is of interest that the cytoplasmic proteins of the etiolated pea leaf are similar in nature to the cytoplasmic proteins of the green leaf.[44]   Even though the etiolated leaf lacks chlorophyll and lacks any considerable amount of chloroplastic protein, still the composition of the cytoplasmic proteins appears to be essentially unaltered.

The cytoplasmic proteins of stems, roots, and other tissues have been still less investigated than have the proteins of leaves.   It is known, however, that the whole cytoplasmic protein of pea stems and of oat seedlings may be prepared according to the general methods used with leaves.[45] In these cases, however, no single bulk protein corresponding to fraction I of leaves is found.   On the contrary, in pea stems, there are at least two principal constituents, each of which makes up of the order of 20% of the total cytoplasmic proteins, while a considerable number of other proteins are included in the residue.

In general summary, the cytoplasmic proteins of leaves and other metabolically active tissues make up some 15% or more of the total dry weight of the tissue.   The cytoplasmic protein of the leaf consists largely of a single electrophoretically homogeneous component which is closely associated with a phosphatase or organic phosphate hydrolyzing enzyme.   In addition to this principal component the leaf cytoplasmic protein includes a variety of minor constituents comprising a considerable number of recognizable enzymes.   Although a very considerable beginning has been made in the preparation of plant cytoplasmic proteins and in the separation of the various cytoplasmic constituents from one another, we still have inadequate information on which to base any detailed picture of the functioning of these constituents in the living plant.

The Nucleoproteins.[46]   Nucleoproteins are constituents of every living cell and appear to be intimately concerned with the reproductive properties of living matter.   It was once thought that nucleoproteins were, as the name implies, confined to nuclear structures of the cell.   We now know, however, that there are two general types of nucleoproteins which differ in chemical nature; one, desoxyribonucleoprotein, is found primarily in the nucleus, while the other, ribonucleoprotein, is found primarily in the cytoplasm of the cell.   In both types, a protein moiety is conjugated with nucleic acid, but nuclear and cytoplasmic nucleoproteins differ in the nature of their associated nucleic acid.   Thus nuclear nucleic acid is characterized by its content of the sugar, desoxyribose, and differs in this respect from cytoplasmic nucleic acid which contains ribose as its sugar constituent.   In the early days of nucleoprotein study, much work was done on nucleic acid of yeast, a ribonucleic acid, and of the thymus gland, a desoxyribonucleic acid, and generalizations were made as to yeast or plant nucleic acid as compared with thymus or animal nucleic acid.   We now know that no such general difference between nucleic acids of plants and animals is to be found. Yeast nucleic acid is merely derived primarily from cytoplasmic nucleoproteins while thymus nucleic acid is derived primarily from nuclear nucleoproteins. The terms plant and animal or thymonucleic acids have now been replaced by the terms ribo- and desoxyribonucleic acid.

Structure of Ribonucleic Acid. The structure of ribonucleic acid is known primarily from chemical work on the material as isolated from yeast. This nucleic acid, on hydrolysis with dilute alkali, yields four separate components called nucleotides, each of which is in turn composed of one molecule each of phosphoric acid, ribose, and a nitrogen base,

TABLE 17-13. The components of nucleoproteins

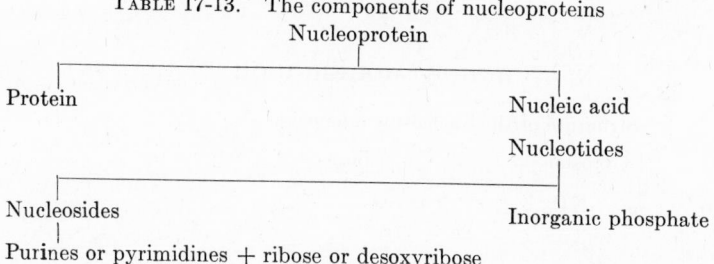

which may be either a purine or a pyrimidine. The phosphoric acid may further be removed from the nucleotide by a more drastic alkaline or enzymatic hydrolysis to yield the ribose-nitrogen base compound which is known as a nucleoside. Ribose may in turn be split from the base by acid hydrolysis. These relations are summarized in Table 17-14. Yeast nucleic acid yields on acid hydrolysis two different purine bases,

TABLE 17-14. Cleavage of yeast nucleic acid into its components by selective hydrolysis

$$\text{Nucleic acid} \xrightarrow[115°,\ 1\ \text{hr.}]{\text{NH}_3} \text{Nucleotides}$$

$$\begin{matrix}\text{Nucleic acid}\\ \text{or}\\ \text{Nucleotides}\end{matrix} \xrightarrow[180°,\ \text{several hr.}]{\text{NH}_3} \text{Nucleosides} + \text{phosphate}$$

$$\begin{matrix}\text{Nucleic acid,}\\ \text{Nucleotides,}\\ \text{Nucleosides}\end{matrix} \xrightarrow{\text{Acid hydrolysis}} \text{Phosphate, ribose, purine, and pyrimidine bases}$$

adenine and guanine, which occur in almost equal amounts. In addition, two pyrimidines, cytosine and uracil, have been isolated from the acid hydrolyzate, and these occur in amounts equivalent to those of the two purines. Yeast nucleic acid contains then as an overall average one molecule each of adenine, guanine, cytosine, and uracil, four molecules of ribose, and four molecules of inorganic phosphate. Since nucleic acids yield nucleotides on alkaline hydrolysis, it is clear that each base is linked to a ribose residue and similarly each nucleoside must be linked to phosphate. Linkage of ribose to purine has been shown both by structure determination[47] and by synthesis[48] to be between nitrogen atom 9 of the base and carbon atom 1 of the ribofuranose residue, while the pyrimidines are linked through nitrogen atom 3 to carbon atom 1 of the sugar. Thus the nucleosides of ribonucleic acid have the general structure:

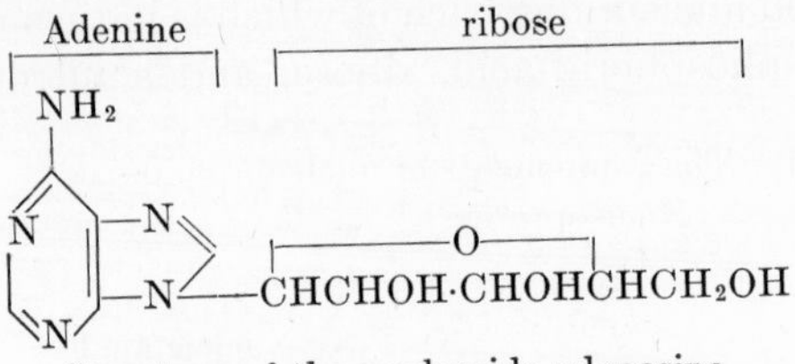

Structure of the nucleoside adenosine

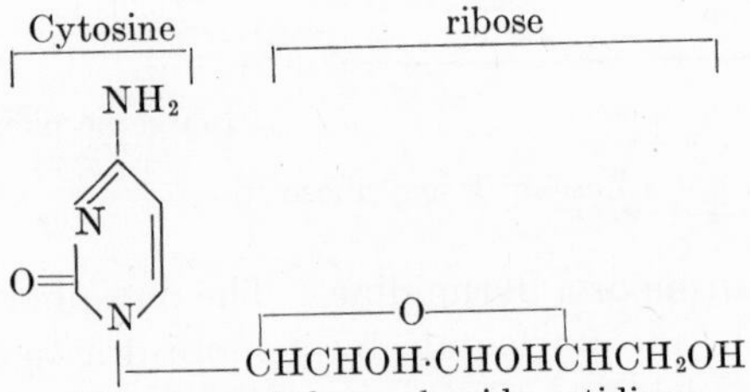

Structure of the nucleoside cytidine

In the nucleotides prepared from yeast nucleic acid, the phosphate is esterified to the ribose residue at position 3.[49] Intact ribonucleic acid is then a polymer of the four nucleotides comprising the four different bases enumerated above, and the internucleotide linkages are through the phosphate residues since each phosphate group of the intact nucleic acid molecule contains only one titratable hydrogen ion.[50] The internucleotide phosphate linkages are thought to be through carbon atom 2 of the ribofuranose residue. We may then picture ribonucleic acid qualitatively, as shown in Fig. 17-7. There are, however, no compelling reasons to believe that the several nucleotides are arranged in the order shown or in fact that they are arranged in units of four nucleotides as depicted here. Nucleic acid is highly polymerized, consists of long chains of such residues, and the several nucleotides may be scattered about the chains in a much more complicated fashion.

Ribonucleic acids similar to yeast nucleic acid have been found and studied in the wheat embryo,[51] rye embryo,[52] pea embryo,[53] and from plant viruses, particularly tobacco mosaic virus.[54] All these nucleic acids yield purines and pyrimidines similar to those found in yeast

ribonucleic acid.   We do not know, however, that all higher plant cytoplasmic nucleic acids are similar to yeast nucleic acid, since isolation and
characterization of the cytoplasmic nucleic acids of higher plants has as yet
been carried out in so few cases.

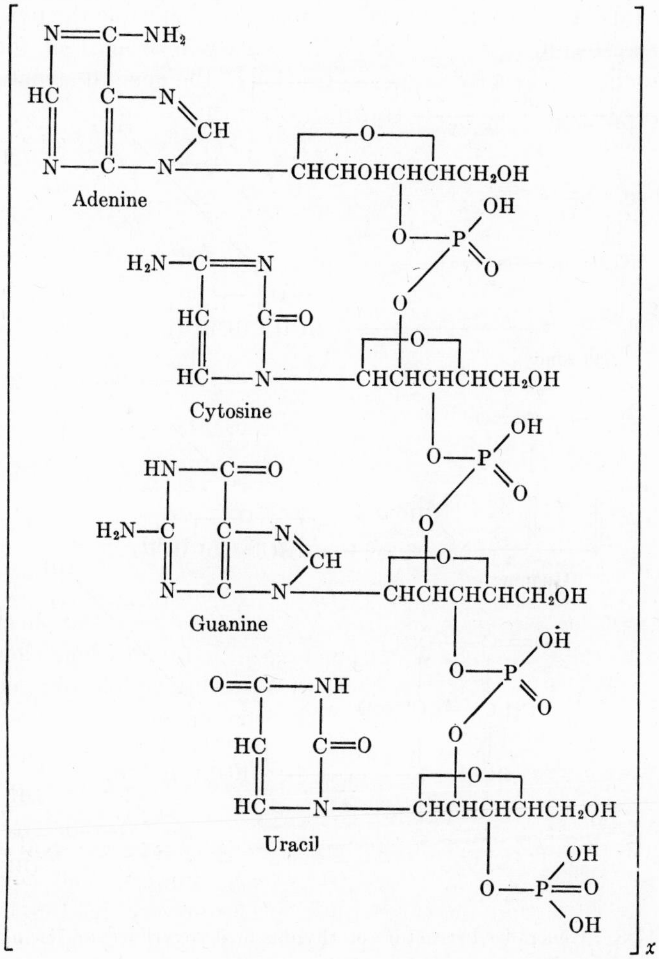

FIG. 17-7.   Generalized structure of yeast or ribonucleic acid.

Structure of Desoxyribonucleic Acid.   Thymus or desoxyribonucleic
acid is distinguished from yeast or ribonucleic acid by the fact that in
thymus nucleic acid the sugar, desoxyribose, replaces ribose.   This sugar
is remarkable for its lability and ease of oxidation, and for this reason
gives a characteristic color reaction, the Feulgen test, which consists of

ability to restore color to a fuchsin-sulfurous acid reagent.[55] The Feulgen test, which is given by desoxyribonucleic acids only after mild acid hydrolysis (needed to free the sugar) is widely used not only for the

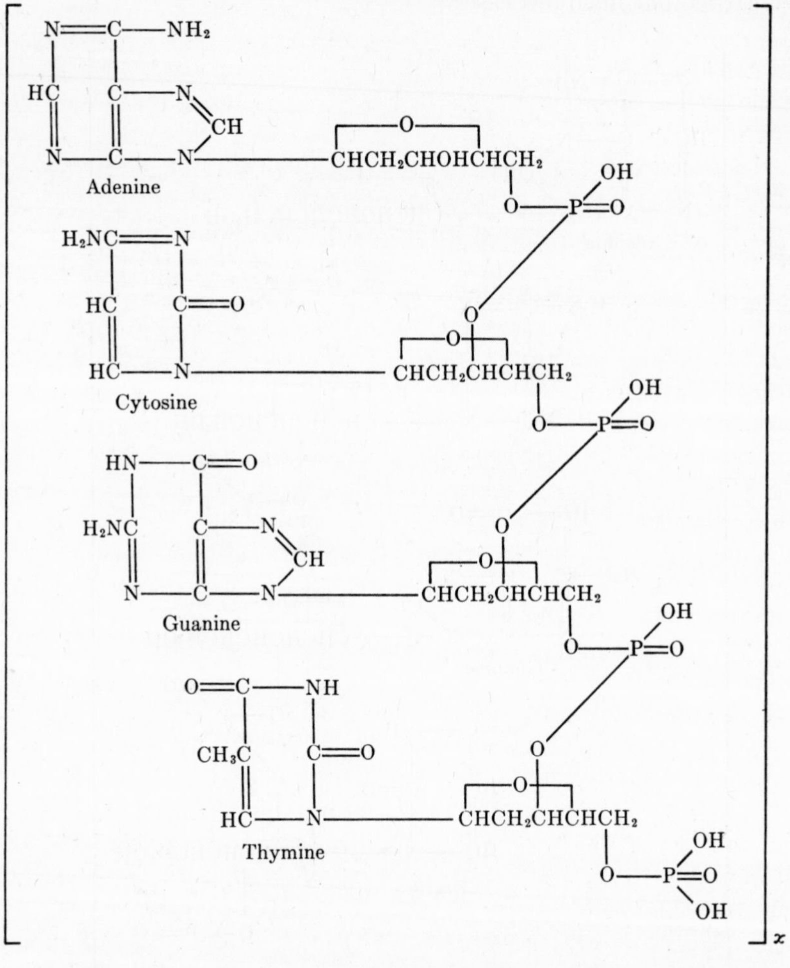

FIG. 17-8. Generalized structure of thymus or desoxyribose nucleic acid.

characterization of isolated nucleic acid preparations but also more particularly for the cytological localization of desoxyribonucleic acids within the cell.

Thymus nucleic acid differs from ribonucleic acid in being more resistant to mild alkaline hydrolysis but much more sensitive to cold acid hydrolysis, which results in liberation of the bases. These include adenine, guanine, and cytosine as in the ribonucleic acids, but uracil is replaced by the pyrimidine thymine. The nucleotide phosphate is bound through position 3 of the desoxyribofuranose residue,[56] but the internucleotide linkages appear to be through carbon atom 5 rather than through carbon atom 2.[57] We may then picture thymus nucleic acid in a general way as shown in Fig. 17-8. Here again there is no compelling reason for adopting any set order or even a basic unit of four nucleotides for the arrangement of the nucleotide residues along the nucleic acid chain.

Molecular Size and Form of Nucleic Acids. Early work on the structure of nucleic acids was carried out with preparations which were extensively degraded and which yielded relatively low molecular weights, as low in fact as 1500, or even less, corresponding to approximately four nucleotide units only. More recent measurements[58] have shown that yeast nucleic acid itself has a molecular weight of the order of 2×10^4 or approximately sixty nucleotide residues per chain, while other ribonucleic acids such as those of tobacco mosaic virus may reach 3×10^5 or roughly three hundred and fifty residues.[59] Thymus nucleic acid has been found to have even higher molecular weights, various values between 5×10^5 and 2×10^6 having been recorded.[60] It is clear in any case that both types of nucleic acid represent high-molecular-weight compounds. How now are these residues arranged geometrically in space along the chain? Astbury[61] has pointed out that the high density of nucleic acids (approximately 1.65 gm./cc.) suggests some form of very close and economical packing of the component nucleotides. This packing must also be along a definite chain axis since double refraction of flow measurements[62] shows that thymus nucleic acid is dispersed in solution as rods about three hundred times as long as thick. From x-ray studies on dried nucleic acid films and fibers, Astbury concludes that the long axis of the molecule is constituted by the repeating chain of ribose-phosphate linkages and that the nucleosides lie packed on one another at right angles to the long axis of the molecule like a stack of hot cakes. The average distance between nucleosides along the axis appears to be 3.3–3.4 Å.

Protein Portion of Nucleoproteins. Little is known concerning the protein constituents of the nucleoproteins of higher plants with the exception of the viruses. In general, however, the cytoplasmic nucleoproteins appear to be present both in solution in the cytoplasm, and in insoluble form as components of particulate matter such as the mitochondria. The nuclear nucleic acid is combined in part with basic proteins of the general nature of protamines or histones. The two components of such a nucleoprotein may be released from one another by a variety of means, treatment with electrolytes at an appropriate concentration being perhaps the simplest. Similar release of nucleic acid from the protein is also accomplished when the nucleoprotein solution is brought to moderately acid pH in which case the nucleic acid is replaced by the stronger acid. These facts suggest that in such cases the binding between nucleic acid and protein may be a simple electrostatic one due to the interaction of positively charged protein with negatively charged nucleic acid. In more complex nucleoproteins such as the viruses, it is necessary to use more drastic methods to liberate nucleic acid from the nucleoprotein, and in general these methods result in denaturation of the protein component. Electrolytes, heat, shaking in solutions of detergents or of chloroform and alcohol, treatment with weak acid or alkali—all effect release of the nucleic acid component to a greater or lesser extent. Here also the evidence favors the view that protein and nucleic acid components are bound through saltlike electrostatic linkages, but with more modification of one moiety by the other than is the case with the simpler nucleoproteins.

Nuclear Nucleoprotein. Great interest attaches to the role of nucleoproteins in the nucleus. That nucleoproteins are actually constituents of the chromosome has been shown directly by Mirsky and collaborators who have isolated chromosomes by fractional centrifugation from red blood cells of fish and chicken as well as from thymus and liver.[63]

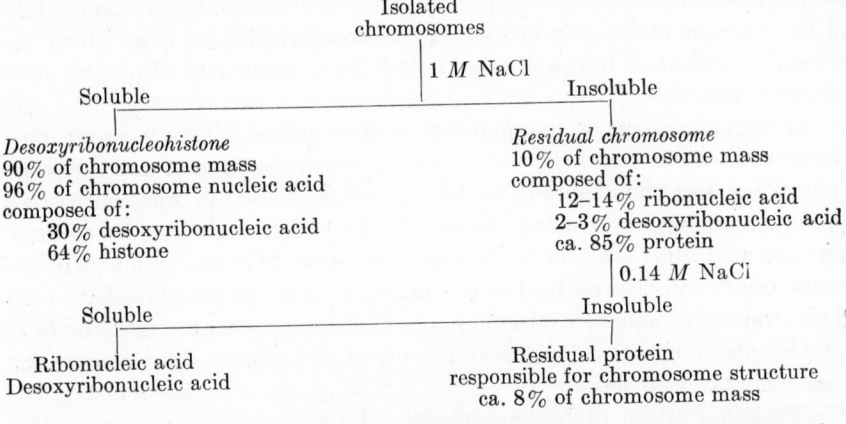

FIG. 17-9. Fractionation of thymus lymphocyte chromosomes according to Mirsky and Ris.[63]

When such chromosomes are suspended in 1 M NaCl they can be separated into a soluble desoxynucleohistone fraction which makes up some 90% of the chromosome, and a coiled threadlike residue, the residual chromosome, which contains 12–14% of ribonucleic acid and 2–3% of desoxyribonucleic acid. The staining characteristics of the chromosome, for example the Feulgen reaction, are due to the NaCl-soluble desoxyribonucleohistone, while the actual shape of the chromosome is apparently due to the insoluble ribonucleoprotein residue. The nucleic acid of this residue may next be removed in 0.14 M NaCl, leaving behind a protein residue which retains the chromosome structure which is thus due to protein alone.

A wealth of information concerning the localization of nucleic acids in the cell has been accumulated by Caspersson and others[64] with the use primarily of a microspectrophotometer which permits of the spectrophotometric determination of purines at their typical absorption in the region of 2600 Å. This work has shown that during the mitotic cycle of the dividing nucleus there is an active series of changes in nucleic acid and nucleoprotein constituents. Beginning in prophase, desoxyribonucleic acid accumulates in the chromosomes, reaches a maximum at metaphase, and largely disappears again by telophase. In telophase and after, however, the nucleoli and a chromosome residue, the heterochromatin, remain, and these contain ribonucleic acid. Caspersson believes that the heterochromatin is responsible for synthesis of the desoxyribonucleoprotein of the metaphase chromosome and is also a center for the production of further ribonucleic acid which escapes into the cytoplasm and contributes to the cytoplasmic nucleoprotein. There is further evidence of a correlational nature that the synthesis of proteins in the cytoplasm may be regulated by the concentration of nucleoproteins present and that protein synthesis may in fact be mediated in some way by nucleoproteins.

Enzymes Which Attack Nucleic Acids. An enzyme, ribonuclease, which attacks yeast nucleic acid is found in animal tissues such as pancreas, liver, and spleen and has been crystallized as an albumin of molecular weight 15,000 by Kunitz[65] from pancreas. This enzyme depolymerizes yeast nucleic acid with the selective liberation of the pyrimidine nucleotides, but the reaction does not proceed to complete hydrolysis of the nucleic acid, and a resistant residue remains. Pancreas and intestinal mucosa also contain a desoxyribonucleic acid depolymerase which acts both on ribo- and desoxyribonucleic acid to liberate mononucleotides.[66] This enzyme occurs in seeds as for example corn, wheat, sunflower, pumpkin, and various beans.[67]

Nucleic acid itself is not attacked by phosphatases but nucleotides are attacked with the liberation of nucleosides and inorganic phosphate. The phosphatases of seeds, seedlings, and leaves are capable of carrying out this nucleotidase reaction.[68]

Nucleosidases which decompose nucleosides to purine or pyrimidine and ribose or desoxyribose have long been known from animal tissues and are widely found in leaves, roots, and seeds.[68] The reaction has been shown by Kalckar[69] to be a phosphorolytic one in which the nucleoside is split to base and sugar phosphate, as is shown below for the case of guanosine. Not only can the nucleoside be decomposed but it can also be

$$\text{Guanine–ribose} + H_3PO_4 \rightleftharpoons \text{Guanine} + \text{ribose–1–}H_2PO_3$$

synthesized from guanine and ribose-1-phosphate in the presence of the enzyme. Nucleosidase may then be more properly termed nucleoside phosphorylase.

We are still far from a full understanding of the exact role of nucleoproteins in the cell. It seems clear, however, that these substances are the carriers of the power of living things to reproduce, to grow and, as components of the genes, to mutate, and to adapt. Whereas the aspects of metabolism with which we have been mainly concerned, carbohydrate metabolism, respiration, nitrogen transformations, etc., concern the maintenance of the individual in the status quo, the nucleoproteins appear to concern the future of the individual, his growth, development, offspring, and evolution.

Virus. The viruses are nucleoproteins which have the property of infectivity, that is, are able to attack a host cell and multiply at the expense of the materials of the host.[70] The higher plant viruses are most commonly spread from plant to plant by particular insects although other modes of transmission are also known, including mechanical transmission through wounds, as is the case with tobacco mosaic virus which may be inoculated into a plant by merely applying a virus solution to a previously abraded leaf surface. Many viruses move readily through the plant, the movement occurring in the vascular tissue, and for this reason, transmission of a virus may also be accomplished by grafting a scion of infected tissue to a receptor which will then receive the virus. With certain viruses, as quick-decline of citrus, transmission through living cells is in fact the only known mode of spread of the disease.

Because of the economic importance of virus diseases of crop plants, a vast amount of work has been done on the transmission, pathology, and other utilitarian aspects of the plant viruses. With the demonstration by Stanley in 1935 that tobacco mosaic virus consists of a single crystallizable protein,[71] biochemical interest in the viruses was aroused, and there is today also a vast body of lore on the chemical and biochemical features of a variety of plant viruses, so much so that they constitute perhaps the best understood group of nucleoproteins. Table 17-15 gives a survey of several properties of a group of plant viruses. The molecular weights of these materials, ranging from 2×10^6 to 40×10^6, are much larger than those of the normal cytoplasmic proteins but are nevertheless much smaller than the particle sizes of the grana of the chloroplasts. It is therefore possible to fractionate virus from whole cytoplasm by frac-

TABLE 17-15. Composition and properties of some plant viruses. (After Greenstein[72])

Virus	Conc. in infected juice gm./liter	Host	Mode of isolation	Form of pure material	Nucleic acid content	Molecular weight approx.	Dimensions
Tobacco mosaic	2.0*	Tobacco	(NH$_4$)$_2$SO$_4$ and centrifugation	Paracryst. needles	5.8%	40 × 10^6	6000 × 150 Å
Ribgrass	0.4	Tobacco	Centrifugation	"	5.8	40 × 10^6	"
Cucumber 4	0.3	Cucumber	"	"	5.8	40 × 10^6	"
Cucumber 3	0.3	Cucumber	"	"	5.8	40 × 10^6	"
Alfalfa mosaic	0.2	Tobacco	"	Amorphous	15	2 × 10^6	Spherical, 165 Å diam.
Potato X	0.1	Tobacco	(NH$_4$)$_2$SO$_4$ and centrifugation	"	5	26 × 10^6	4200 × 98 Å
Tobacco necrosis 1	0.04	Tobacco	(NH$_4$)$_2$SO$_4$	Crystalline plates	15	6 × 10^6	Spherical, 200 Å diam.
Tobacco ring spot	0.012	Tobacco	Centrifugation	Amorphous	40	3 × 10^6	Spherical, 190 Å diam.
Tomato bushy stunt	0.05	Tomato	(NH$_4$)$_2$SO$_4$ and centrifugation	Dodecahedra	15	7 × 10^6	Spherical, 260 Å diam.

* In tobacco.   In tomato 1.3 gm./liter, and 0.15 gm./liter in spinach.

tional centrifugation, combined in some instances with fractional ammonium sulfate precipitation.   Table 17-15 shows also that virus protein occurs in the cytoplasm of the infected cell in concentrations which vary from high in the case of tobacco mosaic (2 gm./liter of expressed juice) to very low (0.01 gm./liter) in the case of tobacco ring spot.   It is not surprising then that tobacco mosaic virus was not only the first to be isolated but has also been used for more detailed study than have the others.

Although the virus particles are not visible in the light microscope, they may readily be seen in the electron microscope, and the measure-

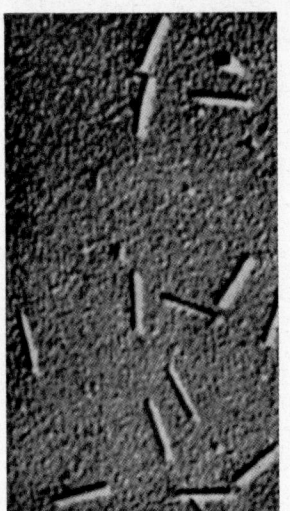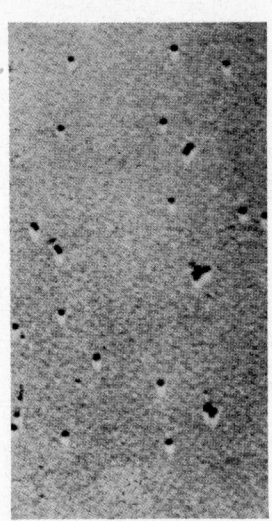

FIG. 17-10.   Electron micrographs of some highly purified viruses.   (After Knight[73])
Left: Tobacco mosaic virus. Right: Tomato bushy stunt virus.

ments of particle size and shape derived from such investigations agree closely with those derived from ultracentrifuge and other physical measurements.

We have seen earlier that the nucleic acid of tobacco mosaic nucleoprotein contains ribose and resembles yeast nucleic acid except for a greater chain length.   The molecular length of nucleic acid from tobacco mosaic is in fact close to the length of the intact nucleoprotein molecule and Cohen and Stanley[74] suggest on this basis and from the nucleic acid content of the virus that in the intact virus molecule approximately eight nucleic acid molecules lie appressed to the protein along its major axis. Less detailed chemical investigations of other plant viruses suggest that all contain nucleic acid of the ribose type.

The protein portion of tobacco mosaic, which composes over 94% of the mass of the virus, is a globulin-like material.   It has been subjected to amino acid analyses particularly by Knight,[75] a selection of whose results are given in Table 17-16. It is of interest that the composition of the virus resembles somewhat the composition of the whole cytoplasmic protein although there are differences particularly in aspartic acid, lysine, and methionine content. The comparison is, however, without great meaning since we are comparing the pure virus with whole cytoplasm which as we have seen is a mixture of many proteins, so that it is not pre-

TABLE 17-16. Amino acid composition of two strains of tobacco mosaic virus as compared with whole cytoplasmic protein. Values in per cent of whole nucleoprotein. (Virus analyses after Knight[73])

	Tobacco mosaic	J14D1	Holmes rib-grass strain	Whole cytoplasmic protein*
Alanine	5.1	4.8	6.4	4.8
Arginine	9.8	10.0	9.9	8.2
Aspartic acid	13.5	13.4	12.6	7.9
Cysteine	0.69	0.64	0.70	....
Cystine	0.0	0.0	0.0	2.2
Glutamic acid	11.3	10.4	15.5	14.3
Glycine	1.9	1.9	1.3	0.3
Histidine	0.0	0.0	0.72	1.5
Isoleucine	6.6	6.6	5.9	} 14.0
Leucine	9.3	9.4	9.0	
Lysine	1.47	1.95	1.51	5.3
Methionine	0.0	0.0	2.2	2.3
Phenylalanine	8.4	8.4	5.4	5.0
Proline	5.8	5.5	5.5	2.8
Serine	7.2	6.8	5.7	....
Threonine	9.9	10.0	8.2	....
Tryptophane	2.1	2.2	1.4	2.2
Tyrosine	3.8	3.9	6.8	5.0
Valine	9.2	8.9	6.2	6.0

* Compiled from Chibnall (1939). These do not apply directly to tobacco but see Table 17-12. The methods used are not quantitatively comparable to those of Knight.

cluded that one of these may resemble the viruses more closely. Table 17-16 also shows, however, that different strains of tobacco mosaic differ in amino acid composition. Strain J14D1 is believed to have arisen from the original by two steps analogous to mutations, and differs from the original in composition by a lower glutamic acid and a higher lysine content. Although strain J14D1 resembles the original strain in size, shape, and physical properties, it produces different symptoms in the host plant, being more lethal than the original. Holmes ribgrass strain also resembles tobacco mosaic physically, although it was originally isolated from ribgrass and is not related in a known genetic way to tobacco mosaic. In tobacco it produces a virus differing markedly in amino acid composition from the usual tobacco mosaic and contains methionine and phenylalanine, which are absent from tobacco mosaic itself.

The appearance of virus protein in tobacco leaf infected with tobacco mosaic virus is not associated with any change in total protein content.

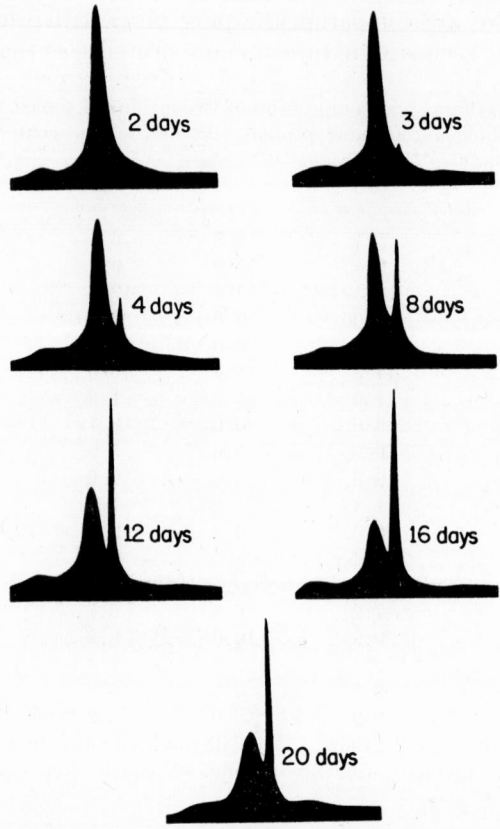

FIG. 17-11. Electrophoretic scanning diagrams of whole cytoplasm of tobacco leaves harvested at various intervals after infection with tobacco mosaic virus. Note appearance of virus component three days after infection. (Courtesy of S. G. Wildman)

On the contrary in the early stages of virus development the total protein of the leaf as well as total cytoplasmic and total chloroplastic protein contents remain essentially constant (Table 17-17). The formation of virus protein is therefore necessarily correlated with the disappearance of a corresponding amount of normal leaf protein. Wildman[76] has shown that it is primarily the cytoplasmic fraction I protein which disappears as tobacco mosaic is formed in infected tobacco leaves. These workers have followed electrophoretically the appearance of virus in the leaf after infection at a remote leaf. Their results (Fig. 17-11) show that within three days and before the appearance of any visible symptoms in the

TABLE 17-17. Distribution of proteins in leaves of *Havana* tobacco during the course of infection with tobacco mosaic virus. (After Wildman *et al.*[76])

Condition of leaves	Days after inoculation	Dry weight of leaves, %	Dry weight extracted as a cell-free juice, %	Total particulate proteins, mg./g. dry wt. of leaves	Total cytoplasmic proteins, mg./g. dry wt. of leaves	Virus protein in total cytoplasmic protein, %
Normal	..	11.0	64.8	146	87	..
Virus-infected	5	10.8	64.4	144	80	0
Normal	..	10.9	65.7	145	81	..
Virus-infected	12	11.4	63.7	136	84	20
Normal	..	11.2	64.1	161	70	..
Virus-infected	17	12.4	64.4	146	78	40

leaf, a new protein appears in the scanning pattern of the leaf proteins. That this is the virus protein can be shown by the fact that when this protein is isolated it is found to be infective. During the period of three to sixteen days after infection the amount of virus protein rises rapidly and this is accompanied by a corresponding decrease in amount of the normal cytoplasmic fraction I. Other proteins of the leaf, chloroplastic proteins as well as fraction II of the cytoplasm, are unaffected during the same period. It is clear therefore that despite the analytical differences between virus strains and between virus and whole cytoplasm, there is a very close relation between virus reproduction and the normal cytoplasmic fraction I protein. This finding implies that virus, in the case of tobacco mosaic at least, may be formed directly or indirectly from a normal protein provided only that virus is present to initiate the transformation.

General References

Schmidt, C. L. A., Chemistry of the Amino Acids and Proteins. Thomas, 2nd ed., 1944.

Advances in Protein Chemistry. Academic Press, 1944 to date.

Chibnall, A. C., Protein Metabolism in the Plant. Yale University Press, 1939.

Symposia of the Society for Experimental Biology, Vol. 1, 1947. A symposium on nucleic acids and nucleoproteins. Cambridge University Press.

Cold Spring Harbor Symposia on Quantitative Biology, Vol. 12, 1947. A symposium on nucleic acids and nucleoproteins.

Greenstein, J. P., Advances in Protein Chemistry. Academic Press, 1944. Vol. 1, p. 209.

References

1. Bull, H. B., Advances in Enzymology. Interscience Publishers, 1941. Vol. 1, p. 1.
2. Anson, M. L., Advances in Protein Chemistry. Academic Press, 1945. Vol. 2, p. 361.
3. Astbury, W. T., Advances in Enzymology. Interscience Publishers, 1943. Vol. 3, p. 63.
   Fankuchen, I., Advances in Protein Chemistry. Academic Press, 1945. Vol. 2, p. 387.
4. Mirsky, A. E., and Anson, M. L., *J. Gen. Physiol.*, **18**, 307 (1935).
5. Edsall, J. T., Advances in Protein Chemistry. Academic Press, 1947. Vol. 3, p. 383.
6. See Sumner, J. B., and Somers, G. F., Enzymes. Academic Press, 2nd ed., 1947.
   Northrop, J. H., Crystalline Enzymes. Columbia University Press, 2nd ed., 1948.
7. Sumner, J. B., *J. Biol. Chem.*, **69**, 435 (1926).
8. Abramson, H. A., Moyer, L. S., and Gorin, M. H., Electrophoresis of Proteins. Reinhold, 1942.
9. Tiselius, A., *Trans. Faraday Soc.*, **33**, 524 (1937).
10. Symposium on Electrophoresis, *Ann. N. Y. Acad. Sci.*, **39**, 105 (1939).
11. For further applications of electrophoresis see Svensson, Advances in Protein Chemistry. Academic Press, 1948. Vol. 4, p. 251.
12. Svedberg, T., and Pedersen, K., The Ultracentrifuge. Clarendon Press, Oxford, 1940.
13. Sumner, J. B., Dounce, A. L., and Frampton, V. L., *J. Biol. Chem.*, **136**, 343 (1940).
14. Mitchell, H. K., and Haskins, F., *Science*, **110**, 278 (1949).
15. Review in Northrop, J. H., Crystalline Enzymes. Columbia University Press, 2nd ed., 1948.
16. Osborne, T., The Vegetable Proteins. Longmans, 1924.
    Blish, M. J., Advances in Protein Chemistry. Academic Press, 1945. Vol. 2, p. 337.
17. Krejci, L., and Svedberg, T., *J. Am. Chem. Soc.*, **57**, 946 (1935).
18. Blish, M. J., Advances in Protein Chemistry. Academic Press, 1945. Vol. 2, p. 337.
19. Schwert, G. W., Putnam, F. W., and Briggs, D. R., *Arch. Biochem.*, **4**, 371 (1944).
20. Block, R. J., and Bolling, D., Amino Acid Composition of Proteins and Foods. Thomas, 1945.
21. Onslow, M., The Principles of Plant Biochemistry. Cambridge University Press, 1931.
22. Watson, C. C., Arrhenius, S., Williams, J. W., *Nature*, **137**, 322 (1936).
23. East, E. M., and Jones, D. F., *Genetics*, **5**, 543 (1920).
    Worzella, W. W., *J. Agr. Research*, **65**, 501 (1942).
24. Svedberg, T., and Eriksson-Quensel, I. B., *Tab. Biol. Period*, **5**, 352 (1935).
25. Smith, E. L., and Greene, R. D., *J. Biol. Chem.*, **167**, 833 (1947).
26. Irving, G. W., Fontaine, T. D., and Warner, R. C., *Arch. Biochem.*, **7**, 475 (1945); **8**, 239 (1945).
27. Bailey, A. E., Cotton-Seed and Cotton-Seed Products. Interscience Publishers, 1948.
28. Payne, D. S., and Stuart, L. S., Advances in Protein Chemistry. Academic Press, 1944. Vol. 1, p. 187.

29. Chibnall, A. C., Protein Metabolism in the Plant. Yale University Press, 1939.
30. Menke, W., *Z. Botan.*, **32**, 273 (1938).
31. Wildman, S. G., and Bonner, J., *Arch. Biochem.*, **14**, 381 (1947).
32. Wildman, S. G., Cheo, C., and Bonner, J., *J. Biol. Chem.*, **180**, 985 (1949).
    Granick, S., *Am. J. Botany*, **25**, 558, 561 (1938).
    Hanson, E. A., *Rec. trav. bot. néerland*, **36**, 183 (1939).
    Hanson, E. A., Barrien, B. S., and Wood, J. G., *Australian J. Exptl. Biol. Med. Sci.*, **19**, 231 (1941).
    Neish, A. C., *Biochem. J.*, **33**, 293, 300 (1939).
33. For reviews of chloroplast structure see Rabinowitch, E., Photosynthesis. Intersci. Publishers, Vol. I, 1945, and Granick, S., in Photosynthesis in Plants, Iowa State College Press, 1949.
34. Menke, W., *Z. Botan.*, **32**, 273 (1938).
    Bot, G. M., *Chron. Bot.*, **7**, 66 (1942).
    Neish, A. C., *Biochem. J.*, **33**, 293, 300 (1939).
    Comar, C. L., *Botan. Gaz.*, **104**, 122 (1942).
35. Smith, E. L., *J. Gen. Physiol.*, **24**, 565, 583, 753 (1941).
36. Brown, D. H., California Institute of Technology. Unpublished report, 1948.
37. Neish, A. C., *Biochem. J.*, **33**, 293, 300 (1939).
38. Menke, W., *Z. physiol. Chem.*, **263**, 100, 104 (1940).
39. Leibich, H., *Z. Botan.*, **37**, 129 (1941).
40. For tea see Li, L., and Bonner, J., *Biochem. J.*, **41**, 105 (1947).
    For chard see Arnon, D., *Nature*, **162**, 341 (1948); *Plant Physiol.*, **1**, 1 (1949).
    For spinach see Laties, G., *Arch. Biochem.*, **27**, 1950, in press.
    Wildman, S. G., and Bonner, J., *Arch. Biochem.*, **14**, 381 (1947).
    Bonner, J., and Wildman, S. G., *Arch. Biochem.*, **10**, 497 (1946).
41. Wildman, S. G., and Bonner, J., *Arch. Biochem.*, **14**, 381 (1947).
    Wildman, S. G., Cheo, C., and Bonner, J., *J. Biol. Chem.*, **180**, 985 (1949).
42. Wildman, S. G., Campbell, J., and Bonner, J., *Arch. Biochem.*, **24**, 9 (1949); also *J. Biol. Chem.*, **180**, 273 (1949).
43. Gest, H., and Kamen, M. D., *J. Biol. Chem.*, **176**, 299 (1948).
44. Wildman, S. G., Parker, M., and Campbell, J. M., California Institute of Technology. Unpublished report, 1948.
45. Wildman S. G., and Campbell, J. M., California Institute of Technology. Unpublished report, 1949.
46. Greenstein, J. P., Advances in Protein Chemistry. Academic Press, 1944. Vol. 1, p. 209.
    Mirsky, A. E., Advances in Enzymology. Interscience Publishers, 1943. Vol. 3, p. 1.
47. Gulland, J. M., In *Symp. Soc. Exptl. Biol.*, **1**, 1 (1947).
    Gulland, J. M., *J. Chem. Soc.*, 1722, 1938.
48. Lythgoe, B., and Todd, A. R., *Symp. Soc. Exptl. Biol.*, **1**, 15 (1947).
49. Levene, P. A., and Harris, S. A., *J. Biol. Chem.*, **98**, 9 (1932).
50. Levene, P. A., and Tipson, R. S., *J. Biol. Chem.*, **109**, 623 (1935).
    See discussion in Gulland, J. M., *Symp. Soc. Exptl. Biol.*, **1**, 1 (1947).
51. Calvery, H. O., and Remsen, D. B., *J. Biol. Chem.*, **73**, 593 (1927).
52. Feulgen, R., Behrens, M., and Mahdihassan, S., *Z. physiol. Chem.*, **246**, 203 (1937).
53. Kiesel, A., and Belozersky, A., *Z. physiol. Chem.*, **229**, 160 (1934).
54. Bawden, F. C., and Pirie, N. W., *Proc. Roy. Soc., London*, **B123**, 274 (1937).
55. Feulgen, R., *Z. physiol. Chem.*, **92**, 154 (1914); **100**, 241 (1917).

56. Gulland, J. M., *J. Chem. Soc.*, 1722, 1938.
57. Gulland, J. M., *Symp. Quant. Biol.*, **12**, 95 (1947).
58. Loring, H., *J. Biol. Chem.*, **128**, lxi (1939).
    Kunitz, M., *J. Gen. Physiol.*, **24**, 15 (1940).
59. Cohen, S. S., and Stanley, W. M., *J. Biol. Chem.*, **144**, 589 (1942).
60. Tennent, H. G., and Vilbrendt, C. F., *J. Am. Chem. Soc.*, **65**, 424 (1943).
61. Astbury, W., *Symp. Exptl. Biol.*, **1**, 66 (1947).
62. Signer, R., Caspersson, T., and Hammarsten, E., *Nature*, **141**, 122 (1938).
63. Mirsky, A. E., and Ris, H., *J. Gen. Physiol.*, **31**, 1 (1947).
64. See review in Caspersson, T., *Symp. Soc. Exptl. Biol.*, **1**, 127 (1947).
65. Kunitz, M., *J. Gen. Physiol.*, **24**, 15 (1940).
66. Levene, P. A., and Medigreceanu, F., *J. Biol. Chem.*, **9**, 375 (1911).
67. Greenstein, J. P., *Federation Proc.*, **1**, 113 (1942).
68. Bredereck, H., *Erg. Enzymforschung*, **7**, 105 (1938).
69. Kalckar, H. M., *J. Biol. Chem.*, **167**, 429, 445, 461, 477 (1947).
    See review in *Symp. Soc. Exptl. Biol.*, **1**, 38 (1947).
70. Bawden, F. C., Plant Virus and Virus Disease. Chron. Bot., 2nd ed., 1943.
71. Stanley, W. M., *Science*, **81**, 644 (1935).
72. Greenstein, J. P., Advances in Protein Chemistry. Academic Press, 1944.
    Vol. 1, p. 209.
73. Knight, C. A., *Symp. Quant. Biol.*, **12**, 115 (1947).
74. Cohen, S. S., and Stanley, W. M., *J. Biol. Chem.*, **144**, 589 (1942).
75. Knight, C. A., *J. Biol. Chem.*, **171**, 297 (1947).
76. Wildman, S. G., Cheo, C., and Bonner, J., *J. Biol. Chem.*, **180**, 985 (1949).
77. Granick, S., in Photosynthesis in Plants. Iowa State College Press, 1949.
78. Frey-Wyssling, A., and Mühlethaler, K., *Vierteljahresschrift Naturforsch. Ges.*, **94**, 3 (1949).

CHAPTER 18

THE PLANT PROTEOLYTIC ENZYMES

General. In the protein molecule, amino acids are linked to one another in long chains through peptide linkages in which the α-amino group of one amino acid is bound as a substituted amide to the carboxyl group of the next amino acid. It is the function of the proteolytic enzymes to cleave these peptide bonds.[1]

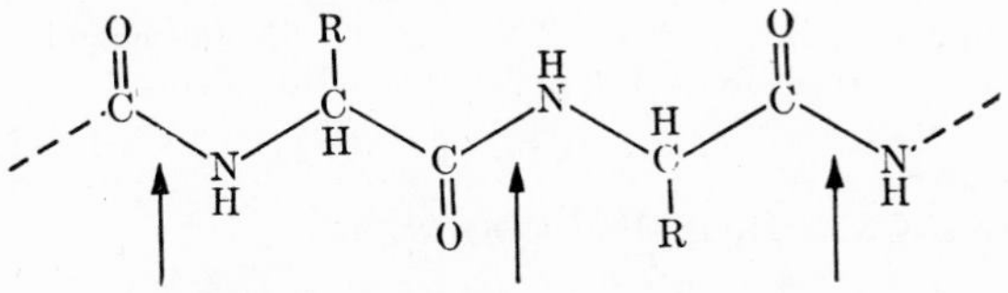

Chain of amino acid residues linked through peptide bonds. Arrows indicate points of attack by hydrolytic enzymes

The proteolytic enzymes may be divided into the peptidases which attack and hydrolyze the peptide bonds of di-, tri-, or polypeptides, and the proteases which are capable of attacking the peptide bonds not only of certain peptides but also of intact protein molecules. Early observations on the existence of proteases in higher plants were made by Vines[2] who found that leaves and other tissues contain enzymes which can liberate tryptophane from plant tissue protein or from peptone. The presence of protease activity in many kinds of plant tissue was also reported by Fisher[3] who determined the release of free amino nitrogen by plant extracts acting on pea meal or on peptone as the substrate. Modern work on plant proteases has principally centered on the study of the enzymes contained in latex of various species, especially of the papaya, *Carica papaya*, and has concerned also, although to a lesser extent, the enzymes of certain seeds.

Proteases. The latex of *Carica*, which is contained particularly in the latex vessels of the leaves and fruit, contains a protease, papain, which has been obtained in crystalline form by Balls *et al.*[4] Papain hydrolyzes a wide variety of proteins to their constituent amino acids and also attacks polypeptides which contain at least two amide bonds, such as leucylglycylglycine or hippurylamide. Papain is inactive in its native condition but is activated by an unknown compound present in papaya latex or by a variety of other compounds including HCN, H_2SO_3, cysteine, ascorbic acid, and other reducing agents. The enzyme appears to exist in two forms, an oxidized inactive form and a reduced active form. This property of activation through reduction, which is not found with the animal proteases pepsin, trypsin, etc. is shared by papain with many other enzymes. The fact that sulfhydryl compounds are effective in the activation suggests that it may be sulfhydryl compounds in the protein which are the centers which must be reduced, the R-SH form of the protein representing the active enzyme, and R-S-S-R or other oxidized forms representing the inactive enzyme. The pH optimum of papain is

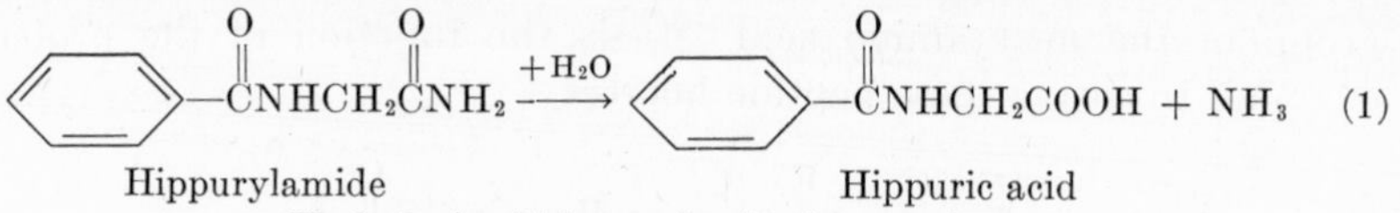

$$\text{Hippurylamide} \xrightarrow{+H_2O} \text{Hippuric acid} + NH_3 \quad (1)$$

Hydrolysis of hippurylamide by papain

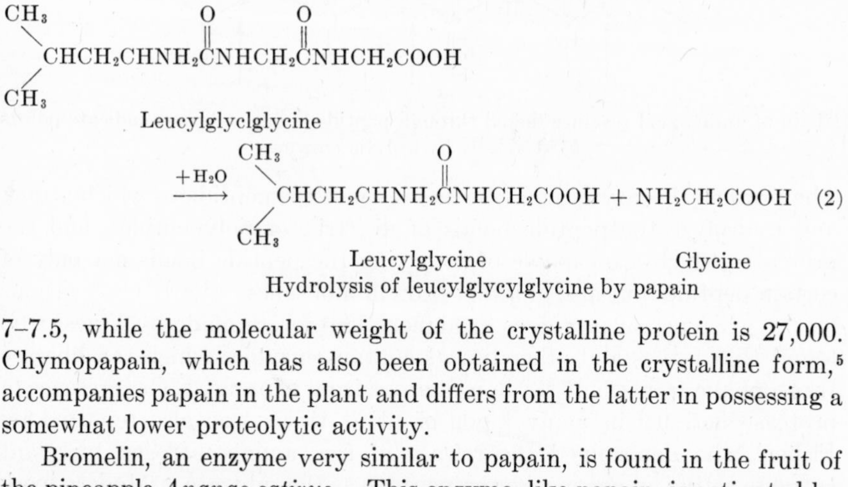

$$\text{Leucylglyclglycine}$$

$$\xrightarrow{+H_2O} \text{Leucylglycine} + \text{Glycine} \quad (2)$$

Hydrolysis of leucylglycylglycine by papain

7–7.5, while the molecular weight of the crystalline protein is 27,000. Chymopapain, which has also been obtained in the crystalline form,[5] accompanies papain in the plant and differs from the latter in possessing a somewhat lower proteolytic activity.

Bromelin, an enzyme very similar to papain, is found in the fruit of the pineapple *Ananas sativus*. This enzyme, like papain, is activated by HCN and cysteine. Still a further and similar enzyme, asclepain, is found in latex of various species of *Asclepias*[6] and the enzyme of *Asclepias syriaca* has been crystallized.[7] The latex of *A. mexicana* yields a related asclepain *m* while that of *A. speciosa* yields the enzyme asclepain *s*. All these enzymes resemble papain in activation characteristics. The euphorbiaceous plant *Hura crepitans* contains in its latex a trypsin-like enzyme, hurain, with a pH optimum of 8 and not activatable by reduction.[8]

The latex of various species of *Ficus*, including *Ficus glabrata* and *Ficus laurifolia*, contains a protease, ficin, which like papain attacks protein with the liberation of amino nitrogen. The pH optimum of ficin is approximately 5 and the enzyme is activated by reducing agents including

TABLE 18-1.   Properties of purified plant proteases.   (After Greenberg and Winnick[6])

Name of enzyme	Source — Common name	Species	Plant part	Crystalline form	Synthetic substrates	pH opt.	Activation by reduction
Papain	Papaya	Carica papaya	Latex of fruit	Needles, hex. plates	Hippurylamide	7–7.5	+
Chymopapain	Papaya	"	"	Needles, plates	....	7	+
Ficin	Fig	Ficus carica F. glabrata	Latex	Hex. plates	Hippurylamide	7	None
Bromelin	Pineapple	Ananas sativus	Fruit	....	Carbobenzoxy-glycylglutamyl-glycylamide	6–7	+
Pinguinain	Maya	Bromelia pinguin	Fruit	....	....	3	
Asclepain	Milkweed	Asclepias species	Latex	Rectangular plates	....	7–7.5	+
Mexicanain	Cuaguayote	Pileus mexicanus	Leaves Fruit	....	....		
Tabernamontain	....	Tabernamontana grandiflora	Latex of fruit	....	....	5–6	+
Euphorbain	Spurge	Euphorbia sp.	Latex	....	....	6	+
Solanain	Horsenettle	Solanum elaeagnifolium	Fruit	....	....	8.5	None
Hurain	Jabillo	Hura crepitans	Latex	....	....	8	None
Arachain	Peanut	Arachis hypogaea	Seed	....	Benzoylarginyl-amide	6.5–7.5	None

cysteine, just as is papain.   Crystalline ficin has been prepared by Walti.[9]

The fruit of *Solanum elaeagnifolium* contains solanain, a protease not affected by oxidation or reduction.[6]

Seeds and seedlings contain active proteases, and the enzymes of wheat and barley, which have been studied particularly,[10] resemble papain in being activated by HCN.   The pH optima of these proteases are lower than those of the latex enzymes and in the case of wheat seedling protease is 4.1.[11]   The protease increases in amount during maturation of the seed, reaches a maximum level, and then falls off with increasing drying out of the grain.[12]   As the seed germinates protease activity increases greatly, the activity being found principally in the tissues of the embryo, including the coleoptile, first leaf, root tips, and scutellum, with the endosperm relatively poor in enzyme.   The great activity of protease in the scutellum during germination suggests that this seed protease may be involved in the mobilization of the reserve protein by the embryonic tissues of the seedling.   The peanut seed contains a highly active trypsin-like enzyme, arachain, which is present particularly in the cotyledons, but is also present although in lesser amounts in the embryo.   The peanut enzyme, like those of cereals, increases in amount during germination of the seed.   Arachain is not influenced by oxidation or reduction and has a pH optimum in the range 6.5–7.5.[13]

The insectivorous plant *Drosera* possesses glandular hairs which function enzymatically in the digestion of entrapped insects.[14]   These glands secrete a protease of high activity, with the extraordinarily low pH optimum of 3.2.   The leaves of *Drosera* yield an extractable protease which differs from the glandular enzyme in having a pH optimum of 4.2–4.3, an optimum range similar to that of the seed enzymes.

Peptidases.   The peptidases,[15] unlike the proteases proper, are restricted in their activities to the hydrolysis of relatively simple peptides containing not more than two or a few amino acid residues.   In addition there appear to be a great many peptidases, each of very considerable specificity as to the substrates which it will attack.   The specificities of the various peptidases have in common the fact that peptides containing D-amino acid residues are either not attacked at all or are split much more slowly than the corresponding peptides containing the natural L-amino acids.   Differences in substrate specificity of the peptidases are based on the nature of the residues attached to the peptide bond or bonds of the substrate and on this basis peptidases may be divided into the general classes shown in Table 18-2.   The dipeptidases hydrolyze peptides such as glycylglycine to their constituent amino acids.   In order that the peptide be attacked it is necessary that it possess a free amino and a free carboxyl group respectively on the two residues flanking the peptide

TABLE 18-2. Classes, specificities, occurrence, and general properties of peptidases

Class of enzyme	Substrate requirement	Typical substrate	Metal required	Occurrence	Remarks
Dipeptidase	Free amino acid carboxyl groups adjacent to peptide bond	Glycylglycine	Co^{++}	Intestinal mucosa, muscle, leaves, seedlings	
Leucylpeptidase	Leucyl residue in peptide linkage	Leucylamide	Mn^{++} or Mg^{++}	Intestinal mucosa, leaves	
Carboxypeptidase	Free carboxyl group adjacent to peptide bond	Carbobenzoxyglycyl-phenylalanine	Mg^{++}	Pancreas	Crystallized by Anson
Prolidase	Proline with free carboxyl group	Glycylproline	Mn^{++}	Intestinal mucosa	
Prolinase	Proline with carboxyl group combined in peptide linkage	Prolylglycylglycine	?	Intestinal mucosa, yeast	
Amino polypeptidase	Peptides with free amino group and three or more residues	Cystinylglycylglycine	?	Fungi, mammalian tissues	
Yeast polypeptidase	Peptide bond adjacent to a free amino group and at least two residues removed from a free carboxyl group	Leucylglycylglycine	?	Yeast	

bond which is to be severed. Secondarily, the structures of the two residues are also of importance in determination of the rate of hydrolysis of the peptide. Glycylglycine dipeptidase is generally distributed in leaves and seedlings[16] as well as in microorganisms and in animal tissues and has been shown in the case of the animal enzyme to be a cobalt pro-

$$NH_2CH_2CONHCH_2COOH \xrightarrow{+H_2O} NH_2CH_2COOH + NH_2CH_2COOH \qquad (3)$$

Glycylglycine Glycine Glycine

Hydrolysis of glycylglycine by dipeptidase

tein. Smith has shown that cobalt possesses the ability of forming complexes with glycylglycine and has suggested that such complex formation is the essential step by which the enzyme and substrate form an enzyme-substrate complex in which the peptide bond becomes unstable and readily hydrolyzed.[17]

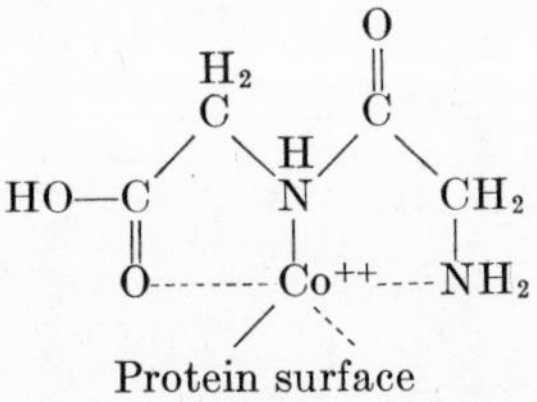

Suggested mode of enzyme-substrate complex formation in the case of dipeptidase. In this complex, glycylglycine is coordinated to cobalt which is in turn coordinated to protein. (After Smith[17])

Leucylpeptidase is restricted in its activity to hydrolysis of peptides in which leucine is bound through its carboxyl group to other amino acids such as in leucylglycine, and in fact, leucylamide is itself split by the

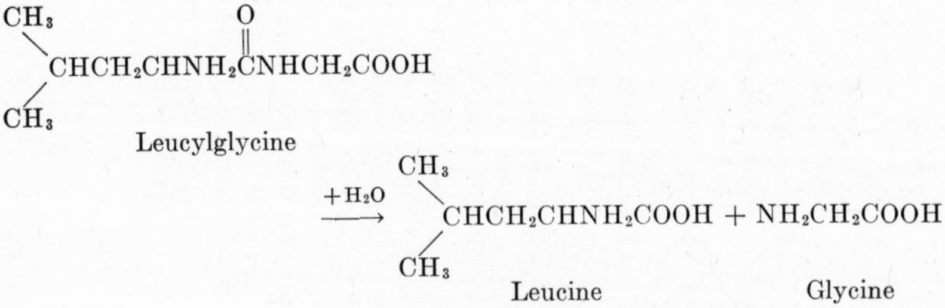

Leucylglycine

Leucine Glycine

Hydrolysis of leucylglycine by leucylpeptidase

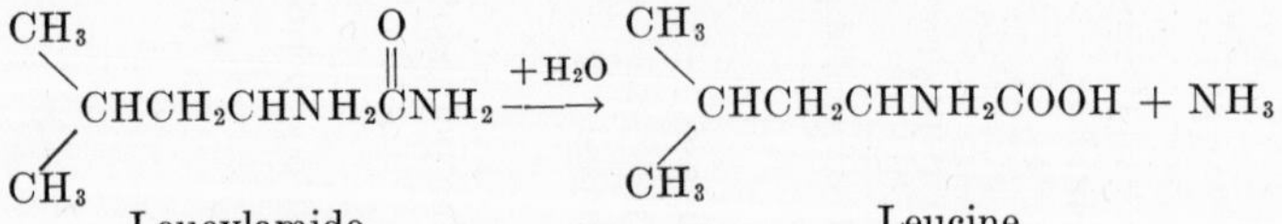

Leucylamide Leucine

Hydrolysis of leucylamide by leucylpeptidase

enzyme. The enzyme is thus an amino peptidase since the presence of a free carboxyl group in the substrate is not required, although a free amino group is essential. This peptidase is widely distributed in animal tissues, is found in leaves,[18] and requires either manganese or magnesium ions for activity.[19] Here also complex formation between substrate and enzyme is undoubtedly mediated through the metal ion.

Carboxypeptidases attack substrates in which a free carboxyl group is adjacent to the peptide bond which is to be split. Thus carbobenzoxyglycylphenylalanine is split by this enzyme but is not split by the dipepti-

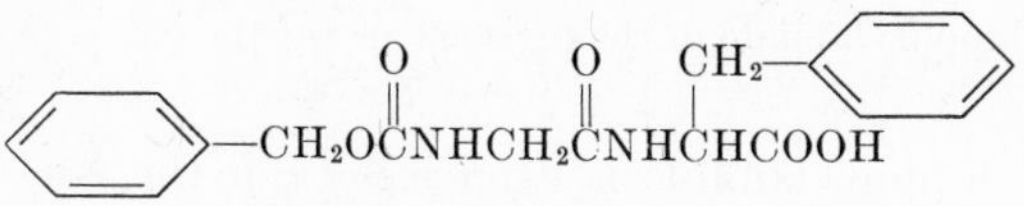

$$\langle\!\!\!\langle\;\rangle\!\!\!\rangle\!-\!CH_2OCNHCH_2CNHCHCOOH$$

Carbobenzoxyglycylphenylalanine

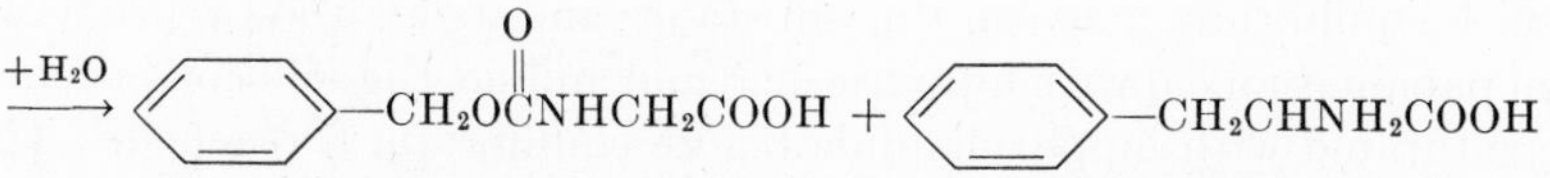

$$\xrightarrow{+H_2O}\;\langle\!\!\!\langle\;\rangle\!\!\!\rangle\!-\!CH_2OCNHCH_2COOH\;+\;\langle\!\!\!\langle\;\rangle\!\!\!\rangle\!-\!CH_2CHNH_2COOH \quad (6)$$

Carbobenzoxyglycine　　　　　　　Phenylalanine
Hydrolysis of carbobenzoxyglycylphenylalanine by carboxypeptidase

dase which requires the further presence of a free amino group. This peptidase has not been characterized from plant tissues but in the case of the animal enzyme is a magnesium protein in which the magnesium is tightly bound to the apoenzyme.[20]

Further peptidases are known from a variety of sources other than higher plants; these are listed in Table 18-2. In general summary, peptidases appear to be metal protein enzymes involving cobalt, manganese, or magnesium as the metal ion. The function of peptidases is entirely obscure.

Peptide Bond Formation. We have seen that there are a great variety of enzymes capable of hydrolyzing peptide bonds, either of intact protein molecules, the proteases, or of the simpler peptides, the peptidases as well as proteases. In the presence of such enzymes the reaction of hydrolysis proceeds to completion, that is, the equilibrium lies far over to the side of splitting of the peptide linkage. It is not therefore ordinarily possible to measure the equilibrium nor is it possible in the usual case to achieve measurable synthesis of a peptide bond by reversal of the hydrolytic reaction. Such a synthesis can, however, be brought about in the special case of synthetic products which are very insoluble and which are removed as formed by precipitation from the reaction mixture. A simple example is that of the synthesis of hippurylanilide from hippuric acid and aniline in the presence of papain.[1]

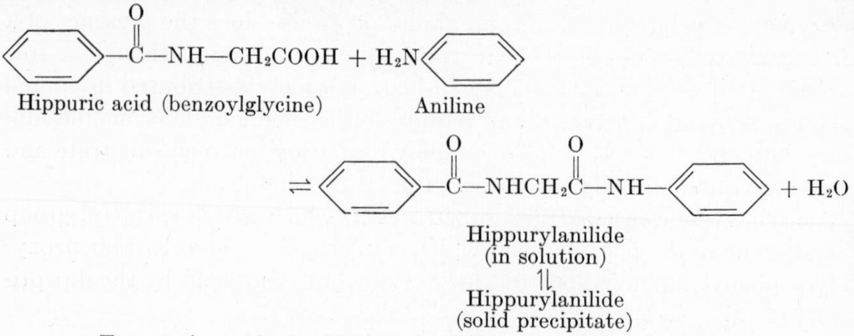

Enzymatic synthesis of hippurylanilide in the presence of papain

In this case the solubility of hippurylanilide is even less than the concentration of hippurylanilide which would need to be present in order to establish equilibrium between this substance and its hydrolytic products. When papain is mixed with hippuric acid and aniline, the solution becomes supersaturated with hippurylanilide before equilibrium is reached. The anilide is deposited as a precipitate, and the synthetic reaction continues. This type of reaction is of great utility in the preparation of anilides and since the synthesis is an asymmetric one in which only peptides of the L-amino acids are utilized, the reaction also serves for the resolution of D,L mixtures.

Peptide bond synthesis by papain (or other proteolytic enzymes) is not restricted to anilide synthesis but can be brought about in other cases in which insoluble products are formed. Thus glycine anilide or glutamic acid anilide may replace aniline itself in the two reactions shown below.

Benzoyl-L-phenylalanylglycine + glycine anilide →
    Benzoyl-L-phenylalanylglycylglycine anilide (insol.)

Carbobenzoxy-L-phenylalanylglycine + glutamic acid anilide →
    Carbobenzoxy-L-phenylalanylglycylglutamic acid anilide (insol.)

Interesting as this type of enzymatic peptide bond synthesis is, it is not known that the mechanism is of importance in peptide bond synthesis *in vivo*. In the synthesis of a protein, for example, the product of a particular synthetic reaction would need to be immediately and nearly completely removed by some second reaction in order to allow this type of synthesis to occur.

The hydrolysis of the peptide bond involves a decrease in free energy of some 3500–4000 cal. per mol.

$$R_1-CHCOOH$$
$$|$$
$$NH$$
$$| \qquad + H_2O \rightarrow R_1CHNH_2COOH + R_2CHNH_2COOH$$
$$C=O \qquad\qquad\qquad\qquad \Delta F = -3500\text{--}4000 \text{ cal.}$$
$$|$$
$$CHNH_2-R_2$$

It would seem probable therefore that the synthesis of peptide bonds, rather than constituting a simple reversal of the hydrolytic reaction, may take a different course in which chemical energy from some source is put into the synthetic reaction. In a general way the problem of peptide bond formation and hydrolysis is analogous to the problem of starch or sucrose synthesis and hydrolysis. In each case, enzymes for the simple hydrolysis are found in nature. With starch and with sucrose, however, we know that synthesis is not a simple reversal of the hydrolysis but involves quite different enzymes and different starting materials, e.g., phosphorylated sugars. It may be anticipated therefore that some such roundabout process may also be involved in peptide bond synthesis.

Little or no study of peptide bond synthesis has been made with plant material, and in fact our knowledge of this process comes principally from work with liver and kidney. Liver slices will carry out the reaction:[21]

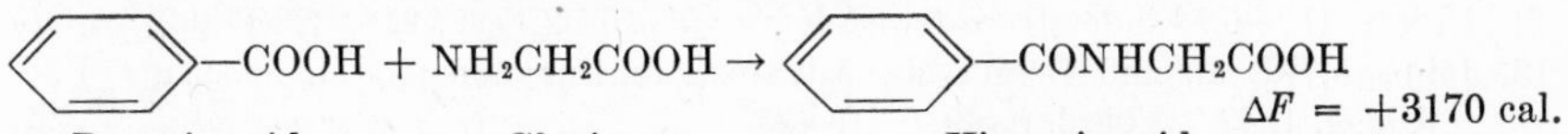

$$\langle\!\!\!\!\bigcirc\!\!\!\!\rangle-COOH + NH_2CH_2COOH \rightarrow \langle\!\!\!\!\bigcirc\!\!\!\!\rangle-CONHCH_2COOH$$
$$\Delta F = +3170 \text{ cal.}$$

Benzoic acid   Glycine   Hippuric acid

para-Amino benzoic acid may be substituted for benzoic acid with the formation of p-amino hippuric acid.[22] This reaction must be supplied with energy in order to proceed to an appreciable extent, and in fact the energy comes from respiratory energy since poisoning of the slices with cyanide stops the reaction, as do other respiratory inhibitors including iodoacetate, fluoride, and azide. Cell-free homogenates, i.e., whole ground tissue, can carry out the above reaction just as do slices, and in fact purified preparations of particulate matter from the cell-free homogenates also carry out the synthesis under appropriate conditions. These conditions include the presence in the system of cytochrome c, fumarate, α-ketoglutarate, or other oxidizable acid, Mg^{++} ions, adenylic acid or ATP, and oxygen. Presumably oxidation of fumarate acts as a continuing source of ATP since under anaerobic conditions a small synthesis can be obtained with the addition of ATP alone. It would appear then that energy-rich phosphate in some manner supplies the energy needed for synthesis of the peptide bond.

General References

Bergmann, M., Advances in Enzymology. Interscience Publishers, 1942. Vol. 2, p. 49.

Bergmann, M., and Fruton, J. S., Advances in Enzymology. Interscience Publishers. 1941. Vol. 1, p. 63.
Sumner, J. B., and Somers, G. F., Enzymes. Academic Press, 2nd ed., 1947.
Smith, E. L., *Ann. Rev. Biochem.*, **18**, 35 (1949).
Greenberg, D. M., and Winnick, T., *Ann. Rev. Biochem.*, **14**, 31 (1945).

References

1. Bergmann, M., and Fruton, J. S., Advances in Enzymology. Interscience Publishers, 1941. Vol. 1, p. 63.
2. Vines, S., *Ann. Botany*, **16**, 1 (1902); **17**, 237, 597 (1903); also series in *idem* to **24**, 213 (1910).
3. Fisher, E., *Biochem. J.*, **13**, 124 (1919).
4. Balls, A. K., and Lineweaver, H., *J. Biol. Chem.*, **130**, 669 (1939).
5. Jansen, E. F., and Balls, A. K., *J. Biol. Chem.*, **137**, 459 (1941).
6. Greenberg, D. M., and Winnick, T., *J. Biol. Chem.*, **135**, 761 (1940); *Ann. Rev. Biochem.*, **14**, 31 (1945).
7. Carpenter, D. C., and Lovelace, F., *J. Am. Chem. Soc.*, **65**, 2364 (1943).
8. Jaffe, W. G., *J. Biol. Chem.*, **149**, 1 (1943).
9. Walti, A., *J. Am. Chem. Soc.*, **60**, 493 (1938).
10. Balls, A. K., and Hale, W. S., *Cereal Chem.*, **13**, 54, 656 (1936); **15**, 622 (1938).
11. Mounfield, J. D., *Biochem. J.*, **30**, 549, 1778 (1936).
12. Bach, A., Oparin, A., and Wähner, R., *Biochem. Z.*, **180**, 363 (1927).
13. Irving, G. W., and Fontaine, T., *Arch. Biochem.*, **6**, 351 (1945).
14. Holter, H., and Linderstrøm-Lang, K., *Z. physiol. Chem.*, **214**, 223 (1933).
15. Johnson, M. J., and Berger, J., Advances in Enzymology. Interscience Publishers, 1942. Vol. 2, p. 69.
16. Ambros, O., and Harteneck, Z., *Z. physiol. Chem.*, **184**, 93 (1929).
    Linderstrøm-Lang, K., and Holter, H., *Z. physiol. Chem.*, **204**, 15 (1932).
    Mounfield, J. D., *Biochem. J.*, **30**, 549, 1778 (1936).
17. Smith, E. L., *J. Biol. Chem.*, **176**, 21 (1948).
18. Linderstrøm-Lang, K., and Sato, M., *Z. physiol. Chem.*, **184**, 83 (1929).
    Sato, M., *Compt. rend. trav. Lab. Carlsb.*, **19**, 1 (1931).
    Berger, J., and Johnson, M. J., *J. Biol. Chem.*, **130**, 655 (1939).
19. Smith, E. L., *J. Biol. Chem.*, **163**, 15 (1946).
20. Smith, E. L., and Hanson, H. T., *J. Biol. Chem.*, **176**, 997 (1948).
21. Borsook, H., and Dubnoff, J., *J. Biol. Chem.*, **132**, 307 (1940); **168**, 397 (1947).
22. Cohen, P. P., and McGilvery, R., *J. Biol. Chem.*, **166**, 261 (1946); **169**, 119 (1947); **171**, 121 (1947).

CHAPTER 19

NITROGEN METABOLISM OF SEEDLINGS

As the seed germinates and grows into a seedling, marked changes in the nature and disposition of the nitrogen compounds of the seed occur. The study of these changes has occupied the attention of many workers and will be considered under the general heading of nitrogen metabolism. These studies have in general been carried out from the standpoint of analysis of the amounts of the several groups of nitrogen compounds in the tissues, rather than from the standpoint of the individual chemical reactions involved.

Proteolysis. The dominating process of nitrogen metabolism in the germinating seed is the hydrolysis of the reserve protein of the seed with the production of amino acids. These amino acids are then believed to be transported to the growing embryonic regions where they are resynthesized to protein. Since the protein of the young plant undoubtedly differs in amino acid constitution from that of the reserve seed protein, it is clear that extensive amino acid breakdown and resynthesis must also occur in the seedling. It has been mentioned in Chapter 18 that active proteolytic enzymes are found in the seed and seedling, and it is presumed that protein degradation in the seed takes place through the agency of these enzymes. The controlling factor responsible for the breakdown of protein is probably the diminution of the free amino acid concentration in the reserve organs of the seed (endosperm or cotyledons) due to mobilization by the growing seedling. Thus the hydrolysis of the reserve protein of the seed may in some cases be diminished in rate by supplying soluble nitrogen as nitrate, or urea, this nitrogen presumably being converted to amino acids by the seedling at the expense of carbohydrate brought from the seed. An example of the protein sparing action of applied nitrogen is given in Table 19-1. Similar results in sparing of protein and decrease

TABLE 19-1. Diminution of protein hydrolysis in endosperm of wheat seedlings as a result of supplying the seedlings with soluble nitrogen. Etiolated seedlings of *Triticum sativum*. Concentrations of nitrogen after 5 days' growth. (After Paech[1])

Nutrient	N in 100 endosperms		N in 100 embryos	
	Protein N	Soluble N	Protein N	Soluble N
Water	35.7 mg.	12.4 mg.	23.4 mg.	14.8 mg.
0.5% NH₄NO₃	47.3	16.7	25.8	27.4
Water	34.0	13.4	22.0	12.3
0.5% urea	45.7	31.7	23.9	41.7

of protein hydrolysis can be achieved with the low carbohydrate seeds of *Lupinus albus* by the supplying of glucose alone to the germinating seedling, as is shown in Table 19-2. The analytical results given in

TABLE 19-2. Influence of glucose in decreasing protein hydrolysis in the cotyledons of seedlings of *Lupinus albus*. Results after 14 days' growth in darkness. (After Paech[1])

	Cotyledons of 6 plants		Embryonic axes of 6 plants	
Nutrient	Protein N	Soluble N	Protein N	Soluble N
N-free	19.8 mg.	23.7 mg.	12.8 mg.	81.5 mg.
N-free + 3% glucose	46.8	36.0	12.0	49.7

Table 19-2 suggest that here sugar may have been converted to amino acids, presumably at the expense of nitrogen already in the seedling. Direct experiments on the regulation of protein hydrolysis by amino acid concentration, while abundantly available for green leaves, are however absent in the case of seedlings.

Amide Formation. The second dominant feature of the nitrogen metabolism of seedlings is the production of the amides, asparagine and glutamine.

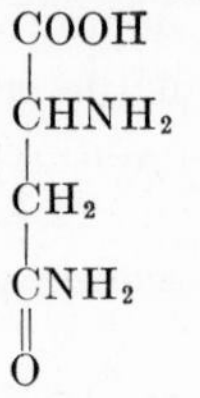

$$
\begin{array}{cc}
\text{COOH} & \text{COOH} \\
| & | \\
\text{CHNH}_2 & \text{CHNH}_2 \\
| & | \\
\text{CH}_2 & \text{CH}_2 \\
| & | \\
\text{CNH}_2 & \text{CH}_2 \\
\| & | \\
\text{O} & \text{CNH}_2 \\
& \| \\
& \text{O}
\end{array}
$$

Asparagine (amide of aspartic acid)　　　Glutamine (amide of glutamic acid)

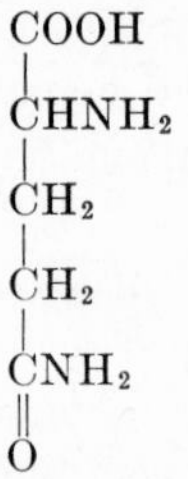

Modern work on this aspect of nitrogen metabolism starts with the investigations of Ernst Schulze, the great Swiss plant chemist. Schulze followed analytically the concentrations of the various nitrogenous compounds in seedlings germinated in the dark and Table 19-3 gives a typical result obtained with lupin. As the seed germinates, the total dry weight and the total protein decrease rapidly. At the same time a marked

TABLE 19-3. Concentrations of various nitrogen compounds found in seedlings of *Lupinus* germinated in the dark. (From Schulze, after Chibnall[2,3])

	Amount per 100 gm. original seeds		
Compound	0 days	7 days	12 days
Total dry weight	100.00 gm.	87.4 gm.	81.7 gm.
Protein	45.1	24.9	11.7
Free amino acids	7.8	15.7	20.4
Asparagine	0.0	9.8	18.2

accumulation of soluble amino acids occurs but, in particular, asparagine is synthesized in large amounts. In other experiments Schulze found asparagine to make up to 25 % of the dry weight of etiolated lupin seedlings. The reasons for and mechanism of this asparagine accumulation have exercised plant physiologists and chemists very considerably.

In lupins grown in the light no accumulation of asparagine such as that found in darkness takes place, and if etiolated lupin seedlings are transferred to the light the accumulated asparagine gradually disappears again. Schulze supposed, following the suggestion of Borodin,[4] that the accumulation of asparagine in etiolated seedlings such as lupin is a response to starvation conditions under which protein is hydrolyzed and the resulting amino acids used as substrates for respiration. Schulze found that the amount of asparagine accumulated greatly exceeded the amount already contained in the reserve protein of the seed and released by protein hydrolysis during germination, and he showed clearly that asparagine synthesis must take place at the expense of other amino acids. Asparagine can, however, be formed directly from ammonia taken up by the seedling from the nutrient solution, as has been shown by Prianischnikov.[5] Pea seedlings placed in NH_4Cl solution took up the ammonia vigorously but accumulated not free ammonia but rather asparagine. Similar results were obtained by Prianischnikov with barley seedlings as is shown in Table 19-4. Prianischnikov's view is then that asparagine is developed

TABLE 19-4. Production of asparagine by seedlings supplied with ammonium salts. Plants germinated in water or in 0.1 % NH_4Cl for 14 days. (Prianischnikov, and Schulow and Prianischnikov[5,6])

	Milligrams/100 seedlings		
	Barley seedlings in NH_4Cl soln.*	Barley embryos in NH_4Cl soln.†	Pea seedlings in NH_4Cl soln.‡
Increase in total N	15.7	79	196
Increase in NH_3 N	0.3	69	...
Increase in asparagine N	19.6	−8	182

* Increases above similar seedlings in N-free solution.
† $CaCO_3$ added to medium.
‡ $CaSO_4$ added to medium.

by the seedling in response to the presence of ammonia in the plant and that asparagine synthesis is in fact a detoxification mechanism. Operation of this mechanism requires, however, that a source of the carbon skeleton and a source of the energy required for the synthesis be available. Prianischnikov has shown that both of these requirements can be filled by hexose. His experiment consisted in depleting carbohydrates from barley or lupin seedlings either, as with lupin, by simply allowing them to germinate in the dark for ten or more days, or in the case of barley by excision of the embryo from the endosperm. Ammonia supplied to such carbohydrate-poor plants is taken up and accumulated as free ammonia, as shown in Table 19-4. When glucose is supplied to such depleted seedlings, however, asparagine formation is increased and ammonia accumulation decreased. Amide synthesis can be induced in seedlings then by conditions of available ammonia together with available carbohydrate or possibly of other non-nitrogenous precursors. Amide formation would appear to be related to the intensive protein hydrolysis of the seedling only through the fact that amide synthesis provides a mechanism for removal of ammonia formed by the deamination of the amino acids produced in proteolysis.

The synthesis and accumulation of asparagine is characteristic of the germination and growth of etiolated seedlings only over a restricted period. With advancing depletion of the reserves of the seedling, the asparagine finally begins to disappear, with the simultaneous appearance of free ammonia.[7] This state is ordinarily soon followed by death of the plant. The asparagine is here presumably utilized as a substrate for respiration after the exhaustion of other reserves.

Glutamine, the amide of glutamic acid, like asparagine is widely distributed in plants and in fact takes over the role of asparagine in many species, accumulating under conditions of proteolysis, etc. In most species glutamine and asparagine are found side by side although in lupin for example asparagine greatly predominates. In other species as in *Ricinus communis* glutamine predominates, while in still other species as *Helianthus annuus*, *Cucurbita* sp. and *Linum usitatissimum*, the two may occur in more or less equivalent amounts.[7,8]

The mechanism of formation of asparagine or glutamine is known only in a general way for plant tissues.[9] The formation of aspartic and glutamic acids from the corresponding keto acids has been discussed in Chapter 16, and it has been shown that glutamic acid may be formed by reaction of α-ketoglutaric acid with ammonia in the presence of glutamic dehydrogenase. Aspartic acid may be formed from glutamic acid and oxaloacetic acid by transamination. The method by which the final amidation is accomplished is, however, obscure. Energy is required for the process (3460 cal. per mol), and the amidation is coupled to energy-yielding reactions of respiration. Krebs has shown[10] that glutamine synthesis in kidney and brain tissue proceeds only aerobically and that inhibition of respiration similarly inhibits amide formation. The synthesis of glutamine in animal tissues appears to involve a phosphorolytic reaction, in which ATP is required, and it is possible that γ-glutamyl-phosphate may be an intermediate in the reaction.[12] The role of respiration in amide formation is therefore in part production of the energy-rich phosphate needed to drive the reaction. The work of Chibnall with leaves, which is discussed in the next section, has shown that α-keto-glutaric acid rather than glutamic acid is the substrate which is primarily consumed during glutamine formation, and both the α-amino nitrogen and the amide nitrogen are therefore derived directly from the ammonia pool.

$$
\begin{array}{ccc}
\text{COOH} & \overset{\displaystyle O}{\underset{\displaystyle |}{C} \sim ph} & \overset{\displaystyle O}{\underset{\displaystyle |}{C} - NH_2} \\
| & | & | \\
CH_2 & CH_2 & CH_2 \\
| & | & | \\
CH_2 + ATP \rightarrow & CH_2 + R \cdot NH_2 \rightarrow & CH_2 + H_3PO_4 + R \\
| & | & | \\
CHNH_2 & CHNH_2 & CHNH_2 \\
| & | & | \\
COOH & COOH & COOH \\
\text{Glutamic acid} & \text{Glutamylphosphate} & \text{Glutamine}
\end{array}
$$

Amidases, enzymes responsible for the hydrolysis of the amide groups of asparagine or glutamine, are found in seedlings of cereals.[11,8] Both reactions have pH optima in the region of pH 8 but there are probably two distinct enzymes involved since an asparaginase without glutaminase activity is known from yeast[8] while glutaminases without asparaginase

$$
\begin{array}{ccc}
\overset{\displaystyle O}{\underset{\displaystyle |}{C} - NH_2} & & \overset{\displaystyle O}{\underset{\displaystyle |}{C} - OH} \\
| & & | \\
CH_2 & \xrightarrow{\text{Asparaginase}} & CH_2 \\
| & + H_2O \longrightarrow & | & + NH_3 \\
CHNH_2 & & CHNH_2 \\
| & & | \\
COOH & & COOH \\
\text{Asparagine} & & \text{Aspartic acid}
\end{array}
$$

$$
\begin{array}{ccc}
\overset{\displaystyle O}{\underset{\displaystyle |}{C} - NH_2} & & \overset{\displaystyle O}{\underset{\displaystyle |}{C} - OH} \\
| & & | \\
CH_2 & & CH_2 \\
| & \xrightarrow{\text{Glutaminase}} & | \\
CH_2 & + H_2O \longrightarrow & CH_2 & + NH_3 \\
| & & | \\
CHNH_2 & & CHNH_2 \\
| & & | \\
COOH & & COOH \\
\text{Glutamine} & & \text{Glutamic acid}
\end{array}
$$

activity are known in animal tissues.[10] The role of these enzymes in amide metabolism is, however, wholly obscure, since cereal seedlings which contain the enzymes are capable of extensive amide accumulation under appropriate circumstances, e.g., during growth in the dark.

General References

Chibnall, A. C., Protein Metabolism in the Plant.   Yale University Press, 1939.

References

1. Paech, K., *Planta*, **24,** 78 (1935).
2. Schulze, E., See complete review in Chibnall's book.
3. Chibnall, A. C., Protein Metabolism in the Plant.   Yale University Press, 1939.
4. Borodin, I., *Botan. Jahresber.*, **4,** 919 (1876); *Botan. Zeitung*, **36,** 802 (1878).
5. Prianischnikov, D., *Ber.*, **40,** 242 (1922).
6. Prianischnikov, D., and Schulow, P., *Ber.*, **28,** 253 (1910).
7. Vickery, H. B., and Pucher, G. W., *J. Biol. Chem.*, **150,** 197 (1943).
8. Schwab, G., *Planta*, **25,** 579 (1936).
9. Mothes, K., *Planta*, **30,** 726 (1940).
   Archibald, R. M., *Chem. Revs.*, **37,** 161 (1945).
10. Krebs, H. A., *Biochem. J.*, **29,** 1951 (1935).
11. Geddes, W. F., and Hunter, A., *J. Biol. Chem.*, **77,** 197 (1928).
12. Speck, J. F., *J. Biol. Chem.*, **168,** 403 (1947).
    Elliot, W. H., *Biochem. J.*, **42,** v (1948).

Chapter 20

NITROGEN METABOLISM OF LEAVES

Behavior of Excised Leaves. The leaf is the great laboratory of the plant and contains a vast array of chemical mechanisms for syntheses of the most varied character. Among the metabolic activities of leaves which are readily available for study, the changes in nitrogenous compounds are outstanding, and a vast amount of effort has gone into studies of the changes in the levels of the various nitrogen fractions in leaves subjected to varying environmental conditions. The nitrogen metabolism of excised leaves in particular has been subjected to such study. Schulze[1] first performed such experiments and found that in excised leaves, kept in the dark with their petioles in water, protein is hydrolyzed and asparagine accumulated much as in seedlings. Similar results have been obtained by many other workers with a variety of species as for example by Chibnall[2] and by Mothes[3] with leaves of *Phaseolus* and *Vicia*.

TABLE 20-1. Changes in N constituents of leaves in *Phaseolus multiflorus* excised and stored with petioles in water. (After Chibnall[2])

	Changes in N constituents as % of total N				
	Protein N	*Nonprotein N*	*Ammonia N*	*2 × amide N*	*Amino N*
After 5 days in light	−13.3	+13.3	+0.07	+7.1	+5.4
After 5 days in dark	−18.6	+18.6	+0.2	+9.9	+4.1

Table 20-1, taken from Chibnall, shows that in excised leaves protein is hydrolyzed and amides and amino acids accumulate. Such changes in constitution would not ordinarily have taken place had the leaves remained attached to the plant, in which case the protein level remains relatively constant as will be discussed below. That protein hydrolysis takes place both in dark where the leaves are depleted of reserves and in light where photosynthetic formation of carbohydrate may occur indicates that protein breakdown is caused primarily by excision rather than by starvation of the leaf.

Table 20-1 shows that a major portion of the nitrogen liberated during protein hydrolysis in the excised leaf may appear as amide nitrogen. This accumulation of amide must be owing to secondary synthesis of the substance, since only relatively small amounts of amides are contained in the leaf protein. Mothes[4] was among the first to study conditions governing amide synthesis in excised leaves, particularly with the hope of discovering the nature of the carbon skeleton from which asparagine is formed. He infiltrated leaves with various compounds structurally related to asparagine such as ammonium aspartate and ammonium succinate and found that all these substances caused large increases in aspara-

TABLE 20-2. Influence of various compounds on the accumulation of asparagine by excised leaves of *Phaseolus multiflorus*. Incubated 30 hours in dark at 28°. (From Mothes[4])

		Amounts in milligrams per sample		
	Fresh wt. of sample, gm.	$NH_3 N$ infiltrated	$NH_3 N$ after 30 hr.	$2 \times$ amide N after 30 hr.
Initial leaves	24.0	0.0	0.5	5.5
Infiltrated with water	19.2	0.0	0.7	7.6
Infiltrated with ammonium succinate	19.5	25.5	1.1	21.2
Infiltrated with ammonium malate	22.4	24.6	2.1	24.8
Infiltrated with ammonium aspartate	23.9	27.5	1.8	30.4

gine, the infiltrated ammonia being almost quantitatively transformed to amide (Table 20-2). There is no evidence, however, that the carbon compounds supplied are actually used in asparagine synthesis, and in fact infiltration with ammonium sulfate is as effective as are ammonium salts of four-carbon acids.[5] It would appear then that the carbon skeleton

TABLE 20-3. Formation of glutamine from various substrates by excised leaves of *Lolium perenne*. Leaves infiltrated in dark at 16–18°C. (From Chibnall[6])

		N in milligrams per 10 gm. fresh leaves		
Substance infiltrated	Time in hr.	Free ammonia	Glutamine amide N	Asparagine amide N
Water	0	0.2	0.9	0.5
	22	0.2	2.3	0.6
Ammonium pyruvate	0	7.6	0.9	0.8
	22	1.0	5.5	0.8
Ammonium phosphate	0	9.9	0.8	0.8
	22	2.3	6.0	0.9

for amide synthesis is already present in the tissue or in any case can be readily formed from an endogenous source and that both N atoms of the amide formed derive from the ammonium supplied. It remained for Chibnall[6] to demonstrate that amide is indeed formed from such an endogenous precursor. Greenhill and Chibnall[7] observed that pot cultures of *Lolium perenne*, the perennial rye grass, after an application of ammonium sulfate, were covered the next day with a white powder.

This powder was glutamine. Chibnall therefore selected this plant as an especially active maker of amide and infiltrated excised leaves of *Lolium* with various possible precursors of glutamine. The results of his experiment, shown in Table 20-3, indicate that rapid and extensive glutamine formation takes place at the expense of either ammonium pyruvate or ammonium phosphate. Infiltration with glutamic acid or ammonium α-ketoglutarate did not increase the rate of synthesis. Chibnall next

TABLE 20-4. Correspondence between formation of glutamine and utilization of α-ketoglutarate in excised leaves of *Lolium perenne*. (After Chibnall[6])

	After 4 hr. of incubation	After 20 hr. of incubation
1. Decrease in α-ketoglutaric acid	17.1 mg./10 gm.	26.2 mg./10 gm.
2. Increase in glutamine N calculated from 1, if all used in synthesis	3.3	5.0
3. Observed increase in glutamine N	3.2	5.3

showed that in leaves infiltrated with ammonium α-ketoglutarate the latter is used up and that a close correspondence exists between keto acid disappearance and glutamine appearance. During the course of the reaction a greater loss of sugar was observed in the leaves supplied with ammonia than those given water alone and Chibnall suggests that this may represent the sugar used in respiratory reactions supplying energy for the glutamine formation. The exact pathway of asparagine formation in leaves remains obscure, however.

TABLE 20-5. Type of amide accumulated in excised leaves of various species in light or in dark

Species of leaf	Principal amide accumulated in:		Author
	Dark	Light	
Tobacco	Asparagine	Asparagine & glutamine	Vickery[9]
Kikuyu grass	Asparagine	....	Wood[10]
Barley	Asparagine & glutamine	....	Yemm[11]
Oat	Asparagine & glutamine	....	Wood[12]
Rhubarb	Glutamine	Glutamine	Vickery[13]
Corn	Asparagine & glutamine	Asparagine	Viets[14]
Perennial rye grass	Glutamine	Glutamine	Chibnall[6]

Tissues other than leaves also respond to high external concentrations of ammonium ions by amide synthesis. Thus beet roots show marked increases in glutamine in as little as three days after application of ammonium fertilizers even in soil culture.[8]

The proportion of asparagine to glutamine in the amides accumulated by starving leaf tissue is characteristic of the species of leaf as well as of the experimental conditions. Table 20-5 summarizes the behavior of a number of species. Tobacco and Kikuyu grass accumulate primarily asparagine during starvation in the dark. Rhubarb and *Lolium* on the contrary accumulate primarily glutamine. Still other species including barley, oats, and corn accumulate both asparagine and glutamine when the detached leaves are maintained in the dark. Amide accumulation may be qualitatively altered in the light. Thus tobacco, which in the dark accumulates primarily asparagine, accumulates both asparagine and glutamine in the light. Corn on the other hand accumulates both amides in the dark, but in the light glutamine synthesis is largely suppressed. These relations must be ultimately owing to the availability of oxaloacetic and α-ketoglutaric acids as precursors of the amides and to effects of light on the relative proportions of these materials in the leaf. The influence of light on amide synthesis may then be due to the interrelations between photosynthetic and respiratory metabolism of organic acids, a subject discussed in Chapter 30.

During recent years several comprehensive series of investigations on the overall metabolism of excised leaves have been carried out. These investigations enable us to compare the changes which take place after excision of the leaf not only with regard to nitrogenous compounds but also with regard to sugars, organic acids, and even respiration. The first of these investigations to be discussed is that of Yemm[11] who used leaves from field-grown barley plants. The leaves were incubated under moist conditions in the dark and samples removed at periodic intervals for analysis. Two principal series of experiments were carried out, the first of which dealt with carbohydrate changes in the leaf and the second with changes in nitrogenous materials. In both cases simultaneous determination of respiratory changes were made (Fig. 20-1). Respiration rate as judged by CO_2 evolution dropped rapidly from an initial high level immediately after excision and became stabilized at a new lower level after approximately forty hours of incubation. Analysis of the sugar changes in the tissue during this period showed that sucrose, which is the principal reserve sugar, was depleted rapidly and at a rate parallel to the dropping respiratory rate. Starch, fructose, and fructosan were similarly depleted, and only glucose showed an increase in concentration up to approximately forty hours. If the total CO_2 evolved is compared with the carbohydrate consumed, it is clear that oxidation of carbohydrate can account for respiratory carbon loss only during the first twenty-four hours or so of incubation. From twenty-four hours on then an increasing fraction of respiration is at the expense of other, noncarbohydrate, substrates.

Changes in protein and in soluble nitrogen in the barley leaves of Yemm's experiment are summarized in Fig. 20-2. From the initiation of the experiment on, protein was hydrolyzed and soluble nitrogenous

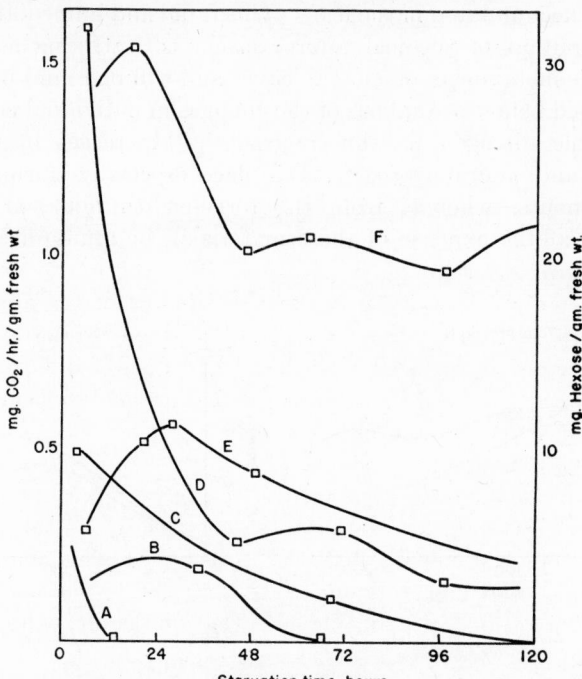

FIG. 20-1. Changes in carbohydrate and in rate of respiration in excised barley leaves in dark. (After Yemm[11])

A. Fructose   B. Fructos an   C. Starch   D. Sucrose   E. Glucose   F. Mean CO_2

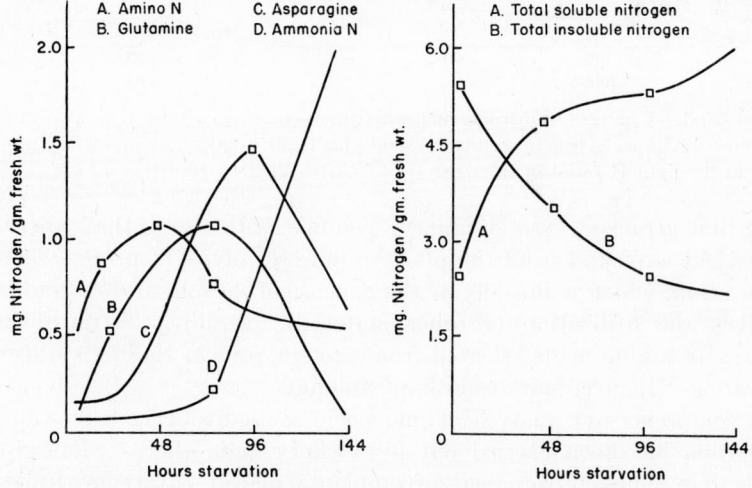

FIG. 20-2. Changes in nitrogenous constituents of excised barley leaves in dark. (After Yemm[11])

constituents accumulated in the leaf. This rapid and immediate decrease in protein content of the leaf after excision is a striking and general feature of the metabolism of excised leaves and will be found in the other cases discussed below. Analysis of the changes in individual components of the soluble nitrogen fraction reveals that increases in glutamine, asparagine, and amino nitrogen take place especially during the first forty-eight hours, whereas from the forty-eighth hour on ammonia accumulates at the expense of these materials, the amino acid nitrogen

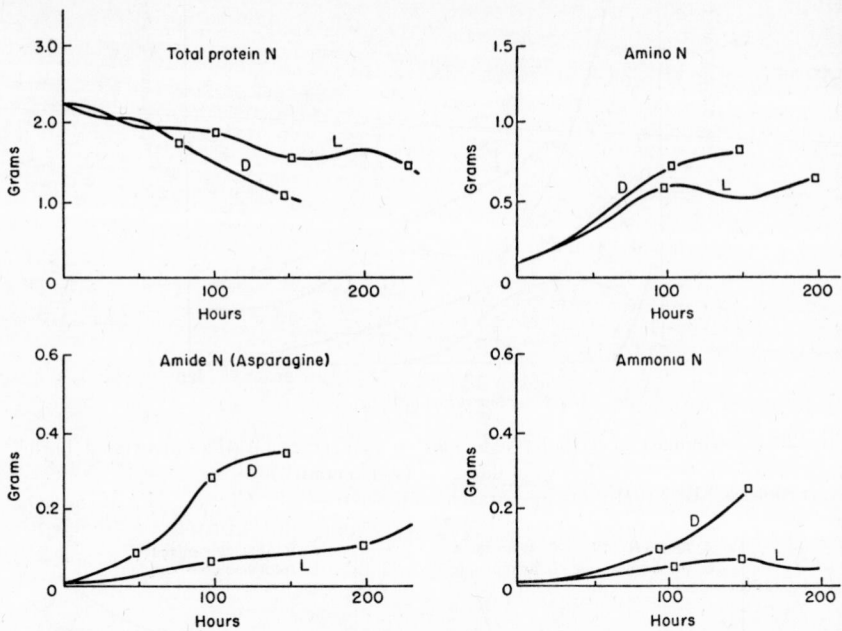

FIG. 20-3. Changes in nitrogenous constituents of excised tobacco leaves in light and dark. Values given in grams N/1000 gm. fresh weight. Samples cultured in water in the light (L) and in the dark (D). (After Vickery et al.[9])

being first attacked. In summary, Yemm's work shows that when the barley leaf is picked from the plant, rapid hydrolysis of protein sets in. Respiration, which is initially at the expense of carbohydrates, gradually involves the utilization of other materials, possibly the deaminated residues of amino acids released from protein, and in the final stages of starvation NH_3 accumulates in large amount.

A comprehensive analysis of changes in excised tobacco leaves during incubation has been carried out by Vickery and others.[9] Full-grown leaves were removed from field-grown plants, placed with their petioles in water or other solution, and cultured either in light or dark. The leaves in light carried on photosynthesis and in fact increased in carbohydrate content and in dry weight during the more than two hundred hours of incubation. Protein hydrolysis took place at a rapid rate both in light and dark, showing again that excision is responsible for a basic alteration in protein metabolism in the leaf and that the hydrolysis of protein is not due solely to depletion of carbohydrate reserves of the leaf. As protein was hydrolyzed, α-amino nitrogen accumulated both in light and dark. With leaves in the light glutamine was synthesized and accumulated, while with leaves in the dark asparagine was the principal amide to accumulate. Appropriate calculations based on the amide content of leaf protein show that the nitrogen of both asparagine and glutamine must

TABLE 20-6. Changes in excised tobacco leaves incubated in the dark. (Taken from Chibnall's[6] summary of the work of Vickery et al.[9])

	Grams/1000 gm. fresh leaves				
	Hours of incubation			Changes during	
Substance or fraction	0	73	143	0–73 hr.	73–143 hr.
Starch	0.43	00.00	00.00	−0.43	0.00
Protein	14.19	10.94	6.25	−3.25	−4.69
Other insoluble organic matter	29.88	29.36	26.80	−0.52	−2.56
Total sugars	2.51	0.67	0.50	−1.84	−0.17
Citric acid	3.26	7.08	9.45	+3.82	+2.37
Malic acid	15.20	6.55	4.48	−8.65	−2.07
Oxalic acid	1.57	1.76	1.64	+0.19	−0.12
Asparagine	0.18	1.50	3.05	+1.32	+1.55
Glutamine	0.12	0.60	0.39	+0.48	−0.21
Nicotine	1.04	0.97	0.87	−0.05	−0.10
Amino acids	0.67	2.26	2.22	+1.59	−0.02
Other nitrogenous organic compounds	2.75	2.30	1.80	−0.45	−0.50
Undetermined	0.70	2.21	−1.30	+1.52	−3.52

have had its origin mainly in secondary deamination of the amino acids set free during the protein hydrolysis. Ammonia itself accumulated only in the latter stages of starvation in the dark and was suppressed in the series kept in light. The changes in organic acids occurring in these excised leaves were also followed, and it was found that in the dark malic acid decreased in concentration rapidly and in a quantity even more than sufficient to account for the asparagine formed. Chibnall in his discussion of this work has made an interesting summary of the alterations in excised tobacco leaves.[6] Chibnall divides the starvation period into two more or less arbitrary periods, an early period from 0 to 73 hours and a second period from 73 to 143 hours. The changes which take place in excised leaves kept in dark are summarized for these two periods in Table 20-6. In tobacco, as in barley, leaves when detached from the plant and placed in dark rapidly consume their reserve of carbohydrate. Conversion of malic acid to other compounds is initiated, while protein is also rapidly broken down. The resulting amino acids are in part deaminated and the ammonia stored as asparagine. During the final period of starvation, protein, amides, and unidentified carbon compounds form the principal respiratory substrate.

In addition to their work on tobacco Vickery and others have done similar starvation experiments with rhubarb.[13] Here again perhaps the most significant feature of the nitrogen metabolism is the rapid and continuous breakdown of the leaf protein from the time of excision on. A

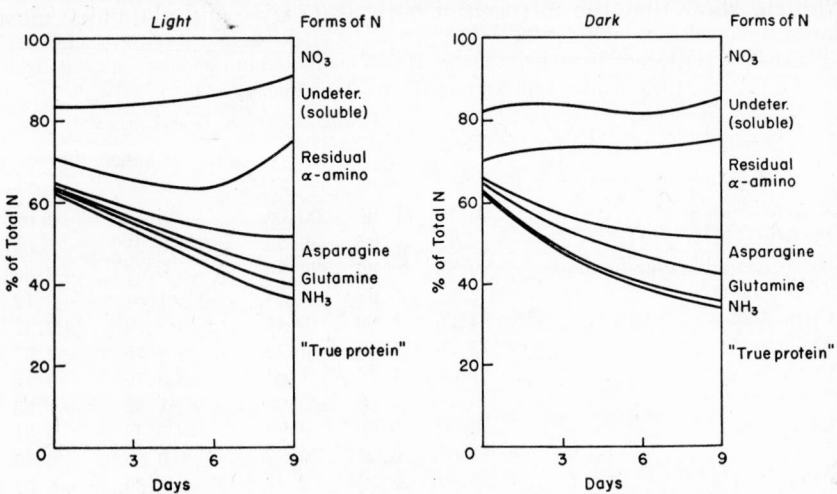

FIG. 20-4. Nitrogen distribution in detached corn leaves cultured for the period indicated in daylight and complete darkness. The vertical distance between two curves represents percentage for that fraction. (After Viets et al.[14])

third extensive series of investigations on the metabolism of starving leaves is that of Wood and co-workers[15] who used leaves of the grasses, primarily of *Avena, Andropogon,* and *Pennisetum* species. With Sudan grass (*Andropogon sudanensis*) and Kikuyu grass (*Pennisetum clandestinum*) as well as with oats (*Avena sterilis*) excised leaves carried on rapid protein hydrolysis with concomitant liberation of amino acids. Protein loss was found to be approximately linear with time up to a limiting point at which 40–60% of the initial protein had been lost. A feature of the metabolism of excised leaves of Kikuyu grass is the fact that during starvation of leaves of this species in a nitrogen atmosphere, protein hydrolysis is greatly retarded and amino acids alone rather than amides and amino acids accumulate. This is no doubt a reflection of the fact that both deamination of the amino acid residues and synthesis of amides are aerobic processes. Other features of the work of Wood will be referred to below. Viets and others[14] have shown with excised corn leaves also that rapid protein loss follows excision and that accumulation of amides and of soluble α-amino nitrogen occurs as in other species.

Thus far protein hydrolysis in the leaf has been treated from the standpoint of the overall protein content of the leaf, including the sum of the cytoplasmic and chloroplastic fractions. That protein loss by the leaf involves the chloroplastic as well as the cytoplasmic fraction was shown in a general way by Michael.[16] Leaves of *Tropaeolum majus* (nasturtium) were detached from the plant and maintained in the dark for varying times, after which they were analyzed for both chlorophyll and total protein. Figure 20-5 shows that there is a direct correlation

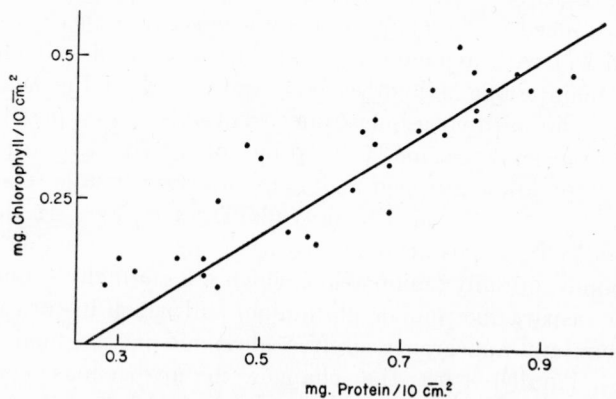

Fig. 20-5. Relation of total protein to chlorophyll in excised leaves incubated in the dark. (After Michael[16])

between these two quantities, leaves of high protein content being similarly high in chlorophyll. Michael concluded that protein hydrolysis of the cytoplasmic and chloroplastic fractions must take place concurrently and that decomposition of the chlorophyll protein complex must lead to total loss of the chlorophyll. It is this process which leads to the yellowing of excised leaves. Carotene and xanthophyll as well as chlorophyll similarly disappear as protein is lost from excised bean leaves.[17] A detailed examination of protein breakdown in leaves has been carried out for Sudan grass and Kikuyu grass,[15] with separate determination of the chloroplastic and cytoplasmic protein fractions. In Kikuyu grass as in *Tropaeolum* the two protein components disappear at the same rate, the disappearance of chlorophyll paralleling the disappearance of total protein. In Sudan grass, however, chloroplastic protein is lost more rapidly than cytoplasmic protein. Here loss of chlorophyll parallels loss of chloroplastic protein rather than loss of total protein. Whether the same leaf proteases are effective in breakdown of the two protein fractions is an interesting question whose answer awaits further study of the leaf enzymes. During starvation in excised leaves of these grasses the chloroplasts gradually shrink. At the time of onset of visible yellowing, disintegration of the chloroplasts has set in and chloroplast fragments are found lying in the cytoplasm. Disintegration of the chloroplast is also associated with the slowing down of protein hydrolysis which occurs in the later stages of leaf starvation (Fig. 20-7).

In general summary, leaves when excised from the plant undergo a rather rapid loss of protein as the outstanding feature of their nitrogen metabolism. This loss of protein appears to be induced primarily by excision, is much more rapid than would have occurred had the leaf remained attached to the plant, and is not offset by either (a) supplying the excised leaf with available nitrogen in the form of ammonium ions or (b) by maintaining a high carbohydrate level in the leaf through exposure to light with consequent photosynthesis. As a possible exception to these general relations it is to be noted that leaves excised from plants grown on a low nitrogen regime are apparently able to synthesize protein rapidly when supplied with available nitrogen.[18] In the excised leaf, protein hydrolysis is attended by accumulation of soluble nitrogenous compounds, initially amino acids, which are gradually supplanted by the amides, asparagine and/or glutamine, and which in turn disappear with concomitant appearance of free ammonia in the final stages of starvation. Parallel with the changes in nitrogenous compounds, changes in carbohydrates and organic acids take place. Respiration of the leaf proceeds primarily at the expense of carbohydrates during the initial period after excision, but amino acids and organic acids are ultimately involved as respiratory substrates and the amides themselves must finally also serve as sources of respiratory metabolites.

Regulation of Protein Level in Leaves. It has been shown that in the germination of the seed, reserve protein is hydrolyzed, the resulting amino acids translocated, in part transformed, and finally resynthesized to protein in the seedling. The leaf is similarly an active site of protein synthesis in the intact plant. We might ask then what are the factors which govern the protein level in leaves. In the normal plant, provided that available nitrogen is not limiting in amount, the leaf protein content is maintained at a relatively constant value over long periods of time. This is shown, for example, by the nearly equal protein concentrations found in successive leaves of barley,[19] tobacco,[20] and many other species. The protein nitrogen level in barley, for example, varies in the experiment of Table 20-7 from 3.23% in the youngest leaves to 4.06% in a mature

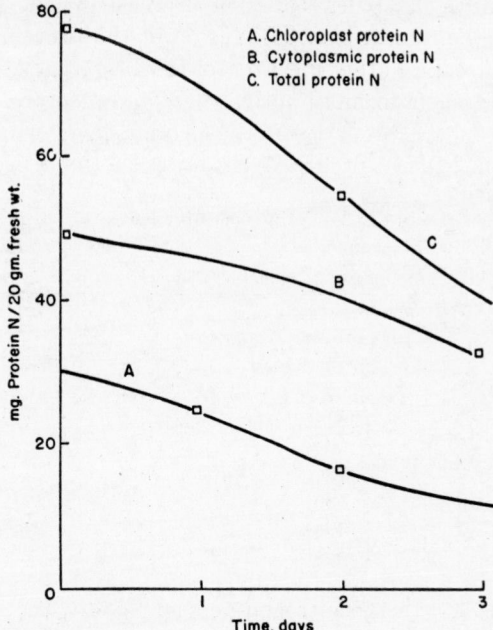

FIG. 20-6. Relative rates of loss of chloroplastic and cytoplasmic protein in excised leaves of Kikuyu grass incubated in the dark. (After Wood et al.[15])

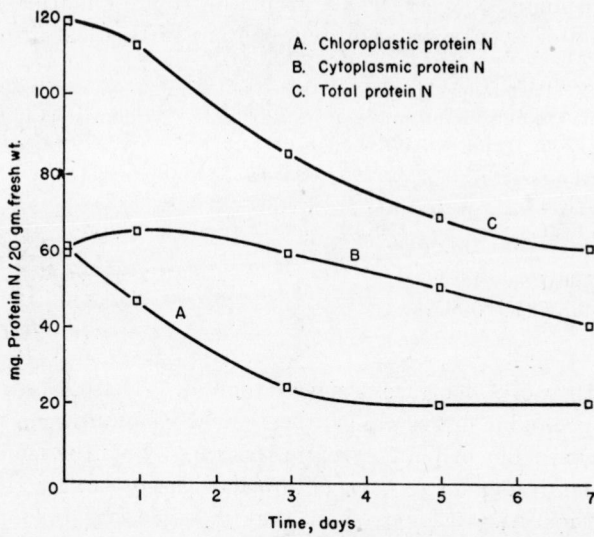

FIG. 20-7. Relative rates of loss of chloroplastic and cytoplasmic protein in excised leaves of Sudan grass incubated in the dark. (After Wood et al.[15])

leaf and 2.52% in the oldest leaves.   The decrease in concentration in the older leaves may be due in part to dilution of the protein by increasing amounts of cellulose and other inert materials; in part it no doubt reflects withdrawal of nitrogen to more actively growing centers.   In nitrogen-deficient plants on the other hand the protein level while high in the young leaves is in general progressively lower in the progressively older

TABLE 20-7.   Concentration of protein in successive leaves of barley plants grown on high and low nitrogen substrates.   (From Gregory and Sen[19])

| | Protein N in per cent of dry leaf | | |
Leaf	Full nitrogen	$\frac{1}{9}$ of full N	$\frac{1}{81}$ of full N
8th (youngest analyzable)	3.23	3.13	1.81
7th	3.87	2.39	1.34
6th	3.94	1.84	1.39
5th	4.06	1.29	0.93
4th	3.70	1.02	0.90
3rd	2.58	1.06	1.06
2nd	2.29	1.18	....
1st	2.52	1.69	....

leaves (Table 20-7).   In the nitrogen-deficient plant, the individual leaf loses protein as it matures, the soluble nitrogen being transported to the actively growing regions.   Walkley[21] has followed the diminution of protein in leaves of nitrogen-deficient barley plants, a particular leaf, the fourth, being analyzed periodically from the time of maturity onward.   Over a period of twenty days, protein content of this leaf dropped from

TABLE 20-8.   Protein level in the fourth leaf of barley plants at various times and with or without the addition of nitrogen to the plant.   All plants decapitated above fourth leaf.   (From Walkley[21])

Days after planting	30	34	38	41	44	50
Mg. protein N/leaf	1.82	1.56	1.31	1.07	0.91	0.19
Mg. soluble N/leaf	0.12	0.12	0.11	0.14	0.11	0.09

Mg. protein N/leaf
after addition of $(NH_4)_2SO_4$
to plants

| | →2.19 | →1.67 | | →1.53 |
| | | →2.79 | →2.43 |

1.82% to 0.19% or a decrease of nearly tenfold.   These leaves are, however, entirely capable of protein synthesis since when nitrogen was added to the nitrogen-deficient plant, protein was rapidly synthesized to a high level.   Similar results have been obtained with cotton.[18]   These and a wealth of other investigations show that nitrogen supply is a primary factor regulating protein level in the leaf of the intact plant.   It is not, however, the sole factor as will be shown below.

Diurnal fluctuations in protein level of the leaf have been observed by several workers.[22,23,24] These fluctuations are small and if present consist of loss of protein at night and gain of protein by day, their magnitude and nature depending on the age of the leaves employed. Young leaves increase in protein continuously, while old or senescent leaves lose protein continuously, especially under conditions of nitrogen deficiency. Only mature but not old leaves show the typical diurnal variation.[25] Exposure of leaves on the intact plant to periods of prolonged darkness induces more extensive protein loss than is found during a single night.[26] In this case also protein is hydrolyzed and soluble nitrogenous compounds transported from the leaf to other tissues.

A further factor regulating leaf protein level is the water stress to which the plant is subjected. Mothes[25] has reported that high water stress causes protein hydrolysis in leaves, and this view is concurred in by Wood and Petrie.[27] The effect is found only when the leaf is attached to

TABLE 20-9. Effect of glucose supply in decreasing protein hydrolysis in excised leaves of *Phaseolus multiflorus*. Leaves in dark throughout. (From Mothes[3])

Treatment	Protein N as per cent of total leaf N		
	Initial	Final	Change
Petioles in water 4 days	81.7	64.7	−17.0
Petioles in 1% glucose 6 days	82.5	70.5	−12.5
Petioles in 2% glucose 6 days	86.7	79.3	− 7.4
Petioles in 2.5% glucose 5 days	77.2	73.2	− 4.0
Petioles in 4% glucose 4 days	80.9	80.5	− 0.4
Petioles in 5–7% glucose 5 days	81.6	83.4	+ 1.8

the plant and may have to do with transport of soluble nitrogen products of protein hydrolysis away from the leaf, i.e., the effect may be primarily on translocation rather than protein synthesis.

Much controversy has arisen over the question of whether the rate of protein hydrolysis by leaves is influenced by the carbohydrate content of the leaf. In the case of excised tobacco leaves as described above, protein loss is not influenced by carbohydrate content where relatively high carbohydrate level is maintained by exposing the excised leaves to light or by supplying them with glucose. Earlier experiments had, however, indicated that in other cases, rate of protein loss might be slowed or even stopped by infiltration of the leaves with sugars. This was found by Paech for leaves of *Lupinus*, *Brassica*, *Phaseolus*, and *Helianthus*, while data taken from Mothes[3] for *Phaseolus* is given in Table 20-9. The supplying of glucose to the leaf in this case completely suppressed protein loss over a period of at least five days. These results are not only at variance with those of Vickery and others with tobacco but also with those of Yemm, which also show protein loss from carbohydrate-rich leaves. A partial resolution of this conflict has come from the work of Wood and others[15] who found some delay of protein hydrolysis in leaves excised from plants pretreated so as to contain two to four times their normal sucrose content. Evidently high carbohydrate level of the leaf gives some protection against protein hydrolysis, but the levels needed are higher than those usually encountered in the leaves used by most investigators.

T. Schulze[29] has reported the existence in leaves of enzyme inhibitors and activators extractable from leaves with acetone and effective in influencing the activity of the leaf proteases. It was supposed that these activators and inhibitors might be compounds capable of entering into oxidation-reduction systems and hence of influencing the reduction-oxidation potential of the reaction mixture. These experiments are open to the serious objection that the influence of the enzyme inhibitors and

TABLE 20-10. Concentration of protein and soluble nitrogen in leaves of barley plants grown with varying concentrations of nitrogen and potassium in the nutrient medium. All figures give mean contents for all leaves of plant. (From Gregory and Sen[19])

Nutrient level						
Nitrogen	1	1	1	$\frac{1}{9}$	$\frac{1}{81}$	$\frac{1}{81}$
Potassium	1	$\frac{1}{9}$	$\frac{1}{81}$	1	1	$\frac{1}{81}$
Per cent protein nitrogen	3.27	3.32	3.32	1.70	1.24	1.21
Per cent amino nitrogen	0.27	0.38	0.46	0.15	0.10	0.09

activators *in vivo* was not demonstrated and has in fact been questioned by Paech,[28] who suggests that they may in fact only work on protease preparations which have been partially oxidized in extraction from the cell.

Still another set of factors of significance for the protein level are the plant nutrients other than nitrogen itself. Gregory and Sen[19] have made a comprehensive investigation of this aspect of protein regulation with barley. Plants were grown with a series of nutrients in which relative nitrogen and potassium levels, 1, $\frac{1}{9}$, $\frac{1}{81}$, were combined in various ways. In each instance the most concentrated nutrient supplied a nonlimiting amount of the element in question. Leaves of the treated plants were harvested and analyzed periodically. Low potassium levels combined with high nitrogen levels, although they leave the protein level unchanged, cause a considerable increase in the soluble nitrogen concentration in the leaf. Gregory and Sen suggest that potassium may be needed for regulation of protein breakdown, that protein synthesis and breakdown may take place along quite different paths, and that protein in the leaf is in a state of flux, a state of steady breakdown and resynthesis. This idea is of course in accord with that developed by Schoenheimer[30] for the proteins of the animal body, which are, as is now known, in a dynamic state, continuous degradation being balanced by continuous resynthesis. That the same concept can be applied to the protein of the leaf is shown by the work of Vickery and others[31] with isotopic N^{15}. A tobacco plant was transplanted to a nutrient containing NH_4Cl in which the nitrogen was tagged with isotopic N^{15} in excess of the amount ordinarily present. After seventy-two hours the plant was dried, ground, and the disposition of the N^{15} in its tissues determined by analyzing the nitrogen of the several fractions in the mass spectrometer. The isotopic N^{15} had found its way into all the tissues of the plant and was recovered not only among the soluble nitrogen constituents but also in the proteins. Since the ratio of N^{15} to N^{14} in the nutrient solution is known, it is possible to calculate from the amount of N^{15} in the plant the amount of total nitrogen

TABLE 20-11. Replacement of N in various compounds and tissues of a tobacco plant during a 72-hr. treatment with nutrient containing isotopic NH_4Cl. (From Vickery et al.[31])

Fraction	Leaf % replacement	Stem % replacement	Root % replacement
Protein	8.2	15.2	18.4
NH₃ N	21.5	22.7	64.3
Amide N	17.9	23.6	....

taken up during the seventy-two hour experimental period. Of the nitrogen absorbed 37% found its way into the protein of the plant, and of all the protein in the plant 10% of the nitrogen was constituted by that taken up during the experimental period. This is taken by Vickery and others to indicate that nitrogen taken up from the nutrient may be exchanged for other nitrogen already in the protein of the plant, which in turn is interpretable on the basis of a protein steady state in which protein hydrolysis is balanced by protein synthesis. The whole protein of the leaves was hydrolyzed and the per cent replacement of the nitrogen in several of the amino acids determined individually. Glutamic and aspartic acids showed higher contents of N^{15} than the remaining acids and would appear therefore to participate more rapidly than the other amino acids in exchange of nitrogen with the NH_3 of the cell. A similar experiment carried out with buckwheat indicated that approximately 6% of the leaf protein nitrogen appeared to have exchanged in forty-seven hours. These preliminary experiments indicate that the amino groups of the plant protein may be readily exchanged with NH_3 freshly introduced from the exterior. This exchange may take place by breakdown of the protein to its individual amino acids, exchange of the nitrogen of the amino acids with nitrogen of NH_3 through the mechanisms outlined in Chapter 16, followed by resynthesis of the protein. The concept of the protein level as a dynamic state, the resultant of rates of synthesis and degradation, is of great importance in the study of the regulation of the protein level since evidently factors influencing either the synthetic or the degradative aspect of protein metabolism may alter the net protein level.

Thus far we have dealt with the protein level of the leaf as it may be studied on the intact plant. As shown earlier, when leaves are excised from nitrogen-rich plants of several species, the protein level drops rapidly, much more rapidly than would be the case had the leaf remained on the plant. This fall in protein content is apparently a result of excision and has lead Chibnall[6] to suggest the possible existence of a

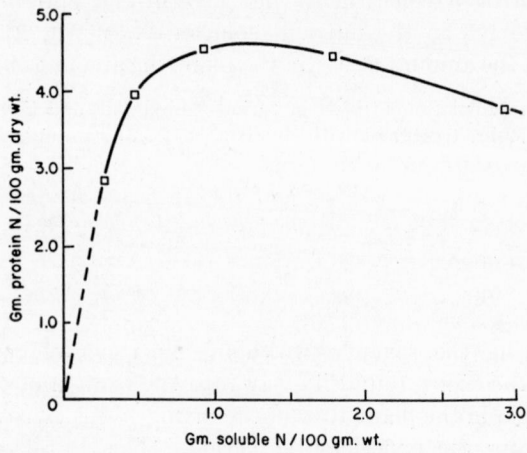

Fig. 20-8. Relation between soluble and protein nitrogen of cotton leaves taken from plants grown under varying environmental and nutritional conditions. (After Phillis and Mason[18])

hormone or correlative factor which would regulate the protein level of the leaf. Such a factor, if it exists, would need to be one supplied to the leaves by the stem or roots and would be a factor which would promote protein synthesis. The absence of this hormone as a result of excision should cause protein synthesis to decrease in rate while protein hydrolysis should continue. Excision of the leaf would, according to this view, merely unmask the protein hydrolysis which takes place continuously in the normal leaf. Excised leaves are, however, capable of extensive and rapid protein synthesis under some conditions, as shown by experiments of Phillis and Mason[18] who grew cotton plants on a low nitrogen regime and then floated leaf disks from these plants on nutrient medium high in nitrate. Protein synthesis took place in such disks as rapidly as in similar leaves left attached to the plant and given nitrate. It would be of interest to compare cotton with tobacco, barley, Sudan grass, etc., from the standpoint of protein loss from the excised leaves of high nitrogen plants. The experiment of Phillis and Mason could be of great value provided that the levels of leaf protein under other conditions were adequately known. Protein synthesis in excised leaves of *Narcissus* has also been reported.[32] The synthesis takes place only in the meristematic portion of the leaf, and the mature portions lose protein as do the majority of excised leaves. In both of these experiments with protein synthesis by excised leaves the rate of synthesis decreases as the leaf matures just as with leaves attached to the plant.[21]

It has frequently been suggested that protein hydrolysis may be controlled simply by the protein-amino acid ratio prevailing in the tissue—a notion derived from the idea that the laws of mass action should be applicable to protein synthesis. This view is supported by Phillis and Mason[18] who grew cotton under a wide variety of nutritional and environmental conditions. Analysis of protein and soluble nitrogen in the leaves of plants from these many different conditions showed a close relation between the two quantities, high soluble nitrogen being correlated with high protein level over a considerable range. This correlation, striking as it is, and though it has been found by other modern workers,[27,33] still cannot make immediately explicable certain results already discussed, as for example the high amino acid level of low potassium barley plants, or the accumulation of amino acids at the expense of protein in excised leaves of tobacco, barley, and other species. A suggestion as to the basis of such discrepancies in the amino acid-protein relationship is offered by the work of Wood and Cruickshank,[12] who have shown that in starved leaves the individual amino acids liberated by protein hydrolysis are not stable in the cytoplasm but are removed from the hydrolyzate at differing rates. During incubation of excised leaves the amounts of free cystine, glutamic acid, arginine, tyrosine, and of tryptophane in the cytoplasm, rise each to its individual maximum only to fall again, indicating release and reutilization or destruction of each amino acid. Furthermore, the amount of each amino acid recovered was found to be less than that contained in the amount of protein hydrolyzed. Such secondary removal of amino acids results in a disproportionality in composition between the free amino acids of the cytoplasm and the amino acids in the protein. Cystine in particular, even though it is contained in the leaf protein hydrolyzed by the starving leaf, does not accumulate at all in the products of hydrolysis, and an equivalent amount of inorganic sulfate appears in its place. Table 20-12 contains data on the disappearance of other amino acids including arginine, glutamic acid, tyrosine, and tryptophane. In each case small amounts of the amino acids are liberated initially, but even these may be transformed so that in the later stages of starvation the concentrations of each amino acid actually drop. Conversely, some amino acids appear in the cytoplasm in larger amounts than those present in the hydrolyzed protein. This is true, for example, of aspartic acid, which must then be synthesized during protein hydrolysis. In this way, according to Wood and Cruickshank, a given protein level may be in equilibrium with varying total amino acid levels, or, in general, protein level may be limited by particular amino acids rather than by the total amino acid level. The most critical amino acids appear to be cystine, glutamic acid, arginine, tyrosine, and tryptophane, in this order.

TABLE 20-12. Disappearance of individual amino acids during the hydrolysis of leaf proteins in darkened excised leaves of Kikuyu grass and of oat (*Avena sterilis*). The figures represent the changes in each amino acid during the particular experimental interval. Data in milligrams N per 20 gm. original fresh weight. The expected amounts of amino acids are based on amino acid analysis of the whole leaf. (After Wood and Cruickshank[12])

Experimental interval hr.	Protein	Glutamic acid		Arginine		Tyrosine		Tryptophane	
		Expected	Found	Expected	Found	Expected	Found	Expected	Found
Kikuyu grass									
0–24	−15.3	+1.23	+0.52	+2.34	+0.35	+0.27	+0.20	+0.35	+0.23
24–48	−16.4	+1.31	0.00	+2.55	+0.08	+0.30	+0.23	+0.37	+0.30
48–72	−12.3	+0.98	0.00	+1.90	−0.28	+0.22	−0.02	+0.27	+0.18
72–96	− 7.2	+0.57	0.00	+1.09	−0.18	+0.14	−0.20	+0.17	−0.16
96–144	− 3.1	+0.25	0.00	+0.48	−0.12	+0.05	−0.17	+0.07	−0.04
Avena sterilis									
0–24	−11.1	+0.89	+1.00	+1.73	+0.68	+0.26	+0.10	+0.20	+0.22
24–32	− 9.2	+0.74	+1.34	+1.40	+0.14	+0.21	+0.01	......	......
32–48	−13.0	+1.04	+1.07	+2.02	−0.34	+0.30	+0.03	......	......
48–72	−12.6	+1.01	−0.39	+1.95	−0.32	+0.29	−0.03	+0.23	−0.06
72–96	− 9.9	+0.79	−0.48	+1.51	−0.14	+0.23	−0.06	+0.18	0.00
96–120	−10.4	+0.83	−0.97	+1.63	−0.14	+0.24	−0.09	+0.19	−0.12

The findings of Wood and Cruickshank suggest that it should prove possible to modify the protein level in leaves by supplying these critical amino acids, perhaps by vacuum infiltration of the leaves. These same findings shed new light also on the suggestion that particular factors may be synthesized in other parts of the plant, transported to the leaves, and be there required for maintenance of the protein level. The work of Wood and Cruickshank suggests that these factors may be the particular amino acids which are unstable in the free form in the leaf. We have seen above that in many species reduction of nitrate and formation of soluble nitrogenous compounds occur in the roots, and in fact isolated roots are known to be capable of themselves synthesizing from nitrate all the amino acids which they require. It is a possibility then that roots or other tissues may export particular amino acids to the leaf, amino acids which the leaf cannot make or can make only in limited quantity. The possibility that particular amino acids may constitute the correlative factors in protein regulation is presented only as a working hypothesis but as one which has elements of plausibility.

In summary, we know of numerous factors which influence the protein level in leaves. All these factors are ones which probably bear only indirect relation to protein level, and as yet no single factor unambiguously and directly connected to protein level has been identified. No simple relation between protein level and amino acid level in the leaf has been found to hold under all conditions. On the other hand, the amino acids which are contained in the cell at any time are not necessarily present in the correct proportion for immediate synthesis into protein, and certain individual amino acids may be present in limiting amounts. This theory, the theory of Wood, would seem to hold promise for future investigation. Maintenance of a given protein level in leaves means

TABLE 20-13. Factors influencing the protein level of leaves

Nitrogen level	High N gives high protein Low N leads to protein hydrolysis
Potassium level	Low K gives high amino acid level but normal protein
Water stress	High water stress gives low protein
Light	Darkness causes protein hydrolysis Light causes protein synthesis (in intact plants)
Excision	Excision causes loss of protein
Sugar content	High sugar delays protein loss in excised leaves to some extent
Oxygen	Protein loss is delayed in absence of O_2

maintenance of a balance between protein synthesis and protein hydrolysis. It is entirely possible that these two processes may go by different paths, differently influenced by most environmental factors, and be quite differently related to quantities such as amino acid level. It will be of evident usefulness in further work to separate the two processes as completely as possible, perhaps by the use of isotopic tracers.

General References

Chibnall, A. C., Protein Metabolism in the Plant. Yale University Press, 1939.
Wood, J. G., *Ann. Rev. Biochem.*, **14**, 665 (1945).
Steward, F. C., and Street, H. E., *Ann. Rev. Biochem.*, **16**, 471 (1947).
McKee, H. S., *New Phytologist*, **48**, 1 (1949).

References

1. Schulze, E., and Bosshard, E., *Z. physiol. Chem.*, **9**, 420 (1885).
2. Chibnall, A. C., *Biochem. J.*, **18**, 395 (1924).

3. Mothes, K., *Planta*, **1**, 472 (1925).
4. Mothes, K., *Planta*, **19**, 117 (1933).
5. Schwab, G., *Planta*, **25**, 579 (1936).
6. Chibnall, A. C., Protein Metabolism in the Plant. Yale University Press, 1939.
7. Greenhill, A. W., and Chibnall, A. C., *Biochem. J.*, **28**, 1422 (1934).
8. Vickery, H. B., Pucher, G. W., and Clark, H. E., *J. Biol. Chem.*, **109**, 39 (1935); *Plant Physiol.*, **11**, 413 (1936).
9. Vickery, H. B., Pucher, G. W., Wakeman, A. J., and Leavenworth, C. S., Conn. Agr. Expt. Sta. Bull. 399, 1937.
10. Wood, J. G., Cruickshank, D. H., and Kuchel, R. H., *Australian J. Exptl. Biol. Med. Sci.*, **21**, 37 (1943).
11. Yemm, E. W., *Proc. Roy. Soc., London*, **B, 117**, 483, 504 (1935); **123**, 243 (1937).
12. Wood, J. G., and Cruickshank, D. H., *Australian J. Exptl. Biol. Med. Sci.*, **22**, 111 (1944).
13. Vickery, H. B., Pucher, G. W., Leavenworth, C. S., and Wakeman, A. J., *J. Biol. Chem.*, **125**, 527 (1938).
14. Viets, F. G., Whitehead, E. I., and Moxon, A. C., *Plant Physiol.*, **22**, 465 (1947).
15. Wood, J. G., Cruickshank, D. H., and Kuchel, R. H., *Australian J. Exptl. Biol. Med. Sci.*, **21**, 37 (1943); **22**, 111 (1944).
    Wood, J. G., Mercer, F. V., and Pedlow, C., *Australian J. Exptl. Biol. Med. Sci.*, **22**, 37 (1944).
    Cruickshank, D. H., and Wood, J. G., *Australian J. Exptl. Biol. Med. Sci.*, **23**, 243 (1945).
16. Michael, G., *Z. Botan.*, **29**, 385 (1935).
17. Jordan, R. C., and Chibnall, A. C., *Ann. Botany*, **47**, 163 (1933).
18. Phillis, E., and Mason, T. G., *Memoirs, Cotton Res. Sta., Trinidad*, **B, 15**, 469 (1943).
19. Gregory, F. G., and Sen, P. K., *Ann. Botany*, **1**(N.S.), 521 (1937).
    See also Gregory, F. G., *Ann. Rev. Biochem.*, **6**, 557 (1937).
20. Hammer, H., Street, O., and Anderson, P., Conn. Agr. Expt. Sta. Bull. 433, 1940.
21. Walkley, J., *New Phytologist*, **39**, 362 (1940).
22. Chibnall, A. C., *Biochem. J.*, **18**, 395 (1924).
23. Mothes, K., *Planta*, **7**, 585 (1929).
24. Mason, T. G., and Maskell, E. J., *Ann. Botany*, **42**, 189 (1928).
25. Mothes, K., *Planta*, **12**, 686 (1931).
26. Wood, J. G., and Barrien, B. S., *New Phytologist*, **38**, 125, 265 (1939).
27. Petrie, A. H. K., and Wood, J. G., *Ann. Botany*, **2**(N.S.), 33, 887 (1938).
28. Paech, K., *Planta*, **24**, 78 (1935).
29. Schulze, T., *Planta*, **16**, 116 (1932).
30. Schoenheimer, R., The Dynamic State of Body Constituents. Harvard University Press, 1942.
31. Vickery, H. B., Pucher, G. W., Schoenheimer, R., and Rittenberg, D., *J. Biol. Chem.*, **129**, 791 (1939); **135**, 531 (1940).
32. Pearsall, W. H., and Billimoria, M. C., *Ann. Botany*, **2**(N.S.), 317 (1938); **3**, 601 (1939).
33. Wood, J. G., *Australian J. Expt. Biol. Med. Sci.*, **20**, 257 (1942).

CHAPTER 21

PURINES AND PYRIMIDINES

THE PURINES

The purines constitute a group of structurally related alkaloids of probably universal occurrence and of general physiological importance in the plant world. They may all be considered as derivatives of purine, imidazole [4–5] pyrimidine, a compound not itself found in nature.

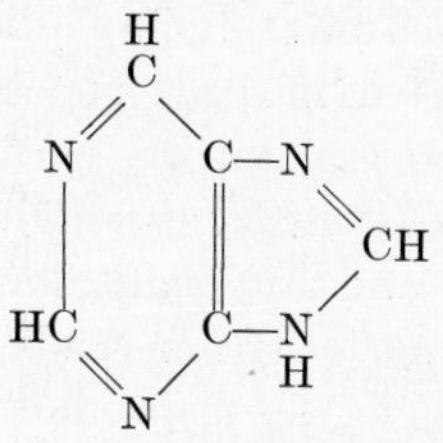

Purine

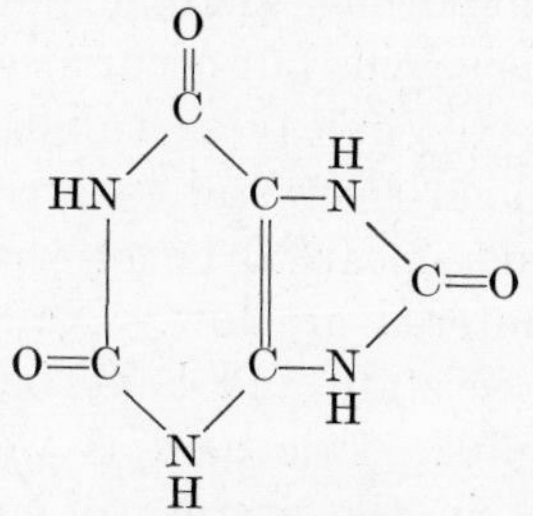

Uric acid

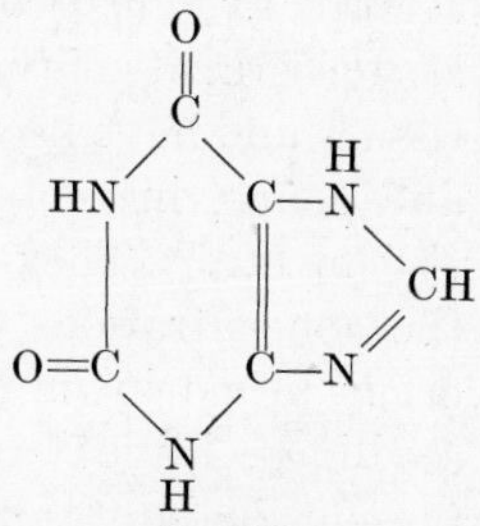

Xanthine

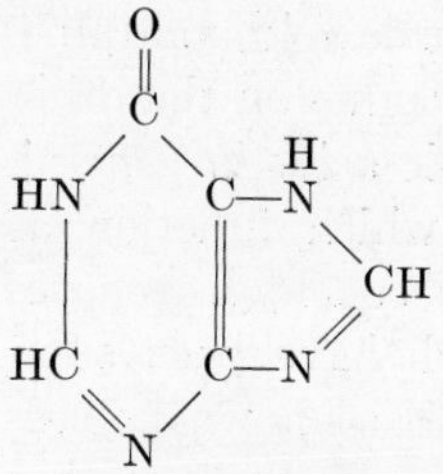

Hypoxanthine

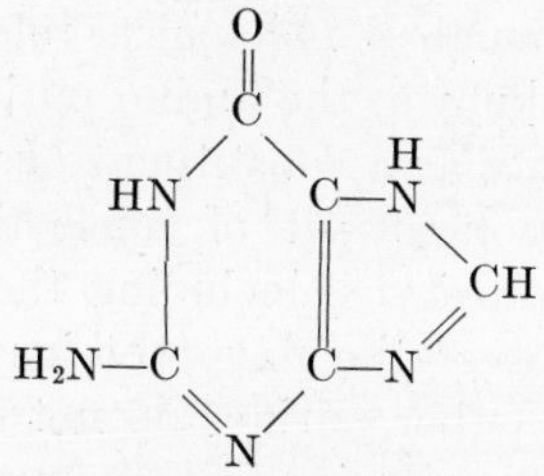

Guanine

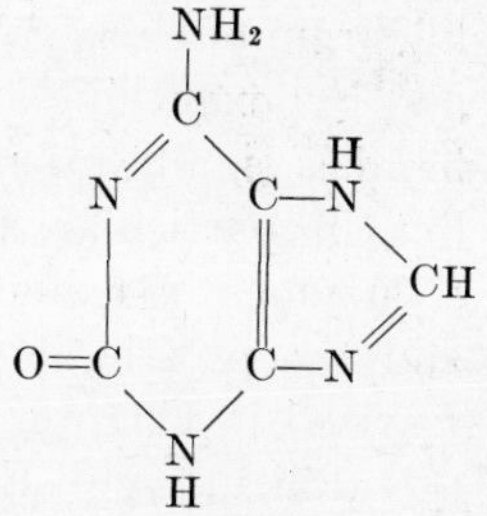

Isoguanine

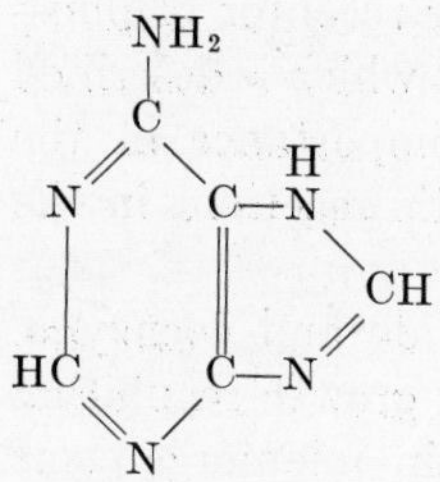

Adenine

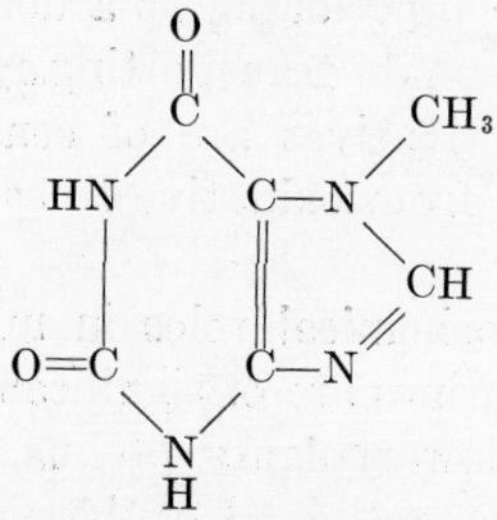

Heteroxanthine

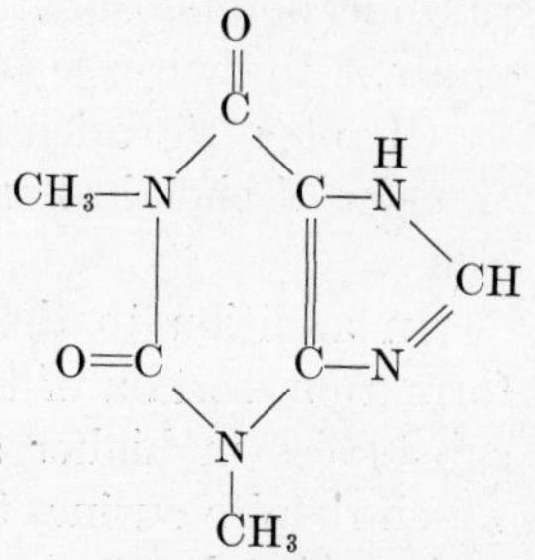

Theophylline

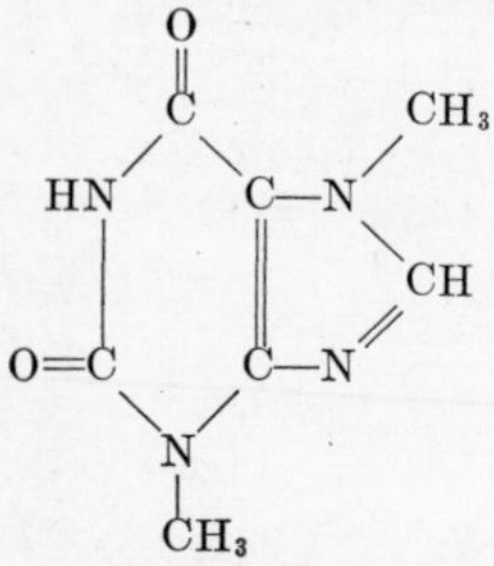

Theobromine 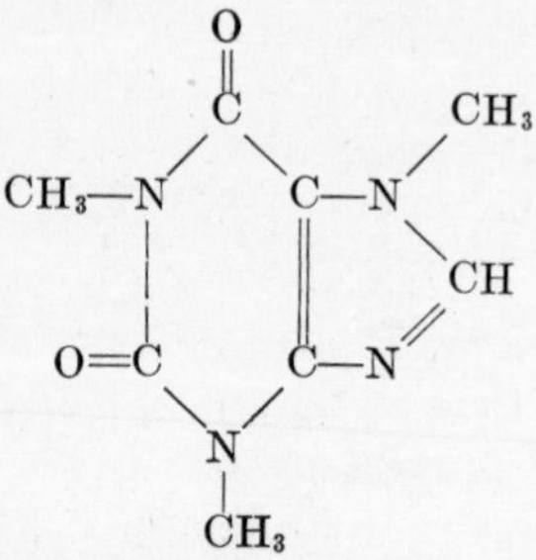   Caffein

The purines may be divided broadly into two groups, the purines proper, of which xanthine, guanine, and adenine are typical, and the methylated purines of which theobromine and caffein are the most important examples. The two amino purines, guanine and adenine, are well known as constituents of the nucleic acids but occur also in the free state in plant tissues and have been detected in leaves as alfalfa and tea, in seeds as pea and coffee, in roots as sugar beet and in green beans, potatoes, corn pollen, etc.[1] They would appear to be of wide distribution although the concentrations encountered are low. Xanthine and hypoxanthine occur primarily in the free state and have been reported from seeds, seedlings, leaves, and fruits. Uric acid is well known as a product present in large amounts in the excreta of animals, but it has been reported also in leaves and seeds in concentrations up to 250 mg. per kilogram of tissue.[2] Although the biological roles of uric acid, xanthine, and hypoxanthine are unknown, the amino purine adenine on the other hand is significant not only as a constituent of nucleic acids in general, but in particular as a constituent of nucleotides which function as coenzymes in oxidative processes, including the two codehydrogenases and the prosthetic groups of the flavoproteins which have been summarized in Chapter 15. Other purines do not so far as known possess the ability to substitute for adenine in these nucleotides. Still another biochemical function of adenine is its property, in the form of the adenine riboside polyphosphates, of functioning as a donor or receptor of phosphate. In their role as phosphate-transporting agents, which is described in Chapter 15, adenine derivatives are of central importance in the transfer of chemical energy from oxidative to synthetic reactions in the cell.

In addition to their biochemical roles in nucleotide and coenzyme formation, certain of the purines are of significance as growth-regulating substances in higher and lower plants. Thus certain microorganisms which require purines for growth have lost the ability to synthesize them and require these substances as exogenous growth factors. Some strains of lactic acid bacteria (*Lactobacillus*) as well as of hemolytic streptococci require external supplies of adenine or of guanine in order for growth to occur.[3] External supplies of guanine or hypoxanthine are partially effective in causing rapid early growth of the fungus *Phycomyces blakesleeanus*.[4] An interesting relation of purines to higher plants has been shown to hold in the case of leaf growth.   Growth of the mesophyll of the leaf, that is, expansion of the young leaf, is dependent on factors supplied from the mature leaves or from the seed.   These factors include, of course, carbohydrates and other nutritive material, but in addition a highly active leaf growth regulating substance isolated from pea seeds and essential for growth of excised leaves of pea and radish was shown to be adenine, replaceable by hypoxanthine.[5] It would appear then that capacity for purine synthesis may not be present in all plant tissues and that purines such as adenine which are required by all tissues for biochemical reasons may correspondingly function as plant growth regulating hormones.

Biosynthesis of Purines. Older concepts of purine synthesis *in vivo* have been based on the reaction of urea with methyl glyoxal[6] followed by reduction and dehydration to yield an intermediate perhaps isomeric with xanthine.   Other purines should then be formed secondarily from xanthine.   Recent work shows, however, that purine synthesis in the

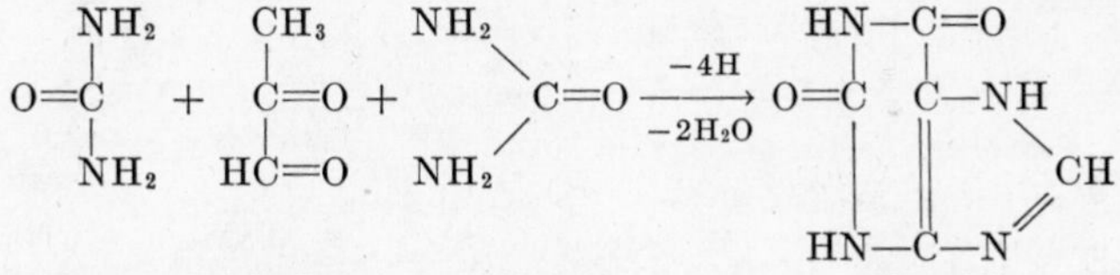

$$
\begin{array}{ccccc}
\text{NH}_2 & \text{CH}_3 & \text{NH}_2 & & \text{HN---C==O} \\
| & | & \diagdown & & | \quad | \\
\text{O==C} \; + & \text{C==O} \; + & \text{C==O} \xrightarrow[-2\text{H}_2\text{O}]{-4\text{H}} & \text{O==C} \quad \text{C---NH} \\
| & | & \diagup & & | \quad \| \quad \diagdown \\
\text{NH}_2 & \text{HC==O} & \text{NH}_2 & & \quad\quad\quad \text{CH} \\
& & & & \text{HN---C---N} \diagup
\end{array}
$$

| Urea | Methyl glyoxal | Urea | Xanthine |

One early suggestion concerning purine biosynthesis

animal at least is probably more complicated.   This has been found in work with formation of uric acid in the pigeon in which precursors containing isotopic carbon were supplied to the intact animal.[7]   The results summarized in the accompanying diagram show that the carbon atoms

The purine nucleus	*Precursor given*	*C atoms in which* *C* appears*
	C^*O_2	Number 6
	CH_3C^*OOH, HC^*OOH	Numbers 2, 8
	$CH_2NH_2 \cdot C^*OOH$	Number 4

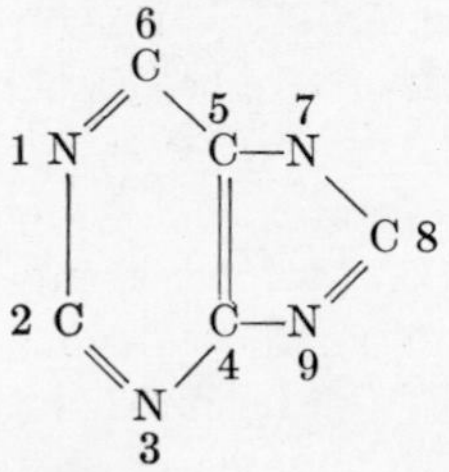

The purine nucleus:

```
       6
       C
     //  \5   7
 1 N      C—N
  |       ‖      \ C 8
  |       ‖      /
 2 C      C—N  //
     \\  /4   9
       N
       3
```

2, 4, 6, and 8 of the nucleus arise from CO_2 and from the carboxyl groups of acetate and glycine. Other evidence indicates further that the α-carbon atom of glycine may constitute carbon atom 5. The carbon of formate is also rapidly incorporated into the purine nucleus even in *in vitro* systems such as liver homogenates.[8] Urea does not, however, appear to be involved as a precursor in the synthesis. The mechanisms involved in purine synthesis in the higher plant are still totally obscure.

Purine Oxidation. In the animal body adenine is oxidatively deaminated by the enzyme adenase to form hypoxanthine. Hypoxanthine is in turn attacked by the flavoprotein xanthine oxidase (which contains adenine as a constituent) to yield xanthine, which is then oxidized to uric acid. The pyrimidine ring of uric acid is opened by the enzyme uricase with the formation of allantoin. Guanine is similarly oxidatively deaminated to xanthine by the enzyme guanase. These enzymes, with the exception of guanase, have not been studied in plants, and it is not known whether purine metabolism, for example in the leaf, parallels that found in animal tissues. On the other hand allantoin has been found in a variety of plants, including young shoots of *Acer*, bark of *Acer*, and

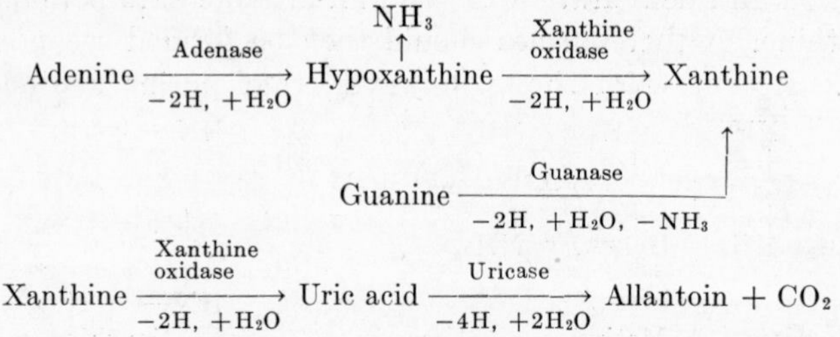

$$\text{Adenine} \xrightarrow[-2H, +H_2O]{\text{Adenase}} \text{Hypoxanthine} \xrightarrow[-2H, +H_2O]{\text{Xanthine oxidase}} \text{Xanthine}$$

$$\text{Guanine} \xrightarrow[-2H, +H_2O, -NH_3]{\text{Guanase}}$$

$$\text{Xanthine} \xrightarrow[-2H, +H_2O]{\text{Xanthine oxidase}} \text{Uric acid} \xrightarrow[-4H, +2H_2O]{\text{Uricase}} \text{Allantoin} + CO_2$$

FIG. 21-1. Outline of oxidative interconversions of purines in the animal body.

wheat seedlings. Allantoic acid, which differs from allantoin by the elements of one molecule of water, has also been reported in leaves of *Acer* and of bean.[9] The presence of these compounds in plant material suggests that in the plant purine breakdown may follow lines generally similar to those found in the animal.

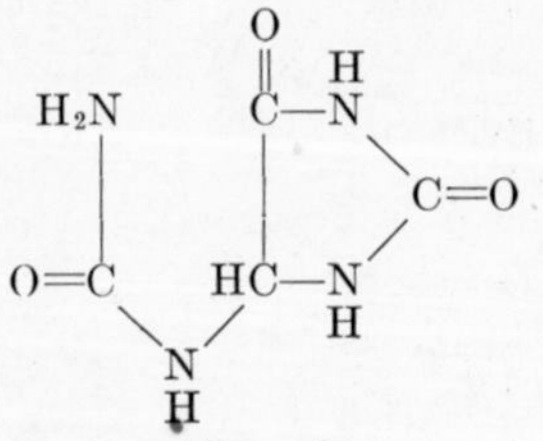

Allantoin 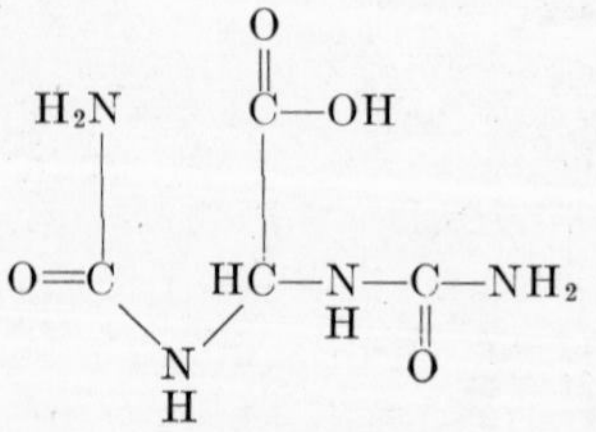   Allantoic acid

The Methylated Purines. The methylated purines heteroxanthine, theophylline, theobromine, and caffein may be regarded as methylated derivatives of xanthine. Heteroxanthine is known from the sugar beet, *Beta vulgaris*, while theophylline is known only in leaves of tea, *Camellia sinensis*. Theobromine and caffein are more widely distributed and are

TABLE 21-1. Distribution of methylated purines in higher plants[10]

Family	Species	Organ of plant	Alkaloids
Rubiaceae	*Coffea arabica*	Leaves	Caffein
		Seed	Caffein
	" *robusta*	Leaves	Caffein
		Seed	Caffein
	" *stenophylla*	Leaves	Caffein
		Seed	None
	" *bengalensis*	Leaves	None
		Seed	None
	" *liberica*	Leaves	Caffein
		Seed	Caffein
Sterculiaceae	*Cola acuminata* (cola)	Leaves	Caffein, theobromine
	Theobroma cacao (cocoa)	Leaves	Caffein, theobromine
		Seed	Caffein, theobromine
Theaceae	*Camellia sinensis* (tea)	Leaves	Caffein, theophylline, etc.
Berberidaceae	*Ilex paraguariensis* (maté)	Leaves	Caffein
		Fruit	None
Sapindaceae	*Paullinia cupana*	Leaves	Caffein, theobromine

found in five widely separated families, as shown in Table 21-1. The alkaloids are always found in the leaves and are absent or essentially absent from roots and wood. Varying amounts are found in the bark and in the flower and fruit. Thus in *Coffea arabica*, *liberica* and *robusta*, caffein accumulates in the seed and may make up to 1.7% of the dry

TABLE 21-2. Distribution of caffein in leaves of tea and coffee. (After Weevers[10])

	Camellia sinensis				Coffea arabica		
Nature of leaf	Gm. dry wt./100 leaves	Per cent caffein	Mg. caffein/ 100 leaves	Nature of leaf	Gm. dry wt./100 leaves	Per cent caffein	Mg. caffein/ 100 leaves
Young	3.0	4.4	131	Young	3.0	1.8	55
Maturing	12.0	2.7	329	Maturing	19.4	1.1	220
Mature	21.8	2.1	469	Mature	30.0	0.8	242
Senescent	29.4	0.0	5	Senescent	28.0	0.55	155

weight. In other species of *Coffea*, caffein is confined to the young leaves and in still others it is absent altogether. In all the species, the concentration of alkaloid on a dry weight basis is highest in young leaves, although the total amount of alkaloid per leaf increases as the leaf matures only to decrease again as senescence sets in. These relations are well illustrated by data of Weevers[10] for tea and coffee. Whether the decrease with age of leaf is owing to withdrawal of the alkaloid from the leaf or to breakdown was not ascertained, although no clear indication of alkaloid transport was obtained in girdling of tea or of *Ilex*.[10,11]

Metabolism of the Methylated Purines. The stored caffein or theobromine of *Coffea* and of *Theobroma* seeds is not utilized when the seeds are allowed to germinate in the dark even over periods of three months.[10] Excised leaves of tea and other species similarly do not decrease in alkaloid content even in the dark over periods of ten days or more. Hence the compounds do not appear to act as sources of reserve nitrogen and are not subject to metabolic attack even in severely depleted tissues.

Biosynthesis of the Methylated Purines. In seedlings of *Theobroma* germinated in the dark, Weevers has shown that a nearly quantitative methylation of theobromine to caffein takes place. It would appear probable then that theobromine may be a general precursor of caffein. Neither methylation of xanthine to form theophylline in tea nor the successive steps in methylation from xanthine through heteroxanthine to theobromine have been sought either *in vitro* or *in vivo*, although such a sequence of reactions would seem to be the probable course in the synthesis of methylated purines. The synthesis of these same purines presents then an excellent opportunity for an attack on the general problem of transmethylation and the formation of labile methyl groups in higher plants (Chapter 16). It will be of interest to discover the nature of these methylation reactions in the leaf.

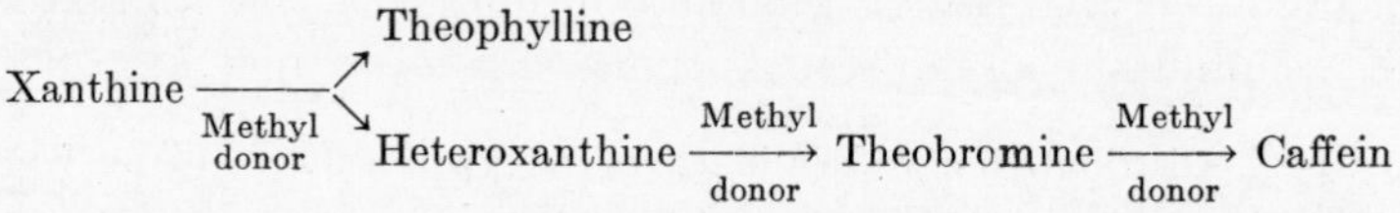

Xanthine → Theophylline / Heteroxanthine (Methyl donor) → Theobromine (Methyl donor) → Caffein (Methyl donor)

Possible steps in the production of methylated purines

THE PYRIMIDINES

The pyrimidines like the purines are found in all higher plants, ordinarily as minor constituents, and in general in bound forms. Uracil

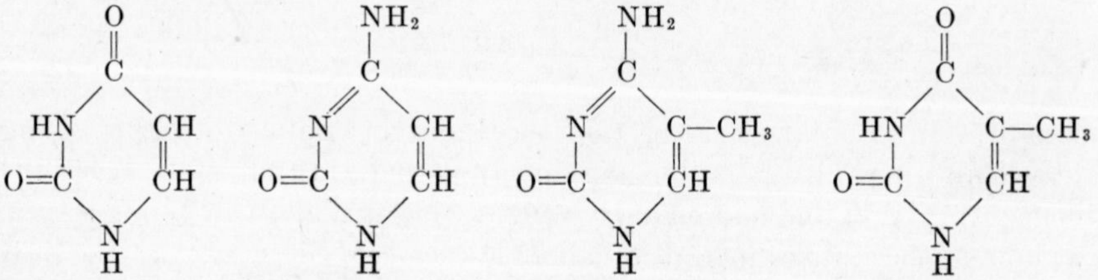

Uracil   Cytosine   5-Methylcytosine   Thymine

and cytosine occur as the pyrimidine components of ribonucleic acid and have been identified not only in yeast but also in rye, wheat, and pea seed nucleic acid as well as in plant virus nucleic acid. Cytosine and thymine occur as the pyrimidine components of desoxyribonucleic acid. 5-Methylcytosine may replace cytosine as reported for nucleic acid of the tubercle bacterium but is not known from higher plants. An unusual pyrimidine glucoside, vicin, is found in seeds of *Vicia faba* and in sugar beets and is made up of a diaminopyrimidine bound through an N glucosidic linkage to D-glucose.

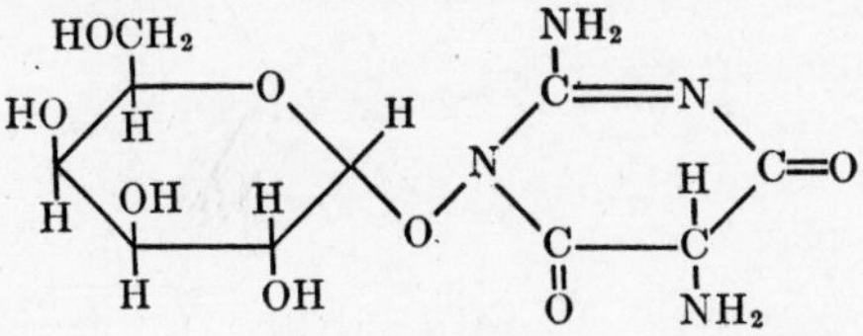

Vicin, a pyrimidine glucoside of *Vicia faba*

Biosynthesis of Pyrimidines. The biosynthesis of pyrimidines has been studied with the aid of pyridimidine-requiring mutant strains of *Neurospora*.[12,13] All these mutants have genetic blocks in their pyrimidine synthesizing mechanism and require the addition of pyrimidines or of pyrimidine precursors to the nutrient medium in order to grow. Since the requirements of several strains can be met either with cytosine or with uracil, it is then probable that these two compounds are interconvertible. In both instances the corresponding ribosides cytidine and uridine are much more effectively utilized than the free pyrimidines themselves, an observation which is not surprising, in view of the fact that it is as the ribosides that these two pyrimidines are found in many nucleic acids. With certain strains of pyrimidine-requiring mutants of *Neurospora*, other simpler substances can fulfill the requirement. Thus oxaloacetate and aminofumaric acid amide act as precursors of pyrimidine for two strains. The relation of these substances to the pyrimidines themselves is indicated by a third strain which cannot produce pyrimidine but which produces large quantities of orotic acid, a carboxyl uracil.[14]

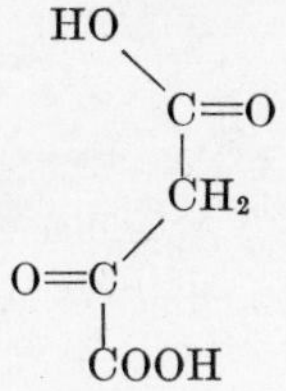

Oxaloacetic acid 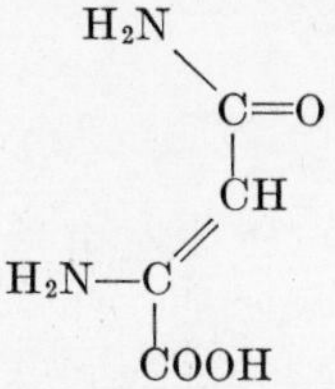 Aminofumaric acid amide 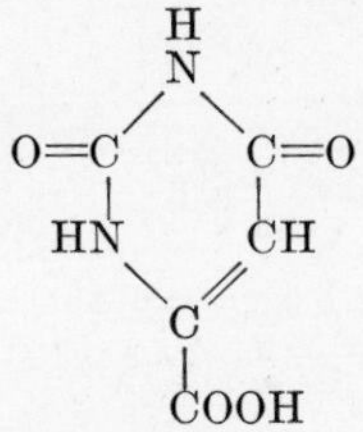 Orotic acid

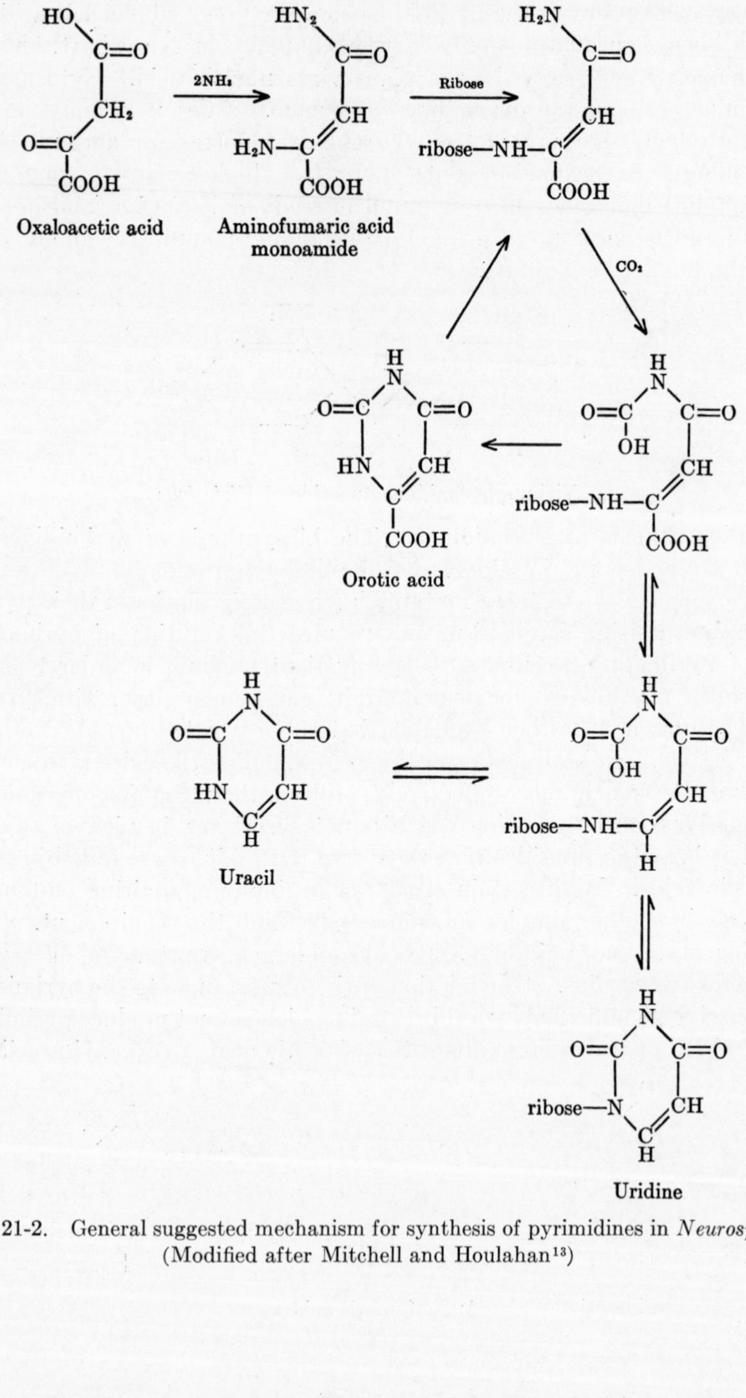

FIG. 21-2.   General suggested mechanism for synthesis of pyrimidines in *Neurospora*.
(Modified after Mitchell and Houlahan[13])

Orotic acid is not apparently an intermediate in pyrimidine synthesis but is rather a side product produced from the immediate pyrimidine precursor when conversion of this precursor to uracil is blocked. Thus orotic acid is not utilized by mutants able to utilize aminofumaric acid amide itself. Mitchell and Houlahan have suggested therefore that in pyrimidine synthesis, introduction of the ribose molecule may occur before actual ring closure, as is shown in Fig. 21-2.[13]

General References

Winterstein, A., and Somlo, F., in Klein, Handbuch der Pflanzenanalyse. Vol. IV, 1: 362, 1933.

Johnson, T. B., in Gilman, Organic Chemistry. Wiley, 1st ed., 1938.

References

1. Distribution of purines and pyrimidines in Klein, Handbuch der Pflanzenanalyse. Vol. IV, 1: 405, 1933.
2. Fosse, R., Graeve, P. de, and Thomas, P., *Compt. rend.*, **194**, 1408 (1932).
3. Knight, B. C. J. G., Vitamins and Hormones. Academic Press, 1945. Vol. 3, p. 105.
4. Robbins, W. J., and Kavanagh, V., *Botan. Rev.*, **8**, 411 (1942).
5. Bonner, D., and Haagen-Smit, A. J., *Proc. Nat. Acad. Sci.*, **25**, 184 (1939).
6. Onslow, M., Principles of Plant Biochemistry. Cambridge, 1931.
7. Sonne, J. C., Buchanan, J. M., and Delluva, A. M., *J. Biol. Chem.*, **166**, 395 (1946); **166, 781** (1946).
8. Greenberg, G. R., *Arch. Biochem.*, **19**, 337 (1948).
9. Fosse, R., *Compt. rend.*, **182**, 869 (1926).
10. Weevers, Th., *Arch. néerland. sci.*, IIIB, **5**, 111 (1930).
11. Weevers, Th., *Proc. Kon. Akad. Weten. Amsterdam*, **32**, 281 (1929); *Rec. trav. botan. néerl.*, **30**, 336 (1932).
12. Loring, H. S., and Pierce, J. G., *J. Biol. Chem.*, **153**, 61 (1944).
13. Mitchell, H. K., and Houlahan, M. B., *Federation Proc.*, **6**, 506 (1947).
14. Mitchell, H. K., Houlahan, M. B., and Nyc, J. F., *J. Biol. Chem.*, **172**, 525 (1948).

CHAPTER 22

ALKALOIDS

NICOTINE

Of the vast number of nitrogenous bases or alkaloids which occur scattered through the plant world only two, aside from the purines and pyrimidines, have been studied extensively from a biochemical and physiological viewpoint. Of these one is the alkaloid nicotine which has been studied in part because of its economic importance as the active principle of tobacco, and in part perhaps because of the amenability of nicotine-producing plants to experimentation. Nicotine, which has the structure shown below, is one of a group of related compounds which occur in species of the genus *Nicotiana*. Nicotine itself is the principal alkaloid of *N. tabacum* and constitutes 0.5% to 8% or more of the dry weight of the leaf, the organ in which the greatest accumulation occurs. Nornicotine, which differs from nicotine by the absence of one methyl group, is also found in tobacco and occurs in particularly high concentrations in varieties which have been selected for low nicotine content. Anabasine, which resembles nornicotine but in which a piperidine ring has been substituted for the pyrrolidine nucleus, is the principal alkaloid

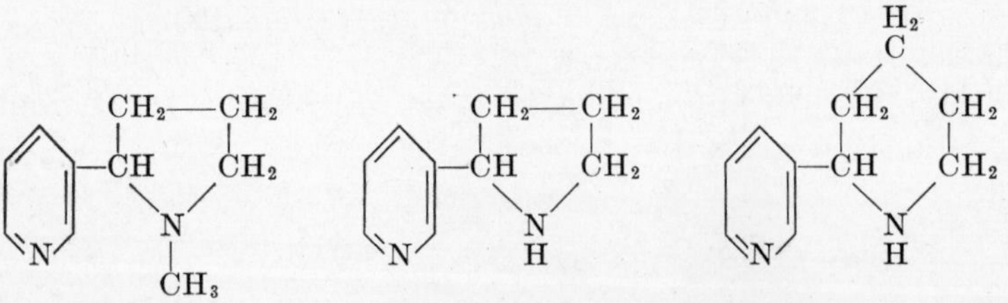

Nicotine   Nornicotine   Anabasine

of *N. glauca* and also occurs in the chenopod, *Anabasis aphylla*. Nicotine and anabasine are both highly toxic central nervous system stimulants while nornicotine is physiologically inactive. In addition to these principal alkaloids, others occur in the tobacco and related plants in much smaller amounts.

Site of Nicotine Synthesis. In the tobacco, nicotine is synthesized exclusively in the roots. This alkaloid is transported upward in the transpiration stream and accumulates passively in the leaves, just as do inorganic salts. Thus when leaves which are rapidly increasing in nicotine content are detached from the plant, accumulation ceases abruptly and completely.[1] If the leaves are placed in sand and allowed to form roots, vigorous nicotine accumulation in the leaf blade is resumed.[2] Similarly leaves of excised tobacco shoots do not form nicotine, and the shoots send out branches which are nicotine-free. Most striking, however, is evidence derived from reciprocal grafts of tobacco with non-nicotine-forming species as tomato. Thus in an experiment of Dawson[3] tobacco shoots which had been grafted to and had developed on tomato stocks contained no more nicotine than that contained in the original scion. Tomato scions grafted on tobacco stocks, on the contrary, contained large amounts of nicotine under the same conditions. The bulk of the nicotine in the tomato appeared again in the leaves, although

TABLE 22-1. Accumulation of nicotine in reciprocal grafts of tobacco and tomato. Weights and amounts of nicotine are expressed in grams or milligrams per plant. (After Dawson[3])

			Nicotine content and fresh weight of plant	
Sample analyzed	Scion	Stock	Fresh wt. (gm.)	Nicotine (mg.)
Whole scion	Tobacco	Tomato	436.2	8.6
Whole scion	Tomato	Tobacco	372.3	570.5

traces were found even in the fruit. These data, which are in agreement with those obtained by others, indicate then that accumulation of nicotine in the leaf depends on the presence in the plant of tobacco roots rather than of tobacco leaves. Still further evidence that the root system is the site of alkaloid synthesis is the fact that aseptic cultures of excised tobacco roots produce and excrete nicotine.[4]

It is of particular interest in connection with the upward transport of nicotine that this movement appears to take place in the xylem rather than in the phloem as is the case with other elaborated substances, notably sucrose. Dawson has shown[3] that the xylem exudate from stumps of decapitated tobacco plants contains nicotine, and other observations confirm the view that transport is indeed in the xylem. Other nitrogenous bases whose translocation has been studied, including particularly the B vitamins such as thiamine and pyridoxine, move exclusively in the phloem and appear to move in the direction of the general nutrient transport, i.e., from leaves toward roots.[5] In the case of nicotine, which moves up the stem from the roots and hence in a direction opposing the principal movement of carbohydrate in the phloem, an entirely different principle, i.e., xylem transfer, appears to govern translocation. It is of interest to determine specifically whether amino acids and other organic forms of nitrogen synthesized in the roots of plants may also move upward in the xylem, at least under conditions where transport in the phloem is primarily in the downward direction.

Biosynthesis of Nicotine. Excised tobacco roots synthesize nicotine from nitrate and hexose and do not appear to require specific nicotine precursors preformed in the leaves. Synthesis of nicotine by the roots of intact tobacco plants depends on an adequate supply of carbohydrate from the leaves and it is well known that decapitation of the tobacco plant, which, by removing the competition of the shoot growing point brings about increased carbohydrate supply to and growth of the remaining leaves and roots, causes particularly high accumulations of nicotine to take place.[6] Attempts to identify particular intermediates in nicotine synthesis have in the past been mainly based on supplying excised tobacco shoots or leaves with the supposed precursor. Such experiments, which were based on the early assumption that the leaf is the site of alkaloid synthesis, have been reported by Klein and Linser[7] who found additions of proline to increase nicotine formation in isolated tobacco leaves. Nicotinic acid has also been reported to cause increased nicotine accumulation in excised tobacco shoots.[8] These results are of very dubious validity since we now know that nicotine is not synthesized in leaves, and actually the small increases in nicotine reported[7,8] may not have been statistically significant. Excised tobacco roots offer good material for critical work on the biogenesis of nicotine, since presumed precursors can be supplied to the tissue under aseptic conditions. It is noteworthy that among the simple compounds structurally related to nicotine and contained in the tobacco plant are nicotinic acid, proline, pyrrolidine, and methylpyrrolidine.

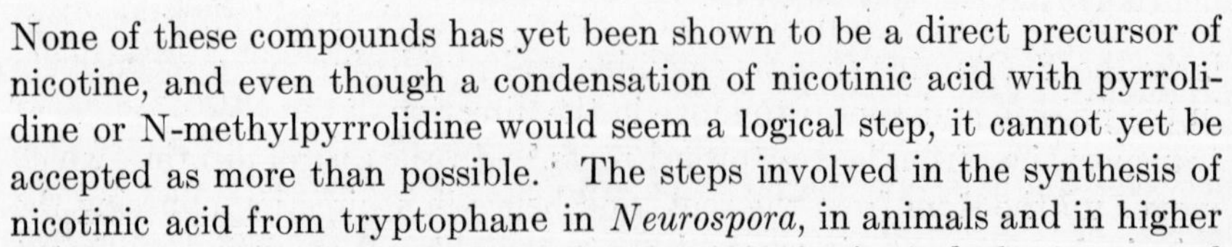

$$CH_2—CH_2$$
$$CH_2 \quad CH—COOH$$

Nicotinic acid   Proline   N-Methyl-pyrrolidine   Pyrrolidine

None of these compounds has yet been shown to be a direct precursor of nicotine, and even though a condensation of nicotinic acid with pyrrolidine or N-methylpyrrolidine would seem a logical step, it cannot yet be accepted as more than possible. The steps involved in the synthesis of nicotinic acid from tryptophane in *Neurospora*, in animals and in higher plants are relatively well understood (see below), and the structural relations involved strongly suggest that pyrrolidine may be derived from proline. It would seem that the time is ripe for a final solution of the problem of nicotine biosynthesis.

Biogenesis of Nicotinic Acid. The synthesis of nicotinic acid proceeds from tryptophane by reactions first elucidated for the rat[16] and for mutant strains of *Neurospora*,[17] but which have been found to apply also to the pea plant[18] and presumably therefore to other plants as well. Tryptophane is first converted to kynurenin by an oxidative opening of the indole nucleus. The kynurenin thus formed is then oxidized to hydroxyanthranilic acid. The conversion of hydroxyanthranilic to

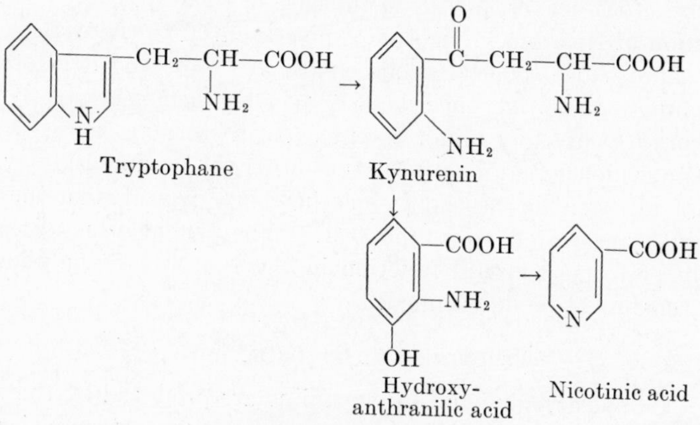

FIG. 22-1.   General course of the reactions involved in biosynthesis of nicotinic acid from tryptophane.

nicotinic acid is the key reaction in this series since at this step the aromatic nucleus is converted to the heterocyclic pyridine ring. The hydroxyl group of hydroxyanthranilic acid is essential to this conversion since anthranilic acid itself is not able to replace hydroxyanthranilic acid as an intermediate. Although the exact enzymatic systems concerned in this biosynthesis are still unknown, knowledge of the course of the reaction is of great importance since it provides us not only with some insight into the mode of production of nicotinic acid but also with a clue concerning the general problem of heterocyclic ring formation.

Relation of Nicotine to Nornicotine and Anabasine. It has been mentioned above that certain tobacco strains selected for low nicotine content show a high content of nornicotine. Roots of the nornicotine-producing strains produce, however, only nicotine,[9] and this alkaloid is transported as such to the leaf where it is demethylated.[10] The difference between the high and low nicotine strains consists in the ability of the leaves of the latter to demethylate nicotine. Evidently a methyl acceptor capable of receiving the methyl group of nicotine is found in such strains of tobacco, and it will be of interest to determine the nature of this transmethylation system.

Anabasine, the principal alkaloid of *N. glauca*, differs physiologically from nicotine in that it is produced not only in the roots but also in the leaves. Thus excised leaves as well as roots of *N. glauca* increase in anabasine content, the leaves only if maintained in the light.[11] *N. glauca* scions on tomato roots contain anabasine, and tomato scions on *N. glauca* roots contain both anabasine and nicotine, suggesting that the roots of *N. glauca* under some conditions may exhibit synthetic powers different from those of the leaves, just as in the case of tobacco.

Function of Nicotine. Nicotine is not essential to the welfare of the tobacco shoot, since tobacco scions on tomato grow and develop in the usual manner. Nicotine once deposited in the leaf is relatively stable and disappears only to a minor extent even in starving excised leaves.[12] There is no evidence that nicotine acts as other than a sluggishly available reserve of nitrogen in the shoot; nicotine once deposited in the leaf is not again translocated from the leaf. Whether nicotine directly or indirectly performs a useful function in the root where the alkaloid is formed, remains to be determined.

ATROPINE AND HYOSCYAMINE

The alkaloids atropine and hyoscyamine are produced by the solanaceous plants *Atropa, Datura, Hyoscyamus* and others. Atropine is related to L-hyoscyamine in that it is the D,L racemization product of the latter. It probably does not occur as such in the plant in more than traces but is produced during isolation of the alkaloid. The alkaloid is formed of two portions, a base, tropine, which is esterified with an acid, tropic acid. Grafting experiments with tomato show that both in *Atropa* and in *Datura* atropine formation appears to take place principally in the roots

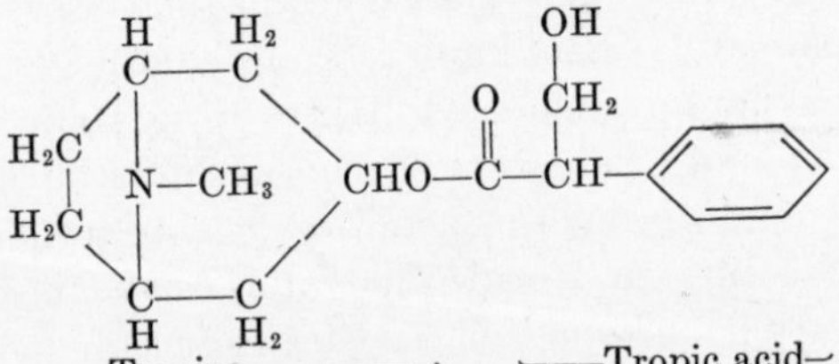

——Tropine——   ——Tropic acid——

Atropine and hyoscyamine

as does nicotine formation in *Nicotiana tabacum*.[13] Here, as in tobacco, transport of the alkaloid appears to take place in the xylem. No alkaloid is found in the seed but formation starts soon after germination, and in the mature plant the highest concentrations are found in the young expanding leaves. In the senescent leaves alkaloid concentration again diminishes, possibly owing to actual breakdown of the material. Even though much of the alkaloid synthesis takes place in the root, it has been claimed that appreciable synthesis may also occur in the leaves. James[14] has supplied detached leaves of *Atropa* with a variety of amino acids and has shown that arginine and ornithine alone have the property of causing small increases in alkaloid content of the leaf. Since arginase is readily demonstrable in the leaf (as it is also in the leaves of other solanaceous plants), it is probable that ornithine may be readily derivable from arginine *in vivo*. The further mechanism of the reaction is not, however, evident from comparison of the structures involved. The synthesis, if it actually occurs by this route, must involve condensation of ornithine with other unknown intermediates to yield, ultimately, tropine, the basic component of hyoscyamine which in turn would yield the complete alkaloid by esterification with the acid portion, tropic acid.

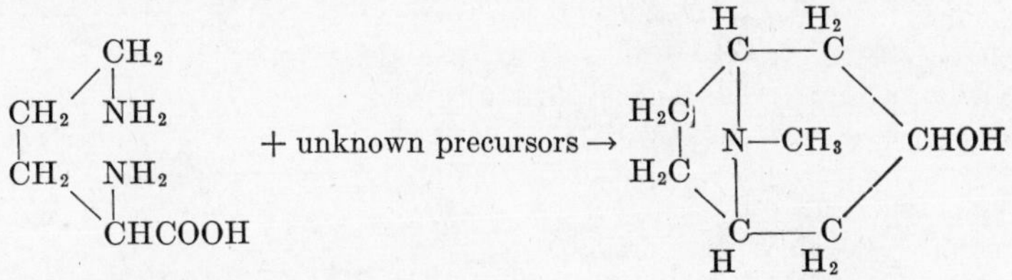

Ornithine   Tropine

In this connection it is of interest that in grafts of tobacco scions on the hyoscyamine-producing solanaceous plant *Duboisia*, tropine alone rather than the complete alkaloid accumulates in the tissues of the scion.[15]

Function of Alkaloids

In general, alkaloids have no known functions in the plant, and it is doubtful if they can play any general biochemical role in view of their limited and irregular distribution in the plant world. Alkaloids have been termed excretion products, a poor term in view of the fact that they may be formed by plants even under conditions of limited nitrogen supply. More probably they may represent by-products of metabolism, secondary products irreversibly formed from essential metabolites in tissues where, for one reason or another, supply of the metabolite outruns immediate consumption. Examples of such secondarily formed exotic and little useful products are common among the mutant strains of *Neurospora* in which genetic blocks in utilization lead to accumulations of metabolites which are then secondarily transformed. It should be borne in mind, however, in assessing the biochemical role of alkaloids in the plant, that the functions of these substances in the organs where they are deposited may be quite unrelated to the functions of the same substances in the organs of synthesis. It is still by no means excluded that alkaloids may in some instances perform significant biological functions in the cells or tissues in which they are actually synthesized.

General References

Henry, T. A., The Plant Alkaloids. Blakiston, 4th ed, 1948.

Dawson, R. F., Advances in Enzymology. Interscience Publishers, 1948. Vol. 8, p. 203.

References

1. Mothes, K., *Planta*, **5**, 563 (1928).
2. Dawson, R. F., *Science*, **94**, 396 (1941).
3. Dawson, R. F., *Am. J. Botany*, **29**, 66 (1942).
4. Dawson, R. F., *Am. J. Botany*, **29**, 813 (1942).
5. Bonner, J., *Am. J. Botany*, **29**, 136 (1942).
6. Dawson, R. F., *Plant Physiol.*, **21**, 115 (1946).
7. Klein, G., and Linser, H., *Planta*, **20**, 470 (1933).
8. Dawson, R. F., *Plant Physiol.*, **14**, 479 (1939).
9. Dawson, R. F., *J. Am. Chem. Soc.*, **67**, 503 (1945).
10. Dawson, R. F., *Am. J. Botany*, **32**, 416 (1945).
11. Dawson, R. F., *Am. J. Botany*, **31**, 351 (1944).
12. Vickery, H. B., Pucher, G. W., Wakeman, A. J., and Leavenworth, C. S., Conn. Agr. Expt. Sta. Bull. 399, 1937.
13. Hieke, K., *Planta*, **33**, 185 (1942).
14. James, W. O., *Nature*, **158**, 377, 654 (1946); **159**, 196 (1947).
15. Hills, K. L., Trautner, E. M., and Rodwell, C. N., *Australian J. Sci.*, **9**, 24 (1946).
16. Rosen, F., Huff, J. W., and Perlzweig, W. A., *J. Biol. Chem.*, **163**, 343 (1946).
17. Beadle, G. W., Mitchell, H. K., and Nyc, J. F., *Proc. Natl. Acad. Sci. U.S.*, **33**, 155 (1947).
    Mitchell, H. K., and Nyc, J. F., *Proc. Nat. Acad. Sci.*, **34**, 1 (1948).
18. Galston, A. W., *Plant Physiol.* **24**, 577 (1949).

Chapter 23

NITRIFICATION AND NITROGEN FIXATION

Nitrification

Nitrogen as it is added to the soil, whether by fixation of atmospheric nitrogen, or by the decomposition of organic matter, is combined in the form of ammonia or other reduced products. We know, however, that nitrate is the principal form of readily available nitrogen in the soil and is also the principal form of nitrogen taken up and utilized by higher plants. The oxidation of ammonia to nitrate in the soil is known as nitrification and like nitrogen fixation is a microbiological process. The first demonstration of the biological nature of this oxidation was given by Schloesing and Munz in 1877. These workers packed a tube with sterile sand and then allowed sewage water containing ammonia to run through the tube. For several days the effluent contained ammonia. On the twentieth day, however, nitrate appeared and ammonia disappeared from the effluent. Sterilization of the sand with chloroform or heat again removed the ability of the sand to convert ammonia to nitrate, but inoculation with a water decoction of soil reinstituted the activity. These experiments then indicated the microbiological nature of ammonia oxidation and actual isolation of the organisms involved was accomplished by Winogradsky.[1]

The nitrifying bacteria isolated by Winogradsky are remarkable in that their growth is inhibited by the presence of organic matter in the culture medium and even gelatin, at that time widely used for preparing plates, proved toxic to them. The successful cultures, obtained with the use of plates solidified with silica gel, contained two types of bacteria, readily separable, the one, *Nitrosomonas*, being responsible for the conversion of ammonia to nitrite, while the other, *Nitrobacter*, further converts nitrite to nitrate. The two organisms are alike in requiring the absence of organic matter, and in requiring the presence of oxygen and of carbonate. With both of these organisms growth and formation of new cell substance is at the expense of CO_2, and the energy required for the reduction of the CO_2 is obtained from the oxidation of ammonia or of nitrite. The nitrifying bacteria are then chemosynthetic organisms which synthesize organic matter from CO_2 at the expense of energy-yielding chemical reactions. The two oxidations which are carried out

335

are both reactions which yield energy, and the amounts produced have been calculated[2] to be as follows:

$$\textit{Nitrosomonas} \quad NH_4^+ + \tfrac{3}{2}O_2 \rightarrow NO_2^- + H_2O + 2H^+$$
$$\Delta F_{298} = -66,500 \text{ cal.}$$
$$\textit{Nitrobacter} \quad NO_2^- + \tfrac{1}{2}O_2 \rightarrow NO_3^- \quad \Delta F_{298} = -17,500 \text{ cal.}$$

In this calculation the concentrations of the reactants have been taken as those which are optimal for the organisms, namely $(NH_4^+) = 0.005M$, $(H^+) = 10^{-8}M$, and $(NO_2^-) = 0.03M$. In the case of *Nitrosomonas* one molecule of CO_2 is reduced for approximately thirty-five molecules of nitrogen oxidized, while with *Nitrobacter* the ratio is closer to 1 to 100. Assuming that all the CO_2 is reduced to glucose (free energy of formation 118,000 cal. per mol of CO_2 reduced), the efficiencies of the two oxidations in supplying energy for CO_2 reduction may be calculated. This calculation shows that for *Nitrosomonas*, 5.9% of the energy liberated is actually used in CO_2 reduction, while for *Nitrobacter* the figure is 7.9%. The rest of the energy is presumably dissipated as heat, and measurements of Meyerhof with *Nitrobacter* show that this is indeed approximately true.[3] It is evident in any case that the oxidation in both instances releases quite sufficient energy to account for the synthesis observed. The mechanism of the transfer of energy from the oxidation to the synthetic reaction is obscure, as is the mechanism of the CO_2 reduction itself.

Ammonia oxidation by *Nitrosomonas* shows a rather narrow pH optimum in the region pH 8.5–8.8, and the pH optimum for *Nitrobacter* is in the same range. Nitrification does proceed, however, at pH's as low as 4. Both organisms are inhibited by conditions which favor the penetration of free ammonia into the cells, this being reflected in the low optimum concentration of NH_4^+ for *Nitrosomonas*, and in marked inhibition of nitrite oxidation by ammonia, especially in more alkaline culture solutions.

The toxicity of organic compounds to the nitrifying bacteria has been studied extensively.[3,4] Glucose in particular depresses growth, concentrations as low as $0.0025M$ causing a detectable inhibition. This concentration of glucose has no effect on rate of ammonia or of nitrite oxidation, and in general, the influence of organic compounds seems to be specifically on growth rather than on nitrification. The inhibitory effects of glucose and other organic compounds are less marked in soil or sand cultures than in liquid media. This undoubtedly bears on the fact that the organisms thrive in soil even though soil ordinarily contains organic matter.

Nitrogen Fixation

Biological nitrogen fixation is the process by which nitrogen of the atmosphere is reduced to ammonia or derivatives of ammonia such as amino acids through the agency of living organisms. The ability to fix molecular nitrogen is restricted to a relatively small number of kinds of organisms, and it is upon the activity of these species that all other living things depend for their nitrogen. The nitrogen of the soil is derived almost exclusively from biological nitrogen fixation, the only further source being that fixed by atmospheric electricity and by man's chemical syntheses. Since nitrogen is continuously lost from the soil through leaching, through the removal of crop residues, and through biological denitrification, it is clear that maintenance of the nitrogen balance requires the continuous replenishment of the soil nitrogen. According to Lipman and Conybeare[5] approximately 24 million tons of nitrogen are lost yearly from the agricultural areas of the United States. Only 3 million tons are replaced by nitrogen contained in manure and synthetic fertilizers, whereas 10 million tons are replaced annually through biological nitrogen fixation. Even though another 3.6 million tons is also added by nitrogen carried down in rainfall, still a yearly deficit of nitrogen is clearly occurring in our agricultural areas taken as a whole. Biological nitrogen fixation is then of great agricultural importance since the amounts of nitrogen added to soil by this process far exceed the amounts which are as yet practicable to add as synthetic fertilizer.

The losses of nitrogen from the soil take place, as noted above, not only through the removal of crop residues from the soil, but also through erosion and leaching. The fixed nitrogen thus lost is carried away in runoff or soil water and is ultimately contributed to the ocean. The marine organisms seem to depend mainly on this source of fixed nitrogen since the ocean appears to support almost no biological nitrogen fixation.[6] Much of the nitrogen contained in human food crops passes ultimately into sewage and is similarly contributed to drainage water. Other smaller losses of fixed nitrogen are due to denitrifying microorganisms, principally anaerobes which reduce nitrate to molecular nitrogen. Man also accomplishes significant amounts of denitrification in the combustion of explosives such as nitroglycerin and trinitrotoluene in which the oxygen of the nitro group is used as hydrogen acceptor for the combustion of reduced organic compounds.

Biological nitrogen fixation is carried on both by free living, non-symbiotic soil microorganisms and by microorganisms which live symbiotically in the roots of leguminous plants. Historically the problem of nitrogen fixation was first approached from the standpoint of the leguminous plants. We, however, shall first discuss the nonsymbiotic nitrogen fixation of free living microorganisms.

Nonsymbiotic Nitrogen Fixation. Jodin in 1862 found that nonsterile solutions containing sugar and containing no initial organic nitrogen could on occasion increase in fixed nitrogen.[7] If the experiment was carried out in a closed system, decreases in both N_2 and O_2 could be demonstrated. In the same way, Berthelot in 1885 showed that increases in fixed nitrogen could occur in moist soil incubated under favorable conditions.[8] The increases did not take place in aseptic soil or in soil incubated at low temperatures. It remained for Winogradsky in 1894 to isolate in pure culture a nitrogen-fixing microorganism from soil.[9] Winogradsky found that inoculation of a sugar nutrient (free of fixed nitrogen) with soil under anaerobic conditions resulted in fermentation of the sugar, mainly to acetic acid, butyric acid, CO_2 and H_2, and in fixation of atmospheric nitrogen. This fixation supplied the nitrogen requirement of the organism. The organism isolated by Winogradsky as responsible for this fixation of nitrogen was *Clostridium pastorianum*, an anaerobic spore former. It is now known that other scattered species of this genus also fix significant amounts of nitrogen.

The most important free living nitrogen-fixing microorganisms of soil are, however, those isolated by Beijerinck, who discovered two further nitrogen-fixing soil microorganisms, *Azotobacter chroococcum*, isolated from soil, and a motile *Azotobacter agile*[10] isolated from canal water. Still other *Azotobacter* species are recognized today including *A. indicum* and *A. vinelandii*. Isolation of these organisms from soil is readily accomplished through the use of enrichment cultures grown on nitrogen-free mannitol media, mannitol being a preferred substrate. *Azotobacter* is characterized by a remarkably rapid rate of metabolism, and its Q_{O_2} of 2000 or more is larger than that of most other known living creatures.[11] The organism is highly aerobic and nitrogen is fixed in relation to oxygen consumed, the consumption of 100 mols of oxygen resulting in the fixation of 0.25–10 mols of nitrogen. Variation in efficiency of nitrogen fixation depends on environmental conditions, especially the oxygen pressure, with low oxygen pressures giving higher efficiencies of nitrogen reduction. Fixation of nitrogen is inhibited when the organism is grown on media containing fixed forms of nitrogen, such as nitrate, or more particularly ammonia. In the presence of fixed nitrogen such as ammonia, the organism continues to grow but uses the fixed nitrogen in preference to carrying on nitrogen fixation. Organic forms of nitrogen such as aspartate or glutamate are much less effective in depressing nitrogen fixation.

The metabolism of *Azotobacter* has been the subject of a great deal of investigation directed primarily at attempts to discover the biochemical system responsible for nitrogen fixation.[12] In general *Azotobacter* appears, however, to possess the usual respiratory enzymes and an orthodox respiratory metabolism. Its cellular material is marked by unusually high contents of the vitamins which participate in coenzyme formation including thiamine, riboflavin, biotin, pantothenic acid, and nicotinic acid. The high content of these materials is doubtless related to the intense respiratory activity of the *Azotobacter* cell.

Of special interest is the enzyme hydrogenase which is found in *Azotobacter* as well as in other bacteria and algae.[13] In the presence of the *Azotobacter* enzyme, molecular hydrogen can bring about the reduction of such hydrogen acceptors as methylene blue and oxygen.

$$H_2 + acceptor \underset{\text{Hydrogenase}}{\rightleftharpoons} Acceptor \cdot 2H$$

Hydrogen acts also as a specific and competitive inhibitor of nitrogen fixation by *Azotobacter*. There is however no direct evidence associating this enzyme with the intermediate reactions of nitrogen fixation, and the fact that nodules of leguminous plants, which also fix nitrogen, lack the enzyme indicates that hydrogenase may not be generally involved as a step in biological nitrogen fixation.

Azotobacter requires molybdenum for growth and nitrogen fixation, 0.1–1.0 ppm. of molybdenum being optimal for nitrogen fixation of the various strains. Although the discovery of this requirement by Bortels in 1930[14] marked the first demonstration of an important role of the element in living things, it is now known that molybdenum is required by other organisms, including higher plants. The molybdenum requirement of *Azotobacter* grown on fixed nitrogen is nevertheless lower than the requirement of the organism when grown with molecular nitrogen. It is possible therefore that molybdenum may play an important role in nitrogen fixation as well as a role or roles in other more general processes.

Nitrogen fixation is carried on by certain blue-green algae, especially *Nostoc muscorum*.[15] The nitrogen fixation of this organism, like that of *Azotobacter*, requires molybdenum, is suppressed in the presence of ammonia, is inhibited by hydrogen and by carbon monoxide and seems to resemble closely the properties of the *Azotobacter* system.[16]

In addition to the biological agents which have been described, claims have been made for nitrogen fixation activity by a host of other organisms, including bacteria, yeasts, fungi, and higher plants. Some of these claims are undoubtedly based on experimental or analytical errors which are particularly important where the amount of nitrogen fixed is small.

In other cases true nitrogen fixation may occur. It is now readily possible to determine with certainty whether utilization of nitrogen occurs in a given case by the use of an atmosphere containing isotopic N^{15}, and this method should be applied before a final decision is made as to the possibility of nitrogen fixation by a given species.[17]

Symbiotic Nitrogen Fixation. Experiments of Bousingault,[18] begun in 1837, in which plants were grown with tap water as the only source of nutrients, indicated clearly that peas and clover were able to grow and to increase in organic nitrogen content beyond that contained in the water supplied, whereas oats and wheat were not able to increase in nitrogen except at the expense of nitrogen in the substrate. The great agricultural chemist, Liebig, however, propounded the view that plants derive their nitrogen from the small amount of ammonia in the atmosphere and denied that molecular nitrogen might be used by any species. Much uncertainty prevailed on the question until the work at the Rothamstead experiment station by Lawes, Gilbert, and Pugh, carried out from 1857 on. These workers showed in field experiments, which were continued over a period of years, that plots continuously cropped to non-legumes steadily decreased in yield until an ultimate stabilization was attained at low level. Plots cropped to legumes, on the contrary, maintained a high productivity over the same period. Legumes appeared also to be able to contribute nitrogen to a succeeding crop in a rotation. The English workers then looked carefully into the possibility that leguminous plants might be able to utilize atmospheric nitrogen. In critical experiments Lawes and Gilbert grew plants under controlled conditions in calcined soil. They determined the initial nitrogen content of seeds and soil as well as that of the water added and of the soil and plant at the end of the experiment. In addition they removed all ammonia from the air supplied to the plants. The results showed unambiguously that neither oats, wheat, barley, peas, beans, nor clover were able to increase in nitrogen at the expense of atmospheric molecular nitrogen alone. We know today that the failure of Lawes and Gilbert to detect nitrogen fixation was due to the calcined and sterile soil which they used, soil which had been freed both of the nonsymbiotic and of the symbiotic nitrogen-fixing organisms. Nevertheless the conclusion of Lawes and Gilbert that plants cannot utilize molecular nitrogen was generally accepted until the work of Hellriegel and Wilfarth in 1888.[19] These workers had planned to study the effects of varying nitrate supply on the growth of pea plants in pot cultures. Their experiments were, however, disturbed by the fact that the pots not supplied with nitrogen showed a very variable growth, some of the plants being yellow and stunted while others grew as though they had received nitrogen from an unknown source. Among explanations for this anomalous and inconsistent behavior, Hellriegel and Wilfarth considered the possibility that airborne bacteria, falling into the cultures, might infect them in such a way as to result in nitrogen fixation. To test this possibility peas were grown from aseptic seeds planted in sterilized sand and furnished with sterilized nutrient. The sand was covered with cotton to ward off airborne contamination. In a portion of the cultures the sand was inoculated with a suspension of garden soil. In still another series, no aseptic precautions were observed. The results of the experiment provided a complete confirmation of the bacterial hypothesis. In the aseptic cultures, the peas failed to grow in the absence of fixed nitrogen, while in the inoculated series, the plants grew luxuriantly in the absence of fixed nitrogen. In the nonaseptic series the results were wholly variable and inconsistent as before. In this experiment Hellriegel and Wilfarth also correlated nitrogen fixation with the presence of root nodules. Only in those plants which fixed nitrogen were nodules present. These nodules are as we know today the site of the nitrogen fixation and contain the symbiotic bacterium whose presence is an essential to leguminous nitrogen fixation. Isolation of the bacterium, known today as *Rhizobium*, from nodules was accomplished by Beijerinck[20] in 1888.

The genus *Rhizobium*, which appears to be allied to *Phytomonas*, has been divided by Fred and others[21] into seven species on the basis of the groups of host plants with which each is associated.

1. Alfalfa group. *Rhizobium meliloti.* Infects *Melilotus, Medicago,* etc.
2. Clover group. *Rhizobium trifolii.* Infects *Trifolium.*
3. Pea group. *Rhizobium leguminosarum.* Infects *Pisum.*
4. Bean group. *Rhizobium phaseoli.* Infects *Phaseolus.*
5. Lupin group. *Rhizobium lupini.* Infects *Lupinus,* etc.
6. Soybean group. *Rhizobium japonicum.* Infects *Glycine.*
7. Cowpea group. *Rhizobium* sp., not well defined. Infects *Vigna, Arachis, Lespedeza, Acacia,* etc.

In general a given *Rhizobium* will infect and thrive only on the roots of a host species within its particular group. Host specificity is, however, not complete and cross inoculation may be obtained experimentally in some instances. Even so, a great measure of specificity does exist and extends even to the varietal level since it is possible to obtain *Rhizobium* species strains which differ from one another in regard to effectiveness of nitrogen fixation with a given host plant. In the cultivation of *Rhizobium* for inoculation of leguminous crops, much emphasis is laid on the matching of *Rhizobium* and legume varieties.

Rhizobium may be readily cultured *in vitro* on a synthetic medium containing inorganic salts, ammonia or nitrate as a source of nitrogen, glucose or mannitol as a source of carbon, and special growth factors which vary somewhat from strain to strain.[22] In general one or more of the vitamins of the B complex including biotin, thiamine, riboflavin, and pantothenic acid exert growth promotive effects on the growth of *Rhizobium* in culture.[22]

Nodules. Invasion of the host plant by the bacteria commences as an active movement of *Rhizobium* cells from the soil toward root hairs of the legume seedling. As the bacteria approach the region of the root, rapid proliferation begins to occur. Many workers have found soil bacteria to be more plentiful in the immediate region of the root than in the soil at large, and for legume roots the difference may amount to fifty-fold. This may be the result of specific substances which are liberated from the root and which influence bacterial growth. Thus it is known that roots of flax and tobacco liberate thiamine and biotin to the surrounding medium,[23] and Wilson has shown that growth factors for *Rhizobium* are given off from roots of seedling clover and alfalfa.[18] After an accumulation of bacterial cells on young root hairs has occurred, the rhizobial cells penetrate the hair; this is followed by a curling of the tip of the root hair. The bacteria now multiply and move as an infection thread down through the hair and through the cortical cells to the endodermis. The inner cortical cells and in many cases the cells of the pericycle next begin to exhibit cell division and cell enlargement activity and from this growth the nodule develops. An interesting cytological aspect of the growth is the fact that the infected cells of the nodule contain double the normal somatic complement of chromosomes.[24] Thus in normal diploid species the nodules contain tetraploid cells, while in tetraploid species the nodules are octoploid, etc.

It is evidently of great interest to ascertain the nature of the interaction between bacteria and host which results in production of the nodule. Thimann has suggested[25] that a growth substance (Chapter 29) liberated by the *Rhizobium* may be the agent responsible for inciting nodule development. *Rhizobium* does in fact produce indoleacetic acid from tryptophane or from pea seedling extract, while in addition local application of indoleacetic acid in high concentrations to pea roots results in formation of nodule-like swellings. Against Thimann's view is, however, to be noted the fact that many other known indoleacetic acid-producing microorganisms do not produce nodules, and it is entirely likely that a more specific agent than indoleacetic acid alone is involved. Thornton has described experiments which indicate that formation of nodules is dependent on factors produced in the leaves of the plant,[26]

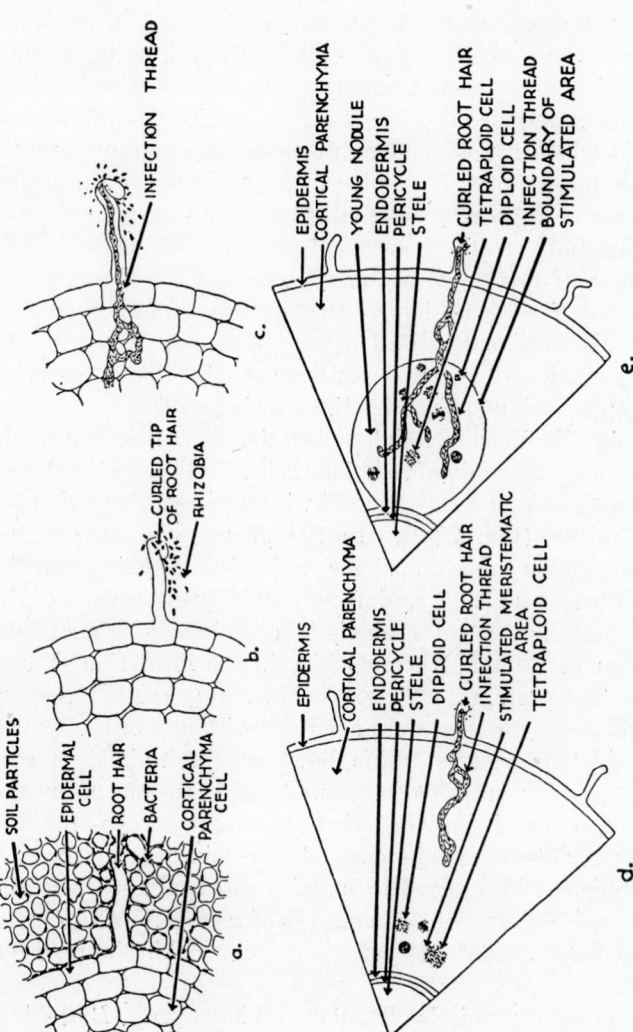

FIG. 23-1. Stages in the formation of a root nodule. (After Wilson[18])
a. Chemotaxis of the organism toward a root hair.
b. Curling of the root hair.
c. Early penetration of the infection thread.
d. Stimulated meristematic region in advance of the infection thread. (The latter has penetrated the outer layer of the cortical parenchyma.)
e. Penetration of the infection thread into the stimulated area.

and many workers have noted the poor formation of nodules in roots which have been excised from the plant. If such roots are grown in aseptic culture through repeated passages *in vitro* and are hence depleted of many of the substances contained in the normal plant, then nodules are not formed in response to *Rhizobium*. It seems entirely likely then that nodule formation may result from interaction between substances produced by the bacteria and other substances contained in the root of the host plant but which are dependent upon the top of the plant for their accumulation in the root. It should be possible to study the nature of the presumptive nodule growth substances by the methods so successfully applied to the study of other aspects of plant growth.

In the last phase of nodule development, the form of the rhizobial cells gradually alters from rod-shaped to irregular bodies known as bacteriods, which are apparently dead or at least inactive cells. The cells of the nodule cease their division, extensive vacuolization takes place, and finally the nodule is sloughed off by the growing root.

Hemoprotein in Nodules. Nodules contain a red hemoprotein similar in color to the hemoglobin of red blood cells.[27] This material may be reversibly oxygenated and deoxygenated with accompanying changes in absorption spectrum (Fig. 23-2). In the presence of oxygen the hematin of the hemoglobin takes up oxygen, which is lost again when the oxygen tension of the system is lowered, for example by evacuation of the system. Carbon monoxide may be similarly combined with the plant hemoglobin, just as with animal hemoglobin, in the formation of carboxyhemoglobin. The protein is not capable of reacting as a cytochrome or cytochrome oxidase *in vitro* and does not react with any of the known plant oxidation-reduction systems. It is, however, able to increase rate of respiration of *Rhizobium* cultures grown under low tensions of oxygen (0.01 atm.) indicating that it may be able to transport oxygen and may serve to make oxygen more available to a system respiring at low oxygen tensions.[27] The pigment is confined to nodules and is absent from the root system proper as well as from other types of nitrogen-fixing organisms such as *Azotobacter*. The role of plant hemoglobin in nitrogen fixation is obscure.

Factors Influencing Symbiotic Fixation of Nitrogen. *Rhizobium* grown in pure culture does not fix nitrogen under any circumstances on any known medium. Application of the isotopic nitrogen method has failed to reveal even a trace of uptake of molecular nitrogen by the free living organism, and it is similarly known with certainty that the legume fails to fix nitrogen in the absence of the bacterium. Nitrogen fixation is thus a property restricted to the symbiotic combination of legume and *Rhizobium*. Of the factors which influence the intensity of fixation, the supply of fixed nitrogen is a most important one. In the presence of fixed nitrogen, growth of the nodule is depressed as is also the number of nodules per plant.[28] Decrease in nitrogen fixation parallels reduction in extent of the nodular tissue as is shown in Table 23-1. Nevertheless, the

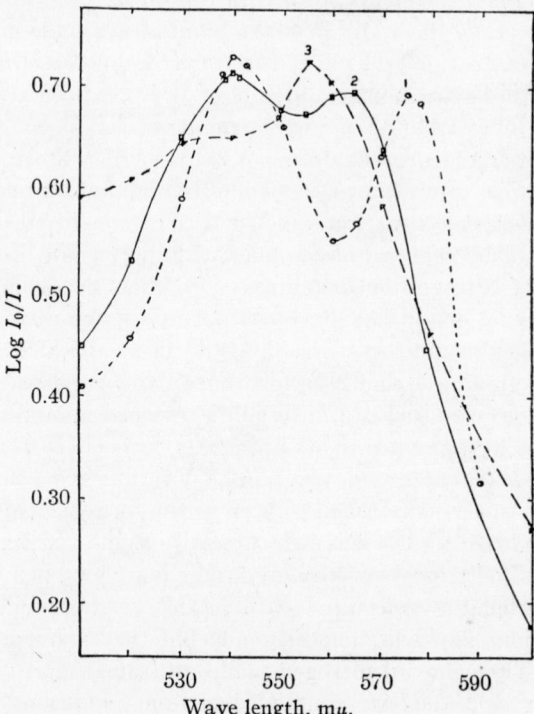

FIG. 23-2. Absorption curves for the hemoprotein from soybean nodules: (1) Oxygenated pigment. (2) CO derivative. (3) Pigment reduced by evacuation. (After Little and Burris[27])

amounts of nitrogen fixed and utilized by a legume under good growing conditions are at least as large as the amounts of nitrogen taken up by the same species grown with an excess of fixed nitrogen.[18]

TABLE 23-1. Influence of nitrate nitrogen on nodule growth in alfalfa inoculated with *Rhizobium meliloti*. (After Thornton and Nicol[28])

Amt. of NO₃ N added/culture (mg.)	Number of nodules produced		Av. length of nodules (mm.)	Vol. of bacteria/ nodule: ($\overline{mm}^3$)
	Per 10 plants	Per gm. dry wt. of roots		
0	496	178	2.2	.024
165	508	145	1.4	.010
330	333	152	1.0	.0055
600	204	100	0.71	.0025
990	69	42	0.55	.0020
1650	68	29	0.59	.0020

A second factor which is related to nodule growth and nitrogen fixation is the carbohydrate-nitrogen status of the host plant and of the nodule. Nodulation and nitrogen fixation both increase over a wide range with increase in plant carbohydrate reserve. Since carbohydrate reserve is lowered by the addition to the plant of available fixed nitrogen, the inhibition of fixation by fixed nitrogen may be merely a reflection of a general effect on the carbohydrate level of the plant. Excessively high carbohydrate in relation to available nitrogen is associated, however, with definite decreases in fixation. This may be shown with young inoculated legumes grown under conditions of high light intensity.[18] The plants grow poorly, a situation which may be remedied by addition of fixed nitrogen or by removal of the plant to a lower light intensity.

Excretion of Nitrogen by Legumes. The roots of leguminous plants may under certain conditions excrete a portion of the nitrogen which is fixed. This phenomenon was first detected in a critical experiment by Lipman[29] who grew peas and oats together in the same pot and showed that the cereal grew more luxuriantly and contained more nitrogen in the presence of the legume than in its absence. This effect persisted when the two species were separated by a porous pot, indicating that the nitrogen derived by the cereal must have been in the form of readily diffusible substances produced by the legume. The excretion of nitrogen by roots of legumes has been most extensively studied by Virtanen,[30] who grew peas and other plants inoculated with *Rhizobium* but under otherwise aseptic conditions and who actually collected the excreted nitrogenous compounds. The excreted nitrogen made up from 10% to almost 80% of the nitrogen fixed, the exact amount depending greatly on the environmental conditions. The conditions for abundant nitrogen excretion are apparently those which favor abundant nitrogen fixation but which discourage growth of the plant. Low temperatures ($55-60°F$) and long days favor excretion while higher temperatures (above $70°F$) in particular diminish the proportion of nitrogen excreted. Much acrimonious dispute[18] has centered about the question of whether or not nitrogen excretion does in fact take place, since in many of the earlier attempts to duplicate the work of Virtanen excretion was not observed, owing undoubtedly to unfavorable circumstances, especially temperature, prevailing in the experiments. Even though excretion of nitrogen can readily be observed under appropriate experimental conditions, it is still questionable whether this excretion is of importance under field conditions, especially as compared to the amounts of nitrogen liberated to the soil from root tissues which have sloughed off and decomposed. This question is discussed in detail by Wilson.[18]

Biochemistry of Nitrogen Fixation. From the biochemical standpoint, nitrogen excretion is of importance since Virtanen and Laine[31] have

FIG. 23-3. Nitrogen excretion by pea plants benefit the growth of associated oats.
(After Virtanen[30])
Pot 1. N-free nutrient soln. Inoculated with *Rhizobium*
Pot 2. N-free nutrient soln. Inoculated with *Rhizobium*
Pot 3. Ca(NO₃)₂ in nutrient soln. Inoculated with *Rhizobium*
Pot 4. N-free nutrient soln. Uninoculated control

collected and identified the excretion products and used their findings as a
basis for a suggested mechanism of nitrogen fixation. Virtanen and
Laine grew sterile *Rhizobium*–inoculated pea plants in cold long days
under low light intensities.   Up to 80% of the nitrogen fixed was excreted
into the sand in which the plants were grown.   The sand was extracted,
and from the extract Virtanen and Laine isolated or otherwise identified
aspartic acid, β-alanine, oximinosuccinic acid, and fumaric acid.   Glutamic acid was later added to this list.

$COOH$	$COOH$	$COOH$	$COOH$	$COOH$
CH_2	CH_2	CH_2	CH_2	CH
CH_2	$CHNH_2$	CH_2NH_2	$C{=}NOH$	CH
$CHNH_2$	$COOH$		$COOH$	$COOH$
$COOH$				
Glutamic acid	Aspartic acid	β-Alanine	Oximino-succinic acid	Fumaric acid

Aspartic and glutamic acids made up the bulk of the nitrogen in the effluvia, aspartic acid alone constituting over half of the total in young cultures. As the culture aged, the per cent of aspartic acid decreased while β-alanine increased in amount, and from this fact Virtanen and Laine inferred that aspartic acid is a primary fixation product, while β-alanine occurs secondarily and as a result of decarboxylation of the aspartic acid. On the Virtanen assumption that aspartic acid is the primary product, the presence of glutamic acid would necessarily be explained as due to transamination of aspartic with α-ketoglutaric acid. The three compounds together in any case made up the bulk, 90% or more, of the total excreted nitrogen. Oximinosuccinic acid, which was also isolated as the crystalline copper salt, made up only 1–2% of the excreted nitrogen and was found primarily in young cultures. On the basis of these findings Virtanen and Laine suggest that nitrogen fixation proceeds through the following steps:

1. Reduction of molecular nitrogen with the production of hydroxylamine, NH_2OH.
2. Reaction of hydroxylamine with oxaloacetic acid to oximinosuccinic acid. This reaction goes rapidly *in vitro*.
3. Reduction of oximinosuccinic acid to aspartic acid.
4. Decarboxylation of aspartic acid to β-alanine. This reaction is carried out *in vivo* by *Rhizobium* cultures as well as by an enzyme system present in crushed nodules.

The central features of the proposal of Virtanen and Laine are then (a) that hydroxylamine is the primary product of nitrogen reduction and (b) that hydroxylamine reacts with oxaloacetic acid. Evidence that hydroxylamine normally occurs in the nodule has been presented by various workers, and the presence of hydroxylamine in cultures of *Azotobacter* has also been reported. The mere presence of hydroxylamine in the organism does not of itself establish this compound as an intermediate in nitrogen fixation, since it has also been detected in cultures of the non-nitrogen fixing *Aspergillus* which apparently forms hydroxylamine as an oxidation product of ammonia. We have seen earlier (Chapter 16) also that hydroxylamine has been found in nonleguminous plants, in which case it has been supposed to be an intermediate in nitrate reduction. Hydroxylamine is highly toxic and even at nontoxic levels cannot be utilized as a source of nitrogen by *Azotobacter* or by leguminous plants. Even if hydroxylamine is an intermediate in nitrogen fixation it must then be one which is present only in minute concentrations and which is quickly transformed further. This transformation would presumably be the reaction of hydroxylamine with oxaloacetic or α-ketoglutaric acids, substances which are known to be present in the roots of leguminous plants. Perhaps the most important and critical point at present supporting the Virtanen and Laine scheme is then the presence of oximinosuccinic acid in the excretion products. The presence of this compound is only readily explained by a reaction between hydroxylamine and oxaloacetic acid. The oximes of oxaloacetic and α-ketoglutaric acids are, as pointed out earlier, metabolizable by higher plants. *Azotobacter*

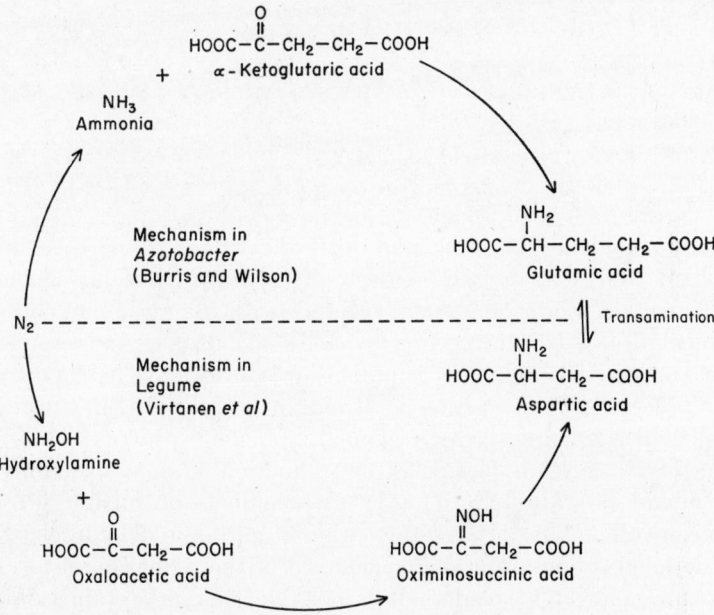

FIG. 23-4. Proposed mechanisms for the reduction of molecular nitrogen by agents of biological nitrogen fixation.

cannot, however, utilize oxime nitrogen, nor does the presence of an oxime depress the fixation of molecular nitrogen as does the presence of ammonia.[32] There seems to be but little question then that neither hydroxylamine nor oximes which arise from hydroxylamine are intermediates in nitrogen fixation by *Azotobacter*. Whether these substances are intermediate in nitrogen fixation by legumes is unclear.

In contrast to the uncertainty as to the role of hydroxylamine in nitrogen fixation is substantial evidence that ammonia is a direct product of nitrogen fixation in *Azotobacter*.[12,33] Studies with isotopic N^{15} have shown that when cells which are rapidly fixing molecular nitrogen are transferred to an atmosphere containing N_2^{15}, the nitrogen fixed is recovered in abundant amounts and within short periods in the glutamic acid of the organism. If similar cells are transferred to a medium containing $N^{15}H_3$, fixation of molecular nitrogen is immediately abolished but the ammonia nitrogen taken up is also rapidly incorporated into the glutamic acid of the organism. In both cases the N^{15} is less rapidly incorporated into aspartic acid and is still less rapidly incorporated into

TABLE 23-2. Comparison of distribution of isotopic N^{15} in *Azotobacter* supplied with N_2 and N_2^{15} or with N_2 and $N^{15}H_3$. All cultures previously grown at the expense of molecular N_2. The atom % excess of N^{15} in the whole cell is taken as 1.00 in each case. (After Wilson and Burris[12])

| | Distribution of N^{15} in cells | |
| | Supplied N_2^{15} | Supplied $N^{15}H_3$ |
Fraction of cells	(after 90 min.)	(after 15 min.)
Whole hydrolyzate of cells	1.00	1.00
Glutamic acid	1.82	2.48
Aspartic acid	1.37	0.79
Arginine	0.67	0.64
Histidine	0.75	0.61

such amino acids as arginine and histidine. The similarities in the metabolism and distribution of isotopic nitrogen whether given as molecular nitrogen or as ammonia, together with the rapid and complete supplantation of nitrogen fixation by ammonia uptake and utilization, strongly suggest that ammonia is in fact an intermediate in the nitrogen fixation process. This position is strengthened also by the fact that glutamic acid is apparently the principal immediate product of nitrogen fixation, since if fixation does result in production of ammonia, this compound should be expected to enter the channels of further nitrogen metabolism principally through the action of glutamic dehydrogenase.

No detailed proposal for the mechanism of the reduction of nitrogen to ammonia can yet be made. It is possible that in certain instances metabolically produced hydrogen may carry out some such reduction as the following:

$$\tfrac{1}{2}N_2 + \tfrac{3}{2}H_2 + H_2O \rightleftharpoons NH_4^+ + OH^- \qquad \Delta F_0 = -170 \text{ cal.}$$

In the case of *Azotobacter* the hydrogen could be produced by the hydrogenase reaction referred to earlier. More probably, however, molecular nitrogen is reduced by hydrogen transfer systems similar to or identical with the respiratory systems but for which molecular nitrogen constitutes a suitable hydrogen acceptor.

In summary there is good evidence (a) that oximes are produced during fixation of nitrogen by legumes and (b) that ammonia is an intermediate in nitrogen fixation by *Azotobacter*. We do not yet understand the intermediary metabolism of the process.

General References

Stephenson, M., Bacterial Metabolism. Longmans, Green and Co., 3rd ed., 1949.
Umbreit, W. W., *Bact. Revs.*, **11**, 157 (1947).

Waksman, S. A., Soil Microbiology. Williams and Wilkins, 2nd. ed., 1932.
Wilson, P. W., The Biochemistry of Symbiotic Nitrogen Fixation. University of Wisconsin Press, 1940.

References

1. Winogradsky, S., *Ann. inst. Pasteur*, **4**, 213 (1890); **5**, 92 (1891).
2. Baas-Becking, L. G. M., and Parks, G. S., *Physiol. Revs.*, **7**, 85 (1927).
3. Meyerhof, O., *Pflügers Arch. ges. Physiol.*, **164**, 353 (1916); **165**, 229 (1916); **166**, 240 (1917).
4. Kingma Boltjes, T. Y., *Arch. Mikrobiol.*, **6**, 79 (1935).
5. Lipman, J. G., and Conybeare, A. B., N. J. Expt. Sta. Bull. 607, 1936.
6. Zobell, C., Marine Microbiology. Chron. Botanica, 1946.
7. Jodin, *Compt. rend.*, **55**, 612 (1862).
8. Berthelot, M., *Compt. rend.*, **101**, 775 (1885).
9. Winogradsky, S., *Compt. rend.*, **118**, 353 (1894).
   For recent work see Rosenblum, E. D., and Burris, R. H., *J. Bact.*, **57**, 413 (1949).
10. Beijerinck, M. W., *Zentr. Bakt.*, II, **7**, 561 (1901).
11. An excellent review is that of Stephenson M., Bacterial Metabolism. Longmans, Green and Co., 3rd ed., 1949.
12. Wilson, P. W., and Burris, R. H., *Bact. Revs.*, **11**, 41 (1947).
13. Stephenson, M., and Stickland, L. H., *Biochem. J.*, **25**, 205 (1931).
14. Bortels, H., *Arch. Mikrobiol.*, **1**, 333 (1930).
    Burk, D., and Horner, C. K., *Soil Sci. Am., Proc.*, **1**, 213 (1936).
15. Allison, F. E., Hoover, S. R., and Morris, H. J., *Botan. Gaz.*, **98**, 433 (1937).
16. Burris, R. H., and Wilson, P. W., *Botan. Gaz.*, **108**, 254 (1946).
17. Burris, R. H., Eppling, F. J., Wahlin, H. B., and Wilson, P. W., *J. Biol. Chem.*, **148**, 349 (1943).
18. These early experiments are summarized in Wilson, P. W., Biochemistry of Symbiotic Nitrogen Fixation. University of Wisconsin Press, 1940.
19. Hellriegel, H., and Wilfarth, H., *Beilageh. Ver. Rübenzucker Ind.*, **1**, 1888.
20. Beijerinck, M. W., *Botan. Zentr.*, **39**, 356 (1888).
21. Fred, E. B., Baldwin, I. L., and McCoy, E., Root Nodule Bacteria and Leguminous Plants. University of Wisconsin Press, 1932.
22. Peterson, W. H., and Peterson, M. S., *Bact. Revs.*, **9**, 49 (1945).
23. West, P. M., *Nature*, **144**, 1050 (1939).
    West, P. M., and Lochhead, A. G., *Can. J. Research*, **18C**, 129 (1940).
24. Wipf, L., and Cooper, D., *Proc. Natl. Acad. Sci.*, **24**, 87 (1938).
25. Thimann, K. V., *Proc. Natl. Acad. Sci.*, **22**, 511 (1936). Also *Third Comm. Intern. Soc. Soil Sci.*, Trans., **A**, 24 (1939).
26. Thornton, H. G., *Proc. Roy. Soc. (London)*, **B104**, 481 (1929).
27. Kubo, H., *Acta Phytochim.*, **11**, 195 (1939).
    Little, H., and Burris, R. H., *J. Am. Chem. Soc.*, **69**, 838 (1947).
28. Thornton, H. G., and Nicol, H., *J. Agr. Sci.*, **26**, 173 (1936).
29. Lipman, J. G., N. J. Agr. Expt. Sta. Bull. 253, 1912.
30. Virtanen, A. I., *Ann. Roy. Agr. Coll. Sweden*, **5**, 429 (1938).
    See also Wyss, O., and Wilson, P. W., *Soil Sci.*, **52**, 15 (1941).
31. Virtanen, A. I., and Laine, T., *Biochem. J.*, **33**, 412 (1939).
    Virtanen, A. I., *Biol. Revs.*, **22**, 239 (1947).
32. Novak, R., and Wilson, P. W., *J. Bact.*, **55**, 517 (1948).
33. Burris, R. H., and Wilson, P. W., *J. Bact.*, **52**, 505 (1946).

PART V. SECONDARY PLANT PRODUCTS

CHAPTER 24

LIPIDS AND LIPID METABOLISM

Classification of the Lipids. The term lipid includes a variety of types of substances which have in common the fact that they yield fatty acids on hydrolysis. The group includes (a) the fats and oils, in which fatty acids are esterified with glycerol, (b) waxes, in which fatty acids are esterified with monohydric, generally long chain, alcohols, and (c) the phospholipids, in which fatty acids are esterified with glycerol or other alcohols and with which phosphoric acid and basic nitrogenous constituents are also combined. Under the wax category certain related waxlike materials and other fatty acid derivatives are also frequently included.

The lipids may be extracted from the plant material by organic solvents such as ether, acetone, alcohol, chloroform, carbon disulfide, and petroleum naphtha. These solvents extract in addition other nonlipid constituents, including resins and resin acids, sterols, and terpenes. The crude lipid extract may and usually does contain then a variety of non-lipid constituents. The lipid fraction may also contain free fatty acids, including the lower acids such as acetic, propionic, and butyric, which are readily volatile in steam, and the higher fatty acids such as those typical of fats themselves.

TABLE 24-1. Typical components of the crude lipid fraction of plant materials

Fats	Fatty acid-containing substances and derivatives		Nonfatty acid-containing substances
	Waxes	Phospholipids	
Esters of fatty acids with glycerol	Esters of fatty acids with monohydric alcohols	Esters of fatty acids including phosphoric acid	1. Resins
			2. Resin acids
			3. Terpenes
	1. Simple aliphatic leaf waxes	1. Esters with glycerol; lecithin, etc.	4. Sterols
	2. Aliphatic hydrocarbons related to fatty acids	2. Esters with carbohydrate alcohols	
	3. Alcohols and ketones related to fatty acids	3. Esters with inositol	

Separation of Lipids.[1] The crude lipid fraction is first prepared from the tissue by continuous extraction of the tissue with a suitable solvent such as ether. With many tissues it is desirable to treat first with alcohol which rapidly extracts the lipids, then to remove the alcohol and extract the residue with ether. The concentrated ether extract may next be extracted with an aqueous carbonate solution to remove acidic substances such as free acids. The ether-soluble fraction can now be

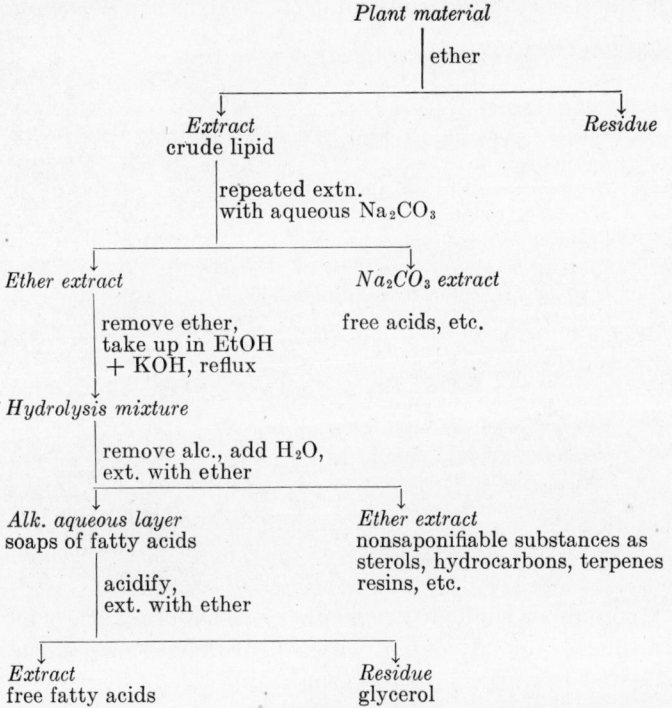

FIG. 24-1. Separation of constituent fatty acids from the crude lipid fraction of plant materials.

saponified, i.e., the fatty acid esters hydrolyzed to free fatty acids, by dissolving the crude lipid (after removal of the ether or other solvent) in ethanol containing KOH and refluxing the solution for several hours. The alcohol is next removed. It is now possible to separate the liberated fatty acids from the unsaponifiable residue by extraction of the aqueous hydrolysis mixture with ether. The neutral substances pass into the ether, leaving the fatty acids in the aqueous phase as their soaps. By acidification and reextraction with ether the free fatty acids can be recovered. Although this method serves to isolate from the tissue those fatty acids which were originally present as fats or as other esters, it does not reveal the exact manner in which each fatty acid molecule was originally combined. To discover this it is necessary to isolate the individual fatty constituents in chemically pure form from the crude lipid extract, a task which is much more complex.

Occurrence of Fats. Fats occur in all portions of the plant including stem, leaf, root, fruit, and flower. In general, however, the vegetative organs contain only low concentrations of fat, and the major concentrations are found in the tissues of the fruit and in the seed, as can be seen in

TABLE 24-2. Fat content of a variety of plant tissues

Tissue	Species	Per cent fat dry wt. basis
Leaf	Dactylis glomerata (cocksfoot grass)	2.2
	Lolium perenne (rye grass)	1.7
	Brassica oleracea (cabbage)	1.7
	Mentha aquatica (peppermint)	5.0
	Spinacia oleracea (spinach)	0.4
Stem (bark)	Tilia cordata (bass wood)	2.3
	Hippophaë rhamnoides (sea buckthorn)	3.0
Root	Beta rapa vulgaris (mangel)	7.0
Fruit (flesh)	Olea europaea (olive)	50
	Persea gratissima (avocado)	20
Seed	Cocos nucifera (coconut)	65
	Ricinus communis (castor bean)	60
	Helianthus annuus (sunflower)	50
	Linum usitatissimum (flax)	35
	Triticum vulgare (wheat)	2.1
	Zea mays (corn)	5.0
	Glycine max (soybean)	20

Table 24-2, which contains a summary of the fat contents of a variety of plant tissues. In the leaf the fat content varies in the few species which have been investigated from 0.4% to 5.0% on a dry weight basis and thus forms an appreciable although not a major constituent. In stems and roots the amounts of fatty material found are of the same order of magnitude as in leaves, with the exception that in unusual cases as the tubers of Cyperus esculentus, the oil content may reach 20–30%. It is in fruits and seeds, however, that fatty materials are most frequently accumulated in high concentration and in these tissues fats often constitute a major component or even the bulk of the tissue, as is shown in Table 24-2. Fats may be accumulated in the pericarp of the fruit proper, as is the case in certain palms, olive, avocado, cocoa, and others. In all these cases the fats are contained primarily in the parenchymatous cells of the fruit itself and can be obtained by expression of the oil from the whole fruit flesh. Fat accumulation occurs also in fruits in which the flesh is made up of tissues of the receptacle, as in fruits of the Rosaceae.

The seeds of many species contain large amounts of fat which may be either in the tissue of the embryo as in soybean and sunflower, in both the tissues of the endosperm and the embryo as in the palm, or primarily in the embryo with only small amounts in the endosperm as in the cereals. According to McNair[2] the seeds of over 88% of the families of higher plants contain fat, and of these three-quarters contain fat to the exclusion of starch as a reserve food for seed germination. On the other hand there is a considerable number of common species whose seeds contain only low concentrations of fat, including, for example, the majority of the cereals as well as peas and some beans.

The lipids are contained in the cell in the form of droplets or globules suspended in the cytoplasm. In vegetative tissue of low lipid content such as leaves, the fat is dispersed as very small droplets. In seeds or fruits which contain a high concentration of lipid, these small drops coalesce to form larger particles. Whether or not fat is also dispersed in the cytoplasm in submicroscopic particles is not readily detectable by such methods as staining with lipophilic dyes or other present histochemical techniques and is still undecided.

The Fatty Acids. The constituent fatty acids of the plant fats and oils may be divided into the saturated and the unsaturated groups. Experimentally, acids of the two groups may be separated in a number of ways, of which a useful method is that recommended by Twitchell.[3] This method is based on the fact that the lead salts of the saturated acids tend to separate out of 95% alcohol as a solid precipitate, whereas the lead salts of the unsaturated acids tend to remain behind in solution. While this separation works well for many fatty acids, it is not equally effective for all. Lower saturated fatty acids (myristic and lower) yield lead salts which are incompletely precipitated and which remain in part in the unsaturated fraction. Certain unsaturated acids such as erucic, elaeostearic, and others tend, on the contrary, to precipitate with the saturated fraction. The lead salt fractionation has therefore definite limitations.

An alternative method for separation of saturated from unsaturated acids is that of fractional crystallization from acetone, ether, or petroleum naphtha at very low temperatures (-20 to $-70°C$). In this process the saturated acids are the most readily precipitated while the most unsaturated (linolenic, etc.) acids are least readily precipitated. The exact solvent and temperature best suited for the separation varies with the composition of the fatty acid mixture under investigation. In the separation of many fatty acid mixtures acetone has proved to be an excellent solvent. Approximately 10 gm. of fatty acid per 100 cc. of acetone are cooled to -20 to $-40°C$ for the precipitation of the saturated fatty acids of cotton seed, corn, soybean, and other important natural oils.

The low temperature crystallization fractionation like the lead salt fractionation does not give complete separation of all saturated fatty acids from all unsaturated acids. It is, however, rapid and convenient in application.

The saturated acids found in the organs of higher plants are all normal (straight chain) acids with even numbers of carbon atoms and include all such acids from C_2 to C_{26}. The lower members of the series are readily volatile with steam and hence form the so-called volatile fatty acids. Actually all fatty acids are steam-volatile to a greater or lesser extent so that the distinction of volatility is at best a quantitative one. While it is doubtful that acetic or butyric acids occur as glycerides in the plant, all

TABLE 24-3. The principal saturated fatty acids of higher plants

Formula	Trivial name	Systematic name	Structure
$C_2H_4O_2$	Acetic acid	Ethanoic acid	CH_3COOH
$C_4H_8O_2$	Butyric acid	n-Tetranoic acid	$CH_3(CH_2)_2COOH$
$C_6H_{12}O_2$	Caproic acid	n-Hexanoic acid	$CH_3(CH_2)_4COOH$
$C_8H_{16}O_2$	Caprylic acid	n-Octanoic acid	$CH_3(CH_2)_6COOH$
$C_{10}H_{20}O_2$	Capric acid	n-Decanoic acid	$CH_3(CH_2)_8COOH$
$C_{12}H_{24}O_2$	Lauric acid	n-Dodecanoic acid	$CH_3(CH_2)_{10}COOH$
$C_{14}H_{28}O_2$	Myristic acid	n-Tetradecanoic	$CH_3(CH_2)_{12}COOH$
$C_{16}H_{32}O_2$	Palmitic acid	n-Hexadecanoic	$CH_3(CH_2)_{14}COOH$
$C_{18}H_{36}O_2$	Stearic acid	n-Octadecanoic	$CH_3(CH_2)_{16}COOH$
$C_{20}H_{40}O_2$	Arachidic acid	n-Eicosanoic	$CH_3(CH_2)_{18}COOH$
$C_{22}H_{44}O_2$	Behenic acid	n-Docosanoic	$CH_3(CH_2)_{20}COOH$
$C_{24}H_{48}O_2$	Lignoceric acid	n-Tetracosanoic	$CH_3(CH_2)_{22}COOH$
$C_{26}H_{52}O_2$	Cerotic acid	n-Hexacosanoic	$CH_3(CH_2)_{24}COOH$

the remaining acids are found in one or more fats or oils. Higher fatty acids in the range of C_{28} to C_{38} are occasionally met with in plant materials but are in general constituents of waxes rather than of fats. Of the saturated group palmitic, lauric, and stearic acids are the most widespread and quantitatively the most abundant. The great bulk of the plant fatty acids belongs, however, to the unsaturated category.

The unsaturated fatty acids present naturally a greater diversity than the saturated acids since they may contain 1, 2, or more ethylenic linkages variously located along the hydrocarbon chain. The unsaturated acids include members from C_{10} to C_{24} and like the saturated acids include, with few exceptions, straight chain molecules of an even number of carbon atoms. Among the unsaturated acids oleic and linoleic are the most widely distributed and the most generally abundant. It has been calculated, on the basis of the world's harvest of plant fats, that oleic acid constitutes some 34 % of the total agricultural fatty acid production.[5] Linoleic acid on the same basis constitutes 29% of the total fatty acids synthesized, while palmitic acid, the most abundant saturated acid,

TABLE 24-4.   Some of the principal unsaturated fatty acids of plants

Unsaturated acids containing one double bond (monoethenoid)

Formula	Trivial name	Systematic name	Structure
$C_{10}H_{18}O_2$	Caproleic	$\Delta^{9,10}$-Decenoic	$CH_2:CH(CH_2)_7COOH$
$C_{12}H_{22}O_2$	Lauroleic	$\Delta^{9,10}$-Dodecenoic	$CH_3CH_2CH:CH(CH_2)_7COOH$
$C_{14}H_{26}O_2$		$\Delta^{5,6}$-Tetradecenoic	$CH_3(CH_2)_7CH:CH(CH_2)_3COOH$
$C_{14}H_{26}O_2$	Myristoleic	$\Delta^{9,10}$-Tetradecenoic	$CH_3(CH_2)_3CH:CH(CH_2)_7COOH$
$C_{16}H_{30}O_2$	Palmitoleic	$\Delta^{9,10}$-Hexadecenoic	$CH_3(CH_2)_5CH:CH(CH_2)_7COOH$
$C_{18}H_{34}O_2$	Oleic	$\Delta^{9,10}$-Octadecenoic	$CH_3(CH_2)_7CH:CH(CH_2)_7COOH$
$C_{18}H_{34}O_2$	Petroselinic	$\Delta^{6,7}$-Octadecenoic	$CH_3(CH_2)_{10}CH:CH(CH_2)_4COOH$
$C_{18}H_{34}O_3$	Ricinoleic	12-Hydroxy-$\Delta^{9,10}$-octadecenoic	$CH_3(CH_2)_5CHOHCH_2CH:CH(CH_2)_7COOH$
$C_{20}H_{38}O_2$	Gadoleic	$\Delta^{9,10}$-Eicosenoic	$CH_3(CH_2)_9CH:CH(CH_2)_7COOH$
$C_{22}H_{42}O_2$	Erucic	$\Delta^{13,14}$-Docosenoic	$CH_3(CH_2)_7CH:CH(CH_2)_{11}COOH$
$C_{24}H_{46}O_2$	Selacholeic	$\Delta^{15,16}$-Tetracosenoic	$CH_3(CH_2)_7CH:CH(CH_2)_{13}COOH$

Unsaturated acids containing one triple bond (ethinoid)

Formula	Trivial name	Systematic name	Structure
$C_{18}H_{32}O_2$	Tariric	$\Delta^{6,7}$-Octadecinoic	$CH_3(CH_2)_{10}C:C(CH_2)_4COOH$

Unsaturated acids containing 2 double bonds (diethenoid)

Formula	Trivial name	Systematic name	Structure
$C_{18}H_{32}O_2$	Linoleic	$\Delta^{9,10,12,13}$-Octadecadienoic	$CH_3(CH_2)_4CH:CHCH_2CH:CH(CH_2)_7COOH$

Unsaturated acids containing 3 double bonds (triethenoid)

Formula	Trivial name	Systematic name	Structure
$C_{18}H_{30}O_2$	Linolenic	$\Delta^{9,10,12,13,15,16}$-Octadecatrienoic	$CH_3CH_2CH:CHCH_2CH:CHCH_2CH:CH(CH_2)_7COOH$
$C_{18}H_{30}O_2$	Elaeostearic	$\Delta^{9,10,11,12,13,14}$-Octadecatrienoic	$CH_3(CH_2)_3CH:CHCH:CHCH:CH(CH_2)_7COOH$
$C_{18}H_{28}O_3$	Licanic	4-Keto-$\Delta^{9,10,11,12,13,14}$-octadecatrienoic	$CH_3(CH_2)_3CH:CHCH:CHCH:CH(CH_2)_4CO(CH_2)_2$-COOH

Unsaturated acids containing 4 double bonds (tetraethenoid)

Formula	Trivial name	Systematic name	Structure
$C_{18}H_{28}O_2$	Parinaric	$\Delta^{9,10,11,12,13,14,15,16}$-Octadecatetraenoic	$CH_3CH_2CH:CHCH:CHCH:CHCH:CH(CH_2)_7$-COOH

Cyclic unsaturated acids

Formula	Trivial name	Systematic name	Structure
$C_{16}H_{28}O_2$	Hydnocarpic	11-$\Delta^{2,3}$-Cyclopentenyl-n-undecanoic	
$C_{18}H_{32}O_2$	Chaulmoogric	13-$\Delta^{2,3}$-Cyclopentenyl-n-tridecanoic	

For Hydnocarpic:

$$CH=CH$$
$$CH(CH_2)_{10}COOH$$
$$CH_2-CH_2$$

For Chaulmoogric:

$$CH=CH$$
$$CH(CH_2)_{12}COOH$$
$$CH_2-CH_2$$

makes up only 11% of the total. Despite this overall predominance of a few fatty acids in the plant world, it is nevertheless characteristic of seed fats in particular that an individual species may contain a large proportion of a fatty acid otherwise of only narrow distribution. This, for example, is the case with the cyclic acids, chaulmoogric and hydnocarpic, which are constituents of chaulmoogra oil contained in the seeds of *Hydnocarpus* and used in treatment of leprosy. The same acids are produced in the oils of other species of the family *Flacourtiaceae*. Similarly, erucic acid is found in the seed fats of crucifers, while a close relative of oleic acid, petroselinic acid, is found in seeds of the *Umbelliferae*. Special attention should be called to tariric acid, which contains an

TABLE 24-5. Distribution of fatty acids in the world's supply of agriculturally produced vegetable fats. (After Boekenoogen[5])

Fatty acid	% of total	Fatty acid	% of total
Oleic	34	Linolenic	6
Linoleic	29	Myristic	3
Palmitic	11	Erucic	3
Lauric	7	Stearic	3

acetylenic linkage, as well as to ricinoleic acid, found in oil of castor seeds (*Ricinus communis*) and which is related to oleic acid but contains an additional hydroxyl group at carbon atom 12. Licanic acid is similarly 4-ketoelaeostearic acid.

In addition to the unsaturated fatty acids given in Table 24-5, there are known in nature further cyclic compounds related to chaulmoogric acid, as well as a number of substituted fatty acids including dihydroxystearic acid (castor oil) and possibly branched chain or methylated acids.

Separation of Fatty Acids. The separation of pure components from a mixture of fatty acids is usually accomplished by fractional distillation of the ethyl or methyl esters. It is desirable first to separate the saturated from the unsaturated acids by the methods outlined earlier in order to decrease the number of components in the distillation mixture since this leads to cleaner and more complete resolution. The two fatty acid fractions are then esterified by refluxing in ethanol or methanol containing 1% to 2% of H_2SO_4 or HCl. In general the ethyl esters are to be preferred since they are more heat stable and since there are available more data for characterization of the separated components. After esterification, the alcohol is partially removed by distillation, the mixture of esters taken up in ether and extracted with water to remove the mineral acid, and the ether then removed by distillation. The mixture of esters may now be fractionally distilled. This is ordinarily carried out *in vacuo* in order to decrease the distillation temperatures and must be done with an efficient fractionating column if good resolution is to be obtained. The

Podbielniak heli-grid type of packing and the Stedman conical packing have both been found very satisfactory for this purpose.[42] The column is insulated against temperature changes, and both column and distilling flask are heated electrically by automatically regulated circuits. In the fractions of distillate collected, the individual esters may be identified by such properties as refractive index, equivalent weight by saponification, melting and freezing points, and by purely physical measurements such as the crystallographic spacings of the pure material as determined by x-ray methods. Properties of derivatives of the fatty acids such as the amides, anilides, and p-bromophenacyl esters may also be used in identification. The unsaturated fatty acids exhibit typical absorption in the ultraviolet in the region of 3200–2400 Å which may be used in determination of these compounds.[43]

The iodine number is widely used as a measure of unsaturation in all work concerning fats and fatty acids. This value is commonly expressed as grams of I_2 absorbed by 100 gm. of fat or fatty acids. The reaction consists in the addition of iodine to the double bond or bonds of the fatty acid in the presence of a catalyst such as mercuric chloride. The thiocyanogen radical, SCN, also adds quantitatively to the double bonds of fatty acids but, unlike iodine, reacts with only one double bond of a diethenoid acid and with two of the three double bonds of a triethenoid acid. The iodine and thiocyanogen values may therefore be used jointly to determine the way in which double bonds are distributed in a fatty acid or fatty acid mixture.

Distribution of Fatty Acids in Plant Fats. The earlier investigations of the plant fats and oils were mainly concerned with the characterization of the materials by various properties of the crude fat, including the saponification number and acid value, which determine the relative amounts of neutral fat and free acid, respectively, and the iodine value which gives an estimate of the average unsaturation of the component fatty acids of the material. Useful as these methods have been in the past, particularly in industrial work, they are now no longer acceptable as methods for the characterization of a fat or oil. It is essential rather to know the component fatty acids in the fat or oil and if possible also to know something of the distribution of the fatty acids in the various glyceride molecules. Just as proteins are characterized by amino acid composition, so fats may be characterized by fatty acid composition.

In general, those fats which are solid at ordinary temperatures are known as fats, whereas those which are liquid at the same temperatures are known as oils. This difference in physical properties is in part due to the varying proportion of saturated to unsaturated fatty acids in the glyceride molecule. Thus the liquid linseed oil of *Linum* seeds contains

35–50% linolenic acid and 25–45% of linoleic acid, while the solid fat of cacao (*Theobroma cacao*) seeds contains about 35% of palmitic and 40% of stearic acid. Various authors have attempted to show that unsaturation is a feature associated with fats of plants grown in temperate or cold climates and that more highly saturated fats are produced by tropical species. This does not, however, appear to be a general rule. The nature of the fatty acids in a particular species is rather genetically controlled without immediate reference to habitat of the species as a whole and in addition many tropical plants form highly unsaturated fats.

When highly unsaturated oils are exposed to oxygen they thicken and/or dry to a solid film owing to oxidation and polymerization of the unsaturated fatty acids and fatty acid oxidation products. Oils which thicken without hardening to a solid film are known as non-drying oils and are those which contain much oleic or other saturated acids as olive oil. Drying oils on the other hand, as those of flax or tung, are those which contain much doubly or trebly unsaturated acid as linoleic, linolenic, elaeostearic, or licanic. The linoleic-linolenic type drying oils find a large-scale application in the manufacture of paint, varnish, and enamel while the edible oils are primarily of the less readily oxidized non-drying type.

Hilditch has arbitrarily divided the fatty acid constituents of a given fat into the major constituents, those acids each of which makes up over 10% of the total, and the minor acids, each of which makes up less than 10% of the total.[4] We may thus speak about the major and minor constituent acids of a given fat or oil. This convention, as well as the data summarized by Hilditch, is followed in the discussion below.

The leaf fats, in so far as investigated, contain linolenic and linoleic as their major components with smaller amounts of oleic acid. The saturated acids make up in general but a small proportion, approximately 10%, of the total. The saturated acids include stearic and palmitic

TABLE 24-6. Component fatty acids of typical leaf fats

| Species | Trivial name | Component fatty acids: % of total | | | | Reference |
		Saturated	Oleic	Linoleic	Linolenic	
Dactylis glomerata	Cocksfoot grass	11	16	31	42	6
Lolium perenne	Rye grass	12	22	26	40	6
Spinacia oleracea	Spinach	Small	30	50	20	8
Brassica oleracea	Cabbage	10	..	Much	Much	7

found in spinach by Speer and others[8] and identified in cabbage by Chibnall and Channon.[7]

The fats contained in fruit tissues other than the seed very uniformly contain palmitic and oleic acids as major components with linoleic generally present in somewhat smaller amounts, as are also stearic and myristic acids. In avocado (*Persea*) pulp oil, as well as in the oil of the patua palm (*Oenocarpus*) and of the olive (*Olea*) fruit, more than 80% of the total fatty acid is oleic accompanied by 10% or less of each of palmitic and linolenic. In the oil palm (*Elaeis*) and in the fruit oil of cacao the proportion of oleic acid is smaller and the amount of palmitic acid correspondingly greater.

Much more is known about the composition of the seed fats than

TABLE 24-7. The component fatty acids of typical fruit fats. (After Hilditch[4])

Species	Trivial name of fat	Component fatty acids: % of total				
		Saturated acids			Unsaturated acids	
		Myristic	Palmitic	Stearic	Oleic	Linoleic
Stillingia sebifera	Chinese vegetable tallow	4	66	1	27	..
Elaeis guineensis	African oil palm	1	43	4	40	11
Oenocarpus pataua	Patua palm	..	9	6	81	4
Theobroma cacao	Cacao fruit oil	..	50	..	35	10
Persea gratissima	Avocado oil	..	7	1	81	11
Olea europaea	Olive oil	1	10	1	80	8

about any other group of plant fats; in fact, this group provides the two foremost oils of the world, those of coconut and peanut as well as the bulk of the other important fats and oils of commerce such as linseed, soybean, maize, and cottonseed oils. Among the seed fats a much greater variation in composition is found than is evident among the leaf or fruit fats. There is nevertheless a certain regularity in composition since, in general, the fat composition of seeds follows phylogenetic lines. The seed fats may be divided on the basis of major fatty acids into the following twelve

TABLE 24-8. Phylogenetic relationships of the seed fats. (After Hilditch[4])

Group	Major fatty acids	Typical families included
1	Linoleic, linolenic, oleic	Coniferae, Juglandaceae, Labiatae, Linaceae
2	Linoleic, oleic	Fagaceae, Papaveraceae, Oleaceae, Compositae
3	Linoleic, oleic, or linolenic; elaeo-stearic, licanic; ricinoleic	Rosaceae, Euphorbiaceae, Cucurbitaceae
4	Palmitic, oleic, linoleic	Anacardiaceae, Tiliaceae, Malvaceae, Solanaceae, Rubiaceae
5	Palmitic, oleic, linoleic	Gramineae
6	Petroselinic, oleic, linoleic	Umbelliferae, Araliaceae
7	Cyclic unsaturated acids	Flacourtiaceae
8	Erucic, oleic, linoleic	Cruciferae, Tropaeolaceae
9	Oleic, linoleic, arachidic, lignoceric	Leguminosae, Sapindaceae
10	Stearic, palmitic, oleic	Sterculiaceae
11	Lauric, myristic	Lauraceae
12	Lauric, myristic	Palmaceae

principal groups. In each group, in addition to varying proportions of the major acids, minor amounts of other acids are also found. Table 24-9 gives a summary of the fatty acid compositions of a selected series of the more important seed fats.

In addition to the seed fats of Table 24-9 there are a number of particular fats to which attention should be called. Petroselinic acid is apparently confined almost exclusively to the seeds of the *Umbelliferae*,

TABLE 24-9. Fatty acid components of some typical seed fats. (Modified after Hilditch[4])

Species	Trivial Name	Group (Table 24-8)	Saturated acids Acid	%	Unsaturated acids Acid	%
Juglans regia	Walnut	1	Palmitic	5.1	Oleic	28.9
			Stearic	2.5	Linoleic	47.6
					Linolenic	15.9
Perilla ocimoides	Perilla	1	Palmitic	6.7	Oleic	10.7
			Stearic		Linoleic	33.4
					Linolenic	49.0
Linum usitatissimum	Flax, linseed	1	Palmitic	5.4	Oleic	9.9
			Stearic	3.5	Linoleic	42.6
					Linolenic	38.1
Aesculus hippocastanum	Horse chestnut	2	Palmitic	4.4	Oleic	67.1
			Stearic	3.6	Linoleic	22.7
					Linolenic	2.2
Olea europaea	Olive	2	Palmitic	6.0	Oleic	83.0
			Stearic	4.0	Linoleic	7.0
Sesamum indicum	Sesame	2	Palmitic	9.1	Oleic	45.4
			Stearic	4.3	Linoleic	40.4
Helianthus annuus	Sunflower	2	Palmitic	3.5	Oleic	34.1
			Stearic	2.9	Linoleic	58.5
Aleurites sp.	Tung	3	Palmitic	4.1	Elaeostearic	79.7
			Stearic	1.3		
Ricinus communis	Castor	3	Stearic	0.3	Oleic	7.2
			Dihydroxy-stearic	1.1	Linoleic	3.6
					Ricinoleic	87.8
Anacardium occidentale	Cashew	4	Palmitic	6.4	Oleic	74.1
			Stearic	11.3	Linoleic	7.7
Gossypium hirsutum	Upland cotton	4	Palmitic	21.9	Oleic	30.7
			Stearic	1.9	Linoleic	44.9
Bertholletia excelsa	Brazil nut	4	Palmitic	14.3	Oleic	58.3
			Stearic	2.7	Linoleic	22.8
Coffea arabica	Coffee	4	Myristic	6.0	Linoleic	38.7
			Palmitic	24.3	Oleic	20.8
Oryza sativa	Rice	5	Palmitic	13.2	Oleic	44.1
			Stearic	1.9	Linoleic	39.4
Zea mays	Maize	5	Palmitic	7.8	Oleic	46.3
			Stearic	3.5	Linoleic	41.8

TABLE 24-9 (Continued)

Species	Trivial Name	Group (Table 24-8)	Saturated acids		Unsaturated acids	
			Acid	%	Acid	%
Apium graveolens	Celery	6	Palmitic	3.0	Oleic	26.0
					Linoleic	20.0
					Petroselinic	51.0
Brassica campestris	Rape	8	Palmitic	1.0	Oleic	32
					Linoleic	15
			Lignoceric	1.0	Linolenic	1
					Erucic	50
Arachis hypogaea	Peanut	9	Palmitic	6.3	Oleic	61.1
			Stearic	4.3	Linoleic	21.8
			Arachidic, etc.	5.9		
Glycine max	Soybean	9	Palmitic	6.8	Oleic	33.7
			Stearic	4.4	Linoleic	52.0
			Arachidic	0.7		
Theobroma cacao	Cacao	10	Palmitic	24.4	Oleic	38.1
			Stearic	35.4	Linoleic	2.1
Cocos nucifera	Coconut	12	Caprylic	7.9	Oleic	5.7
			Capric	7.2	Linoleic	2.6
			Lauric	48.0		
			Myristic	17.5		
			Palmitic	9.0		
			Stearic	2.1		

being found in addition only in certain *Araliaceae* and *Simarubaceae*. The acetylenic acid, tariric acid, is known from species of *Picramnia*, a member of the *Simarubaceae* in which it is a major component. The drying oil of the tung seed (*Aleurites*) is characterized by large amounts of the triply unsaturated elaeostearic acid.

The fats of fruit and seed of the same species may resemble one another in composition or they may differ widely, as in the case of the oil palm in which lauric acid, which constitutes 47% of the seed fatty acids, is present in negligible amounts in the fruit fat (Table 24-10). Unfor-

TABLE 24-10. Comparison of the fatty acid composition of fruit and seed fats in the same species. (After Hilditch[4])

Species	Common name	Tissue	Component acid: % of total				
			Lauric	Palmitic	Oleic	Linoleic	Others
Elaeis guineensis	Oil palm	Fruit	.....	35–40	40–50	5–11	....
		Seed	47	9	18	1	14% myristic
Laurus nobilis	Laurel	Fruit	3	20	63	14	....
		Seed	43	6	32	18	....
Olea europaea	Olive	Fruit	7–15	70–85	4–12	......	....
		Seed	.....	6	83	7	4% stearic
Theobroma cacao	Cacao	Fruit	.....	50	35	10	....
		Seed	.....	24	38	2	35% stearic

tunately, a direct comparison of these fats with the leaf fat of the same species is not available.

Structure of the Fats. The fats of the higher plants, and indeed of all organisms, consist, according to Hilditch,[4] of fatty acids esterified with glycerol on the basis of maximum heterogeneity of acid composition per molecule. Thus in a fat containing glycerol, G, and equal amounts of fatty acids F_1, F_2, and F_3, the molecular species represented most abundantly is $GF_1F_2F_3$. Molecules of the type $GF_1F_1F_2$, etc. are present in

TABLE 24-11. The component glycerides of cocoa butter (seed fat of *Theobroma cacao*) and of Borneo tallow (seed fat of *Shorea sp.*, *Dipterocarpaceae*). Fractions obtained by acetone partition. (After Hilditch[4])

Component glyceride	Amount of glyceride present in whole fat	
	Cacao butter (mol. %)	Borneo tallow (mol. %)
Fully saturated		
Tripalmitin	...	1
Dipalmitostearin	2	2
Palmitodistearin	...	1
Monooleoglycerides		
Oleodipalmitin	6	8
Oleopalmitostearin	52	31
Oleodistearin	19	40
Dioleoglycerides		
Palmitodiolein	9	3
Stearodiolein	12	13
Triolein	...	..
Total mol. % accounted for	100	99

Fatty acid composition of above fats

Fatty acid	Cacao butter (mol. %)	Borneo tallow (mol. %)
Palmitic	26.2	19.5
Stearic	34.4	42.5
Oleic	37.3	36.9

lower amounts, and those of type $GF_1F_1F_1$, etc. in still smaller quantities. Indeed formation of the simple triglyceride $GF_1F_1F_1$ appears to occur extensively only in cases where one component acid is present much in excess of the others. This conclusion has been arrived at primarily on the basis of selective crystallization of the triglycerides from acetone at low temperature, the fractions least soluble in acetone being those which contain the smallest content of unsaturated acids, with the progressively more soluble fractions being those containing higher proportions of unsaturated acids. Distribution of the component fatty acids of a fat among the component glycerides on the basis of maximum heterogeneity does not mean that they are distributed randomly. Thus if a particular fatty acid makes up one-third of the total, this fatty acid will tend to be represented once in each glyceride molecule. The number of glycerides in which it will be represented twice is actually much lower than would be expected on a basis of purely random distribution. If a particular fatty acid makes up two-thirds or more of the total, then essentially all the glyceride molecules will contain two molecules of this acid, and the simple triglyceride will also occur, its amount being larger the greater the excess of the particular fatty acid above two-thirds of the total. Thus in the case of a particular sample of olive oil in which oleic acid made up 76% of the total fatty acids, triolein attained the relatively high value of 29% of the total glycerides. In a different sample of olive oil in which oleic acid

TABLE 24-12. Component glycerides of some fats of general interest. Only major fatty acids are considered in this table. (After Hilditch[4])

	Component fatty acids		Component glycerides	
Species	Acid	Mol. %	Glyceride	Mol. %
Olea europaea (olive)	Palmitic	11.8	Monosat'd-diolein	45
	Oleic	76.4	Linoleo-diolein	26
	Linoleic	6.3	Triolein	29
Arachis hypogaea	Palmitic	10.1	Monosat'd-diolein	11
(peanut)	Oleic	58.0	Monosat'd-oleo-linolein	45
	Linoleic	22.8	Linoleo-diolein	24
			Triolein	19
Gossypium hirsutum	Palmitic	24.4	Palmito-oleo-linolein	41
(cotton)	Oleic	24.9	Palmito-dilinolein	18
	Linoleic	46.7	Oleo-dilinolein	28

made up only 67% of the total, triolein made up but 5% of the total glycerides.[4]

Examples of composition of the solid fats of different seeds with reference to the proportions of the several types of glycerides present are given in Table 24-11. The examples in this table are chosen as representing fats containing three major fatty acids in relatively equal proportions. It may be seen that in all the cases listed in Table 24-11 the simple triglycerides form only a small proportion of the total glycerides and that glycerides of the form $GF_1F_2F_3$ and $GF_1F_1F_2$ form the majority of molecules.

In fats which contain fatty acids in widely disparate proportions the distribution of these acids among the component glycerides is more complex than shown in the relatively simple cases of Table 24-11. The general principle of maximum heterogeneity is, however, preserved. Table 24-12 gives the distribution of component glycerides in a variety of important natural plant oils.

The Plant Waxes.[9] The waxes, as pointed out above, resemble the fats in frequently containing fatty acids as constituents. Unlike the fats, the waxes contain the fatty acids esterified to monohydric (occasionally dihydric) alcohols containing 24-36 carbon atoms. In addition, waxes frequently contain free fatty acids as well as free alcohols, hydrocarbons of high molecular weight, and high-molecular-weight ketones.

The alcohols contained in the plant waxes are solids at ordinary temperatures and differ from the related fatty acids by being, in general, less soluble in organic solvents. The normal C_{24} alcohol is soluble at room temperature in benzene and chloroform, and is soluble in other solvents at elevated temperatures. The normal C_{36} alcohol, on the other hand, is insoluble in cold solvents and only slightly soluble in hot benzene or chloroform. The hydrocarbons included in the wax component are remarkable in containing an odd number of carbon atoms ranging in number from 25 to 37. Shorter hydrocarbons such as n-heptane have been found in essential oils as in turpentine of *Pinus ponderosa* but not in the waxes proper. These paraffins are solids at ordinary temperatures, the melting points ranging from 55° for the C_{25} hydrocarbon to 75° for the C_{37} hydrocarbon. The melting points are thus spread over only a small range and cannot be used as a criterion of purity, particularly since a mixture of two hydrocarbons differing by two C atoms shows no depression in melting point. The wax paraffins are soluble in petroleum naphtha and in carbon disulfide at room temperature, the solubility decreasing with increasing chain length. In addition to alcohols, esters, and hydrocarbons, the plant waxes contain long chain ketones and secondary alcohols, which are soluble in hot solvents such as benzene and chloroform and somewhat soluble in hot methyl or ethyl alcohol.

The waxes occur in the plant in three different forms: (a) as leaf cuticle waxes, (b) as waxes of fruit cuticle and seed coats, and (c) as waxes which occur dispersed in the cell in the same manner as the fats. Leaf waxes, which occur superficially, may be obtained by immersing the leaves in hot water where the wax melts and floats to the surface. This group includes important waxes of commerce as carnauba wax of *Copernica cerifera* (*Palmaceae*). Carnauba wax[9,10] contains primarily esters of C_{26}, C_{28}, C_{30}, C_{32}, and C_{34} alcohols in combination with saturated fatty acids having the same series of numbers of carbon atoms. The C_{27} n-hydrocarbon is also present in small amounts. Commercial candelilla wax (*Euphorbia* sp.) contains on the other hand 50–60% of a paraffin, n-hentriacontane ($C_{31}H_{64}$), 15% of a mixture of C_{30}, C_{32}, and C_{34} acids, and 5% of a similar mixture of alcohols.

Fruit and seed-coat waxes which have been investigated include those of apple and of cotton fiber. Chibnall[11] has shown that apple cuticle wax contains hydrocarbons, acids, and alcohols. The paraffins include n-nonacosane ($C_{29}H_{60}$) and n-heptacosane ($C_{27}H_{56}$), while the alcohols

TABLE 24-13. Summary of the composition of some plant waxes.   (After Chibnall and Piper[12])

	n-Fatty acids						Primary alcohols						Hydrocarbons							n-Nonacosan		
	C_{24}	C_{26}	C_{28}	C_{30}	C_{32}	C_{34}	C_{24}	C_{26}	C_{28}	C_{30}	C_{32}	C_{34}	C_{25}	C_{27}	C_{29}	C_{31}	C_{33}	C_{35}	C_{37}	10-ol	15-ol	15-one
Leaf cuticle																						
Carnauba	−	+	+	+	+	+	−	+	+	+	+	+	−	+	−	−	−	−	−	−	−	−
Candelilla	−	−	+	+	+	+	−	−	+	+	+	+	−	−	−	+	+	−	−	−	−	−
Fruit cuticle																						
Apple	−	+	+	+	+	−	−	+	+	+	−	−	−	+	+	−	−	−	−	+	−	−
Cytoplasmic																						
Brussels sprouts	−	+	+	−	−	−	−	+	+	+	+	−	−	+	+	−	−	−	−	−	+	+
Cabbage	−	−	−	−	−	−	−	−	−	−	−	−	−	+	+	−	−	−	−	−	−	+
White mustard	−	+	+	+	−	−	−	−	+	+	+	−	−	−	−	−	−	−	−	−	−	−
Tobacco	−	−	−	−	−	−	−	−	+	−	−	−	+	+	+	+	+	−	−	−	−	−
White clover	−	+	+	+	−	−	−	−	+	+	−	−	−	−	−	−	−	−	−	−	−	−
Cocksfoot grass	−	−	−	−	−	−	−	+	−	−	−	−	−	−	−	−	−	−	−	−	−	−

include n-triacontanol ($C_{30}H_{61}OH$), n-octacosanol ($C_{28}H_{57}OH$), and n-hexacosanol ($C_{26}H_{53}OH$). The acids include a mixture of C_{26}, C_{28}, C_{30}, and C_{32} saturated fatty acids. The wax of cotton fibers, which constitutes 0.4–0.97% of the dry weight of the fiber, contains a mixture of esters of the C_{24}–C_{32} primary alcohols and acids together with paraffin constituents.[12]

The waxes contained in the leaf have been investigated particularly in the *Cruciferae*.[12] In these cases the plant material is ground, filtered free of cell wall constituents, and the cytoplasmic portion subjected to fractionation, so that the waxes obtained cannot derive from the cuticle but must actually constitute cytoplasmic constituents. An example of such an isolation is that in which 220 kg. of cabbage leaves (*Brassica oleracea*) yielded 0.1% of ether-soluble material which was then precipitated from ether by acetone and redissolved in hot acetone. The 35 gm. of material obtained were saponified, recrystallized, and fractionally distilled *in vacuo*. This product (27 gm.) was fractionated by distillation and crystallization to yield a ketone, n-nonacosan-15-one ($C_{29}H_{58}O$), and two hydrocarbons, n-nonocosane ($C_{29}H_{60}$), and hentriacontane ($C_{31}H_{64}$). Brussels sprouts (*Brassica oleracea*) yielded a cytoplasmic wax containing n-nonacosane, n-hentriacontane, n-nonacosan-15-one, n-nonacosan-15-ol, and a mixture of primary alcohols in the range C_{26}–C_{32} as well as C_{26} and C_{28} fatty acids. Seeds of *Simmondsia californica* contain as a reserve material a fatty material which is not a fat at all but a wax in which eicosenoic and docosenoic acids are esterified with eicosol ($C_{20}H_{39}OH$) and docosenol ($C_{22}H_{43}OH$).[13]

The waxes with the exception of that of *Simmondsia* play no recognizable part in the metabolism of the plant and do not appear to be utilized once they have been formed. Chibnall regards them as "end products" of metabolism.

The Biosynthesis of Fats and Fatty Acids. The study of the biosynthesis of fats and fatty acids in plants has proceeded along two main lines: (*a*) the investigation of the accumulation of lipids in fruits and seeds and (*b*) the study of the actual synthesis of the fatty acid molecules. A mass of evidence from higher animals has shown that lipids can be synthesized from carbohydrate in the animal organism. The evidence from higher plants suggests that the same may be true in fruits and seeds also. Thus du Sablon[14] found that in the walnut and in the almond sugars disappear from the developing seed during the period of fat accumulation and similar data have been obtained for the sunflower and for the cotton seed. Data such as that obtained by du Sablon cannot, however, be accepted without caution as signifying a simple transformation of sugars to fat in the seed, as is shown by the work of Thor and Smith[15] with the pecan (*Hicoria pecan*). In this case sugar, as expressed on a percentage basis, decreases during the filling of the seed with fat. The amount of sugar which disappears cannot, however, account for more than 5% of the oil which appears in the same period. Evidently material translocated to the fruit is used immediately in fat synthesis in this case. That fat is formed in the organ itself, rather than transported into it from the leaf, is indicated not only by the different composition of leaf and fruit or seed fats (see above) but also by the fact that in the olive, as well as in flax, fat formation persists for a short time after removal of the immature organ from the plant.[16,17]

A feature of oil formation which should be noted is the fact that it proceeds at an extraordinarily rapid pace for a short period in the develop-

TABLE 24-14. Depletion of carbohydrate from seeds of walnut and almond during ripening and fat accumulation. (After Du Sablon[14])

Date of harvest	Composition of seed; % of weight			
	% oil	% glucose	% sucrose	% starch, etc.
		Walnut		
July 6	3	7.6	0.0	21.8
Aug. 1	16	2.4	0.5	14.5
Aug. 15	42	0.0	0.6	3.2
Oct. 4	62	0.0	1.8	2.6
		Almond		
June 9	2	6.0	6.7	21.6
July 4	10	4.2	4.9	14.1
Aug. 1	37	0.0	2.8	6.2
Oct. 4	46	0.0	2.5	5.3

ment of the fruit or seed. The oil content of the flax seed may rise from 2% to over 30% over a two-week period starting about ten days after flower opening.[17] Similar observations have been made on cotton,[18] on pecan,[15] on castor seed,[19] and on other oil-bearing species.

In the case of flax it is clear that free fatty acids are first manufactured and that formation of glycerides follows as the seed matures, as shown in Table 24-15. The degree of unsaturation of the oil also increases with maturity of the seed. Thus the oil of immature flax seed taken less than 8–10 days after flowering may have an iodine value of 100 or less. During development the iodine value rises rapidly until it reaches 180–200 in oil of the fully mature seed.[17] This effect is reflected also in the changing physical nature of the fat which is semi-solid as extracted from immature seeds, gradually becoming more and more liquid as the seed develops. This may well be due to the deposition of increasing amounts of the more unsaturated acids during the latter stages of seed maturation.

Investigation of the respiratory gas exchange in developing seeds has confirmed the conclusions drawn above (a) that fat is synthesized in the seed or fruit itself and (*b*) that fat is synthesized from an oxygen-rich precursor such as carbohydrate. This mode of approach to the problem is based upon the consideration that the transformation of an oxygen-rich to an oxygen-poor material must be attended by the evolution of excess CO_2, i.e., a portion of the oxygen-rich substrate must be burned to

TABLE 24-15. Accumulation of oil and decrease of free fatty acid during development of flaxseed. (After Eyre[17])

Age of seed: days after flowering	Dry wt. per 1000 seeds gm.	Oil %	Free acid calc. as mg. oleic per gm. oil
12	1.10	2.20	43.4
16	1.59	5.53	41.4
20	1.99	16.01	20.7
25	2.83	29.23	4.9
30	3.28	34.76	3.2
32	3.62	36.03	3.0
38	4.19	38.89	2.8
46	4.73	38.18	3.0
76	4.79	37.44	2.4

CO_2, using a further portion as a hydrogen acceptor. In the case of the conversion of hexose to fat the process should approximate:

$$2C_6H_{12}O_6 + 3O_2 \rightarrow 6CO_2 + 6H_2O + (-CH_2-)_6$$

Hexose — Carbon at reduction level of fatty acid

In this reaction two CO_2 molecules should be evolved for each O_2 molecule consumed and the R.Q. of the gas exchange of a tissue carrying on such an overall process should be 2, or much higher than 1, the value for the normal respiratory oxidation of hexose. Measurements of the gas exchange of ripening olive fruits, as well as ripening seeds of almond and *Ricinus*, show that during the period of rapid fat formation the R.Q. is well above 1 and may reach a value as high as 1.5.[16] Burr and Miller[19] have investigated the gas exchange of the ripening castor seed in detail, using a method in which a chamber was fixed over the fruit while the latter remained attached to the plant. They found, as shown in Fig. 24-2, that the respiratory quotient of the fruit just after pollination starts at the low level of 0.7 but increases steadily with time. In the period 10–20 days after pollination a rapid burst of fatty acid formation occurs, the concentration of total fat going from 6% to 26% during this period. During the same period also, the R.Q. remains at 1 or above. With the slowing down of fat accumulation the R.Q. again sinks to a low level.

Investigation of the mode of fatty acid formation has thus far been confined to microorganisms and higher animals. This work has revealed the important fact that acetate is an intermediate in fatty acid formation. Thus in the relatively simple case of butyric acid formation by *Clostridium acetobutylicum*[20] the overall reactions appear to be:

$$2CH_3C^*OOH \rightarrow CH_3C^*CH_2C^*OOH \xrightarrow{(H)} CH_3C^*H_2CH_2C^*OOH$$
$$\underset{\text{O}}{\|}$$

Acetate Acetoacetate Butyric acid

One molecule of acetate condenses with another to form acetoacetate which is then further reduced to butyric acid. The incorporation of isotopic carbon into the carboxyl group of the acetate, as is shown in the

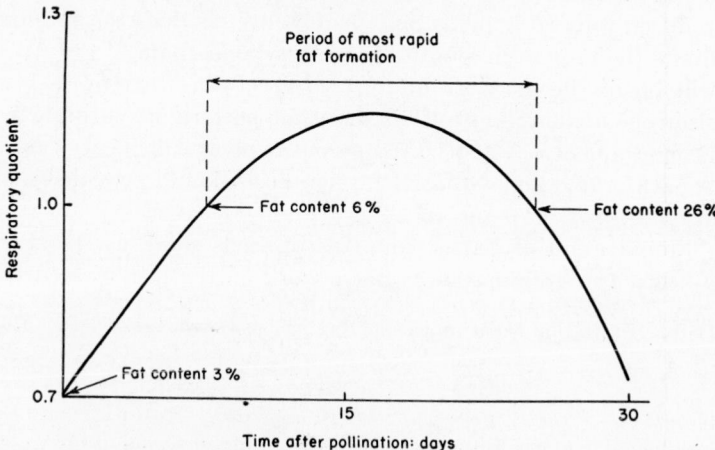

FIG. 24-2. Respiratory quotient of developing castor fruits in relation to fat accumulation. (After Burr and Miller[19])

reaction above, confirms the mechanism of the condensation of acetate. Using an enzyme preparation from *Clostridium*, Stadtman and others[21] have been able to synthesize *in vitro* not only butyric acid but also the six-carbon caproic acid. The reaction appears to proceed from acetylphosphate rather than from acetate itself since while acetylphosphate is readily converted to the higher acids, acetate is utilized only when a source of energy-rich phosphate such as ATP is simultaneously supplied. This evidence suggests then that fatty acid formation proceeds along the general pathway shown below:

$$CH_3COO \sim ph + CH_3COOH \rightarrow CH_3C:OCH_2COOH + ph$$
$$CH_3C:OCH_2COOH \xrightarrow{\text{Reduction}} CH_3CH_2CH_2COOH$$
$$CH_3CH_2CH_2COO \sim ph + CH_3COOH \rightarrow CH_3CH_2CH_2C:OCH_2COOH + ph$$
$$CH_3CH_2CH_2C:OCH_2COOH \xrightarrow{\text{Reduction}} CH_3CH_2CH_2CH_2CH_2COOH$$

That condensation of acetate with the four-carbon acid proceeds as shown above has also been directly demonstrated by isotope experiments. A similar mechanism for synthesis of higher fatty acids in the animal is indicated by the work of Rittenberg and Bloch[22] who fed $CH_3C^{13}OOH$ to rats and recovered the labeled carbon in the fatty acids of the depot and liver fats. The content of labeled carbon in the carboxyl group of the fatty acid was twice the average labeled C content of the molecule as a whole, consistent with the view that the labeled carbons are at alternate positions through the molecule. The fact that acetate is the basic building block for the synthesis of fatty acids presumably accounts for the observation that only fatty acids with even numbers of carbon atoms occur in the natural fats. The mechanism by which double bonds are introduced into fatty acids remains obscure, as does the mechanism which limits the growth in length of the fatty acid chain.

It will be of the greatest interest to ascertain whether fatty acid synthesis in the higher plant follows this same pattern of synthesis from a two-carbon fragment. The seed tissues of the plant which carry out such extraordinarily rapid and extensive fat synthesis should provide excellent material both for *in vitro* and for *in vivo* studies.

The Mobilization of Fats. The fats of seeds serve as reserve food and are utilized in germination. Early work on the fate of seed fat by

TABLE 24-16. Utilization of fat in germinating seeds of sunflower. (After Miller[24])

Days after germination:	0 days		3.5 days		7 days		14 days	
	Coty-ledons	Embry. axis	Coty-ledons	Embry. axis	Coty-ledons	Embry. axis	Coty-ledons	Embry. axis
Dry wt./100 pl.	4.9	....	3.2	0.60	2.3	1.21	1.6	2.17
Fat %	54	....	53	9.9	36	3.8	9	1.8
Fat gm./100 pl.	2.6	....	1.7	0.06	0.8	0.05	0.1	0.04
Loss of dry wt. from cotyledons	.....	....	1.7	....	2.6	.....	3.3	....
Loss of fat from coty-ledons	.....	....	0.9	....	1.8	.....	2.5	....
% which fat loss is of total loss	.....	....	53	....	69	.....	76	....
Total sugars %	4.1	4.1	1.1	9.2	3.1	18.6	2.1	7.7
Total sugars gm./100 pl.	0.28	0.02	0.06	0.07	0.12	0.43	0.05	0.16

Hellriegel in 1855 and others was succeeded by the quantitative investigations of Ivanov[23] and Miller.[24] Thus Ivanov found that during eight days of germination the oil content of flaxseed may decrease from 34% to 16%. Data of Miller (Table 24-16) show that the loss of fat from the cotyledons of germinating sunflower seeds commences within the first four days and that the fat disappearing from the cotyledons does not reappear in the growing embryonic axis but is presumably transformed into other materials. In the sunflower this material does not accumulate as sugar but must constitute the structural components of the plant since the increase in dry weight of seedling, apart from its sugar content, can only be accounted for by the loss of fat from the cotyledons. The same is true of germinating seeds of *Cucurbita pepo* in which the bulk of the dry weight of the growing embryonic axis can be accounted for only by the loss of fat from the cotyledons.[25] In the castor seed, on the other hand, an appreciable proportion of the fat lost from the endosperm may appear as sugar in the seedling since this plant accumulates as much as 40% of sugars in the seedling hypocotyl.

We should now inquire as to whether the conversion of fats to other compounds occurs in the reserve tissues or whether the fats or fatty acids are translocated to the growing portions of the seedling and there converted. This question has been considered by Murlin and others.[26] The conversion of the ricinoleic acid of castor seed to sucrose, according to the overall equation would have a respiratory quotient of 0, whereas

$$2C_{18}H_{34}O_3 + 14O_2 \rightarrow 3C_{12}H_{22}O_{11} + H_2O \qquad R.Q. = 0$$
Conversion of ricinoleic acid to sucrose

the complete oxidation of the fatty acid to CO_2 and H_2O would have a respiratory quotient of approximately 0.72. The data of Murlin given in Table 24-17 show that the young seedling axis of *Ricinus* has an R.Q.

TABLE 24-17. Respiratory quotient of whole germinating *Ricinus* seed compared with that of the young plant minus endosperm. (Data of Murlin[26])

Tissue used	Length of hypocotyl at stage used	Respiratory quotient CO_2/O_2
Whole seedling	3 mm.	0.30
Seedling, endosperm removed	3 mm.	0.97
Whole seedling	12 mm.	0.34
Seedling, endosperm removed	12 mm.	0.99
Whole seedling	30 mm.	0.33
Seedling, endosperm removed	30 mm.	0.93

close to 1, as would be expected for carbohydrate oxidation, whereas the

$$C_{18}H_{34}O_3 + 25O_2 \rightarrow 18CO_2 + 17H_2O \qquad R.Q. = 0.72$$
Complete oxidation of ricinoleic acid

whole seedling has a respiratory quotient of slightly over 0.3. It must be concluded, therefore, that the fat of the endosperm is in part burned and in part converted to carbohydrate in the endosperm and that the respiration of the seedling is carried on at the expense of carbohydrate rather than of fat. This conclusion is confirmed by calorimetric experiments in which it has been found that the oxygen-poor fatty acids of the endosperm are converted into oxygen-rich compounds during germination, whereas the actively growing embryonic axis consists during its entire life of oxygen-rich substances with a combustion R.Q. of 0.93–0.97, closely approximating that of carbohydrate.[27]

The fatty materials of the cotyledon undergo marked changes in composition with germination.[23,24,28] These changes consist of an increase in the free fatty acid content and an increase in the proportion of saturated fatty acids present in the fat. Ivanov suggests that the latter represents preferential utilization of the unsaturated acids.

Pathways of Fat Metabolism. The first stage in fat metabolism, i.e., the hydrolysis of the glyceride to fatty acids and glycerol, is effected by the enzyme lipase. This enzyme is probably present in all plant tissues but is particularly well known from seeds such as those of *Ricinus communis*, which contain an unusually high lipolytic activity. The enzyme is contained in the endosperm, is water insoluble or only slightly soluble, and is obtained from defatted seeds by finely grinding the dry ether-insoluble residue. The enzyme has been purified to some extent by Takamiya.[29] The pH optimum of castor seed as well as of cottonseed lipase is 4.7–5.0 and the enzyme acts on fat emulsions with the liberation of the free fatty acids. Different natural fats are hydrolyzed at different rates by lipase of castor seed, peanut oil being most rapidly saponified, while castor seed, corn, cottonseed, soybean, ripe olive, and linseed oils are attacked with decreasing rapidity in the order named.[30] Lipase also attacks simple esters such as glycerol triacetate and ethyl butyrate. A second water-soluble enzyme of the castor seed is said, however, to have an even greater activity on ethyl butyrate than lipase.[31] Wheat seed lipase unlike castor seed lipase is fully soluble and rapidly attacks both simple esters and glycerides. Unlike castor seed lipase also, the pH optimum of the wheat seed enzyme is in the neighborhood of 7.4.[41]

The oxidation of fatty acids in the plant is but little understood and work on this subject is confined to the study of the enzyme lipoxidase,[32] discovered in 1928 by Bohn and Haas in the soybean and found since in leaves as in alfalfa. This enzyme adds oxygen to the double bonds of unsaturated fatty acids with the formation of a peroxide.

$$R—CH=CH—R + O_2 \rightarrow R—CH—CH—R$$
$$\underset{\text{O——O}}{}$$

Double bond in	Peroxide formed at
fatty acid	double bond

The peroxide formed is able to carry out secondary nonenzymatic oxidations such as the destruction of carotene, and for this reason lipoxidase is an important factor in the destruction of carotene in the drying or storage of vegetable products. Lipoxidase attacks one double bond of linoleic and of linolenic acid. It does not attack oleic, erucic, or elaidic acids and does not attack saturated acids. Lipoxidase has been isolated in crystalline form from soybean seeds[33] and is a metal-free protein.[35] Although the action of lipoxidase may represent an initial step in fatty acid breakdown in the plant, this has not been shown to be the case. More detailed knowledge of fatty acid oxidation obtained in animal tissues suggests that quite a different pathway may be operative.

The cornerstone of our knowledge of fatty acid oxidation is the work of Knoop who in 1904 reported experiments which led him to propose the concept of β-oxidation.[34] In this work Knoop took advantage of the fact that when substituted fatty acids are fed to animals they are not completely oxidized but are excreted by the body in modified form. Thus phenylacetic acid is excreted in the form of a substituted amide, and the same is true of benzoic acid which is excreted as hippuric acid.

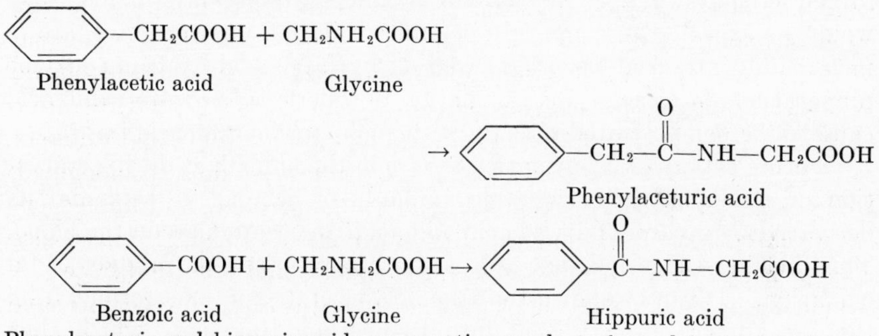

Phenylaceturic and hippuric acids as excretion products formed from phenylacetic and benzoic acids *in vivo*

Knoop then fed animals phenyl-substituted higher fatty acids and found that while such fatty acids with even numbers of carbon atoms in the fatty acid chain result in excretion of phenylaceturic acid, phenyl-substituted fatty acids with odd numbers of carbon atoms in the fatty acid chain result in the excretion of hippuric acid.

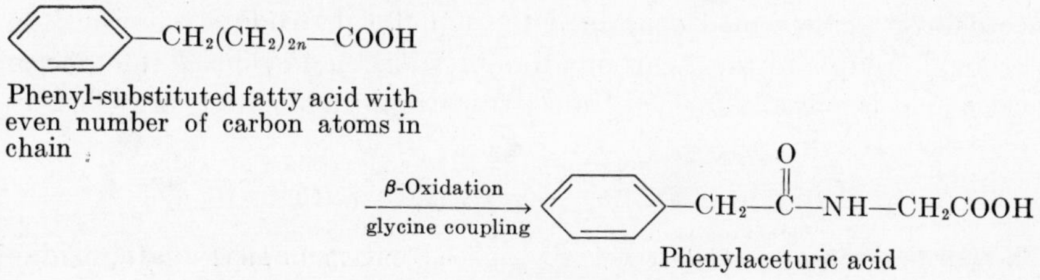

$$\langle\!=\!\rangle\!-\!(CH_2)_{2n}\!-\!COOH \xrightarrow[\text{glycine coupling}]{\beta\text{-Oxidation}} \langle\!=\!\rangle\!-\!\overset{\overset{\textstyle O}{\|}}{C}\!-\!NH\!-\!CH_2COOH$$

Phenyl-substituted fatty acid   Hippuric acid
with odd number of carbon
atoms in chain

Excretion products of phenyl substituted fatty acids containing even (above) or odd (below) numbers of carbon atoms in fatty acid chain

The concept of β-oxidation expresses the fact that the fatty acid chain is degraded by an even number of carbon atoms in all instances, and this is generally held to suggest that the oxidation is by removal of successive two-carbon fragments. The principal objection to the β-oxidation concept has been the fact that oxidation of the higher fatty acids might be expected to result in formation of appreciable amounts of fatty acids of intermediate chain lengths which are not actually found. This may, however, only signify that once a fatty acid molecule is attacked the oxidation is completed before the molecule is released from the active enzyme complex. The general principle of β-oxidation has been confirmed in many ways. Thus β-hydroxy and β-carbonyl fatty acids which would be expected to constitute possible oxidation intermediates, are in fact readily attacked *in vivo* as well as by fatty acid oxidizing enzyme preparations *in vitro*.[35] Beta-oxidation of valeric acid to propionic acid can also be demonstrated directly with such enzyme preparations.[36]

During extensive fat oxidation *in vivo* in the animal, as during starvation or in diabetes, considerable amounts of acetoacetic acid and its derivatives, β-hydroxybutyric acid and acetone, accumulate in the blood. This fact has suggested that acetoacetate may arise as a product of fat oxidation. It can be shown that the acetoacetate is in fact formed from the breakdown of fatty acid into two-carbon fragments which then

$$\underset{\substack{\beta\text{-Hydroxybutyric}\\ \text{acid}}}{\overset{\overset{\textstyle OH}{|}}{CH_3CHCH_2COOH}} \xleftarrow[-2H]{} \underset{\text{Acetoacetic acid}}{\overset{\overset{\textstyle O}{\|}}{CH_3C\!-\!CH_2COOH}} \rightarrow \underset{\text{Acetone}}{\overset{\overset{\textstyle O}{\|}}{CH_3CCH_3}} + CO_2$$

recombine to form the four-carbon compound. This experiment has been done by Weinhouse and others,[37] who allowed liver slices to oxidize *n*-octanoic acid in which the carboxyl group contained isotopic C^{13}. The acetoacetic acid formed contained C^{13} equally distributed between the carboxyl group and the β-carbon atom. Additional evidence that acetoacetic acid is actually formed from two-carbon fragments is provided by

$$\underset{\substack{\text{Octanoic acid}\\ \text{(carboxyl labeled)}}}{CH_3(CH_2)_6C^*OOH} \xrightarrow[O_2]{\text{Liver slices}} \underset{\substack{\text{Acetoacetate}\\ \text{(carboxyl and }\beta\text{ labeled)}}}{\overset{\overset{\textstyle O}{\|}}{CH_3C^*\!-\!CH_2C^*OOH}}$$

the experiments with *Clostridium acetobutylicum* cited earlier which show that this organism can synthesize acetoacetate from acetate. It might be suspected, therefore, that acetate or a closely related two-carbon compound may be the initial breakdown product of fatty acid oxidation. That an acetylphosphate rather than acetate may be the actual oxidation product is indicated by the fact that ATP is essential to the oxidation of fatty acids by liver homogenates (cell-free ground liver preparation) according to Lehninger.[38] It has been suggested that ATP activation of

TABLE 24-18. Dependence of fatty acid (octanoate) oxidation by liver homogenate on the presence of ATP. (After Lehninger[38])

System	O_2 uptake	Octanoate oxidized
Enzyme + substrate + ATP	7.0 micromols	1.7 micromols
Enzyme + substrate alone	1.6 micromols	−0.1 micromol

oxidation of the fatty acid may possibly be due to formation of fatty acid acylphosphates. Activation may also be achieved by the addition of α-ketoglutarate to the system, the oxidation of this substrate resulting in the production of the required energy-rich phosphate. The enzyme system involved in fatty acid oxidation is an insoluble one and occurs in the particulate matter of the cytoplasm. All the required enzymes are apparently incorporated in this one active unit.

The further oxidation of the acetate or closely related breakdown product of fatty acid oxidation by animal tissue involves the conversion of this compound into the metabolites of the Krebs cycle.[35] Thus, when acetate containing C^{13} in the carboxyl group is oxidized by kidney homogenate, the C^{13} is recovered in α-ketoglutarate, succinate, and fumarate. Similarly, the oxidation of acetoacetate by kidney homogenate involves the production of α-ketoglutarate and fumarate.[39] Quantitative studies with liver homogenate also show that when octanoate is oxidized in the presence of fumarate, at least half of the carbon atoms of the octanoate oxidized can be recovered in the form of citrate, isocitrate, and other acids of the Krebs cycle.[38] With enzyme preparations of tissues such as muscle, which do not accumulate acetoacetate, the fatty acid also is oxidized through the Krebs cycle and can be quantitatively recovered as succinate, provided that the oxidation of this compound is blocked by malonate.[40] It would seem then that in fatty acid oxidation the initial step consists in the production of a reactive two-carbon compound, acetate or an acetylphosphate, a second step consists in the incorporation of this compound into a metabolically active plant acid, and the final step consists in the oxidation of this acid through the usual pathway of carbohydrate metabolism.

Strange as it may seem, no information on fatty acid oxidation by the higher plant analogous to that summarized here for animal tissues is available. It will be an important problem in plant biochemistry to illuminate this significant phase of lipid metabolism.

THE PHOSPHOLIPIDS

The phospholipids are fatty compounds in which phosphate is also bound in ester linkages. It is probable that these materials occur in all living cells to the extent of a few tenths to a few per cent, tissues high in fat tending to be also rich in phospholipid. They make up then, at most, a small part of the total crude lipid fraction of the plant. The phospholipids of the higher plant comprise three general types, the phosphatidic acids, the phosphatides, and the lipositols.

Phosphatidic Acids. The phosphatidic acids are glycerides in which two of the hydroxyl groups are esterified with fatty acids while the remaining hydroxyl group is esterified with phosphoric acid. In the plant the material is not present as the free acid but as the salt, combined

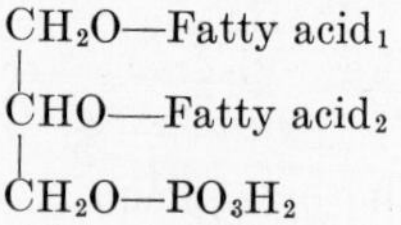

$$CH_2O\text{---Fatty acid}_1$$
$$CHO\text{---Fatty acid}_2$$
$$CH_2O\text{---PO}_3H_2$$

Phosphatidic acid

with the Ca, Mg, or K ions of the cytoplasm. Phosphatidates have been isolated from (1) leaves, (2) seeds, and (3) latex of *Hevea brasiliensis*, and the component fatty acids of material isolated from all these sources have been investigated. Thus the phosphatidate of cabbage leaves contains mainly linoleic and linolenic acids with only small amounts of palmitic and stearic acids while phosphatidic acid of leaves of *Dactylis glomerata* also contains a high proportion of unsaturated fatty acids. In general, the fatty acid composition of the phosphatidate closely parallels that of the phosphatide of the same source but may be appreciably different from the composition of the related fat.

Lecithin and Cephalin. In the phosphatides proper, the phosphoric acid residue of the phosphatidic acid is further esterified with a nitrogenous base. In the case of lecithin, the base involved is choline, while in cephalin it is ethanolamine. Animal tissues contain a further cephalin-

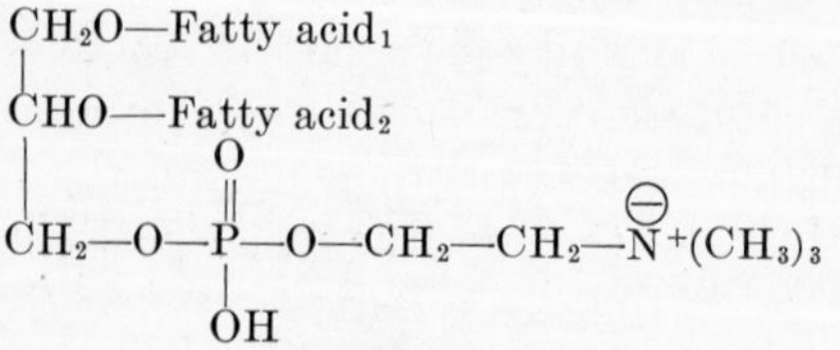

$$CH_2O\text{---Fatty acid}_1$$
$$CHO\text{---Fatty acid}_2$$
$$CH_2\text{---O---P}(=O)\text{---O---CH}_2\text{---CH}_2\text{---N}^+(CH_3)_3 \quad | \quad OH$$

Lecithin

$$CH_2O\text{---Fatty acid}_1$$
$$CHO\text{---Fatty acid}_2$$
$$CH_2O\text{---P}(=O)\text{---O---CH}_2\text{---CH}_2\text{---NH}_2 \quad | \quad OH$$

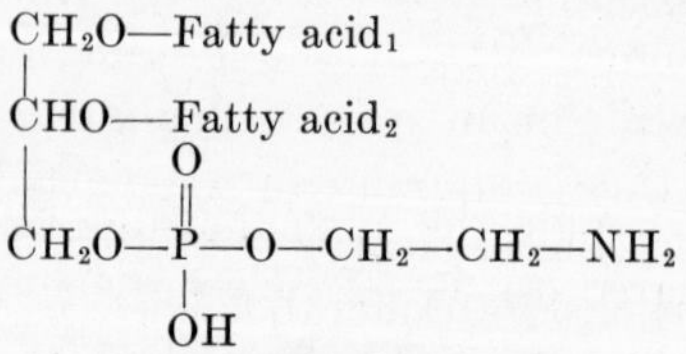

Cephalin

like substance in which the base involved is the amino acid serine rather than ethanolamine. This phosphatidyl serine has been but little studied in higher plants.[47]

Lecithin and cephalin are of extremely wide distribution and are the dominant phospholipids of the great majority of plant tissues.[48] In cabbage leaves,[44] however, both lecithin and cephalin appear to be replaced by phosphatidic acid, while the latex of *Hevea* contains lecithin and phosphatidates with but little cephalin.[46] In seeds the ratio of lecithin to cephalin varies only over a narrow range, as is shown in Table 24-19. The fatty acid composition of seed phosphatides has been

TABLE 24-19. Ratio of lecithin to cephalin in seeds of various plant species

	Per cent of total phospholipid		
Species	Lecithin	Cephalin	Authority
Peanut	36	64	(48)
Cottonseed	29	71	(48)
Linseed	36	64	(48)
Sunflower	52	41	(48)
Lupin	74	26	(49)
Wheat	75	25	(45)

studied extensively (Table 24-20) and is of interest in comparison to the composition of the fat of the same species. In general, there is a marked parallelism in composition, but there are also characteristic differences in that (a) the seed phosphatides tend to be higher in saturated fatty acids than the seed fat and (b) the phosphatides contain small amounts of highly unsaturated C_{20} and C_{22} fatty acids not found in the related fat.

TABLE 24-20. Comparison of the fatty acid composition of seed phosphatides and glycerides. P = phosphatide fraction. G = glyceride fraction. (After Hilditch and Zaky[50])

	Cotton		Sunflower		Peanut		Linseed	
Acids	P	G	P	G	P	G	P	G
Palmitic	17.3	23.4	14.7	5.6	16.2	8.3	11.3	5.4
Stearic	7.3	1.1	5.1	2.2	2.8	3.1	10.6	3.5
Arachidic	2.8	1.3	9.5	0.9	....	....	....	0.6
Hexadecenoic	1.5	2.1	....	....	....	....	3.5	....
Oleic	20.3	22.9	19.3	25.1	47.1	56.0	33.6	18.8
Linoleic	44.4	47.8	45.9	66.2	22.7	26.0	20.4	24.2
Linolenic	....	....	....	....	....	....	17.4	47.3
C_{20-22} unsat'd	6.4	....	5.5	....	4.1	....	3.2	....

Although the phosphatides differ in fatty acid composition from the related fat, nevertheless the composition of the several phosphatides in a given seed is closely similar, as has been shown for soybean.[51]

Lipositols. The lipositols are more complex than the phosphatides in structure and are phospholipids in which inositol, rather than glycerol, constitutes the alcohol. The lipositol of soybean[52] contains per mol of inositol one mol of galactose, one mol of phosphoric acid, a mixture of fatty acids including palmitic and stearic, and one mol of ethanolamine tartrate. Of the whole phosphatide fraction of soybean, approximately 40% is made up of lipositols, while lecithin makes up approximately 29% and cephalin 31%.[53]

Separation of Phospholipids. Phospholipids are removed from tissues by the usual lipid solvents, although alcohol or alcohol-ether mixtures are most effective. Phosphatides and phosphatidates are insoluble in acetone. Lecithin is less soluble then cephalin in cold alcohol and lipositol is less soluble still. Phosphatidates may be separated from the bases by partitioning into ether from aqueous HCl.[44]

Distribution and Metabolism. The phospholipids do not necessarily occur in the plant in the same manner as the true fat, i.e., dispersed as droplets. There is evidence, on the contrary, that phospholipids may be present in the tissue in a bound state, combined perhaps to protein.[54] The phospholipid of the leaf appears to be largely confined to the chloroplasts and in fact to the grana (see Chapter 17). The seed phosphatides disappear from the cotyledons or endosperm during germination, as do the fats, and may then represent reserve material, particularly since the disappearance of the reserve phosphatide is not balanced by the appearance of any corresponding amount of phosphatide in the growing seedling.[45,55] A detailed study of the distribution of phosphatides in the growing bean plant (*Phaseolus multiflorus*) by Jordan and Chibnall[45] has shown that in general rapidly growing tissues are low in phosphatide and that the material is accumulated as the tissue matures. The phosphatide of the mature leaf is, however, much less readily depleted than is the fat during starvation of the tissue in the dark. In general, then, although phosphatides appear to serve in part at least as reserve material in seeds, the distribution and sluggish metabolism of these materials in leaves suggest that they here serve some other function.

Enzymes Attacking Phosphatides. Lecithin is attacked by at least four different enzymes: lecithinases A and B, glycerophosphatase, and choline phosphatase. Lecithinase A yields as a product one free fatty acid and lysolecithin, which has the structure shown below.[56] Lecithinase B, on the contrary, removes both fatty acid residues from lecithin and also attacks lysolecithin with the liberation of the remaining fatty acid. Both of these enzymes also attack cephalin with production of the corresponding compounds, lysocephalin, etc. Lecithinases have been found in seeds of rice,[57,58] but their main interest derives from the fact that they are contained in the venoms of snakes (*Colubridae* and *Viperidae*), bees, wasps, and scorpions. The lysolecithin produced by the action of lecithinase A is a strong hemolytic agent which hemolyzes both red and white blood cells. A similar action of rice seeds has not, however, been reported. Choline phosphatase, which converts lecithin or cephalin to the corresponding phosphatidic acid has been but little investigated as to distribution but is found in takadiastase (prepared from *Aspergillus*) and

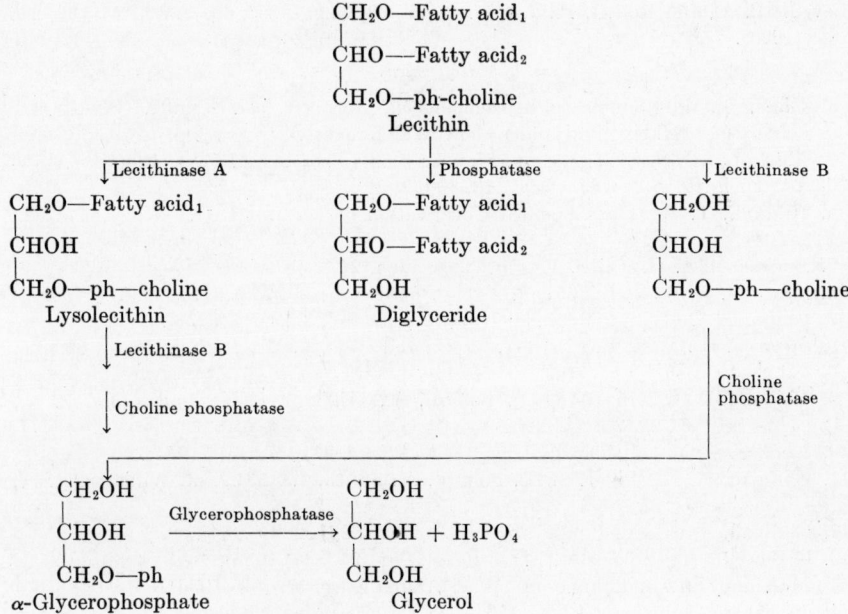

FIG. 24-3.   Enzymes attacking lecithin.

in the rice seed[58] as well as in snake venoms. Takadiastase also contains an enzyme which splits only the phosphate-glycerol bond, thus liberating choline phosphate and the diglyceride. Glycerophosphate, generated by the action of lecithinase B and choline phosphatase, is hydrolyzed to inorganic phosphate and glycerol by glycerophosphatase, an enzyme of wide distribution in plant tissues.

Quite obviously no general picture of the metabolism of the phospholipids can be drawn at present. Information concerning the mode of synthesis of these materials, as concerning their function, is almost totally lacking. The widespread occurrence of phospholipids indicates possible important functions in the plant which should justify a more intensive investigation.

General References

Jamieson, G. S., Vegetable Fats and Oils. Reinhold, 2nd ed., 1943.

Hilditch, T. P., The Chemical Constitution of Natural Fats. Wiley, 2nd ed., 1947.

Ralston, A. W., Fatty Acids and Their Derivatives. Wiley, 1948.

Markley, K. S., Fatty Acids. Interscience Publishers, 1947.

Bloor, W. R., The Biochemistry of the Fatty Acids. Reinhold, 1943.

MacLean, H., and I. Smedley-MacLean, Lecithin and Allied Substances, the Lipins. Longmans, 1927.

Chargaff, E., Advances in Protein Chemistry. Academic Press, 1944. Vol. 1, p. 1.

Levene, P. A., and Rolf, I. P., *J. Biol. Chem.*, **62**, 759 (1925); **65**, 545 (1925); **68**, 285 (1926).

References

1. Klein, G., Handbuch der Pflanzenanalyse. Springer, Vol. II, 1932. See also the book of Markley cited under General References.
2. McNair, J. N., *Am. J. Botany*, **17**, 662 (1930); *Botan. Rev.*, **11**, 1 (1945).
3. Twitchell, E., *Ind. Eng. Chem.*, **13**, 806 (1921).
4. Hilditch, T. P., The Chemical Constitution of Natural Fats. Wiley, 2nd ed., 1947.
5. Boekenoogen, H., Olien, Vetten, Oliezaaden, **26**, 143 (After 4).
6. Smith, J. A. B., and Chibnall, A. C., *Biochem. J.*, **26**, 218, 1345 (1932).
7. Chibnall, A. C., and Channon, H. J., *Biochem. J.*, **21**, 479 (1927).
8. Speer, J. H., Wise, E. C., Hart, M. C., and Heyl, F. W., *J. Biol. Chem.*, **82**, 105, 111 (1929).
9. Chibnall, A. C., Piper, S. H., Pollard, A., Williams, E. F., and Sahai, P. N., *Biochem. J.*, **28**, 2189 (1934).
10. Koonce, S. D., and Brown, J. B., *Oil and Soap*, **21**, 167, 231 (1944).
11. Chibnall, A. C., Piper, S. H., Pollard, A., Smith, J. A. B., and Williams, E. F., *Biochem. J.*, **25**, 2095 (1931).
12. Chibnall, A. C., and Piper, S. H., *Biochem. J.*, **28**, 2209 (1934).
13. Greene, R. A., and Foster, E. O., *Botan. Gaz.*, **94**, 826 (1933). McKinney, R., and Jamieson, G. S., *Oil and Soap*, **13**, 289 (1936).
14. du Sablon, L., *Compt. rend.*, **123**, 1084 (1896). *Rev. gén. botan.*, **9**, 313 (1897).
15. Thor, C. J. B., and Smith, C. L., *J. Agr. Research*, **50**, 97 (1935).
16. Gerber, C., *Compt. rend.*, **125**, 658, 732 (1897).
17. Eyre, J., *Biochem. J.*, **25**, 1902 (1931).
18. Caskey, C. C., and Gallup, W. D., *J. Agr. Research*, **42**, 671 (1931).
19. Burr, G. O., and Miller, E. S., *Botan. Gaz.*, **99**, 773 (1938).
20. Wood, H. G., Brown, R. W., and Werkman, C. H., *Arch. Biochem.*, **6**, 243 (1945).
21. Stadtman, E. R., Stadtman, T. C., and Barker, H. A., *J. Biol. Chem.*, **178**, 677 (1949).
22. Rittenberg, D., and Bloch, K., *J. Biol. Chem.*, **154**, 311 (1944).
23. Ivanov, S., *Beih. botan. Zentr.*, **28**, 159 (1912).
24. Miller, E., *Ann. Botany*, **24**, 693 (1910); **26**, 889 (1912).
25. Henmann, W., *Planta*, **34**, 1 (1944).
26. Murlin, J. R., *J. Gen. Physiol.*, **17**, 283 (1933).
27. Daggs, R. G., and Halcro-Wardlaw, H. S., *J. Gen. Physiol.*, **17**, 303 (1933).
28. Matthes, E., *Botan. Arch.*, **19**, 79 (1927).
29. Takamiya, E., *Proc. Imp. Acad. Tokio*, **12**, 73 (1936).
30. Longenecker, H. E., and Haley, D. E., *J. Am. Chem. Soc.*, **57**, 2019 (1935).

31. Falk, K. G., and Sugihara, K., *J. Am. Chem. Soc.*, **37**, 217 (1915).
32. Bergstrom, S., and Holman, R. T., Advances in Enzymology. Interscience Publishers, 1948. Vol. 8, p. 425.
33. Theorell, H., Holman, R. T., and Akeson, A., *Arch. Biochem.*, **14**, 250 (1947).
34. A complete review of early work on fatty acid oxidation may be found in Breusch, F. L., Advances in Enzymology. Interscience Publ., 1948. Vol. 8, p. 343.
35. Grafflin, A. I., and Green, D. E., *J. Biol. Chem.*, **176**, 95 (1948).
36. Atchley, W. A., *J. Biol. Chem.*, **176**, 123 (1948).
37. Weinhouse, S., Medes, G., and Floyd, N. F., *J. Biol. Chem.*, **153**, 689 (1944); **155**, 143 (1944); **157**, 411 (1945); **158**, 35 (1945).
38. Lehninger, A. L., *J. Biol. Chem.*, **161**, 413 (1945); **164**, 291 (1946).
39. Buchanan, J. M., Sakami, W., Gurin, S., and Wilson, D. W., *J. Biol. Chem.*, **157**, 747 (1945).
40. Lehninger, A. L., *J. Biol. Chem.*, **165**, 131 (1946).
41. Singer, T. P., and Hofstee, B. H. J., *Arch. Biochem.*, **18**, 229, 245 (1948).
42. Norris, F. A., and Terry, D. E., *Oil and Soap*, **22**, 41 (1945).
    For complete discussion see Markley, K. S., Fatty Acids. Interscience Publishers, 1947.
43. Beadle, B. W., *Oil and Soap*, **23**, 140 (1946).
44. Chibnall, A. C., and Channon, H. J., *Biochem. J.*, **21**, 233 (1927).
    Chibnall, A. C., and Sahai, P. N., *Ann. Botany*, **45**, 489 (1931).
    Smith, J. A. B., and Chibnall, A. C., *Biochem. J.*, **26**, 218, 1345 (1932).
45. Channon, H. J., and Foster, C., *Biochem. J.*, **28**, 853 (1934).
    Jordan, R. C., and Chibnall, A. C., *Ann. Botany*, **47**, 163 (1933).
46. Tristram, G. R., *Biochem. J.*, **36**, 400 (1942).
47. Folch, J., and Schneider, H. I., *J. Biol. Chem.*, **137**, 51 (1941).
48. Thierfelder, H., and Klenk, E., Die Chemie der Cerebroside und Phosphatide, Springer, 1930. See also Rewald, B., *Biochem. J.*, **36**, 822 (1942); *Oil and Soap*, **20**, 212 (1943); and Klein, G., Handbuch der Pflanzenanalyse. Springer, Vol. II, 1932.
49. Diemair, W., and Weiss, K., *Biochem. Z.*, **302**, 112 (1939).
50. Hilditch, T. P., and Zaky, Y. A. H., *Biochem. J.*, **36**, 815 (1942).
51. Thornton, M. H., Johnson, C. S., and Ewan, M. A., *Oil and Soap*, **21**, 85 (1944).
52. Woolley, D. W., *J. Biol. Chem.*, **147**, 581 (1943).
53. Scholfield, C. R., Dutton, H. J., Tanner, F. W., and Cowan, J. C., *J. Am. Oil Chemists' Soc.*, **25**, 368 (1948).
54. Chargaff, E. Advances in Protein Chemistry. Academic Press, 1944. Vol. 1, p. 1.
55. Houget, J., *Compt. rend.*, **216**, 821 (1943).
56. Belfanti, S., Contardi, A., and Ercoli, A., *Ergeb. Enzymforschung*, **5**, 213 (1936).
57. Iwata, M., *Biochem. Z.*, **224**, 430 (1930).
58. Contardi, A., and Ercoli, A., *Biochem. Z.*, **261**, 275 (1933).

CHAPTER 25

ESSENTIAL OILS

Introduction to the Terpenes. The terpenes and their derivatives make up a group of naturally occurring plant materials which have in common the fact that they may be conceived of as derivatives of the five-carbon compound isoprene, C_5H_8.

$$CH_2{=}\overset{\overset{\textstyle CH_3}{|}}{C}{-}CH{=}CH_2$$

Isoprene

This group includes components of such diverse materials as essential oils, resins, carotenoids, and rubber. In all these materials, the carbon skeleton may be thought of as built up of two or more isoprene units condensed either in open chains or in closed systems containing one or more rings. In addition to the unsaturated hydrocarbons which constitute multiples of C_5H_8, there are also found in nature a wide variety of reduction and oxidation products of these same compounds. Such alteration may consist either in addition of hydrogen with reduction of double bonds, in the removal of hydrogen with formation of additional double bonds, or in the introduction of oxygen with the formation of alcohol, carbonyl or carboxyl derivatives. Because of the diversity of chemical alterations to which a given carbon skeleton is apparently subject in the plant, an enormous number of terpenes and terpene derivatives are known as natural products. This discussion will not take up individual compounds except as examples of general classes or of general principles, and no attempt will be made to cover all the known terpenes.

Broadly speaking, the terpenes may be divided into three groups, the lower terpenes, the carotenoids, and the polyterpenes rubber and gutta percha. This differentiation is based more on differences in physical properties than on basic differences in chemical nature, since all are structurally related to isoprene. The lower terpenes, ten-, fifteen-, twenty-, and thirty-carbon compounds, make up the bulk of the essential oils and resins although these materials may also contain non-terpene components. The carotenoids are solid materials characterized by a particular arrangement of the component isoprene units and containing

384

additional double bonds. Rubber and gutta percha are high polymer derivatives of isoprene.

Table 25-1 gives a brief summary of the general classes of terpenes. It seems probable that most if not all plants possess the ability to synthesize one or more compounds of the terpene family although in most

TABLE 25-1. Relations between the several classes of terpenes

Terpene class	Empirical formula	Occurrence	Oxygen-containing derivatives
Isoprene	C_5H_8	Does not occur in nature	Isovaleraldehyde
Mono-terpenes	$C_{10}H_{16}$	Essential oils	Monoterpene alcohols, aldehydes, ketones
Sesquiter-penes	$C_{15}H_{24}$	Essential oils, resins	Alcohols, ketones
Diterpenes	$C_{20}H_{32}$	Essential oils, resins	Phytol, vitamin A
Triterpenes	$C_{30}H_{48}$	Resins, latex	Triterpene alcohols
Tetrater-penes	$C_{40}H_{64}$	Carotenes	Xanthophylls
Polyterpenes	$(C_5H_8)_n$	Rubber, gutta percha	None

species this is restricted to carotenoids and to phytol, a constituent of chlorophyll. Ability to form mono-, sesqui-, di-, tri-, and polyterpenes and their derivatives, on the other hand, is scattered irregularly through the plant kingdom.

Essential Oils. Essential oils are produced by some 2000 species (of perhaps 400,000 total species) scattered through about sixty of the families of higher plants and are particularly characteristic of such families as the *Pinaceae, Umbelliferae, Myrtaceae, Lauraceae, Rutaceae, Labiatae,* and *Compositae.* The oil is ordinarily formed in special cells or groups of cells which may be present as scattered individual glandular cells or as glandular hairs such as are found in many leaves and stems. In these cases the oil is contained as a globule or globules in the cell. The oil may also be excreted from cells which line schizogenous ducts or canals. Thus the oleoresin of *Pinus* species (from which turpentine is obtained by steam distillation) is formed by the cells which line the resin canals of the bark, and the oil is obtained by tapping of these canals. Essential oil may be present in such glandular cells or ducts in any or all organs of the plant: root, stem, leaves, buds, flowers, and fruits. Commonly, however, the oil is concentrated in one particular organ as leaves, bark, or fruit. Even in species which contain oil in various organs it is common to find that the oil of different parts possesses different individual chemical components.

Over 500 different chemical compounds have been identified in the essential oils, including terpenes, sesquiterpenes, higher terpenes in small amounts, and a variety of derivatives of these substances, as well as non-terpene compounds. Such non-terpenes include, for example, the allylisothiocyanate of the *Cruciferae*, various organic sulfides and mercaptans, the indole and anthranilic acid esters which contribute the odor of jasmine and of orange blossoms, as well as normal hydrocarbons such

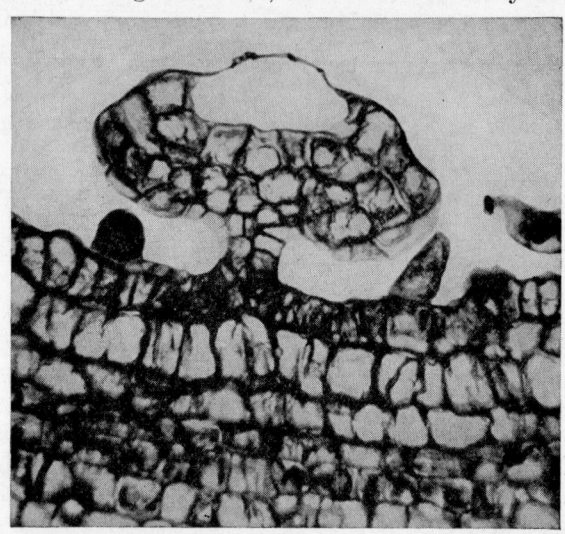

FIG. 25-1. Lysigenous oil sac of *Rubus rosaefolius*. Oil is secreted into the sac by the surrounding glandular cells. (After Engard, courtesy of A. J. Haagen-Smit)

as *n*-heptane, which is the major constituent of the turpentine of *Pinus jeffreyi*.

$$CH_2{=}CH{-}CH_2{-}N{=}C{=}S \qquad CH_2{=}CH{-}CH_2{-}\underset{\underset{O}{\|}}{S}{-}S{-}CH_2{-}CH{=}CH_2$$

Allylisothiocyanate   Allylsulfinyl-allylsulfide, allicin (anti-
(crucifer oil)   bacterial principle of garlic)

$$CH_3{-}CH_2$$
$$CH{-}S{-}S{-}CH{=}CHCH_3$$
$$CH_3$$

secondary Butylpropenyldisulfide
(active principle of *Ferula asafoetida*)

Indole   Methyl anthranilate

Active principles of orange and jasmine flowers
Some non-terpene constituents of essential oils

Nevertheless from a quantitative point of view it is the lower terpenes which are characteristic of essential oils.

Monoterpenes. Among the simplest of the monoterpenes are the aliphatic compounds myrcene and ocimene which consist of two C_5H_8 units united in a single open chain.

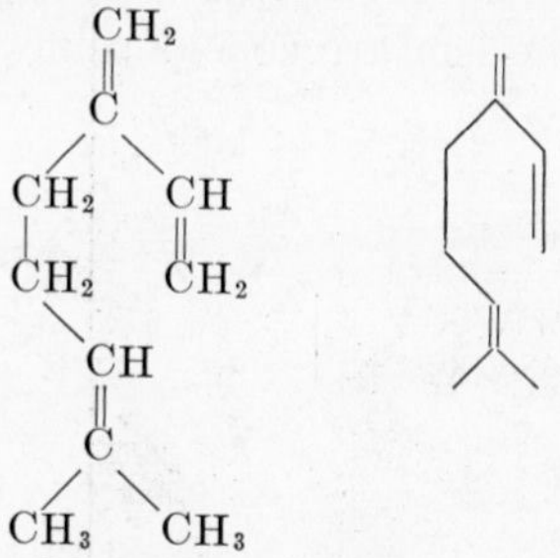

Myrcene 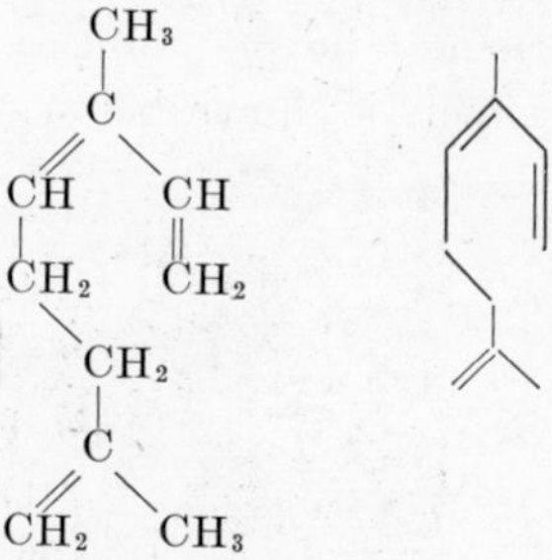 Ocimene

Both compounds contain three double bonds, and they differ only in the arrangement of these bonds. A convenient presentation of the structures of these and other terpenes is also illustrated above for myrcene and ocimene. By this convention, all the atoms of the structure are omitted just as in the usual notation for the aromatic nucleus.

Derivatives of the aliphatic monoterpenes include alcohols, aldehydes, ketones, and acids. Citronellol, $C_{10}H_{19}OH$, and geraniol, $C_{10}H_{17}OH$, are two of the widely distributed alcohols. Both of these compounds are hydrogenated as compared to the monoterpene hydrocarbon, citronellol containing only one double bond and geraniol two double bonds. The related aldehydes, citronellal, and citral are also found in a wide variety of oils, as is the aliphatic acid, geranic acid.

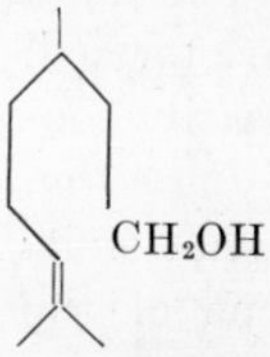

Citronellol 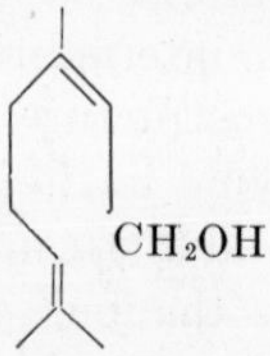 Geraniol 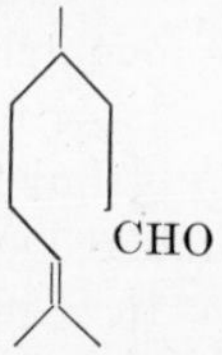 Citronellal 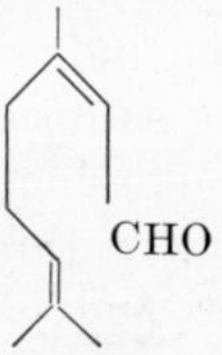 Citral 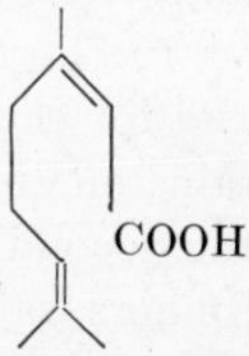 Geranic acid

Two aliphatic ketones are found in *Tagetes* sp., tagetone, $C_{10}H_{16}O$, and a related compound containing one less double bond, dihydrotagetone.

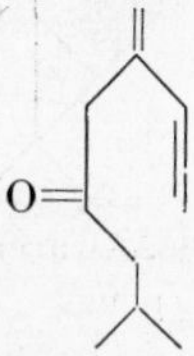

Tagetone 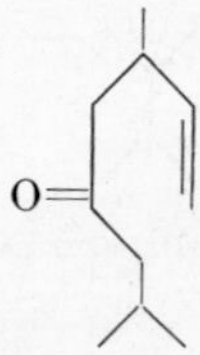 Dihydrotagetone

In other monoterpenes, the chain may be closed to form monocyclic structures. Thus limonene, which has the structure shown below, is a typical monocyclic monoterpene. This compound is found as a component of a wide variety of oils, including those of citrus. In the monocyclic terpenes the possibilities for isomerism are considerable; a series of such isomers is shown below.

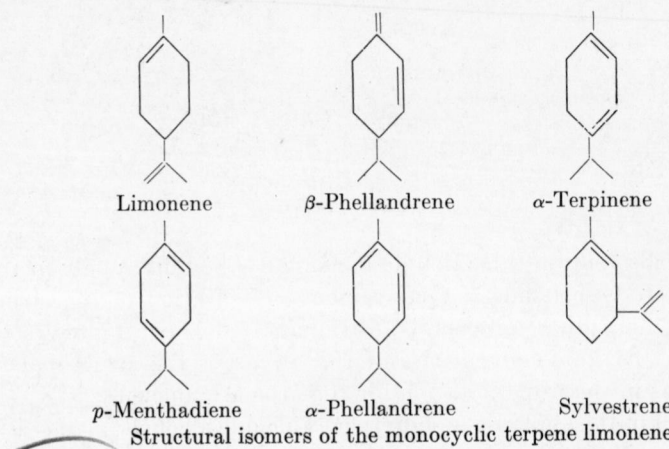

Structural isomers of the monocyclic terpene limonene

Figure 25-2 illustrates graphically the many modifications of the monocyclic terpene structure which take place as the result of oxidation or reduction and which are found among the components of essential oils. The alcohols include the well-known menthol, component of mint oils (*Mentha piperita*), while among the ketones are menthone (peppermint oil), carvone of caraway (*Carum*) and piperitone of *Mentha pulegium*.

In the bicyclic and tricyclic monoterpenes, internal rings are formed, and for each ring so formed the basic terpene structure necessarily contains one less double bond. In α-thujene, found in the oil of *Boswellia*, the second ring is a three-membered one, and the same is true of its isomers sabinene and carene. In α-pinene, the bulk terpene of the

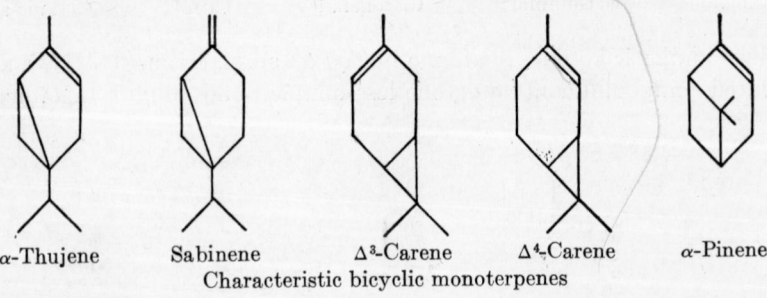

Characteristic bicyclic monoterpenes

essential oil of *Pinus palustris* and others, a four-membered ring is found.

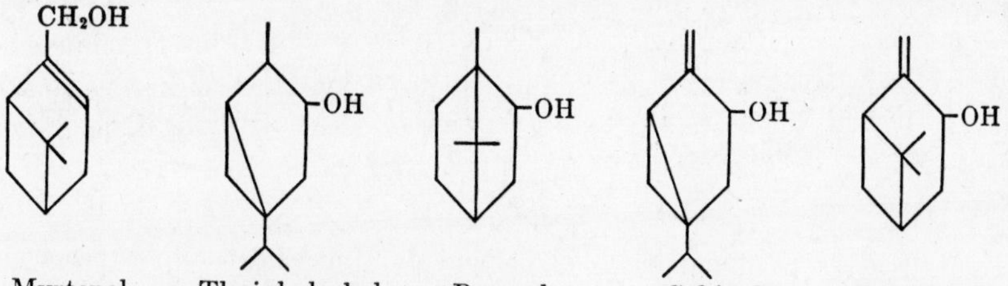
CH₂OH
OH
OH
OH
OH

Myrtenol Thujyl alcohol Borneol Sabinol Pinocarveol

Characteristic bicyclic monoterpene alcohols

Bicyclic terpene alcohols include myrtenol of *Myrtus communis,* thujyl alcohol of *Artemesia absinthium,* borneol of hemlock oil, sabinol of

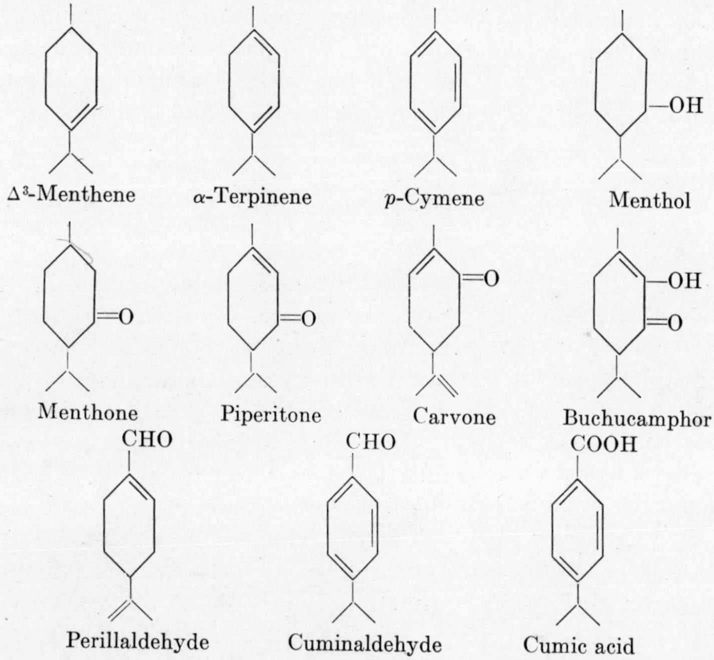
OH
Δ³-Menthene
α-Terpinene
p-Cymene
Menthol
O
O
O
OH
O
Menthone
Piperitone
Carvone
Buchucamphor
CHO
CHO
COOH
Perillaldehyde
Cuminaldehyde
Cumic acid

FIG. 25-2. Various stages in the oxidation of monocyclic monoterpenes. (Modified after Haagen-Smit)

Juniperus species and pinocarveol of *Eucalyptus.* The one commonly known bicyclic terpene aldehyde is myrtenal of *Hernandia* oil. Camphor is the bicylic terpene ketone contained in *Cinnamomum camphora* and many other species. Other such ketones are thujone of *Thuja* and *Salvia officinalis,* umbellulone of *Umbellularia californica,* and verbenone of *Verbena.*

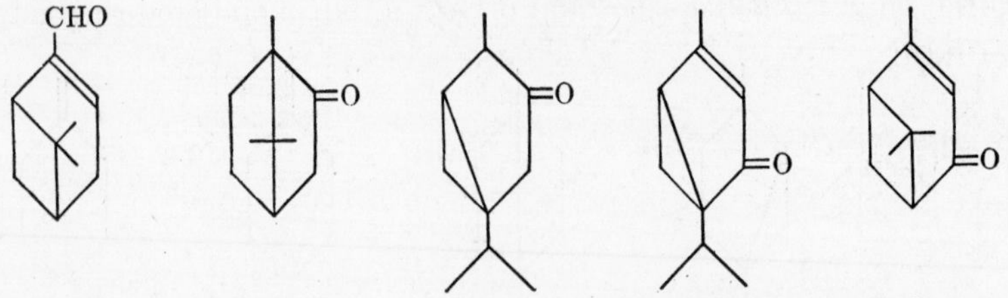

Myrtenal Camphor Thujone Umbellulone Verbenone
Characteristic bicyclic monoterpene aldehydes and ketones

Sesquiterpenes. The sesquiterpenes may be thought of as consisting of three isoprene units and have as their basic empirical formula $C_{15}H_{24}$. Like the monoterpenes they occur in essential oils and are found also in a variety of reduced and oxidized derivatives. Aliphatic sesquiterpenes are not known from higher plants, but the aliphatic sesquiterpene alcohol farnesol occurs in various oils including rose oil and oil of citronella. Nerolidol is likewise a component of peru balsam and of neroli oil.

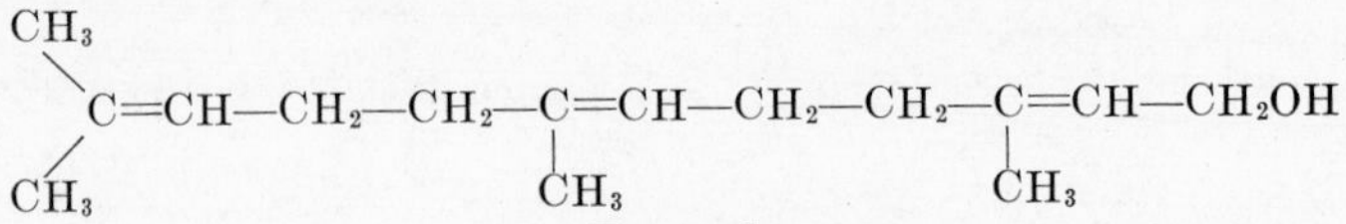

Farnesol

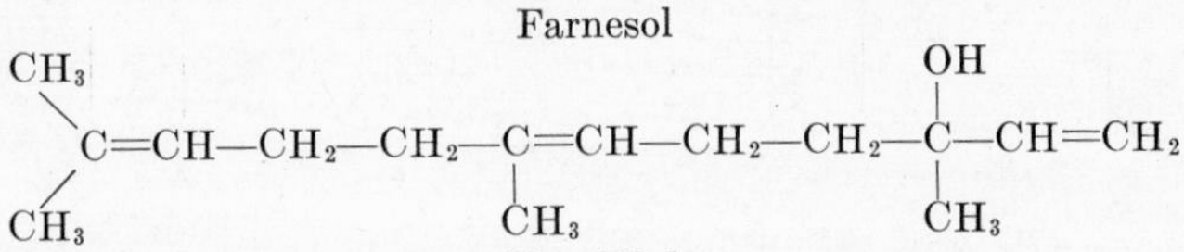

Nerolidol

Characteristic monocyclic sesquiterpenes are bisabolene (oil of bergamot) and zingiberene of ginger, *Zingiber officinale.*

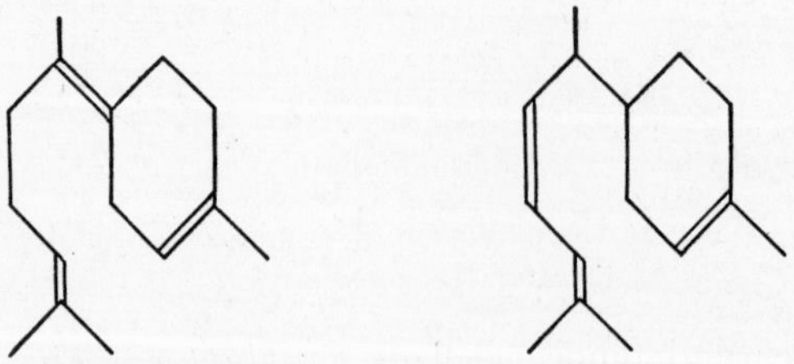

γ-Bisabolene Zingiberene

The bicyclic sesquiterpenes are represented by a variety of ring structures. Beta-selinene, a component of celery oil, contains two six-membered rings, as does cadinene, a widely distributed sesquiterpene found in oils of cubeb, galbanum, guayule, and others. Representative

β-Selinene 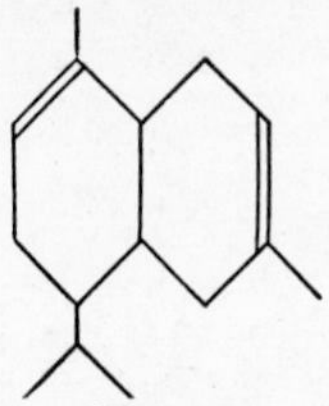   Cadinene 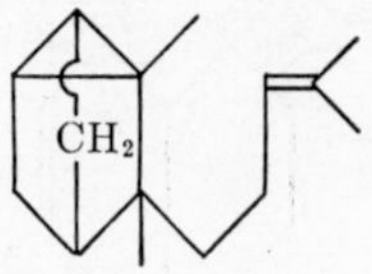   α-Santalene

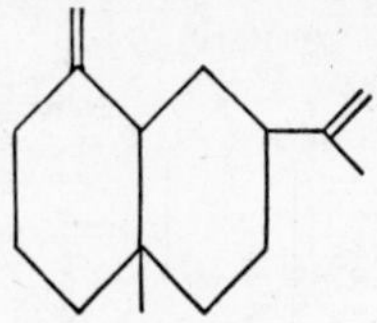

of the tricyclic sesquiterpenes found in nature is α-santalene from *Santalum album*. Oxygenated cyclic sesquiterpenes are represented by cadinol, a derivative of cadinene found in cubeb and in sandalwood oil, and by the eudesmol of the *Eucalyptus* oils.

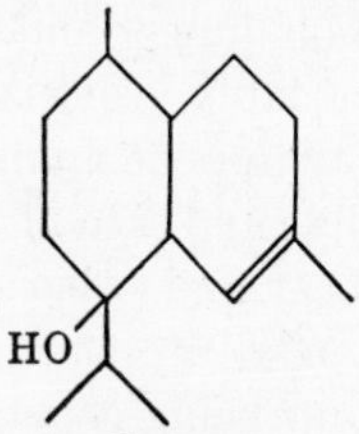

Cadinol 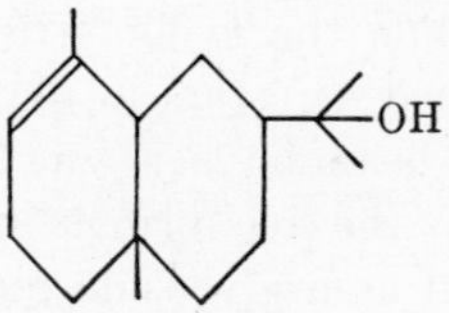   α-Eudesmol

Diterpenes. The diterpenes, $C_{20}H_{32}$, are represented in nature by relatively few compounds. Alpha-camphorene, of *Cinnamomum camphora* oil is a monocyclic diterpene while the *Pinaceae* and *Taxaceae* contain cryptomerene, dacrene, phyllocladene, and other diterpenes. Phytol, $C_{20}H_{40}O$, the alcohol with which chlorophyll is esterified, may be regarded as a hydrogenated diterpene alcohol. Vitamin A, which is

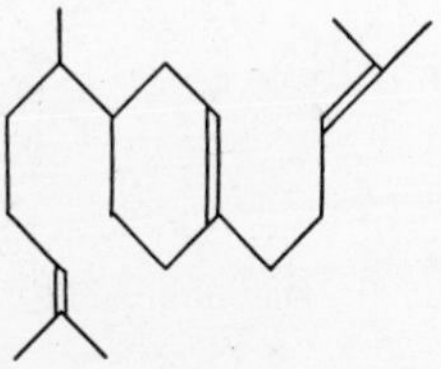

α-Camphorene 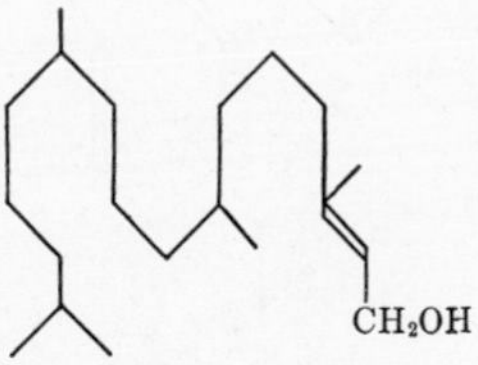   Phytol

not found as such in plants but which accumulates in large quantities in fish liver oils, is similarly a monocyclic diterpene alcohol.

The nonvolatile portion of the whole oil of many plants contains acidic diterpenes. Thus the oil of *Pinus palustris*, of which turpentine constitutes the volatile portion, contains in the nonvolatile residue abietic and D-pimaric acids.

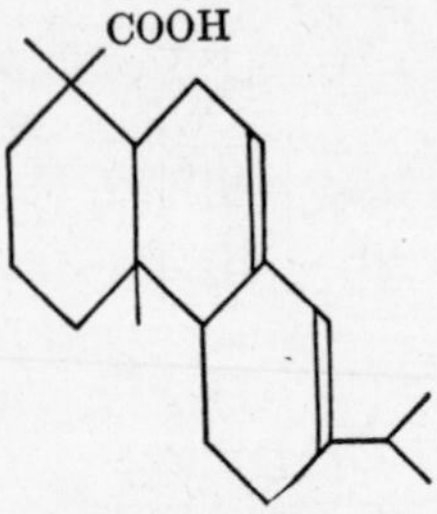

Abietic acid

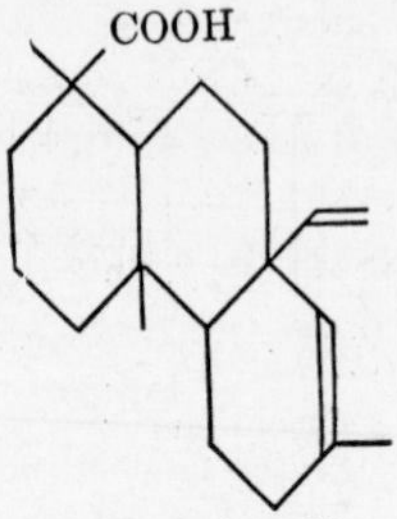

D-Pimaric acid

Triterpenes. Triterpenes, containing six of the elementary isoprene units, are of still less frequent occurrence than the lower terpenes. The aliphatic triterpene squalene has been reported to be present in higher plant oils, while the cyclic triterpene alcohols lupeol and β-amyrin are found combined as esters in many species of latex, especially those low in rubber. Oleanolic acid, a pentacyclic triterpene hydroxy acid,[7] is found both in the free form as in leaves of olive and skin of the grape and also combined as a glycoside in the sugar beet and other plant tissues. Such triterpene hydroxy acid glycosides constitute one group of the highly surface-active saponins, a second group being constituted by the sterol glycosides.

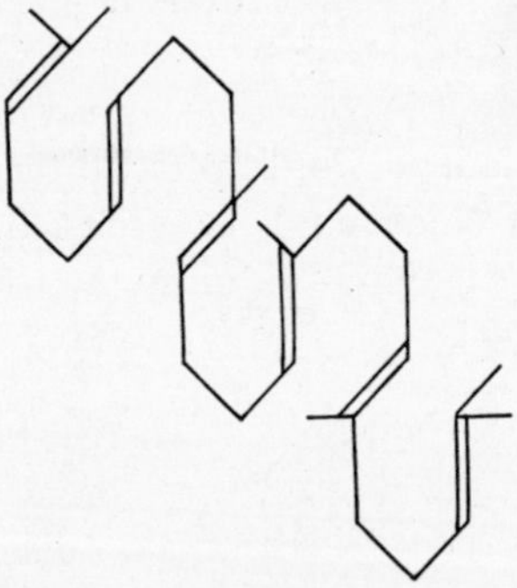

Squalene

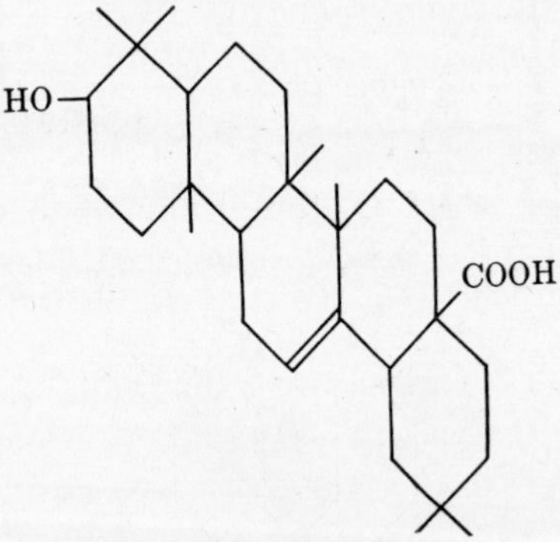

Oleanolic acid

Extraction and Separation of the Lower Terpenes. The essential oils are ordinarily removed from the plant by steam distillation of the whole fresh plant tissue. The tissue is suspended in water and steam passed through the whole, the distillate being collected through a condenser. The oil floats on the surface of the water and may be readily separated. Steam distillation removes in general compounds up to and including the diterpenes; triterpenes, if present, are to be sought in the residue from the distillation. Further separation of an essential oil into its components may be accomplished by fractional distillation and Fig. 25-3 gives an example of such a fractionation conducted on oil of peppermint. At temperatures of 20–150°C (at one atmosphere pressure) acetaldehyde, acetone, isovaleric acid, and isoamyl alcohol first distil over in the order named. These are followed by the monoterpene hydrocarbons in the range of 150–200°C. At slightly higher temperatures, 200–230°C, menthone and menthol which make up the bulk of this oil are next distilled over. Sesquiterpenes and their oxygenated

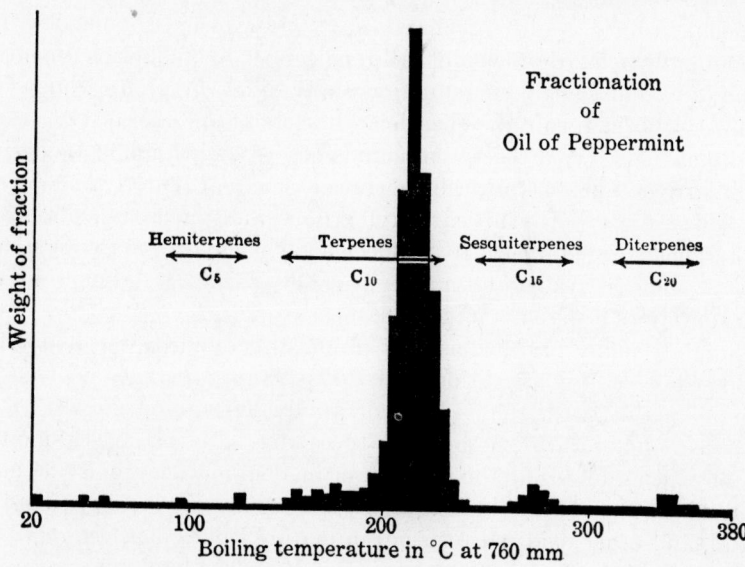

FIG. 25-3. Fractionation of oil of peppermint. (After A. J. Haagen-Smit, in Guenther, E., The Essential Oils. Van Nostrand, Vol. I, 1948.)

derivatives appear over the range 250–300°C, followed by the diterpenes at approximately 350°C. Because of the high temperatures involved in distillation at atmospheric pressure it is of course desirable to distil the sesqui- and diterpenes under reduced pressure.

Metabolism of the Lower Terpenes. Since in general a wide variety of terpenes and their derivatives occur together in a single oil, it is logical to assume that they must have in common a precursor which by varied modes of condensation and cyclization may yield the varied carbon skeletons. We must distinguish according to this view three steps in the formation of the individual terpene: the formation of the common building block, the condensation reactions leading to the appropriate carbon skeleton, and the further transformations such as oxidation or reduction which lead to the final products. But little is known concerning any of these steps in terpene formation, and no understanding of the nature of the ultimate building block has come from the study of terpene chemistry. It does not in any case appear to be isoprene itself, since this compound has never been discovered in plant material. The aliphatic monoterpenes such as citronellol readily cyclize *in vitro*, and other similar instances of cyclization are known. This has led Kremers[1] to the suggestion that citral, a widely distributed monoterpene derivative, may represent an intermediate in the formation of the cyclic terpenes. Citral itself would first need to be reduced, with the loss of one double bond, to citronellal, which readily cyclizes *in vitro* to form isopulegol. This compound by oxidation would yield pulegone, by reduction, menthol. On the other hand citral by reduction at the aldehyde group would yield geraniol which in turn by cyclization with loss of one mol of H_2O would yield limonene. From these compounds others in turn might be derived by similar reactions of known biochemical type, particularly since it has been shown that both the carboxyl groups and the double bonds of monoterpenes are subject to reduction by dehydrogenase systems.[2] Kremers' general proposal for the interconversion of terpenes in two species of mints is given in Fig. 25-4.

There are many indications that oil once secreted by the plant does not possess a fixed constitution, but that chemical changes continue to occur. Thus in citrus, oil is formed rapidly during the growth of the shoot but is not formed in the stem thereafter. Nevertheless limonene gradually accumulates at the expense of linaloöl and geraniol.[3] Similar evidence indicates that β-phellandrene and cymene may yield phellandral, cuminal, and other oxidized terpenes in mature leaves of *Eucalyptus*.[4]

Function of the Lower Terpenes. The terpenes have no known role in the plant. Since they are volatile, it is inevitable that a portion will be lost by direct evaporation into the air from the living plant. Thus a single plant of *Juniperus* may lose as much as 30 gm. of oils per day in this way.[5] Further evidence that oils may be lost in large quantities is the fact that oil-bearing plants frequently yield enough odor to be smelled at appreciable distances. This loss is evidently greater during the day than during the night since higher contents of leaf oil are recorded during the night than during the day.[6] There is no evidence that the simple terpenes once formed enter into metabolism again or that they are, for example, used as reserve food materials. The fact that the oils are accumulated in specialized glandular cells or ducts would also seem to preclude their reutilization in metabolism. The simple terpenes may on the contrary represent compounds formed by aberrant metabolism perhaps of the five-carbon precursor, which must be synthesized by all plants for the formation of carotenes, phytol, etc. It is possible also that certain

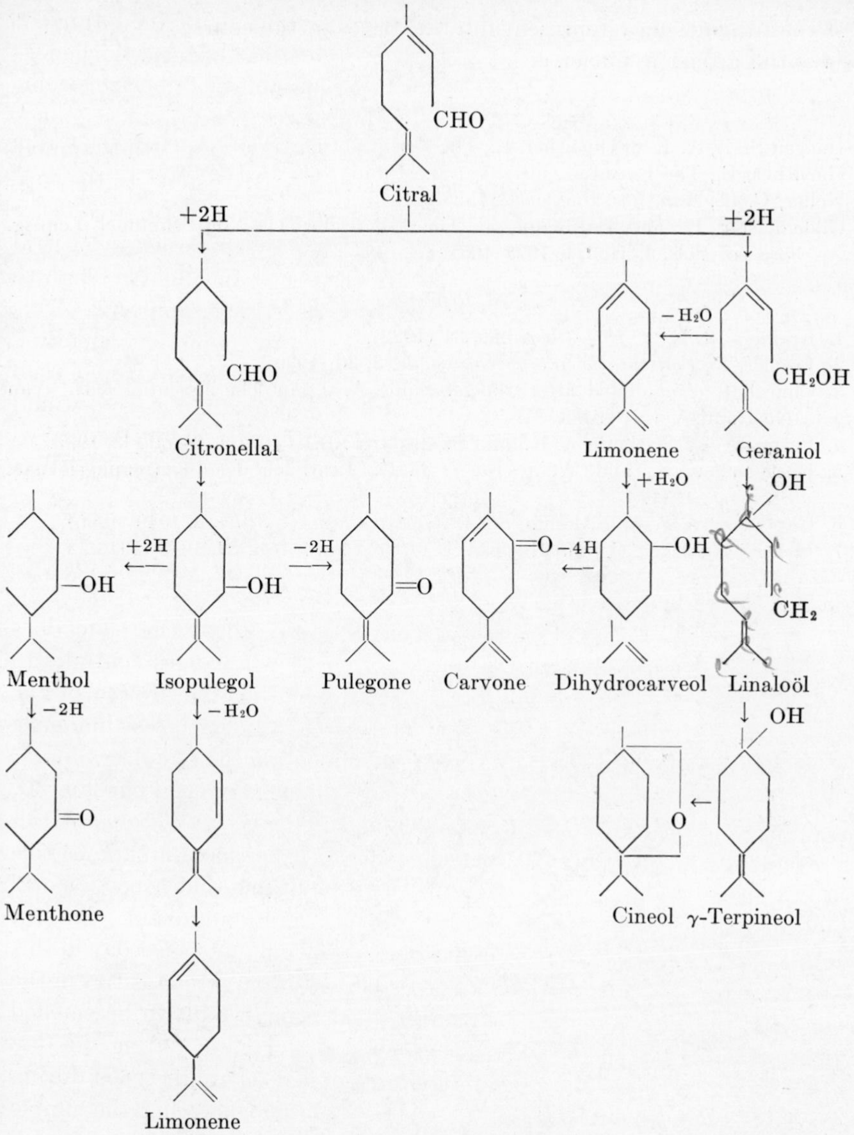

Fig. 25-4. Possibilities for formation of various terpenes and terpene derivatives from citral. On left, peppermint; on right, spearmint. After Kremers (1).

of the terpenes may represent intermediates in the course of synthesis of essential aromatic nuclei.

General References

Haagen-Smit, A. J., in Guenther, E., The Essential Oils. Van Nostrand, Vol. I, 1948.
Guenther, E., The Essential Oils. Van Nostrand, Vols. II, III, 1949.
Noller, C. R., *Ann. Rev. Biochem.*, **14**, 383 (1945).
Gildemeister, E., and Hoffmann, F., Die Ätherischen Öle-Von Schimmel, Leipzig. 3rd ed. Vols. I, II, III, 1928–1931.

References

1. Kremers, R. E., *J. Biol. Chem.*, **50**, 31 (1922).
2. Fischer, F., *Fortschr. chem. org. Naturstoffe*, **3**, 30 (1939).
3. Charabot, *et al.*, cited after Haagen-Smit, A. J., in The Essential Oils. Van Nostrand, Vol. I, 1948.
4. Berry, P. A., Macbeth, A. K., and Swanson, T. B., *J. Chem. Soc.*, 1443, 1937.
5. Kostytschev, S., and Went, F. A. F. C., Lehrbuch der Pflanzenphysiologie. Springer, 1931.
6. Gaponenkov, T., and Aleshin, S., *J. Applied Chem. (USSR)*, **8**, 1049 (1935).
7. Meisels, A., Jeger, O., and Ruzicka, L., *Helv. Chim. Acta*, **32**, 1075 (1949).

CHAPTER 26

THE CAROTENOIDS

Chemistry of the Carotenoids. The carotenoids constitute a ubiquitous or nearly ubiquitous group of tetraterpene pigments of both higher
and lower organisms. Carotenoids of the higher plant are found in roots,
stems, leaves, flowers, and fruits, deposited as plastid pigments in all
instances. Thus in the special case of the green leaf, the carotenoids are
intimately associated with chlorophyll in the chloroplasts. The carotenoids may be divided into two general groups, the carotenes and the
xanthophylls. The carotenes are hydrocarbons, typically soluble in
petroleum ether and but little soluble in alcohol. The xanthophylls,
oxygen-containing derivatives of the carotenes, are typically insoluble in
petroleum ether but soluble in alcohol.

There are many methods of extracting the carotenoids from plant
tissue, and if the carotenoids alone are to be extracted, use is frequently
made of solvent extractions on the dried tissue. Most frequently however, it is desired to extract and determine at the same time the chlorophyll of the tissue. For the extraction and separation of chlorophyll
and the carotenoids the following general procedure based on the early
work of Willstätter and Stoll[1] may be used. The fresh tissue is ground
in aqueous acetone, in which all the pigments are soluble. To the
aqueous acetone layer ether is now added. The chlorophyll and the
carotenoids pass into the ether layer, which is then separated from the
aqueous acetone layer. A concentrated solution of KOH in methyl
alcohol is next added to the ether layer. The KOH saponifies the
chlorophyll and on the addition of water the chlorophyll remains in the
aqueous methyl alcohol while the carotenoids pass into the ether layer.
The chlorophyll may be determined spectrophotometrically in the chlorophyll fraction. The ether layer is next evaporated and the residue par-
titioned between 85% methyl alcohol and petroleum ether. The carotenes are found in the petroleum ether layer, the xanthophylls in the
methyl alcohol layer. This method while appropriate for the separation
and determination of the principal pigments is hardly adaptable for the
large-scale preparation of carotenes or xanthophylls. For the latter
purpose it is in general more practical to dry the initial material and then
to extract it at once with the appropriate solvent.

For the separation of the individual carotenoids the chromatographic technique, originally proposed by Tswett in 1907, is a powerful and widely used tool. This technique is based on the fact that even closely related chemical structures differ significantly in their affinity for appropriate adsorbents. Thus when a solution of carotenes in petroleum ether or benzene is passed through a column of calcium hydroxide or calcium hydroxide mixed with alumina, the mixture separates into a number of sharply differentiated zones, each consisting of one chemically individual carotene. The zones may be separated and the individual pigments eluted in acetone or acetone mixtures. Separation of the xanthophylls may be similarly achieved by passing a suitable solution through a column of calcium carbonate. Details of the chromatographic method may be found in the works of Zechmeister and Cholnoky, and of Strain cited in the general references.

Chemical work on the structure of the carotenes dates from the observation of Willstätter and Meig in 1907 that the carotene of carrots and of green leaves possesses the empirical formula $C_{40}H_{56}$. Zechmeister and Cholnoky[2] in 1928 then showed that the material possesses eleven double bonds and in 1930 Karrer and others[3] proposed the structure for β-carotene which is recognized today. This structure consists of two C_9 six-membered rings which are attached to the ends of a 22-carbon chain.

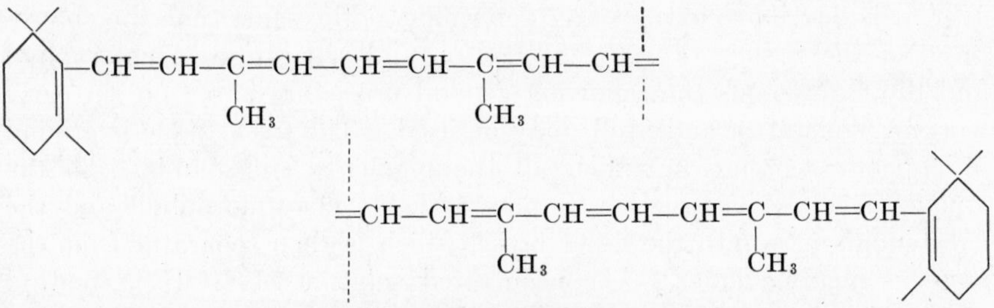

Structure of β-carotene

It may be seen that the structure of carotene is clearly that of a terpene and may be thought of as composed of eight isoprene-like units. Beta-carotene further possesses a center of symmetry at the center, the molecule consisting of two identical portions linked together at this point.

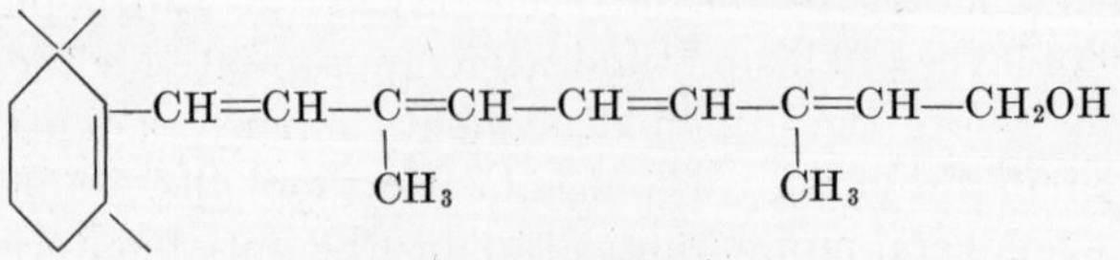

Structure of Vitamin A

Oxidative splitting of the β-carotene molecule at the center yields two molecules of vitamin A, a compound not found in higher plants. Beta-carotene is however able to act as a source of vitamin A for the animal body and may hence be referred to as a provitamin A.

The eleven double bonds of the β-carotene molecule are conjugated with respect to one another. This conjugated system is reponsible for the deep color of the compound, the crystalline material being in fact red, although β-carotene in solution exhibits three absorption maxima in the blue region of the spectrum. The position of these absorption maxima are highly characteristic for a carotenoid and may be used to identify and characterize a given compound.

Beta-carotene is the principal carotene of leaves and occurs in these tissues in the concentrations of 0.02–0.1%. Two other carotenes which are found in appreciable quantities in many cases are α- and γ-carotene whose structures are given in Table 26-1. Alpha-carotene like β-carotene possesses 11 double bonds but only ten of these are conjugated, the last being isolated in one ring. Alpha-carotene may constitute a negligible portion of the leaf carotene as in spinach or may make up to 35% of the total as in other leaves. Gamma-carotene possesses twelve double bonds of which eleven are conjugated with one isolated from the rest. Unlike the α and β isomers, γ-carotene contains but one ring, the other end of the chain being an open one. This isomer, while it makes up only a small part of leaf carotenes, is found as a substantial constituent of flower and fruit carotenes, e.g., flowers of *Mimulus longiflorus*.

In addition to the three hydrocarbons mentioned above, a fourth, lycopene, $C_{40}H_{56}$, is widely distributed in plants. Lycopene, which has the structure shown in Table 26-1, has an open chain structure, both halves resembling the open chain portion of γ-carotene. It contains, hence, eleven conjugated and two isolated double bonds. Lycopene constitutes the red pigment of tomato and many other red fruits but is ordinarily not found in leaves.

A polyene differing somewhat in character from those described above is phytofluene, which has the composition $C_{40}H_{64}$ and contains approximately seven double bonds. Phytofluene is colorless in visible light but possesses strong absorption maxima in the ultraviolet region.[4] It is widely distributed in flowers, fruits, stems, and roots but appears to be absent from green leaves. Phytofluene is the only carotenoid which possesses the empirical formula $C_{40}H_{64}$ expected for a true tetraterpene and hence represents an important connecting link between the lower terpenes and the carotenoids.

A wide variety of oxygenated carotenoids or xanthophylls is found in

TABLE 26-1. The principal carotenes of higher plants

Name	No. of conjugated double bonds	Structure
α-Carotene	10	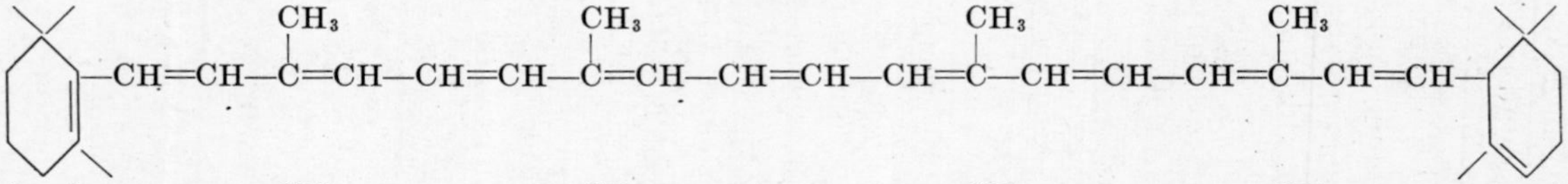 (cyclohexene ring)—CH=CH—C(CH₃)=CH—CH=CH—C(CH₃)=CH—CH=CH—CH=C(CH₃)—CH=CH—CH=C(CH₃)—CH=CH—(cyclohexene ring)
β-Carotene	11	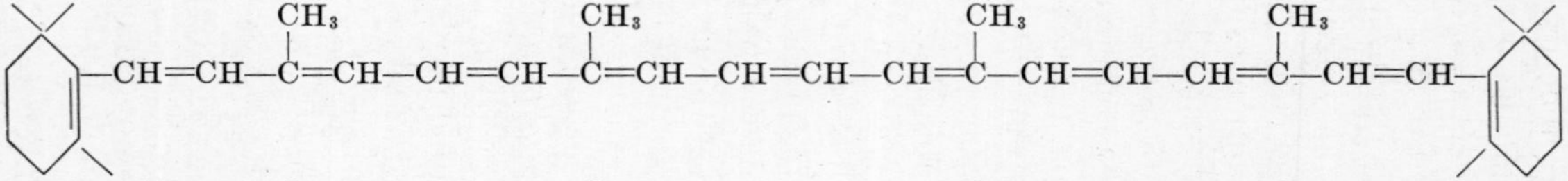 (cyclohexene ring)—CH=CH—C(CH₃)=CH—CH=CH—C(CH₃)=CH—CH=CH—CH=C(CH₃)—CH=CH—CH=C(CH₃)—CH=CH—(cyclohexene ring)
γ-Carotene	11	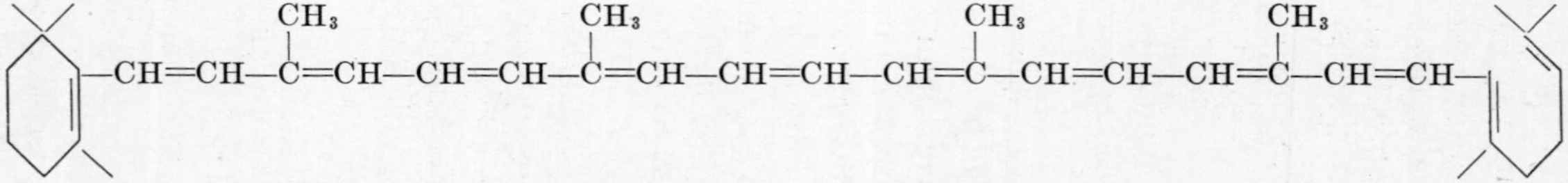 (cyclohexene ring)—CH=CH—C(CH₃)=CH—CH=CH—C(CH₃)=CH—CH=CH—CH=C(CH₃)—CH=CH—CH=C(CH₃)—CH=CH—(open chain)
Lycopene	11	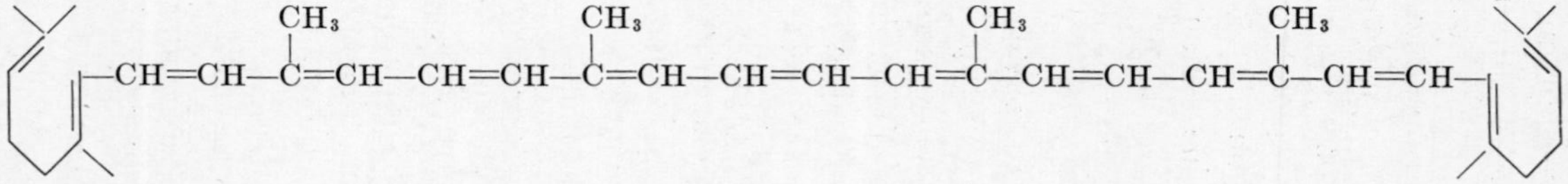 (open chain)—CH=CH—C(CH₃)=CH—CH=CH—C(CH₃)=CH—CH=CH—CH=C(CH₃)—CH=CH—CH=C(CH₃)—CH=CH—(open chain)

leaves, fruits, flowers and other plant organs. The major constituent of the leaf xanthophyll of many leaves is lutein (Table 26-2) which has the structure of α-carotene with an additional hydroxyl group in each terminal ring. Lutein is also the principal xanthophyll of egg yolk. Other leaf xanthophylls are zeaxanthin, a dihydroxy-β-carotene; lycoxanthin, a hydroxylycopene; lycophyll, a dihydroxylycopene; rubixanthin, a hydroxy-γ-carotene; and hyptoxanthin, a hydroxy-β-carotene. In all of these oxidation products, the hydroxyl groups occupy the same positions in the structure, i.e., the 3 position, although in other cases additional OH groups may occur in other positions as in capsanthol

TABLE 26-2. Summary of some of the important structural features of the principal carotenoid pigments. (After Strain[18])

Pigment	Formula	Double bonds		Conjugated $C=O$ groups	Hydroxyl groups	Rings
		Total	Conjugated			
α-Carotene	$C_{40}H_{56}$	11	10	0	0	2
β-Carotene	$C_{40}H_{56}$	11	11	0	0	2
γ-Carotene	$C_{40}H_{56}$	12	11	0	0	1
Lycopene	$C_{40}H_{56}$	13	11	0	0	0
Cryptoxanthin	$C_{40}H_{56}O$	11	11	0	1	2
Lutein	$C_{40}H_{56}O_2$	11	10	0	2	2
Zeaxanthin	$C_{40}H_{56}O_2$	11	11	0	2	2
Rhodoxanthin	$C_{40}H_{50}O_2$	14	12	2	0	2

(Table 26-3). These same xanthophylls are of wide occurrence in tissues other than leaves. Thus zeaxanthin is also the principal pigment of corn and other yellow seeds. Lycoxanthin is found in the tomato fruit, while lycophyll is found in berries of *Solanum dulcamara*.

In addition to hydroxyl derivatives, the xanthophylls include esters, ketones, and hydroxy ketones. Helenien, the yellow pigment of flowers of the *Helenieae*, is the dipalmitate ester of lutein while physalien, the pigment of yellow ground cherry (*Physalis*) fruits, is the dipalmitate of zeaxanthin. Rhodoxanthin, the red pigment of berries of *Taxus baccata*, is a diketoxanthophyll. The pigments of the red pepper *Capsicum annuum* include capsanthin and capsorubin. The latter contains two hydroxyl and two keto groups (Table 26-3).[5]

Stereoisomerism in the Carotenoids. The presence of numerous aliphatic double bonds in the carotenoid molecule suggests at once the possibility of *cis-trans* isomerism, and such isomers have in fact been found. Certain of the double bonds of the carotenoid molecules are stable only in the *trans* configuration due to stereochemical factors. Only those double bonds can assume the *cis* configuration in which the structure shown below is present and in which X and X^1 are both hydrogen atoms.

TABLE 26-3. Important naturally occurring xanthophylls of plants. R signifies the common $C_{20}H_{24}$ central portion of the molecule. R' signifies a tautomeric form in which position of the double bonds is shifted

Name	Structural relation	Structure
Alcohols		
Lycoxanthin	3-Hydroxylycopene	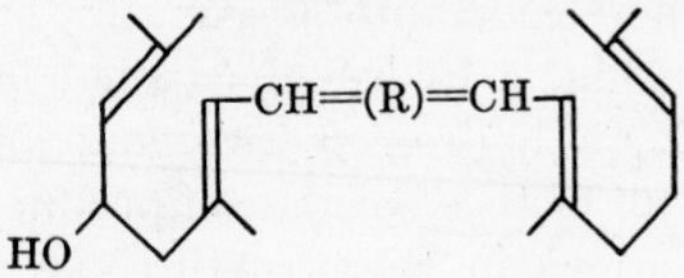
Lycophyll	3,3'-Dihydroxylycopene	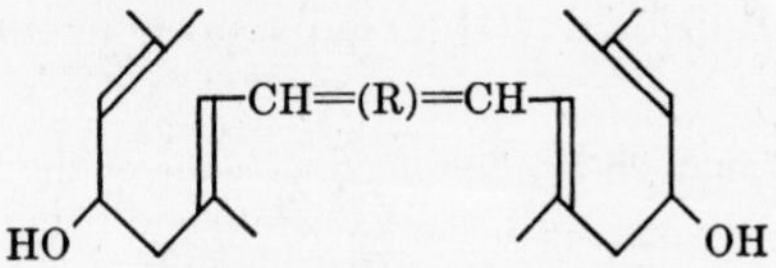
Rubixanthin	3-Hydroxy-γ-carotene	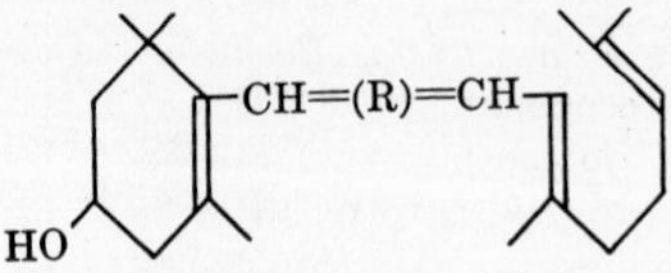
Cryptoxanthin	3-Hydroxy-β-carotene	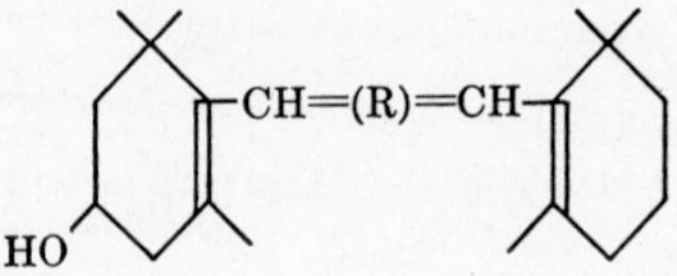
Zeaxanthin	3,3'-Dihydroxy-β-carotene	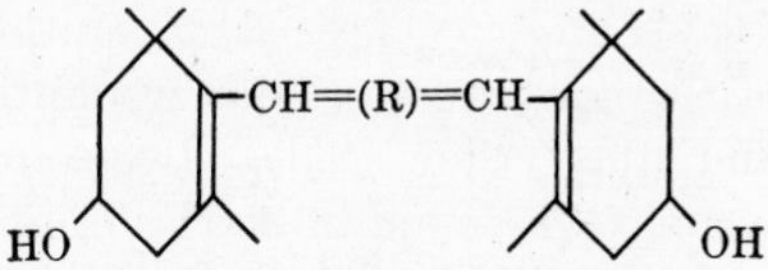
Xanthophyll (lutein)	3,3'-Dihydroxy-α-carotene	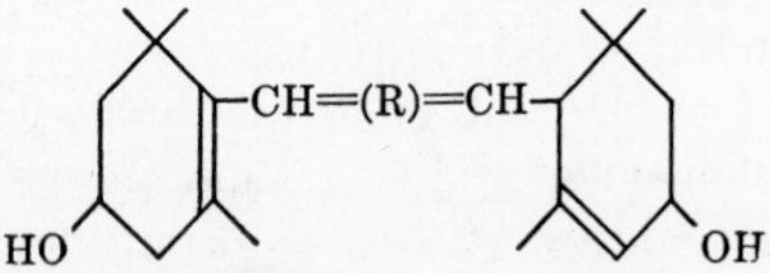
Capsanthol		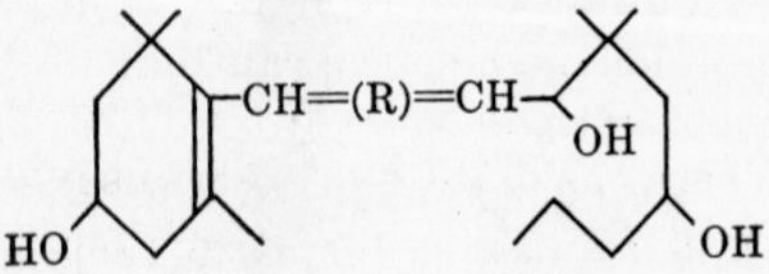
Ethers		
Rhodoviolascin		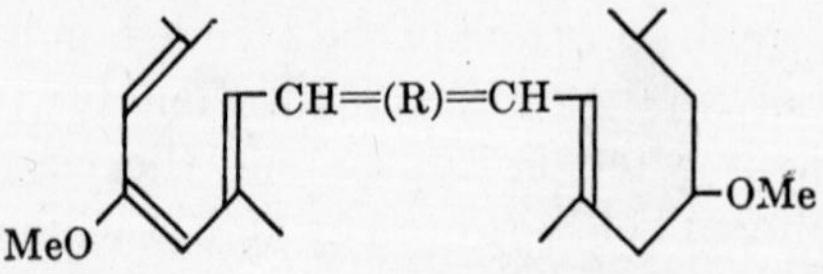

TABLE 26-3 (*Continued*)

Name	*Structural relation*	*Structure*

Esters

Helenien — Xanthophyll dipalmitate

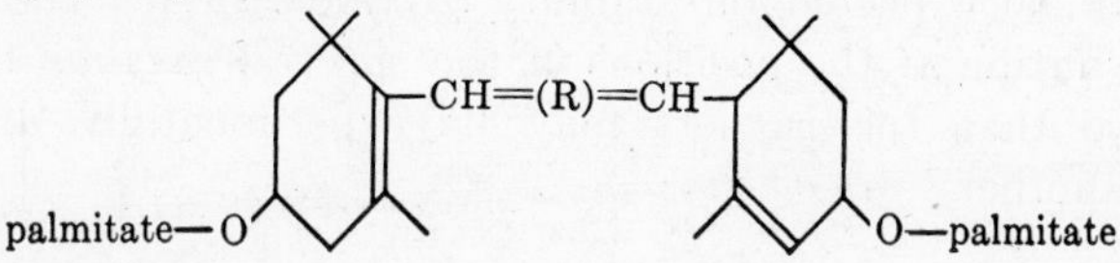

Physalien — Zeaxanthin dipalmitate

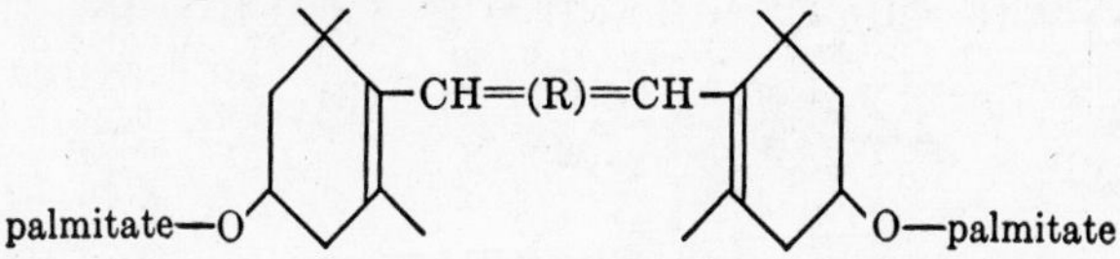

Ketones and hydroxyketones

Myxoxanthin

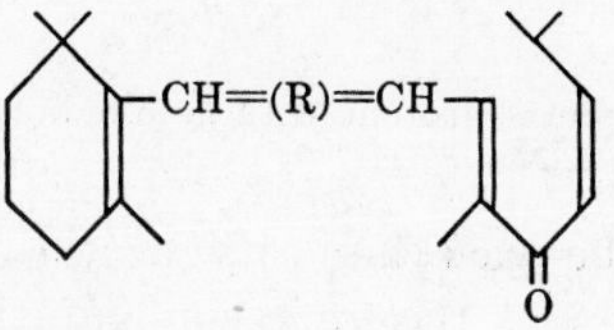

Rhodoxanthin — Diketoxanthophyll (tautomeric form)

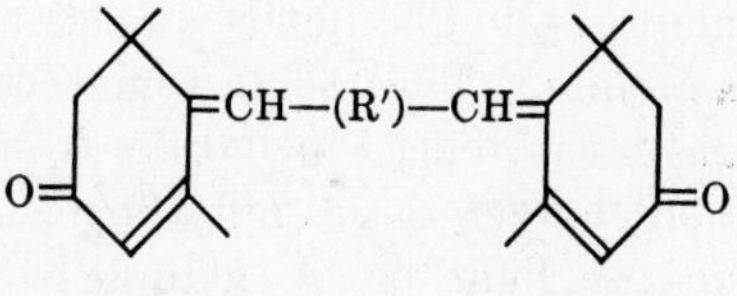

Capsanthin

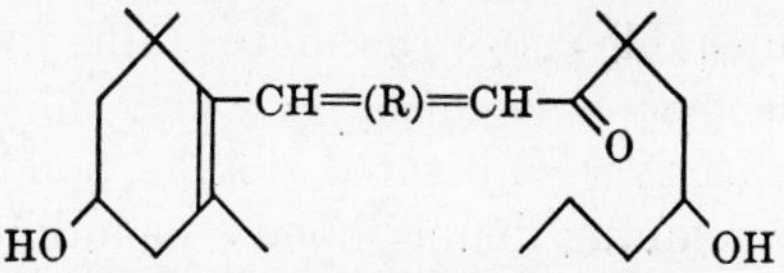

Capsorubin

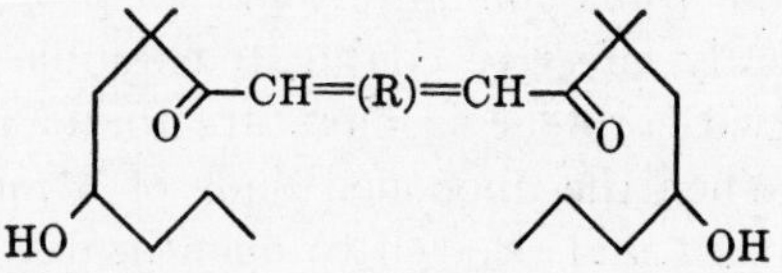

Where one of the two positions is filled by a methyl group, steric hindrance

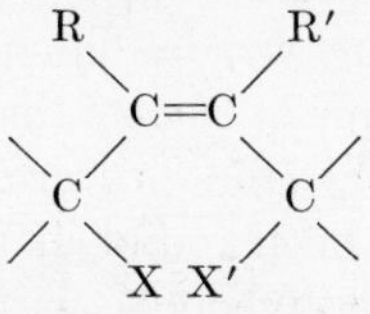

renders the *cis* structure unstable.[6] In β-carotene, for example, of the nine double bonds contained in the central chain, only five are potentially capable of assuming the *cis* configuration so that of the maximum of 272 *cis-trans* isomers possible with this asymmetric molecule only twenty are to be expected on stability grounds. Each *cis* double bond involves shifting of the positions of the spectral maxima to shorter wavelengths so that the isomers may be experimentally distinguished from one another.[7]

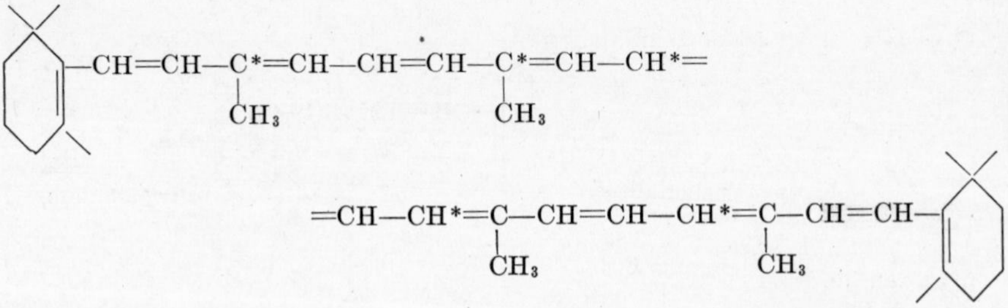

Beta-carotene with double bonds capable of assuming *cis* configuration marked*

The vast majority of the double bonds found in native carotenoids possess the *trans* configuration and although isomers containing one or more *cis* double bonds do occur naturally, these are found only in small amounts. Thus some tomatoes contain two pigments, prolycopene and pro-γ-carotene, which differ from lycopene and γ-carotene in their absorption characteristics but which may be converted to the latter *in vitro* by intense light or by iodine, isomerizing treatments which convert the unstable *cis* to the more stable *trans* double bonds.[8] Prolycopene is an isomer of lycopene in which all but one of the seven stereochemically effective conjugated double bonds possess the *cis* configuration. The remaining double bonds, including in this case the central one, possess the *trans* configuration. In pro-γ-carotene, five of the six stereochemically effective conjugated double bonds are in the *cis* position. These two carotene isomers are found primarily in seeds, fruits, and petals.[7] When the unopened buds of *Mimulus longiflorus* are removed from the plant and allowed to open in diffuse light, procarotene and prolycopene are found in the petals in substantial amounts.[9] If the flowers open under normal conditions in full sunlight only the all-*trans* isomers are found. Apparently then, the *cis* double bonds are isomerized to the more stable *trans* configuration *in vivo* as they are *in vitro*.

In addition to prolycopene and procarotene, further *cis* isomers of lycopene and carotene have been found in nature.[10] The tomato fruit contains small amounts of neo-lycopene with two *cis* double bonds and neo-γ-carotene, likewise a di-*cis* isomer. Neo-γ-carotene P of *Pyracantha* fruits is a mono-*cis* isomer of γ-carotene. Other examples of *cis-trans* isomerism are also found in the xanthophylls.[11]

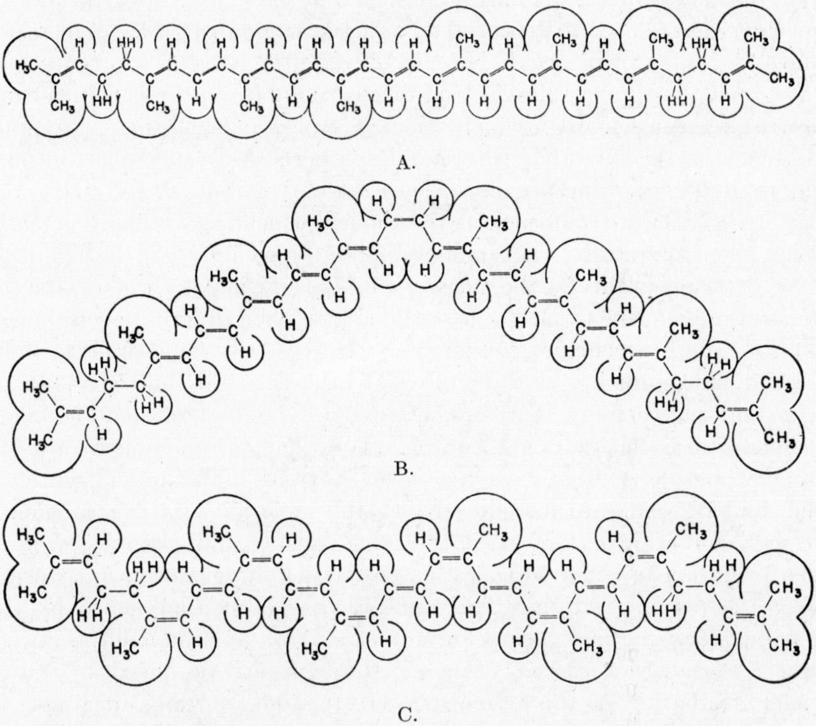

FIG. 26-1. A. Model of all-*trans*-lycopene. B. Model of central mono-*cis*-lycopene. C. Model of prolycopene (a hexa-*cis*-lycopene). (After Zechmeister[29])

Biogenesis of the Carotenoids. The carotenoids may be formed from the same branched chain, five-carbon precursor, discussed above (Chapter 25) in relation to the lower terpenes. As with the terpenes there is no specific evidence as to the nature of this precursor although work on rubber to be discussed in the following chapter indicates that a five-carbon acid may be involved. Since in all the carotenoids the direction of polymerization appears to be reversed at the center of the chain, it seems reasonable to presume that polymerization takes place to a C_{20} intermediate or intermediates and that formation of the final carbon skeleton proceeds further by combination of two such halves. According to this view β-carotene would be formed by condensation of the two appropriate vitamin A-like structures; lycopene would be formed by condensation of the two appropriate C_{20} open chain molecules, while γ-carotene would be formed by one molecule of β-carotene precursor combining with one molecule of lycopene precursor, etc. The appropriate C_{20} intermediates have not, however, been detected in plant material, and indeed the only well-known C_{20} compound which may be considered as related to the carotenoids is the alcohol, phytol. Willstätter and Mieg[12] first suggested that phytol might represent a precursor in carotenoid synthesis, and it is true that carotenoid synthesis is in general active in green leaves in which much phytol is being formed. Against this view is the fact that carotenoid synthesis also takes place in non-chlorophyll-containing organs such as the carrot root, many seeds, and flowers. Even in fruits such as the tomato which are normally green in color when immature, the amount of phytol which could be liberated during disappearance of the chlorophyll is insufficient to account for the amount of lycopene formed. It would appear therefore that even though phytol were a precursor in carotenoid synthesis, it would necessarily be formed independently of chlorophyll synthesis. It is entirely possible also that phytol may represent a reduction product of some common precursor of both phytol and carotenoids rather than the precursor itself.

It would be of interest to know at what point in carotenoid synthesis the double bonds are introduced into the molecule of the carotenoids. If, as would appear reasonable, the same basic C_5 unit is involved as in the formation of other terpenes, extensive dehydrogenation must occur after the polymerization. This question has been approached by the genetic block method, using mutant strains of a carotenoid-producing yeast, *Rhodotorula rubra*.[13] Among the mutants produced by ultraviolet irradiation of this organism were found (a) strains incapable of producing any carotenoid, (b) one strain which produced the colorless phytofluene as well as small amounts of carotenes, (c) strains which produced large amounts of carotenes, and (d) the original parent strain which produced mainly torulene, a xanthophyll. Although no final conclusion may be drawn from this type of data, still it may be suggested that carotenoid synthesis in *Rhodotorula* may follow the general pathway shown below in which colorless precursors are transformed first to the true tetraterpene phytofluene (seven double bonds) thence to carotenes (eleven and twelve double bonds in this case) and thence to the oxygen-containing torulene, which probably contains thirteen conjugated double bonds.

Colorless precursor → Phytofluene → Carotene → Xanthophyll
(Tetraterpene) $\left(\begin{smallmatrix} \beta\text{-Carotene} \\ \gamma\text{-Carotene} \end{smallmatrix}\right)$ (Torulene)

Suggested course of carotenoid formation in *Rhodotorula*

Function of the Carotenoids in Plants. The carotenoids of the green leaf are found as universal components of the chloroplast. Both carotenes and xanthophylls are bound to or intimately associated with protein and in fact appear to be components of the chlorophyll-protein complex which has been discussed in Chapter 17. Despite the fact that carotenoids are contained in the chloroplast, no clear indication of a general relation of carotenoids to photosynthesis is known, with the reservation that in the diatom *Nitzschia* and in the green alga *Chlorella* some of the light absorbed by the carotenoids appears to be available for photosynthesis.[14,15] No specific role in photosynthesis is, however, known for the carotenoids, and no physiological role for the pigments contained in other plant organs is known. Frey-Wyssling[16] has expressed the view that many of the carotenoids, and in particular the xanthophylls, may represent functionless metabolic byproducts, as seems to be the case with other terpenes. It should be noted, however, that though the xanthophylls do not as yet appear to have physiological functions they may in many cases play a definite role in the plant by virtue of the coloration which they impart to flowers, fruits, and seeds. These colors undoubtedly are in many instances factors in fertilization of the flower and in seed dispersal by animals.

Physiology of the Carotenoids. All the carotenoids appear to be synthesized *in situ* in the organ in which they are found. Thus carotenoids are synthesized in excised roots and in excised fruits such as the tomato. Data on carotenoid formation in excised leaves appear, however, to be meager. Experiments designed to test whether or not carotenoids are translocated in the stem of tomato plants revealed no evidence of any accumulation of any carotenoids in girdled stems in which sucrose and various vitamins were actively accumulated above the girdle.[17] Leaves of etiolated seedlings contain carotenoids, but the amounts present are increased by exposing the plant to light.[18] The etiolated plant contains, however, generally the same pigments as the green, light-grown plant.

Of the factors influencing carotenoid production, temperature is an important one in the tomato. The optimum temperature for lycopene formation in the tomato appears to be approximately 19–24°C. Isolated tomato fruits do not develop normal lycopene when stored at temperatures above 30°C, whereas normal pigmentation is developed at temperatures lower than 30°C. This phenomenon has been investigated in detail and the influence of temperature on individual pigments is shown in Table 26-4. High temperatures depress not only the formation of lycopene, the principal pigment, but also the formation of the other, minor, pigments. Carotenoid formation is, however, resumed when the fruits are removed from high temperature storage and restored to lower temperatures. This temperature factor does not operate in a similar manner for the watermelon, which forms lycopene even at high temperatures.[19] Oxygen is necessary to lycopene formation both in tomato and in watermelon.

Numerous investigations have been made concerning the influence of nutritional status of the plant in relation to carotenoid formation. The β-carotene content of carrots is but little influenced by major element

TABLE 26-4. Effect of temperature on the development of carotenoid pigment in isolated tomato fruits. (After Went et al.[20])

Temperature of storage	Content of pigment in milligrams per 100 gm. dry fruit				
	Xanthophyll	Lycoxanthin	Lycopene	γ-Carotene	β-Carotene
26.5°C	5.1	92.0	270.0	3.4	9.6
33.0°C	0.0	0.0	17.1	0.6	6.2
33°C, then moved to 26.5°C	1.2	2.2	55.0	1.0	5.0

nutrition over a wide range, and the same is true of tomatoes.[21] Iron deficiency, which interferes with chlorophyll synthesis and results in chlorosis, is also associated with a greatly decreased carotenoid content of the chlorotic tissue.[22]

Genetic control of carotene content of roots, leaves, seeds, and fruits has been demonstrated for cabbage, beans, corn, and many other species. A detailed investigation of the genetic control of carotenoid formation in tomato has been made by LeRosen and others.[23] The color of the epidermis of the tomato fruit is due to a pigment formed under the influ-

TABLE 26-5. Genetic control of fruit color in the tomato. (After LeRosen, Went, and Zechmeister[23])

Genetic constitution	Skin color (alkali sol. pigment)	Flesh color (carotenoids)	Individual carotenoids in flesh (mg./100 gm. dry wt.)		
			Lycopenes	Xanthophylls	β-Carotene
yr	White	Yellow	0	4.5	1.5
Yr	Yellow	Yellow	0	2.9	1.8
yR	White	Red	219	6.3	20
YR	Yellow	Red	202	10.3	19

ence of a dominant gene Y. This pigment, which is not a carotenoid, is not formed in the homozygous recessive yy. Production of lycopene in the flesh of the fruit is controlled primarily by a dominant gene R. The double recessive rr results in yellow flesh, and the total absence of lycopene. Although plants which are rr in constitution produce appreciable amounts of carotenes and xanthophylls, these constituents are increased in amount by some ten times in the presence of R. Similarly, the color of the endosperm of yellow corn seeds, which is mainly due to zeaxanthin, is controlled by a gene Y for which a recessive allele y is known. Plants having the endosperm composition yyy are white whereas those having the composition YYY are deep yellow. Not only the amount of yellow pigment but also the amount of provitamin A (β-carotene) seems to be affected by genetic constitution with respect to this gene, as shown in Table 26-6. Many genes are known which affect chlorophyll formation

TABLE 26-6. Effect of genetic composition of endosperm on color and provitamin A content of maize seeds. (After Mangelsdorf and Fraps[24])

Factorial composition of endosperm	No. of genes for yellow	Color of endosperm (mainly zeaxanthin)	Provitamin A β-carotene: microgm./gm.
yyy	0	White	0.03
yyY	1	Pale yellow	1.35
yYY	2	Dilute yellow	3.00
YYY	3	Deep yellow	4.50

in leaves, particularly in corn, barley, and other cereals. These genes in some instances must also influence carotenoid formation since genetically albino (nongreen) corn seedlings may contain as little as 1% of the normal carotenoid.[25]

Changes in coloration of fruits from green to yellow or orange during ripening are associated either with decreases in chlorophyll, increases in carotenoids, or both. In the tomato, as we have seen, the green chlorophyll color of the immature fruit disappears on maturation, this process being accompanied by a synthesis of carotenoid pigments, and the same is true of other fruits. Artificial hastening of ripening by ethylene treatment brings about not only chlorophyll destruction but also the usual increase in polyenes.[19,26] The development of coloration in flowers during maturation has also been investigated in the pumpkin by Zechmeister and others.[27] These workers gathered partially mature flowers in which the upper two thirds were yellow and the lower third still green. The yellow part contained four times as high a concentration of carotene and three times as high a concentration of xanthophyll as the green portion, indicating that here, as in the tomato, attainment of color on maturation involves not only loss of chlorophyll but also new formation of polyene pigments.

The coloring of leaves in the fall such as takes place with certain species, notably of *Acer* and *Quercus*, is primarily due to loss of chlorophyll and consequent unmasking of the yellow plastid pigments. These pigments are then converted to alkali-soluble red pigments of unknown structure.[28]

General References

Zechmeister, L., Carotinoide. Springer, Berlin, 1934.

Karrer, P., and Jucker, E., Carotinoide. Birkhäuser, Basel, 1948.

Strain, H. H., Leaf Xanthophylls. Carnegie Institution of Washington, Pub. 490, 1938.

Zechmeister, L., and Cholnoky, L., Principles and Practice of Chromatography. Wiley, 1943.

Strain, H. H., Chromatographic Absorption Analysis. Interscience Publishers, 1942.

References

1. Willstätter, R., and Stoll, A., Untersuchungen ü. Chlorophyll. Springer, 1913.
2. Zechmeister, L., and Cholnoky, L., *Ber.*, **61**, 1534 (1928).
3. Karrer, P., Helfenstein, A., Wehrli, H., and Wettstein, A., *Helv. Chim. Acta*, **13**, 1084 (1930).
4. Zechmeister, L., and Sandoval, A., *J. Am. Chem. Soc.*, **68**, 197 (1946).
5. The distribution of carotenoids in plant materials is treated extensively in the works of Zechmeister, Strain, and Karrer, cited under general references.
6. Pauling, L., *Fortschr. chem. organ. Naturstoffe*, **3**, 203 (1939).
7. Zechmeister, L., *Chem. Revs.*, **34**, 267 (1944).
8. Zechmeister, L., LeRosen, A. L., Went, F. W., and Pauling, L., *Proc. Natl. Acad. Sci.*, **27**, 468 (1941).
9. Schroeder, W. A., *J. Am. Chem. Soc.*, **64**, 2510 (1942).
10. Zechmeister, L., and Pinckard, J., *J. Am. Chem. Soc.*, **69**, 1930 (1947).
11. Zechmeister, L., and Lemmon, R. M., *J. Am. Chem. Soc.*, **66**, 317 (1944).
12. Willstätter, R., and Mieg, W., *Liebigs Ann. Chem.*, **355**, 1 (1907).
13. Bonner, J., Sandoval, A., Tang, Y. W., and Zechmeister, L., *Arch. Biochem.*, **10**, 113 (1946); **21**, 455 (1949).
14. Dutton, H. J., and Manning, W. M., *Am. J. Botany*, **28**, 516 (1941).
15. Emerson, R., and Lewis, C. M., *Am. J. Botany*, **30**, 165 (1943).
16. Frey-Wyssling, A., Die Stoffausscheidungen der höheren Pflanzen. Springer, 1935.
17. Gorham, P., Unpublished results of California Institute of Technology, 1943.
18. Strain, H. H., *Plant Physiol.*, **13**, 413 (1938). Strain, H. H., *J. Am. Chem. Soc.*, **70**, 588 (1948).
19. Vogele, A. C., *Plant Physiol.*, **12**, 929 (1937).
20. Went, F. W., LeRosen, A. L., and Zechmeister, L., *Plant Physiol.*, **17**, 91 (1942).
21. Ellis, G. H., and Hamner, K. C., *J. Nutrition*, **25**, 539 (1943).
22. Sideris, C. P., and Young, H. Y., *Plant Physiol.*, **19**, 52 (1944).
23. LeRosen, A. L., Went, F. W., and Zechmeister, L., *Proc. Natl. Acad. Sci.*, **27**, 236 (1941).
24. Mangelsdorf, P., and Fraps, G., *Science*, **73**, 241 (1931).
25. Strain, H. H., Leaf Xanthophylls. Carnegie Institution of Washington, Pub. 490, 1938.
26. Rosa, J., *Proc. Am. Soc. Hort. Sci.*, **22**, 315 (1925).
27. Zechmeister, L., Béres, T., and Ujhelyi, E., *Ber.*, **68**, 1321 (1935).
28. Zechmeister, L., Carotinoide. Springer, 1934.
29. Zechmeister, L., *Ann. N. Y. Acad. Sci.*, **XLIX**, 220 (1948).

Chapter 27

RUBBER

Natural Distribution of Rubber. There are over 2000 species of plants which form rubber in greater or lesser quantity. Of these, about 500 produce rubber in sufficient quantity to have figured in the literature as rubber-bearing plants. Among these 500 species, rubber content may vary from over 20% on a dry weight of plant basis as in guayule or kok saghyz to 1% or less on a dry weight of plant basis as in many lactiferous plants, which do, however, produce latex containing a high percentage of rubber. The vast majority of rubber-producing plants contain only small amounts of rubber, concentrations of a few tenths of a per cent being usual. Thus rubber formation is a quantitative character ranging from high rubber content on the one extreme to complete absence of rubber on the other, with all possible intermediate situations.

Rubber formation is a property which is possessed by widely scattered plant species, particularly in the families *Moraceae*, *Euphorbiaceae*, *Apocynaceae*, *Asclepiadaceae*, and *Compositae*. No monocotyledenous plant, no gymnosperm, and no lower plant is known to accumulate rubber. The vast majority of rubber-forming species are tropical, and temperate zone species which accumulate sufficient rubber to be considered as potential rubber sources are few, including primarily the guayule (*Parthenium argentatum*) and *Taraxacum kok saghyz* and its allies. Gutta, closely related to rubber, is produced by a small number of tropical trees, especially *Palaquium gutta* and *Mimusops balata* as well as by temperate zone *Euonymus* and *Eucommia* species. Chicle which has been used as the base for chewing gum is a mixture of gutta and triterpenols and is produced primarily by the Central American *Achras sapota*.

Latex.[1] Rubber, in general, occurs in the plant in the form of microscopic particles suspended in a liquid serum, the whole forming a latex which is in turn contained in specialized latex cells or vessels. The milky character of latex depends simply on the presence in the serum of many suspended particles of a refractive index greatly different from that of the dispersion medium. The presence in a plant of latex does not guarantee the presence of rubber and in fact numerous latices contain mainly suspended matter other than rubber, as protein (*Ficus callosa*), waxes (*Brosimum galactodendron*), or triterpenol esters (stems of *Cryptostegia madagascariensis*, many temperate *Euphorbias* and *Compositae*).

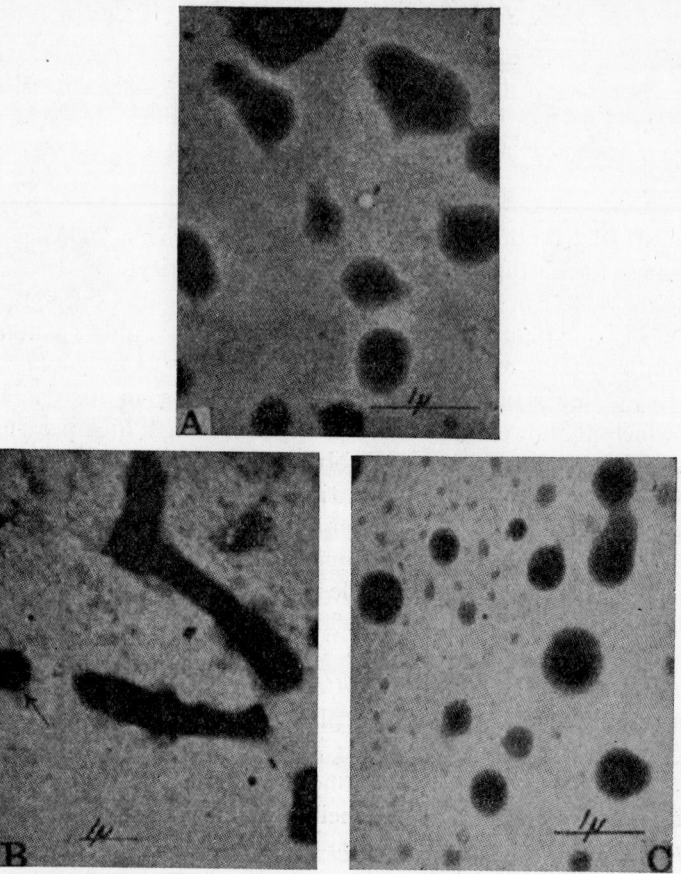

FIG. 27-1. Electron micrographs of various latex particles.[2]

A. *Hevea brasiliensis*. Large particles approximately 0.6 micron in diameter. Several pear-shaped particles show evidence of coalescence into small spheres.

B. *Manihot glaziovii*. Rod-shaped particles several microns in length and about 0.5 micron in diameter. Round particle having same diameter as rods shown by arrow. These particles can also be seen as parts of some of the rods.

C. *Taraxacum kok saghyz*. All sizes of particles, showing growth by coalescence. The smallest particles contain only a few rubber molecules.

Latex particles vary from 0.01 to 50 or more microns in diameter among various species. In any one latex, particles of varying size may be found, the smallest containing but one or a few rubber molecules, and the larger being formed by union or repeated fusion of the smaller particles.[2] Latex particles also vary in shape from species to species. Thus, in *Hevea* rubber particles are pear shaped, in *Manihot* rod shaped, and in kok saghyz globular.

The latex cell or vessel is a living cell, or cells, in which the rubber particles occur throughout the vacuole and cytoplasm. There is, in fact, no sharp differentiation between vacuole and cytoplasm, but rather a continuous intergradation of the two. The rubber molecule appears to be formed entirely in the cytoplasm of the latex vessel, and in *Hevea* and in kok saghyz the latex vessel appears to be able to carry out the entire synthesis starting with carbohydrate, since as rubber is formed reserve carbohydrates are mobilized from adjacent cells. Rubber is also produced from carbohydrate reserve by excised kok saghyz roots in storage.[3] The latex particles move freely in the latex vessel of *Hevea* or kok saghyz in a longitudinal direction. Latex particles found in leaf latex vessels are not, however, transported to stem or root, as has been shown in experiments utilizing stocks and scions having differently shaped latex particles.[4]

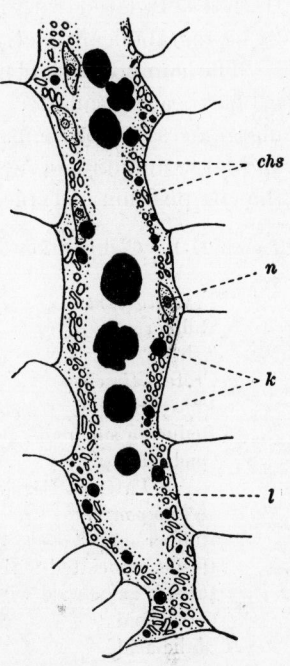

FIG. 27-2. Formation of rubber in the latex vessels of *Ficus carica*. (After Frey-Wyssling[22])

In addition to rubber and water, latex contains an assortment of other materials. *Hevea* latex contains 20–60% rubber by weight, 0.3–0.7% ash, 1–2% nitrogenous compounds, approximately 2% resins and 1–2% sugars and quebrachitol. The ash includes inorganic salts, especially KCl. The nitrogenous fraction includes 0.3–0.5% protein of which a portion is adsorbed on the surface of the latex particles and contributes to the stability of the colloidal system. Flocculation of the rubber in the latex may be achieved by denaturation of this protein layer, in particular by acids. The remainder of the latex proteins are contained in the rubber-free latex serum and comprise one principal protein and six electrophoretically distinguishable further components.[5] These proteins include a peroxidase since peroxidase activity is found in *Hevea* latex serum. Other latices also contain striking enzyme activity, peroxidase being found in latex of *Ficus*, *Cryptostegia*, and others, proteolytic activity in latex of *Ficus*, *Carica papaya*, *Asclepias*, and others, and catalase activity in still other latices. In addition to protein, *Hevea* latex contains a variety of amino

n, nucleus, forming spindles; *c*, chondriosomes (white); *k*, globules of rubber (black); *l*, latex in the vacuole with large rubber particles.

acids, all in small amounts.[6] Further trace constituents include lecithin (0.03% of the whole latex), sterols, resin acids, fatty acids, wax, reducing sugars (up to 0.35%), and quebrachitol. The latter compound which makes up 0.5–2.0% of the latex undergoes variations in concentration correlated with the physiological activity of the tree and appears to be metabolically active as a reserve food.

The composition of *Hevea* latex is better known than that of other rubber-bearing plants. *Cryptostegia* latex, like *Hevea* latex, contains inorganic salts, proteins, α-amino compounds (presumably amino acids) and large amounts of a phenol of unknown constitution.[7] Summaries of the composition of typical latices are given in Table 27-1.

TABLE 27-1. Composition of latex of *Hevea brasiliensis* and *Cryptostegia grandiflora*

Hevea latex (various sources)

Component	% of whole latex
Rubber	20–60
Ash	0.3–0.7
Protein (free)	0.3–0.5
Phospholipids	0.03
Reducing sugars	0.35
Quebrachitol	0.5–2.0

Latex of *Cryptostegia grandiflora*. (After Stewart *et al.*[7])

Component	% of latex solids	% of serum solids
Rubber	57.1	....
Resins associated with rubber	7.2	....
Protein associated with rubber	2.1	....
Citric acid	0.3	0.8
Malic acid	0.1	0.3
Serum resins	2.3	6.9
Phenol	19.5	59.0
KCl	4.6	14.0
Amino acids	3.7	11.2
Protein of serum	3.0	9.1

Chemistry of Rubber. Rubber and the related gutta are polyterpenes in which long chains of isoprene residues are linked together head to tail (Fig. 27-3). The number of residues per polyisoprene chain is of the order of 500 to 5000 or more in rubber, while in gutta the number is less, perhaps of the order of 100 residues. These long chains appear to be

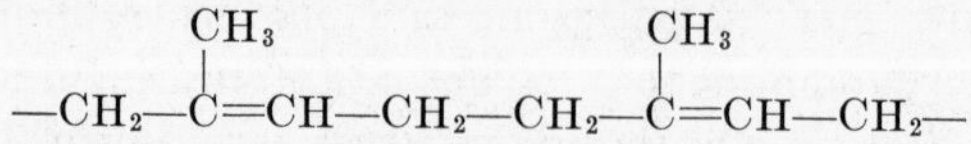

$$-CH_2-\underset{\underset{CH_3}{|}}{C}=CH-CH_2-CH_2-\underset{\underset{CH_3}{|}}{C}=CH-CH_2-$$

FIG. 27-3. Generalized form of polyisoprene chains linked together in head-to-tail union.

unbranched, except for the methyl groups, and, in fact, the vulcanization process by which rubber may be caused to harden and to set to a definite shape consists in the introduction of sulfur cross linkages between iso-prene chains. In the unstretched state rubber is amorphous, but it can be caused to crystallize by stretching or by prolonged cooling at 0°. In the stretched form rubber yields x-ray interference diagrams so that its crystalline structure may be determined just as described earlier for fibers of cellulose.[8] Such studies show that the identity period along the chain axis is 8.1Å, comprising two isoprene residues and that all the double bonds of the polyisoprene chain are in the *cis* configuration. Gutta percha unlike rubber is plastic rather than elastic at ordinary tempera-tures and is microcrystalline, i.e., consists of very small crystals. Gutta differs from rubber structurally in that the identity period along the chain axis is 4.8Å, comprising one isoprene unit, and all the double bonds of the polyisoprene chain are in the *trans* configuration. The structures of rubber and of gutta are shown in Fig. 27-4. Mixed polyisoprene chains containing both *cis* and *trans* double bonds do not occur in nature,[9] although rubber may be partially isomerized to a form containing a mixture of *cis* and *trans* bonds by treatment with iodine, a treatment described above for the isomerization of the carotenoids.

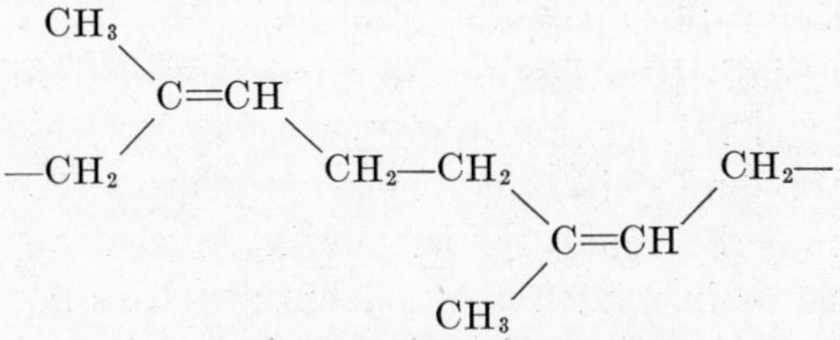

(*a*) Polyisoprene chain with double bonds in *cis* position as in rubber.

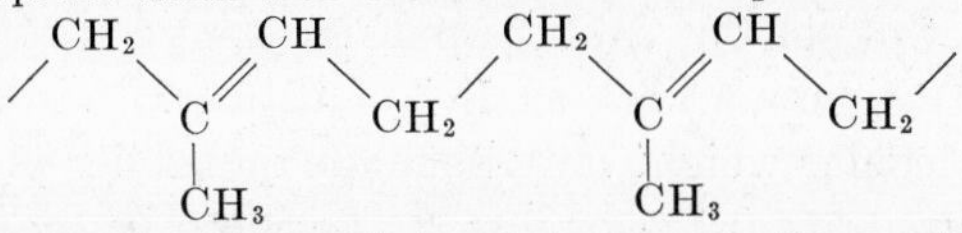

(*b*) Polyisoprene chain with double bonds in *trans* position as in gutta percha.

FIG. 27-4. The two isomers of the polyisoprene chain as they occur in rubber and gutta percha.

Rubber is amorphous at ordinary temperatures because the thermal motions of the atoms of the polyisoprene chain are continuously undoing the tendency of the chains to crystallize into orderly arrays. Thus at low temperatures where the thermal motions are decreased, rubber does indeed crystallize. These thermal motions are largely rotational and vibrational in directions at right angles to the axis of the molecule, since the forces which determine the carbon-carbon bond distances and angles are large compared to the forces which restrict transverse motions.

Rubber therefore tends to expand at right angles to the chains, and since this tendency is necessarily accompanied by tensions along each chain axis, the expansion continues until the chains are snarled up in a disorderly fashion such that the tendency for expansion is equal in all directions. When rubber is stretched, the molecules are oriented along the tension axis. The above outlined considerations cause the rubber to tend to return to its original disorderly state. This makes rubber elastic.[10] In gutta on the other hand, steric factors associated with the more stable *trans* configuration seem to minimize the magnitude of thermal motions and the material is crystalline even at ordinary temperatures. If the magnitude of the thermal motions is increased by heating above the melting point, gutta becomes elastic, although less so than rubber.

Rubber is typically soluble in benzene, petroleum ether, ether, carbon disulfide, etc., but is insoluble in acetone, methyl alcohol, etc. Rubber may in fact be purified by fractional precipitation from benzene solution by acetone or methyl alcohol, a method which has been used to obtain pure hydrocarbon of the composition $(C_5H_8)_n$ free of accompanying oxidation products and other impurities. Fractional precipitation also reveals that native rubber is made up of molecules which vary greatly in size. A great deal of attention has been paid to the determination of the molecular weight of rubber, and satisfactory use has been made of both osmotic pressure and viscosity measurements.[11] Osmotic pressure measurements may be made on rubber in benzene, chlorobenzene, or toluene solutions, or in mixtures of solvents such as benzene containing 15% of methyl alcohol.[12] This mixture is particularly to be recommended since in this solvent rubber appears to act as a nearly ideal solute even at concentrations as high as 1% of rubber. Viscosimetric determinations of molecular weight may be made on rubber provided that the concentrations used are 0.1% or less. The viscosity must be measured for at least two concentrations and the specific viscosity at infinite dilution determined by extrapolation. This quantity is the intrinsic viscosity, η_i.

$$\eta_i = \left(\frac{1}{c} \cdot \eta_c\right)_{\to \text{Infinite dilution}} \tag{1}$$

From the intrinsic viscosity, the molecular weight is obtained from relation 2. If c, the concentration of rubber in equation 1, is given in grams rubber per 100 cc. of solvent, and if the solvent is benzene at 25°C,

$$\text{Molecular weight} = K(\eta \text{ intrinsic}) \tag{2}$$

then K, the constant of equation 2, assumes the value[13] 6×10^4.

Application of these methods to fractions precipitated from a pure preparation of *Hevea* rubber shows that the fractions most soluble in acetone are those of lowest molecular weight. In the sample for which data is given in Table 27-2, molecular weights ranged from 30,000 to over 300,000, with the bulk of the rubber in the fraction of 324,000 molecular weight.[14] The same general techniques have shown that rubber of young plants or of young parts of mature plants is predominantly of low molecular weight whereas that of mature parts of mature plants is predominantly of high molecular weight. Similar data on the distribution

TABLE 27-2. Distribution of molecular sizes in fractions obtained from crepe rubber. (After Bloomfield and Farmer[14])

Comp. of pptn. mixture: Petroleum ether/acetone (by vol.)	Yield of pptd. material % of whole rubber	Mol. wt. by viscosity
50/50	$\left\{ \begin{array}{c} \\ 1.5 \\ \\ \end{array} \right\}$	30,000
52/48		60,000
60/40	15.0	216,000
62.5/37.5	70.0	324,000
85/15	12.3	Undetermined

of molecular weights have been reported for a variety of other rubbers.[15]

Biosynthesis of Rubber. The problem of the course of the biosynthesis of rubber is closely related to the problem of essential oil and of carotenoid synthesis since presumably in all these cases the same basic precursor may be involved. Indirect evidence indicating that rubber formation is in fact related to the synthesis of other terpenes has been obtained in *Cryptostegia* and in guayule. In the case of *Cryptostegia* there are two related species, *C. grandiflora* which produces latex containing rubber, and *C. madagascariensis* which produces latex containing little rubber, but which accumulates a fatty acid ester of the triterpene alcohol lupeol.[16] In hybrids of the two species rubber formation is dominant over triterpenol formation, while in the F_2 generation triterpenol formation segregates essentially as a simple recessive character. This result suggests that the two terpenes, rubber and triterpenol, may be formed from a common precursor and that in the presence of the appropriate enzyme, whose production is genetically controlled, the precursor may go to rubber, while in the presence of a second enzyme, also genetically controlled, the same precursor may go to triterpenol.

In the case of *Parthenium argentatum*, the guayule, the plant forms not only rubber, but also considerable amounts of an essential oil. This oil, of which the principal component is α-pinene,[17] is contained primarily in the leaves whereas the rubber is produced and accumulated in parenchymatous cells of the stem and roots. Leaf oil is found in high concentration during the high temperatures of spring and summer when rubber formation is at a minimum and decreases to lower concentrations during the low temperatures of fall and winter when rubber formation is proceeding actively.[18] In addition, it is known that in the guayule, rubber accumulation in the tissues of the bark and root is dependent directly on the presence of leaves. If the leaves are removed, rubber formation ceases abruptly even though the stem tissues contain much reserve carbohydrate. This behavior is in contrast to that of *Hevea* and of kok

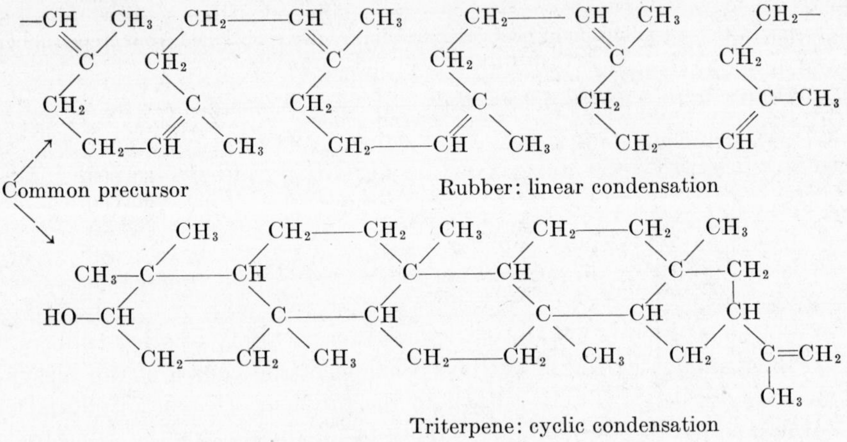

Common precursor

Rubber: linear condensation

Triterpene: cyclic condensation

FIG. 27-5. Alternative pathways for condensation of a precursor common to formation of rubber, a linear polyterpene, and lupeol, a cyclic triterpene alcohol. (After Wildman *et al.*[16])

saghyz in which the rubber-accumulating tissue is able to form rubber directly at the expense of carbohydrate. In guayule, on the contrary, factors essential to rubber formation, possibly rubber precursors, are evidently formed in the leaves and translocated to the site of rubber synthesis. It has been suggested[18] that the precursor common to lower terpenes and to rubber may be formed in the leaves and that under conditions (low temperature) favorable to rubber formation this precursor is utilized for rubber formation in the stem and roots, whereas under other conditions (high temperature), this precursor may be used for terpene synthesis in the leaves themselves.

The fact that precursor formation and rubber synthesis appear to be confined to different tissues in the guayule have made it possible to attack experimentally the nature of the rubber precursor in this plant.[19] Excised stem tissue of guayule can be grown in tissue culture on suitable synthetic nutrient medium. Under these conditions, the tissue grows rapidly, but fails to accumulate rubber. If an extract of guayule leaves is added as a supplement to the nutrient medium, the stem sections not only grow vigorously but also produce rubber. Thus, the guayule leaf contains some material essential to rubber formation by the stem. The active principle of leaf extract can be replaced by acetate, and in fact when isotope labeled acetate is added to the nutrient medium used for culture of excised guayule stems, or even of intact guayule plants, the isotopic carbon of the acetate is found to be rapidly incorporated into the rubber. Acetate does not, of course, bear any obvious relation to the isoprene structure. It appears possible, however, that acetate may be converted to β-methylcrotonic acid, a naturally occurring compound,[20] which is also capable of supporting rubber formation in excised stem sections. Beta-methylcrotonic acid may possibly be formed from acetate and acetone, the latter having been shown by isotope experiments to be also utilized in rubber formation. The path of rubber formation indicated by these experiments is shown below:

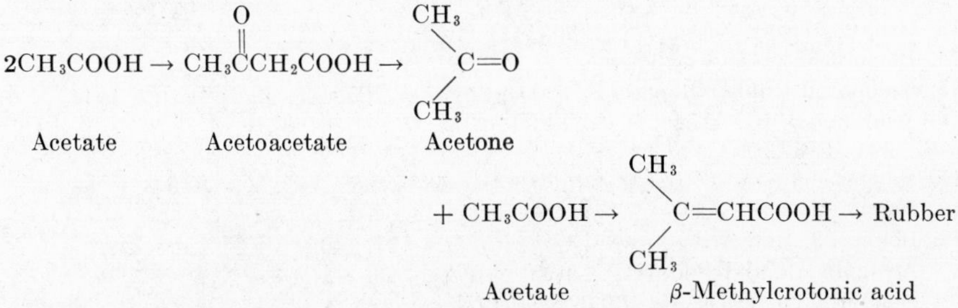

$$2CH_3COOH \rightarrow CH_3\overset{O}{\underset{\|}{C}}CH_2COOH \rightarrow \quad \overset{CH_3}{\underset{CH_3}{\diagdown\diagup}}C=O$$

Acetate   Acetoacetate   Acetone

$$+ CH_3COOH \rightarrow \quad \overset{CH_3}{\underset{CH_3}{\diagdown\diagup}}C=CHCOOH \rightarrow \text{Rubber}$$

Acetate β-Methylcrotonic acid

Possible pathway of rubber synthesis in the guayule

According to this view, the formation of rubber would constitute an analogy to that of fatty acid formation with β-methylcrotonic acid rather than acetate taking part as the repeating unit.

The Role of Rubber. It has been suggested frequently in the past that rubber may function as a reserve food. The evidence available at present does not, however, support this view. Rubber once deposited in the plant, whether in *Hevea*, in guayule, or in kok saghyz, does not appear to be again mobilized even under severe conditions of starvation such as prolonged etiolation[21] or continued defoliation. Enzymes capable of breaking down rubber are not known in higher plants, although rubber is attacked by microorganisms. It seems possible that rubber may represent a nonfunctional byproduct of cellular metabolism.

General References

Memmler, K., The Science of Rubber. Reinhold, Eng. ed., 1934.

Prokofiev, A. A., Formation of Rubber in Plants (Russian), *Bull. acad. sci. URSS*, 909 1939.

Bonner, J., and Galston, A. W., *Botan. Rev.*, **13**, 543 (1947).
Krotkov, G., *Botan. Rev.*, **11**, 417 (1945).

References

1. Moyer, L. S., *Botan. Rev.*, **3**, 522 (1937).
2. Hendricks, S. B., Wildman, S. G., and McMurdie, H., *India Rubber World*, **110**, 297 (1944).
3. McGavack, J., and Faulks, P., *Rubber Age*, **58**, 204 (1945).
4. Hauser, E. A., *Kautschuk*, **3**, 357 (1927).
   Prokofiev, A. A., *Compt. rend. acad. sci. URSS*, **44**, 162 (1944).
5. Roe, C. P., and Ewart, R. H., *J. Am. Chem. Soc.*, **64**, 2628 (1942).
6. For work on composition of *Hevea* latex see series of papers by Altman, R. F. A., in *Arch. Rubbercultuur*, Vols. 23–25, 1940–41.
7. Stewart, W. S., Bonner, J., and Hummer, R., *J. Agr. Research*, **76**, 105 (1948).
8. Bunn, C. W., *Proc. Roy. Soc. (London)*, **A180**, 40 (1942).
9. Hendricks, S. B., Wildman, S. G., and Jones, E. J., *Arch. Biochem.*, **7**, 427 (1945).
10. Treloar, L., *IRI Trans.*, **18**, 256 (1942); *Trans. Faraday Soc.*, **39**, 241 (1943).
    Bunn, C. W., *Proc. Roy. Soc. (London)*, **A180**, 82 (1942).
11. Staudinger, H., and Bondy, H. F., *Ber.*, **63**, 734 (1930).
12. Gee, G., *Trans. Faraday Soc.*, **36**, 1162 (1940).
13. Gee, G., *Trans. Faraday Soc.*, **36**, 1171 (1940).
14. Bloomfield, G. F., and Farmer, E. H., *IRI Trans.*, **16**, 69 (1940).
15. Goldenrod Rubber Report, U. S. Dept. of Agr., So. Reg. Lab., Part 2, 1944.
16. Wildman, S. G., Abegg, F. A., Elder, J. A., Hendricks, S. B., *Arch. Biochem.*, **10**, 141 (1946).
17. Haagen-Smit, A. J., and Siu, R., *J. Am. Chem. Soc.*, **66**, 2068 (1944).
18. Bonner, J., and Galston, A. W., *Botan Rev.*, **13**, 543 (1947).
19. Bonner, J., and Arreguin, B., *Arch. Biochem.*, **21**, 109 (1949).
    Arreguin, B., and Bonner, J., *Arch. Biochem.*, **26**, 178 (1950).
    Bonner, J., *J. Chem. Ed.*, **26**, 628 (1949).
20. Asahina, Y., *Arch. Pharm.*, 355, 1913.
21. Bobilliof, W., Anatomy and Physiology of *Hevea brasiliensis*. Zürich, 1930.
    Prokofiev, A. A., *Bull. acad. sci. URSS*, 589, 1940.
22. Frey-Wyssling, A., Die Stoffausscheidungen der höheren Pflanzen. Springer, 1935.

CHAPTER 28

THE ANTHOCYANINS AND ANTHOXANTHINS

The vast majority of the water-soluble blue, red, and yellow colors of the higher plants are compounds belonging to the group of anthocyanin and anthoxanthin pigments. In contrast to the carotenoids which are water insoluble and which are contained in particulate form in the cytoplasm, the anthocyanins and anthoxanthins are freely soluble in water and are present in the cell primarily in the vacuolar sap. The anthocyanins, which comprise red, violet, and blue pigments, are present in the plant as glycosides and yield on acid, alkaline, or enzymatic hydrolysis a sugar, or sugars, together with an aglycon or sugar-free residue. The aglycons which are derived in this manner from the anthocyanins are known as anthocyanidins. The anthoxanthins are also, in the main, glycosides and yield sugar-free aglycons on hydrolysis. Although relatively little is known concerning the biochemistry of the anthocyanin and anthoxanthin pigments, a great deal is known not only about their chemistry but also about the genetic control of pigment production. In fact, the knowledge of the genetic control of anthocyanin production constitutes perhaps the most detailed study in the physiological genetics of higher plants.

Chemistry of the Anthocyanidins. The anthocyanidins, whose structures were largely elucidated through the early work of Willstätter[1] and of the Robinsons,[2] are all derivatives of 2-phenylbenzopyrylium salts (flavylium salts) and, in fact, are all derivatives of polyhydroxyflavylium salts. The ring oxygen atom of this structure is known as an oxonium group and is the ionic group of the flavylium ion. The flavylium ion is commonly written in the form of its chloride since anthocyanins are ordinarily isolated as chloride salts. In the plant however the flavylium ion is undoubtedly combined with other anions, probably in a large measure with those of plant acids. The flavylium nucleus is then common to all anthocyanidins, which differ among themselves mainly in their substituents in the 2-phenyl ring. Three principal groups of anthocyanidins can be differentiated on this basis, namely the pelargonidin, cyanidin, and delphinidin pigments. In the pelargonidin pigments, one phenolic hydroxyl group is found in the 2-phenyl ring and this in the *para*

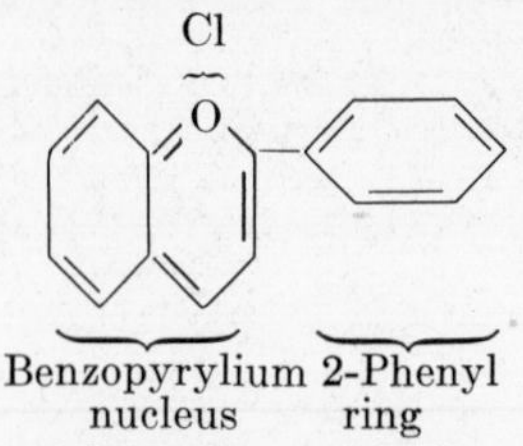

2-Phenylbenzopyrylium chloride (flavylium chloride)

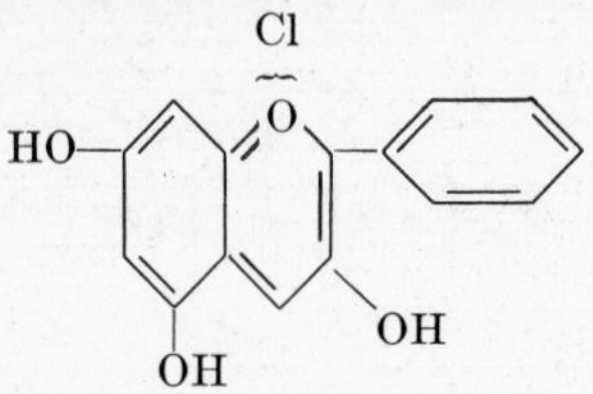

Trihydroxyflavylium chloride (parent nucleus of principal anthocyanidins)

position. The cyanidin pigments, on the contrary, contain two such phenolic hydroxyl groups, while the delphinidin pigments contain three.

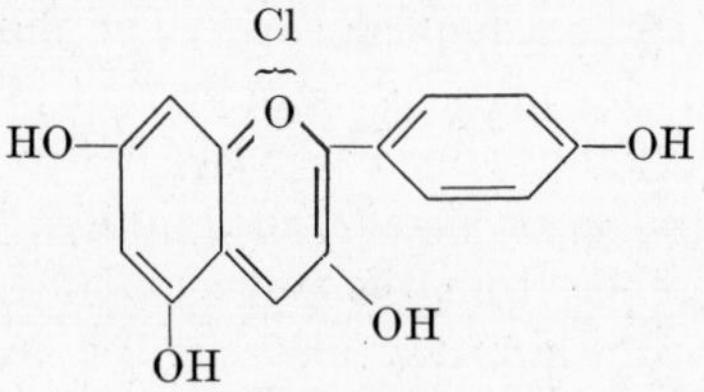

Pelargonidin

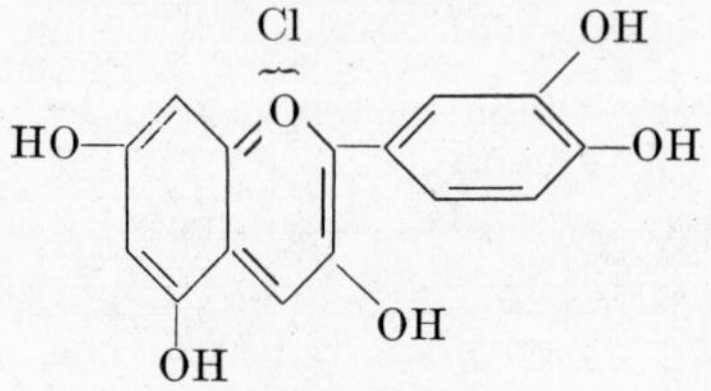

Cyanidin

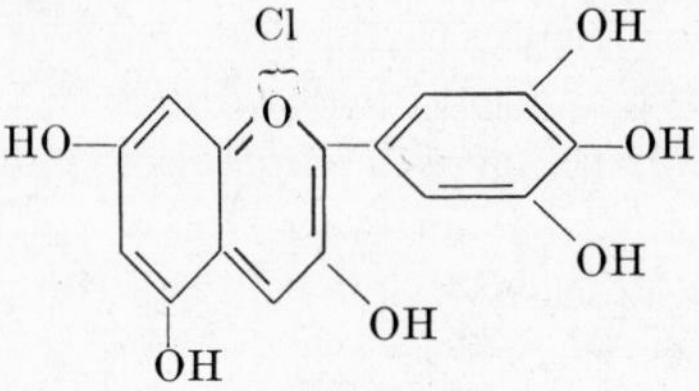

Delphinidin

A simple technique suggested by Karrer[3] for the establishment of the type of pigment involved in a particular case is based on hydrolysis of the anthocyanidin with 10% $Ba(OH)_2$ in an inert atmosphere, under which conditions the pigment is cleaved to form phloroglucinol and the appropriate phenolic acid. Thus pelargonidin yields *p*-hydroxybenzoic acid under these conditions, while cyanidin yields the diphenolic protocatechuic acid, and delphinidin yields the triphenolic gallic acid.

Further variation in the basic anthocyanidin structure is afforded by the fact that one or more of the phenolic groups may be methylated. Thus peonidin resembles cyanidin but for the methylation of the phenolic group at the *meta* position of the 2-phenyl ring. Similarly malvidin resembles delphinidin in which both *m*-phenolic groups have been

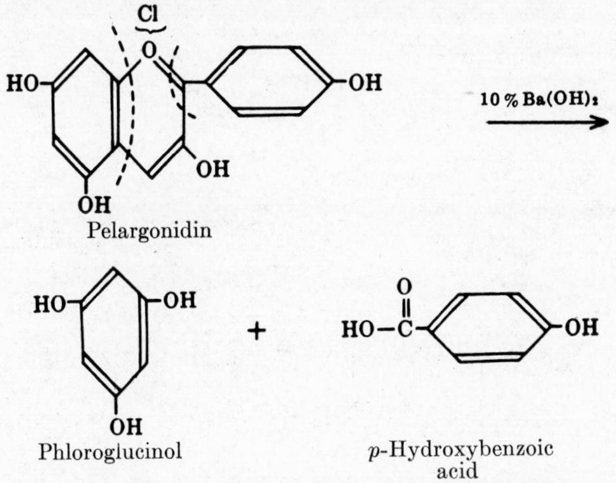

FIG. 28-1. Type reaction for establishment of substituents in anthocyanidin nucleus.

methylated. Finally in the hirsutidin group of anthocyanidins one of the phenolic groups in the benzopyrylium nucleus itself is methylated.

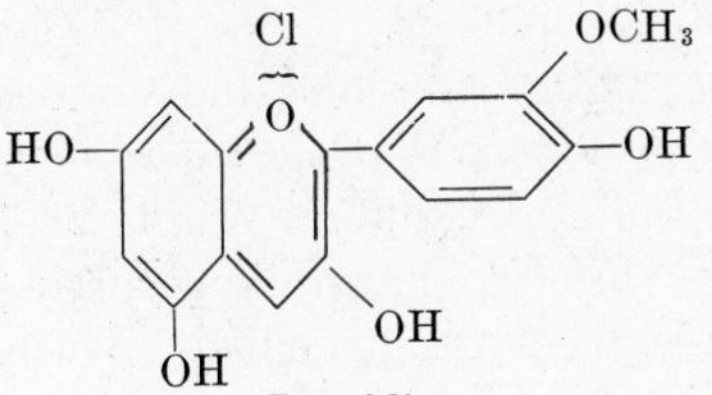

Peonidin

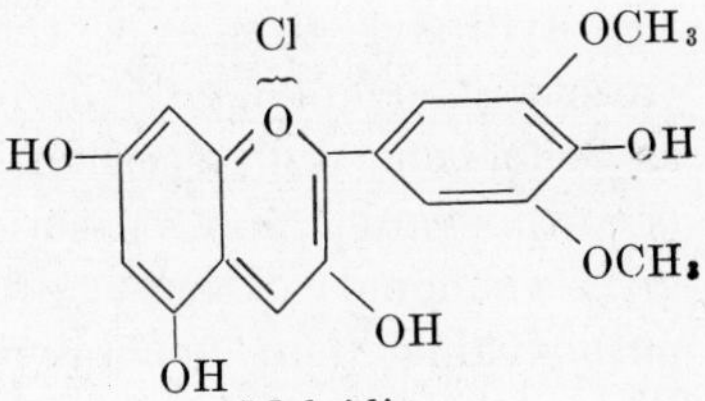

Malvidin

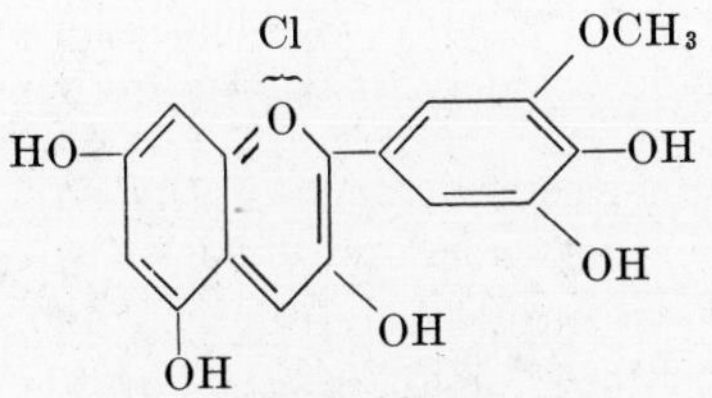

Petunidin

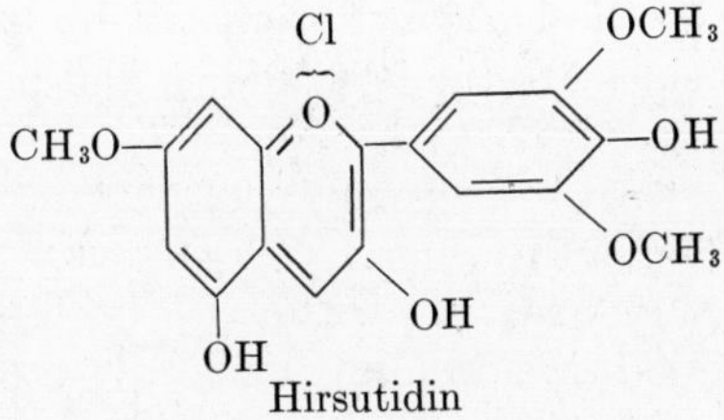

Hirsutidin

The anthocyanidins are indicators and change color according to the hydrogen ion concentration of the solution, the acid salts being in general red and the salts formed in basic solution blue. At intermediate acidities the free violet or purple color base exists as such in solution. The structural changes which cause these color alterations probably include the transformation of the 2-phenyl ring to a quinoid form as shown below.

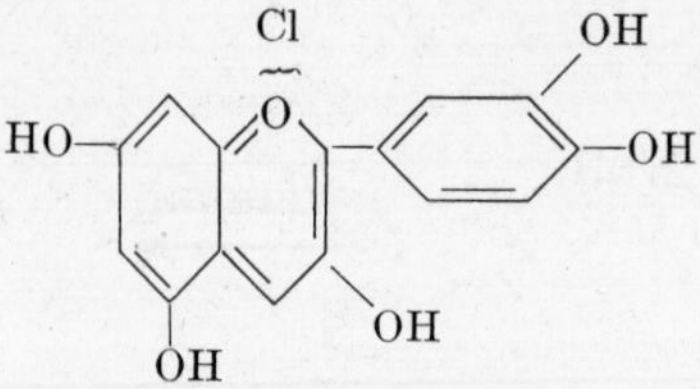

Oxonium salt of cyanidin
(acid solution)
Red

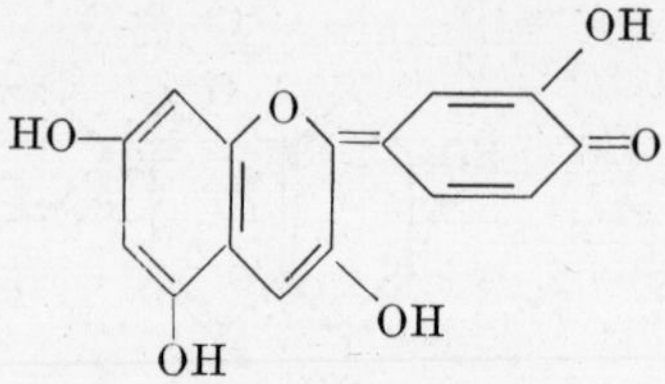

Free color base of cyanidin
(neutral solution)
Violet

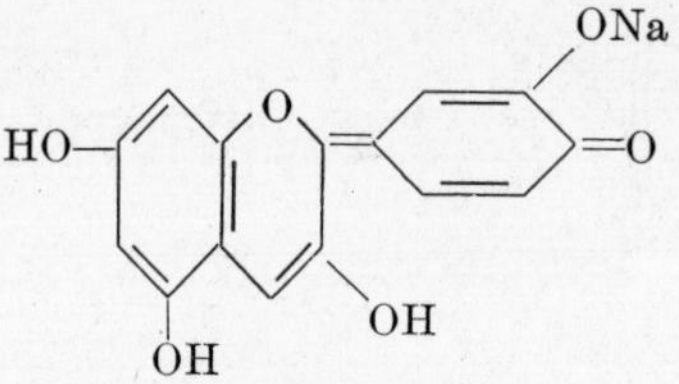

Salt of color base of cyanidin
(alkaline solution)
Blue

Chemistry of the Anthocyanins. The anthocyanins are, as stated above, glycosides of the anthocyanidins, the sugars being combined to the hydroxyl groups of the benzopyrylium nucleus. Since a great variety of sugars may be involved and since both mono- and diglycosides as well as mixed diglycosides may occur, it is evident that a great variety of anthocyanins may be derived from each anthocyanidin. Table 28-1 gives a summary of the sugars occurring in a variety of naturally occurring anthocyanins. In general, the sugar residues are attached to the hydroxyl group at position 3 (see numbering system below) or more often to the hydroxyl groups at both position 3 and position 5. The diglucosides are the best known and possibly the most widely distributed of the

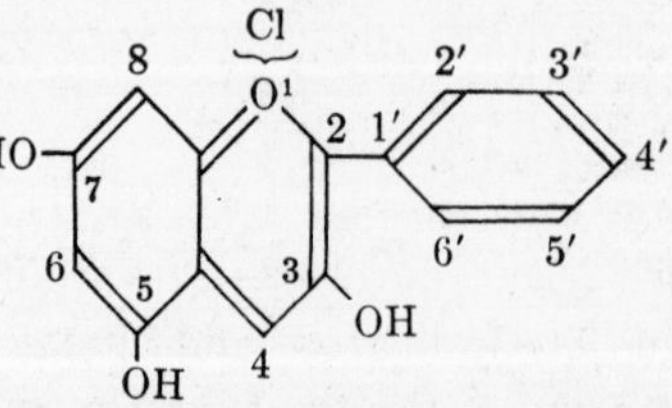

Numbering system for the anthocyanin nucleus

anthocyanins, although, as can be seen in Table 28-1, both the mono-glucosides and glycosides involving sugars other than glucose are also of common occurrence.

Qualitative tests enable a partial identification of the nature of an unknown anthocyanin even without isolation and structure determination on the pure material.[4] Among the most useful of these tests are behavior toward alkaline oxidation, color produced with ferric chloride in amyl alcohol, and partitions between various solvents. For the alkaline oxidation test, a dilute solution of the pigment is shaken in air with dilute NaOH. Petunidin and delphinidin pigments are decolorized whereas pigments of other groups are more stable. For the ferric chloride

TABLE 28-1. Glycosidal nature of naturally occurring anthocyanin pigments

Anthocyanidin	Anthocyanin	Glycosidal nature	Typical occurrence
Pelargonidin	Pelargonin	Diglucoside	Scarlet pelargonium
	Callistephin	Monoglucoside	Callistephus
	Fragasin	Monogalactoside	Fragaria vesca
Cyanidin	Cyanin	Diglucoside	Red rose, blue corn flower (Centaurea)
	Keracyanin	Rhamnoglucoside	Black cherry
	Idaein	Monogalactoside	Cranberry
	Chrysanthemin	Monoglucoside	Scarlet winter aster
Delphinidin	Delphinin	Diglucoside	Delphinium
	Violanin	Rhamnoglucoside	Viola tricolor
	Gentianin	Monoglucoside (also contains p-OH-cinnamic acid)	Gentiana
Peonidin	Peonin	Diglucoside	Red peony
Malvidin	Malvin	Diglucoside	Primula viscosa, Epilobium
	Oenin	Monoglucoside	Blue grape
Hirsutidin	Hirsutin	Diglucoside	Primula hirsuta

test, the pigment is extracted into amyl alcohol in the presence of acetate and ferric ions. Cyanidin gives a marked blue color, delphinidin a less intense color; members of other groups give less color or none at all. The distribution tests may be carried out by partition between dilute aqueous HCl (10%) and an anisole phase containing 20% of an ethyl isoamyl ether-picric acid (50%) mixture. Malvidin, pelargonidin, and peonidin are wholly extracted into the nonaqueous layer, but delphinidin is not extracted and cyanidin only partially so. Still another solvent partition test is based on distribution between aqueous HCl (10%) and a mixture of 80% toluene and 20% cyclohexanol. Pelargonidin and peonidin are extracted into the organic solvent layer while delphinidin, malvidin, and cyanidin are essentially not extracted. A summary of these and other properties of the anthocyanidins is given in Table 28-2.

These classical methods for the qualitative identification of anthocyanin and related pigments will undoubtedly be supplemented or replaced by methods based on the separation of pigments by chromatography on paper.[17] In this method, the anthocyanin-containing solution is allowed to dry on one end of a strip of paper. The edge of the paper is then dipped in a suitable solvent (in this case ethyl acetate, phenol, or others) and as the solvent boundary moves along the paper, each pigment is carried along its own characteristic distance. The ratio of distances traveled by the solvent and solute, or Rf value, is itself frequently a useful criterion in judging the identity of a particular pigment, although the usual qualitative tests may also be applied to the separated compounds.

TABLE 28-2. Qualitative differentiation of anthocyanidins

	Pelar-gonidin	Cyanidin	Delphi-nidin	Peonidin	Malvidin
Color in H^+ soln.	Red	Violet-red	Blue-red	Violet-red	Violet-red
Color in OH^- soln.	Blue	Blue	Blue	Blue	Green-blue
Stability toward NaOH in air	Stable	Stable	Destroyed	Stable	Stable
Ferric chloride test	None	Deep blue	Blue	Trace of color	None
Extraction into anisole-ether	Completely extd.	Partially extd.	Not extd.	Completely extd.	Completely extd.
Extraction into toluene-cyclo-hexanol	Extd.	Slightly extd.	Not extd.	Extd.	Not extd.

Factors Affecting Color. The chemical structure of the anthocyanidin affects its color markedly. Thus pelargonidin which contains one phenolic hydroxyl group in the 2-phenyl ring is redder than cyanidin or delphinidin which contain more hydroxyl groups. Methylation on the other hand causes increasing redness, with malvidin (two methoxyl groups) being redder than delphinidin (no methoxyl groups). In general also, the diglycosides tend to be bluer than the corresponding monoglycosides. In addition to these effects, however, other substances present in the vacuolar sap may cause intensification of the anthocyanin color, anthoxanthins and tannins being common agents responsible for such copigmentation effects. Acidity of the cell also is of importance in regulating anthocyanin color, and in flowers which show a diurnal change in color, changes in pH may be of importance. The freshly opened flowers of *Ipomoea leari*, for example, are pink and have a pH of approximately 6.0. During the course of the day the color changes to blue, and the pH of the cell contents may simultaneously go as high as 7.8.[5] In *Hydrangea*, a classical case, red flowers also tend to be somewhat more acid than blue ones.

Factors Affecting Anthocyanin Production. Temperature is of importance in regulation of anthocyanin production. In seedlings of red cabbage, for example, production of anthocyanin was found to be maximal over the temperature range 20°–30°C and to be markedly decreased at 10°C.[6] More specific effects of temperature have been observed with developing buds, in which treatment during a particular critical period may modify the anthocyanin content of the flower when the latter is subsequently allowed to open at ordinary temperatures. Thus in the yellow-flowered *Dahlia variabilis*, production of red flowers is induced by treating the buds at 30° for a period of about three weeks ending ten days before full opening of the flower.[7] In other species, as *Calceolaria*, high

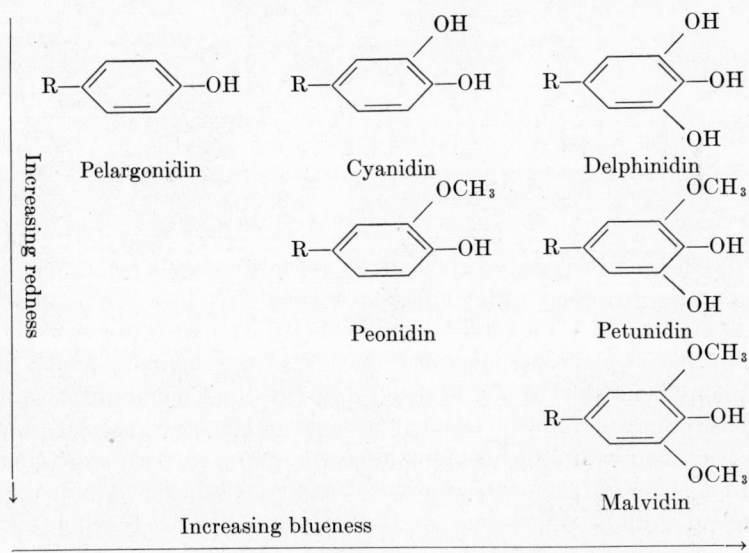

FIG. 28-2. Effect of structure on color of anthocyanidins.

temperature treatment suppresses anthocyanin production in the developing flower.[8]

In many cases anthocyanin production is independent of light and takes place, for example, in etiolated seedlings of red cabbage, as well as in roots of many plants. In other cases, however, including fruits of peaches, apples, pears, and apricots, anthocyanin production is favored by light, and full pigmentation occurs only in the light.[9] This effect is specifically brought about by radiation in the wavelength range of 3600–4500 Å, indicating that pigments other than chlorophyll are involved in the light-absorbing part of the reaction.

Many nutrient deficiencies, including particularly those of nitrogen and phosphorus, are associated with increased anthocyanin production, and in fact extensive anthocyanin production is often a typical symptom of deficiency for these elements. The exact manner in which this effect is brought about is unknown, although it is commonly ascribed to the high carbohydrate status of nitrogen- and phosphorus-deficient plants. Direct experiments fail, however, to reveal any general parallel between sugar content and anthocyanin content.[6]

Chemistry of the Anthoxanthins. The anthoxanthins include the widely distributed flavone and flavonol pigments, both of which are derivatives of 2-phenylbenzopyrone.

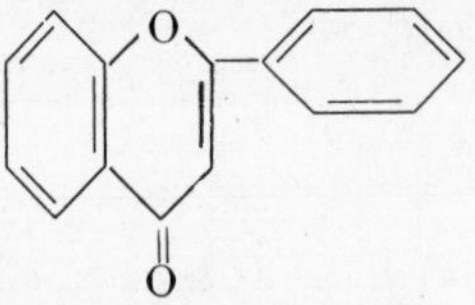

2-Phenylbenzopyrone
(flavone)

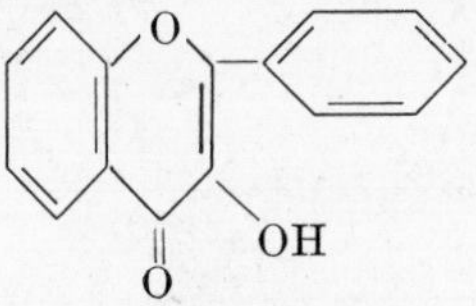

3-Hydroxyflavone
(flavonol)

The flavones and flavonols, which are yellow or orange in color, may occur either in the free form or in combination as glycosides. They are found in flowers, fruits, leaves, bark, and roots, and appear to be of general distribution in plant tissues. The nucleus of the flavone pigment series is 2-phenylbenzopyrone and is then related to the general anthocyanidin nucleus, differing from the latter in absence of the hydroxyl group at the 3 position and containing a quinoid oxygen group at position 4. In the flavonol series of pigments, on the contrary, both the 3-hydroxy and 4-quinoid groups are present in the parent nucleus. Flavone itself is found in nature as the yellow deposit on stems, leaves, and seed capsules of *Primula* but, in general, the naturally occurring flavones contain phenolic hydroxyl groups in the 5 and 7 positions as do the anthocyanins. Additional hydroxyl groups may frequently occur in other positions, especially in the 3' and 4' positions as in the anthocyanins. A series of typical flavone pigments are listed in Table 28-3, which also includes two examples of the flavanone pigments in which the double bond of the chromene ring is reduced with formation of a chroman.

The flavonol pigments combine the general structure of the flavones with an additional hydroxyl group in the 3 position of the chromene nucleus as in the anthocyanidins. The structures of the flavonols found in nature are often otherwise homologous with the structures of known flavone pigments, as is noted for several cases in Table 28-4. Flavonols, like flavones, occur in nature primarily as glycosides and the same individual flavonol may in fact be found combined with quite different sugar residues in different species. This is strikingly the case for the flavonols kaempferol and quercitin, which are included in Table 28-4. Dihydro-flavonols or flavanols are likewise known in nature, as for example the compound fustin which occurs in combination as one of the coloring matters of *Quebracho* wood.

TABLE 28-3. Typical flavone and flavanone pigments

Class	Pigment	Structure	Occurrence
Flavone	Chrysin	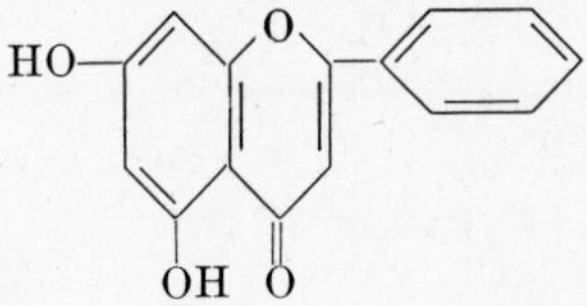	Buds of *Populus*
Flavone	Apigenin	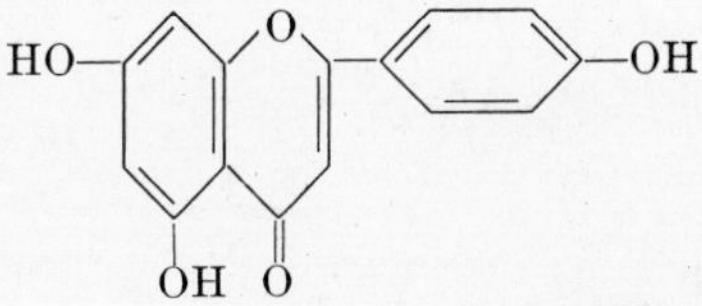	Yellow *Dahlia*. As 7-glucoside in flowers of *Cosmos*. As di-glycoside with glucose and apiose in parsley.
Flavone	Luteolin	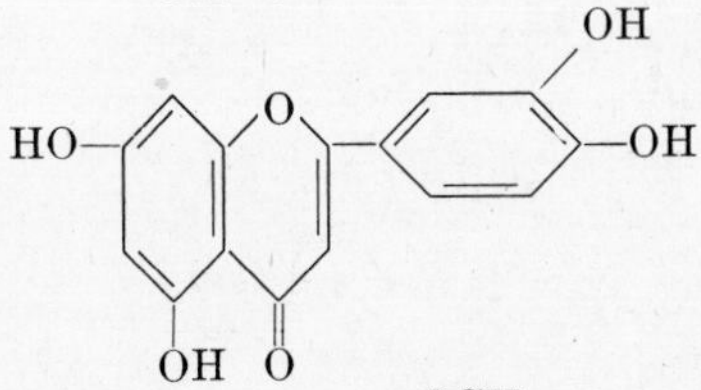	*Reseda luteola Genista tinctoria*.
Flavanone	Citronetin	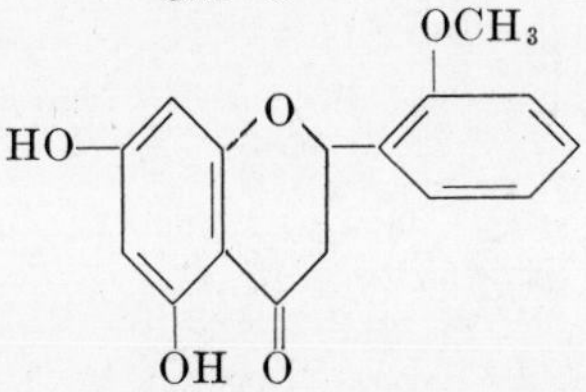	Peel of citrus fruit as glycoside.
Flavanone	Naringenin	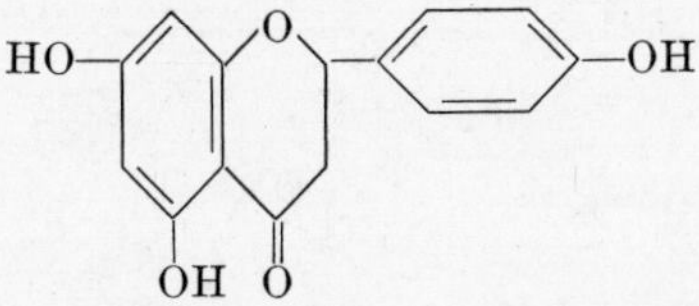	Peel of citrus fruit as rhamnoside.

Résumé of Structural Relations of Anthocyanin and Related Pigments. We have seen that there are in nature a variety of types of pigments related to the anthocyanins. Each general type consists of numerous individual pigments which differ in number and arrangement of phenolic hydroxyl groups and in nature of the sugar components with which the aglycon is combined. The principal types of pigments of the

TABLE 28-4.　Typical flavonol and flavanol pigments

Class	Pigment	Structure	Occurrence
Flavonol	Galangin (3-OH chrysin)	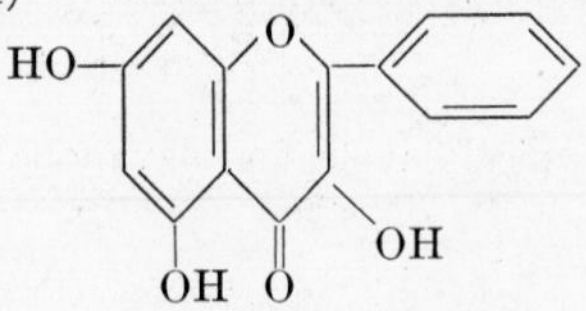	Rhizomes of *Alpinia*.
Flavonol	Kaempferol (3-OH apigenin)	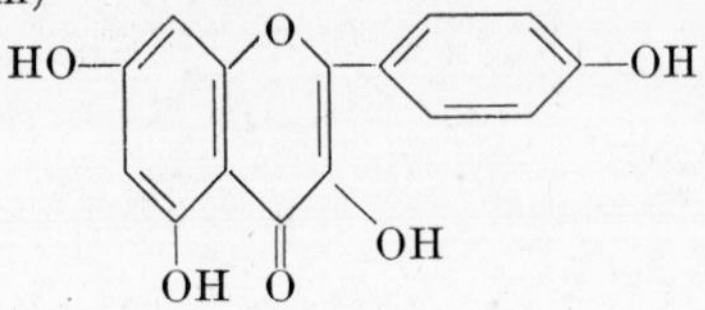	Flowers of *Delphinium*; 3-robinoside in *Robinia*, 3-rhamnoside in *Rosa* and *Indigofera*; diglucoside in *Cassia*.
Flavonol	Quercitin (3-OH luteolin)	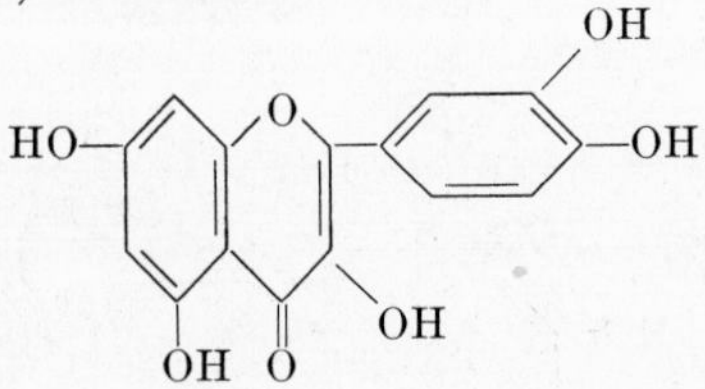	3-Rhamnoside in bark of *Quercus*; 3-glucoside in *Zea*; 3-galactoside in apples.
Flavonol	Myricetin	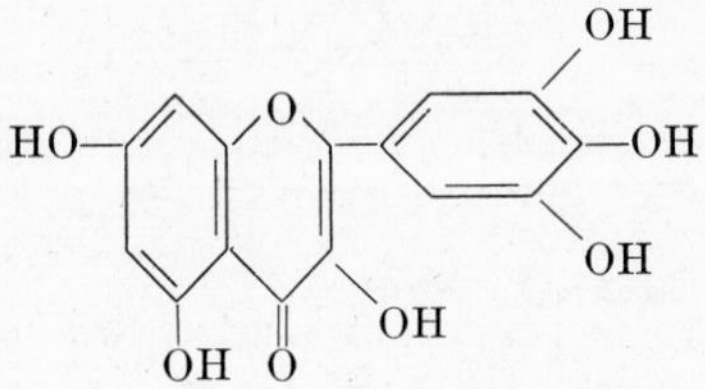	3-Rhamnoside in *Myricaceae* and *Rhus*.
Flavanol	Fustin	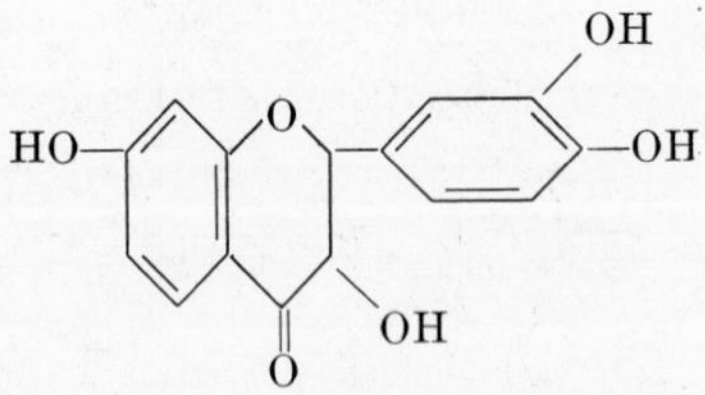	Glycoside in wood of *Quebracho colorado*.

anthocyanin and anthoxanthin series are, however, characterized by differences in nuclear structure. These differences are summarized in Fig. 28-3 for the anthocyanidin, flavone, flavanone, flavonol, flavanol, and isoflavone series.

The catechins, colorless substances of wide occurrence in woods and leaves, are related to the anthocyanins as reduction products. Thus

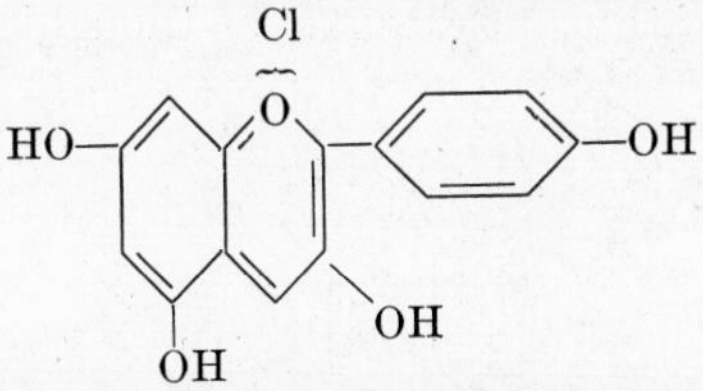

Anthocyanidin (pelargonidin)

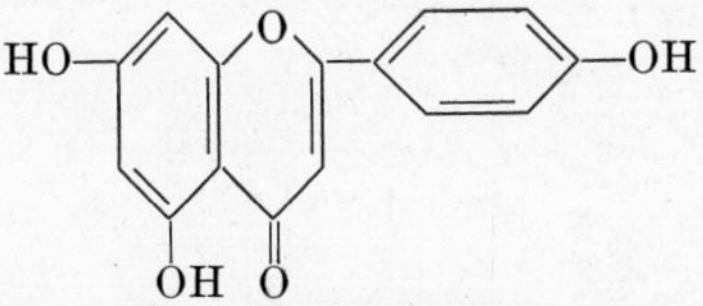

Flavone (apigenin)

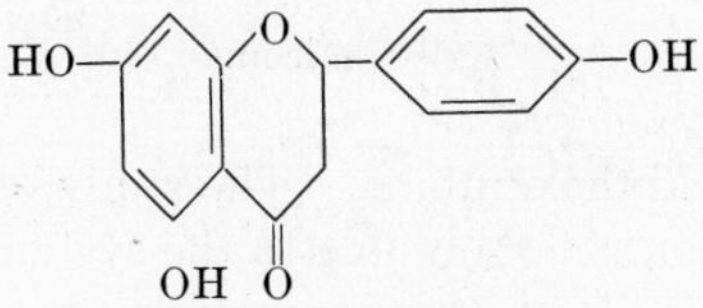

Flavanone (naringenin)

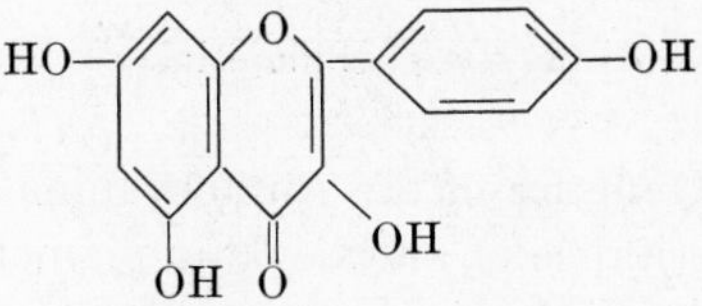

Flavonol (kaempferol)

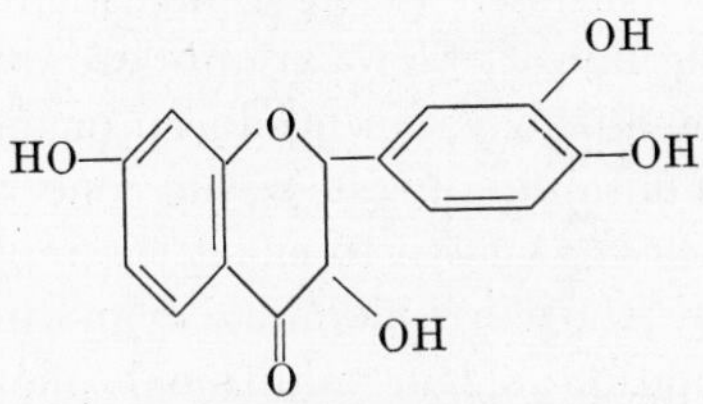

Flavanol
(fustin; note that characteristic
structural change is that on
pyrane ring)

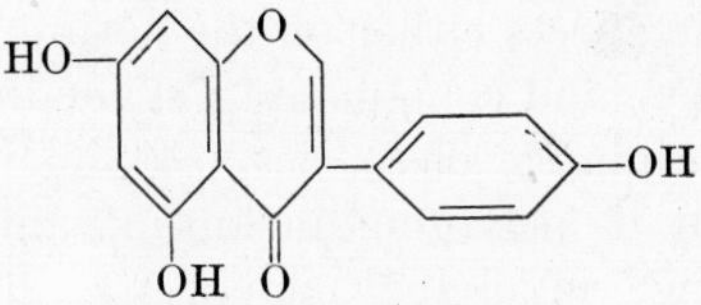

Isoflavone (genistein)

FIG. 28-3. Structural relationships among the principal groups of pigments related
to the anthocyanins.

epicatechin, which may be regarded as a reduction product of cyanidin, is found particularly in *Acacia* wood, while tea catechin of tea leaves is similarly related to delphinidin.

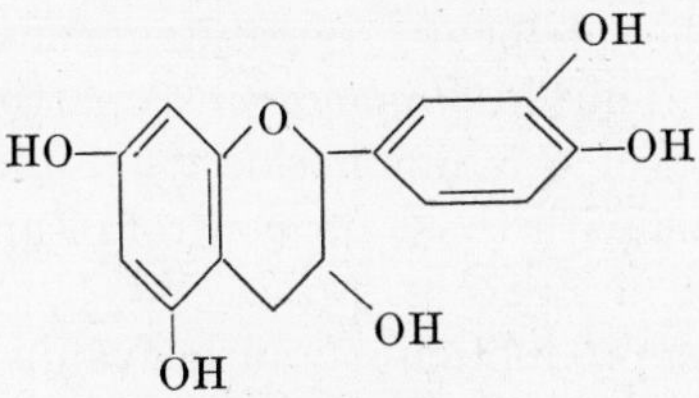

Epicatechin

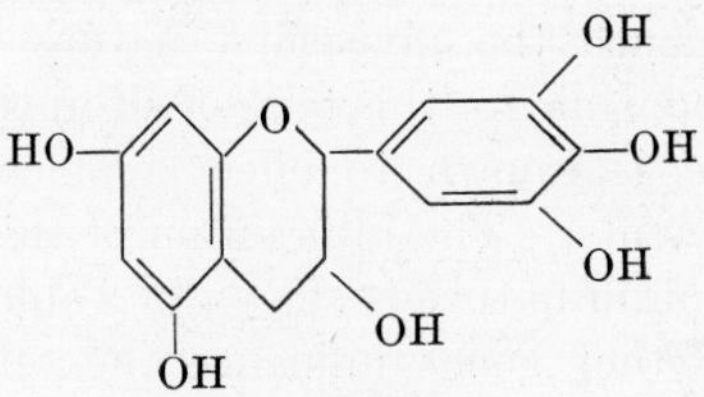

Tea catechin

Still another group of colorless compounds closely related to the anthoxanthin pigments is represented by phloretin, in which the pyrone ring has been opened at the oxygen atom. Phloretin, which occurs in barks as the glycoside phloridrin, is in turn a direct derivative of chalcone, to which indeed all anthocyanins and anthoxanthins may be thought of as related.

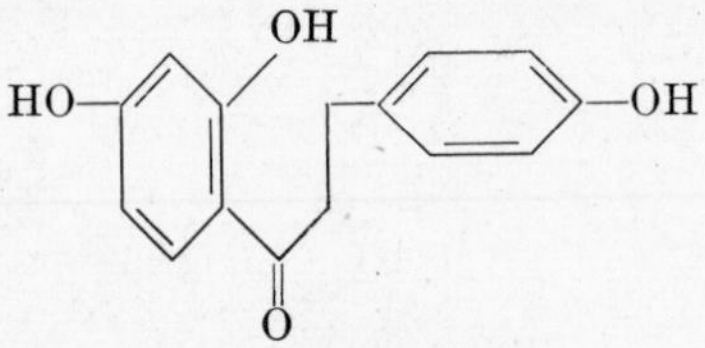

Phloretin

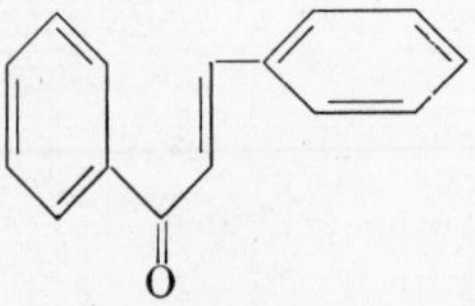

Chalcone

Genetics of the Anthocyanins and Anthoxanthins. The study of the inheritance of anthocyanin pigmentation is largely due to the systematic work of Scott-Moncrieff and associates with flower colors[10] although earlier important work was done by Onslow.[11] These investigations have shown that many of the heritable flower color variants are due to genic effects either on the nature of the anthocyanin produced or on the acidity and copigmentation conditions obtaining in the tissue concerned. In a number of species, particular genes are known to effect the replacement of pelargonidin anthocyanins by the corresponding cyanidins, a change involving merely the introduction of an additional phenolic hydroxyl group, the remainder of the molecule, including its glycosidal arrangements, being left unaffected. These instances are summarized in Table 28-5. In all instances, the allele for production of the cyanidin is the dominant one. In *Primula*, on the other hand, the replacement of pelargonidin by the dimethoxylated malvidin is effected by a single dominant gene. The same conversion in *Lathyrus* requires the intervention of two genes, one being responsible for replacement of pelargonidin by peonidin and the second for the replacement of peonidin by malvidin.

In none of the instances cited in Table 28-5 does the gene concerned influence the glycosidal nature of the pigment. Separate genes are responsible for this type of alteration. Thus in *Streptocarpus* a particular gene is known to effect replacement of 3,5-dimonoside by 3-bioside pigments. The interaction of this gene with those for control of anthocyanidin nature is shown in Table 28-6.

Other genes influence the relative production of anthocyanin and anthoxanthin pigments. In *Lathyrus*, recessive genes c and r produce anthoxanthin at the expense of anthocyanin, which is only produced in the presence of the dominant alleles C and R.[12] In leaves of *Zea mays* isoquercitin, an anthoxanthin, is produced in the presence of the recessive gene a, while anthocyanin is produced only in the presence of the dominant allele A. In still other cases the production of anthocyanin may be

TABLE 28-5. Genetic control of anthocyanin structure. (After Scott-Moncrieff[10])

Organism	Pigment formed by homozygous recessive form	Pigment formed by dominant form	Structural change effected
Papaver	Pelargonidin-3-bioside	Cyanidin-3-bioside	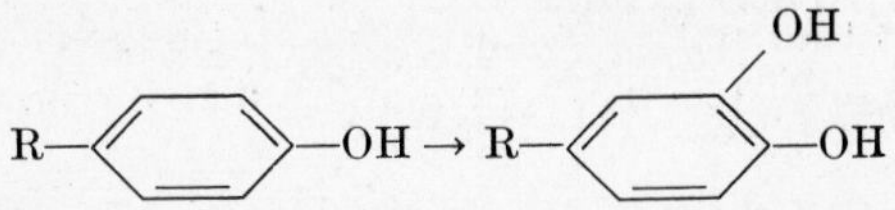 $R{-}C_6H_4{-}OH \rightarrow R{-}C_6H_3(OH){-}OH$
Tropaeolum	Pelargonidin-3-bioside	Cyanidin-3-bioside	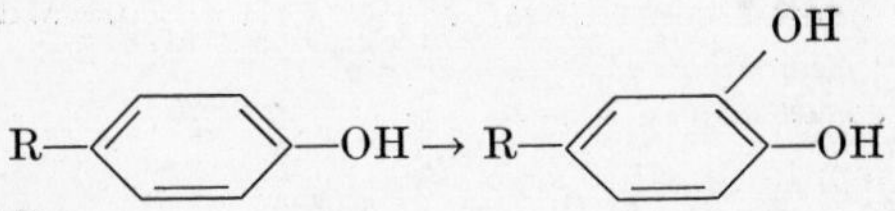 $R{-}C_6H_4{-}OH \rightarrow R{-}C_6H_3(OH){-}OH$
Antirrhinum	Pelargonidin-3-rhamnoside	Cyanidin-3-rhamnoside	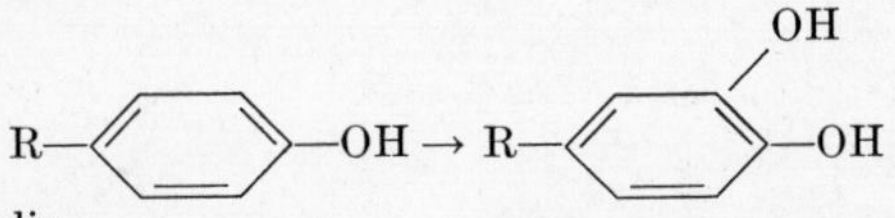 $R{-}C_6H_4{-}OH \rightarrow R{-}C_6H_3(OH){-}OH$
Cheiranthus	Pelargonidin-3,5-dimonoside	Cyanidin-3,5-dimonoside	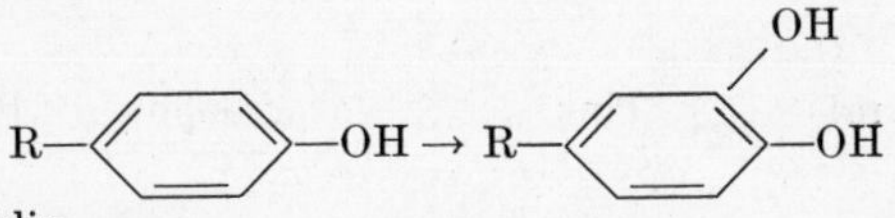 $R{-}C_6H_4{-}OH \rightarrow R{-}C_6H_3(OH){-}OH$
Rosa	Pelargonidin-3,5-dimonoside	Cyanidin-3,5-dimonoside	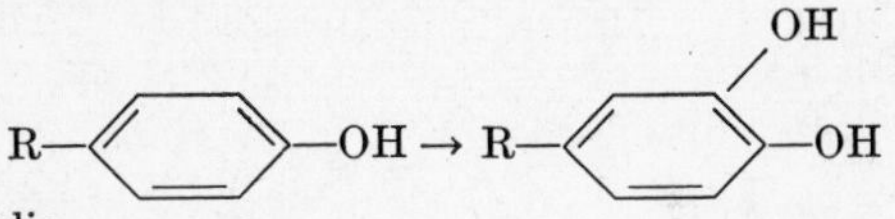 $R{-}C_6H_4{-}OH \rightarrow R{-}C_6H_3(OH){-}OH$
Primula	Pelargonidin-3-monoside	Malvidin-3-monoside	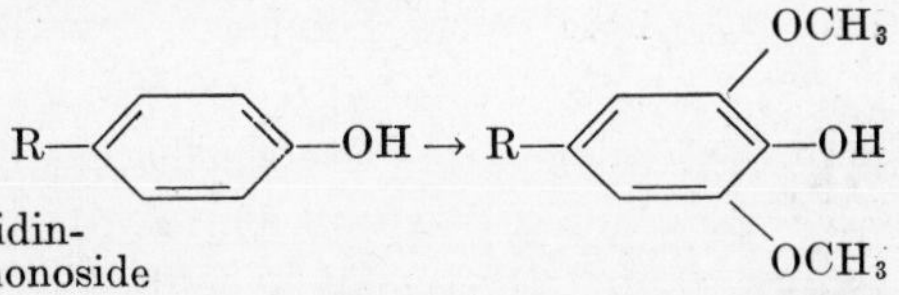 $R{-}C_6H_4{-}OH \rightarrow R{-}C_6H_2(OCH_3)(OCH_3){-}OH$
Pharbitis	Pelargonidin-3,5-dimonoside	Peonidin-3,5-dimonoside	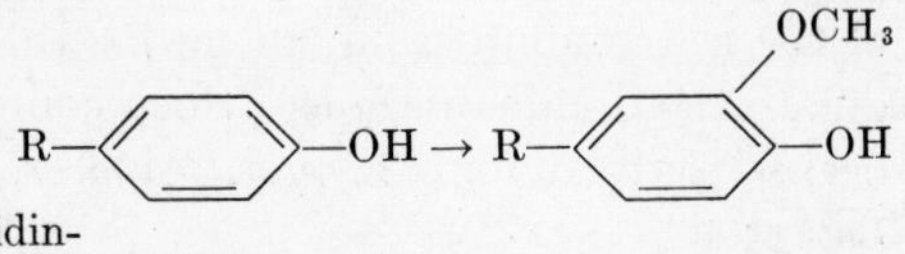 $R{-}C_6H_4{-}OH \rightarrow R{-}C_6H_3(OCH_3){-}OH$
Lathyrus	Pelargonidin dimonoside	Peonidin dimonoside	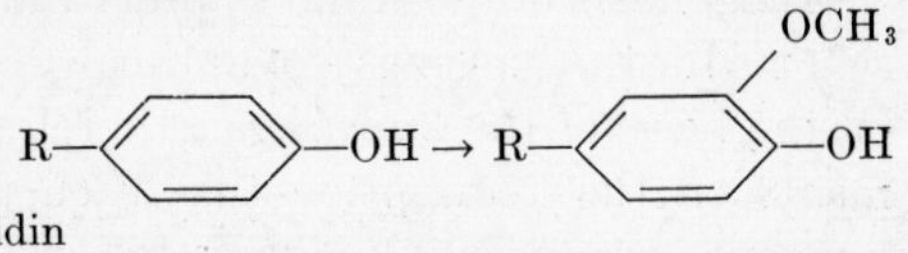 $R{-}C_6H_4{-}OH \rightarrow R{-}C_6H_3(OCH_3){-}OH$
Lathyrus	Peonidin dimonoside	Malvidin dimonoside	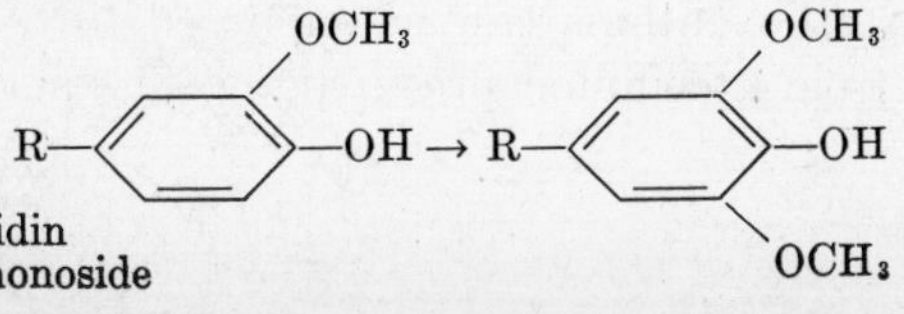 $R{-}C_6H_3(OCH_3){-}OH \rightarrow R{-}C_6H_2(OCH_3)(OCH_3){-}OH$

recessive to anthoxanthin production, as with certain genes of *Lathyrus*.[12] This reciprocal relation suggests that anthocyanin and anthoxanthin may be formed from a common precursor which can be directed into one or the other pathway according to the genetic situation.

A last type of genic effect on flower color is that in which the *p*H

TABLE 28-6. Pigment composition as affected by genetic constitution in a hybrid *Streptocarpus*. The cross involved is *S. rexii* (genetic constitution ROD) by *S. dunnii* (genetic constitution rod). (After Scott-Moncrieff[10])

Genetic constitution	Color	Pigment	Structure	Nature of glycoside
ROD	Blue	Malvidin	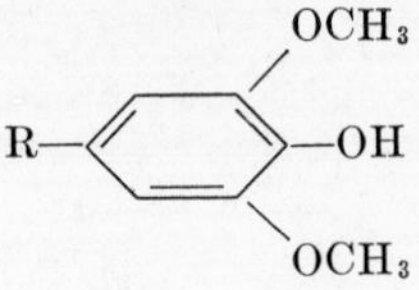 R—⟨ ⟩—OH (OCH_3, OCH_3)	3,5-Dimonoside
RoD	Magenta	Peonidin	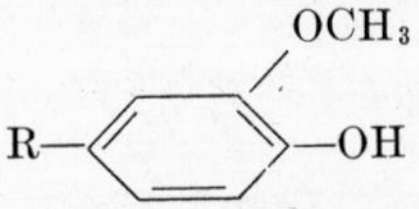 R—⟨ ⟩—OH (OCH_3)	3,5-Dimonoside
roD	Pink	Pelargonidin	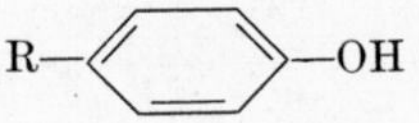 R—⟨ ⟩—OH	3,5-Dimonoside
ROd	Mauve	Malvidin	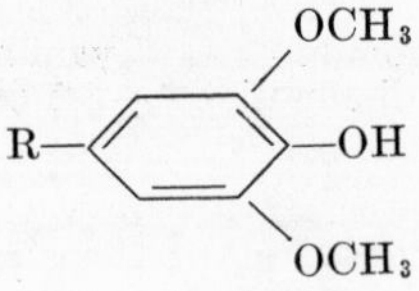 R—⟨ ⟩—OH (OCH_3, OCH_3)	3-Bioside
Rod	Rose	Peonidin	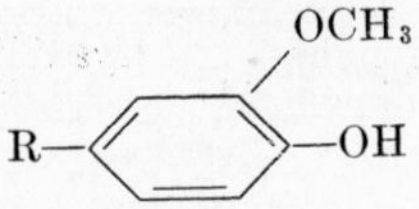 R—⟨ ⟩—OH (OCH_3)	3-Bioside
rod	Salmon	Pelargonidin	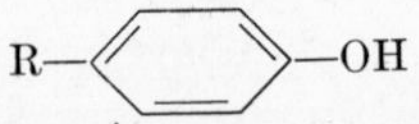 R—⟨ ⟩—OH	3-Bioside

of the cell contents is influenced. In all instances, the more acid *p*H is due to the dominant gene. Such control of cellular *p*H of the flower has been described for *Primula*, *Papaver*, and *Tropaeolum*[13] as well as for *Lathyrus*.[12]

Biogenesis of the Anthocyanins and Anthoxanthins. A purely formal scheme for the biogenesis of anthocyanins and anthoxanthins can be made on the basis of the known genetic relationships between the various pigments on the assumption that each gene controls a particular step in pigment production. The relationships between anthocyanins and anthoxanthins can be supposed to be owing to the derivation of both from a common precursor. Thus in *Lathyrus* genes *c* and *r* both block

production of anthocyanin, and anthoxanthin accumulates, while genes
m and *k* block production of anthoxanthin so that anthocyanin accumulates.   In other species, as *Antirrhinum* and *Pharbitis*, there are recessive
white forms in which the recessive gene blocks formation of both anthocyanin and anthoxanthin.   In *Pharbitis* the gene *Ca* is needed for even
small amounts of pigment production, and the gene *C* is also necessary
if full production is to be achieved.   Absence of these two genes blocks
formation of the common precursor.   As we have seen already, a further
series of genes affects the final disposition of glycosidic groups in the
anthocyanin itself.   These relations are summarized in Fig. 28-4.

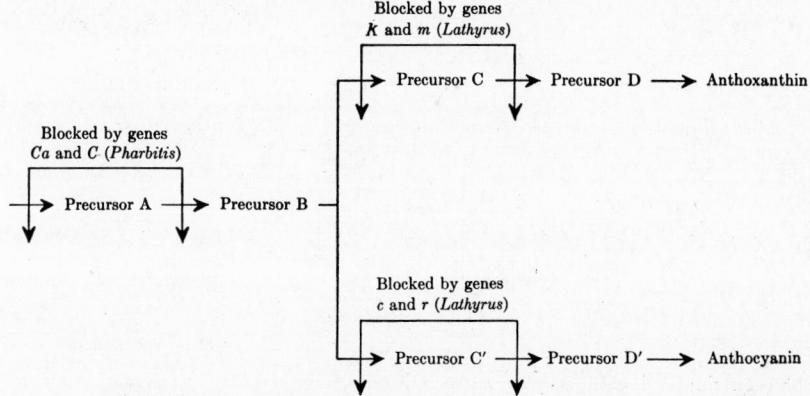

FIG. 28-4.   Diagrammatic representation of the course of anthocyanin and anthoxanthin formation on the basis of genetic evidence.

The nature of the common precursor of anthocyanins and anthoxanthins has been approached by genetic means in the case of flowers of
cotton (*Gossypium*) by Stephens.[14]   Here a red spot at the base of the
petal, due to a glycoside of cyanidin, is formed only in the presence of
dominant genes *G* and *S*.   Plants of the genotype *gs* accumulate a
quercetin glycoside, whereas plants of the genotype *Gs* accumulate an
essentially colorless substance which may be converted to cyanidin by
reduction *in vitro*.   A similar colorless substance is known to occur as an
intermediate in the reduction of quercetin to cyanidin *in vitro*.[15]   This
compound may then represent the common intermediate, capable of
yielding either cyanidin or quercetin.

The steps in the biosynthesis of the general anthocyanidin-antho-
xanthin structure are still obscure, since but little work of a biochemical
nature has been done on the problem.   The best-known suggestion is
that of Robinson,[16] who proposes on organic chemical, rather than bio-

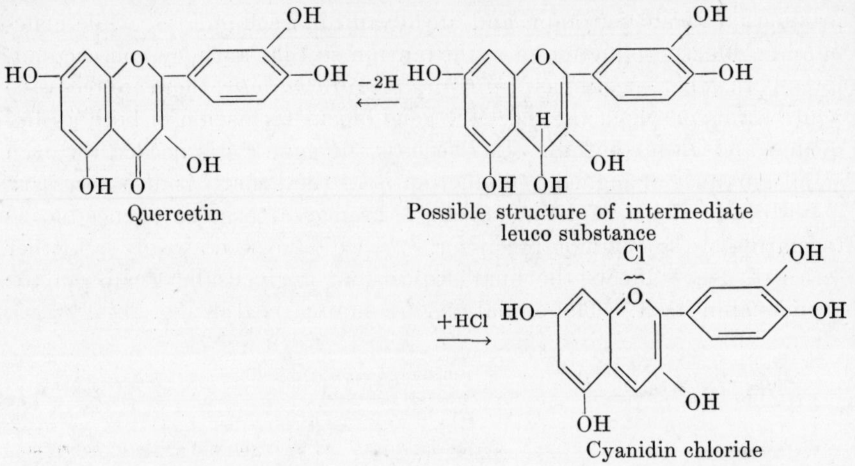

FIG. 28-5. Possible course of flavonol and anthocyanidin synthesis from a common precursor.

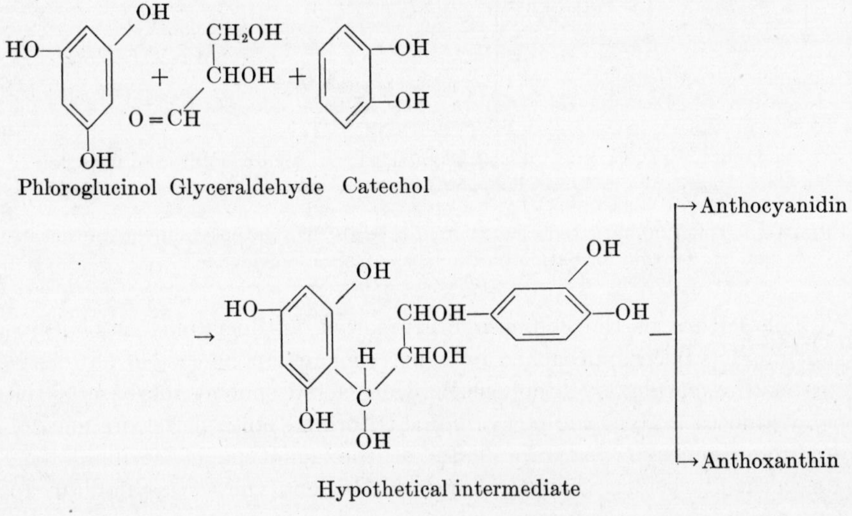

FIG. 28-6. General course of anthocyanidin synthesis as suggested by Robinson.[16]

chemical, grounds that, in general, phloroglucinol, catechol, and some three-carbon compound, such as glyceraldehyde, may join to produce the C_6-C_3-C_6 carbon skeleton through some such intermediate as that shown in Fig. 28-6. This intermediate, by ring closure and reduction, would yield the leuco substance described above as a possible intermediate in both anthocyanidin and anthoxanthin synthesis. It is entirely likely, however, that appropriate biochemical investigation will reveal that other pathways are actually involved.

General References

Link, K. P., in Organic Chemistry, H. Gilman, Ed., Vol. II, 2nd ed. Wiley, 1943.
Mayer, F., and Cook, A. H., The Chemistry of Natural Coloring Matters.   Reinhold, 1943.
Blank, F., *Botan. Rev.*, **13**, 241 (1947).
Onslow, M. W., The Anthocyanin Pigments of Plants, 2nd ed.   Cambridge University Press, 1925.

References

1. Willstätter, R., and co-workers, a series of 18 papers in *Ann.*, Vol. 401, 1913 to Vol. 412, 1917.   Summary in *Naturwissenschaften*, **20**, 612 (1932).
2. Robinson, R., *Nature*, **135**, 732 (1935).
3. Karrer, P., in Klein, G., Handbuch der Pflanzenanalyse.   Vol. III, Springer, 1932.
4. Robinson, G. M., and Robinson, R., *Biochem. J.*, **25**, 1687 (1931); **26**, 1647 (1932); **27**, 206 (1933).
    Lawrence, W. J. C., Price, J. R., Robinson, G. M., and Robinson, R., *Biochem. J.*, **32**, 1661 (1938).
5. Philip Smith, E., *Trans. Bot. Soc. Edinb.*, **30**, 230 (1931).
6. Frey-Wyssling, A., and Blank, F., *Ber. Schweiz. botan. Ges.*, **53A**, 550 (1943).
7. Harder, R., and Döring, B., *Nachr. Ges. Wiss. Göttingen VI*, **2**, 89 (1935).
8. Floren, G., *Flora*, **35**, 65 (1941).
9. Arthur, J. M., in Biological Effects of Radiation.   Vol. II, McGraw-Hill, 1936.
10. Scott-Moncrieff, R., *Ergeb. Enzymforsch.*, **8**, 277 (1939).
    Lawrence, W. J. C., and Price, J. R., *Biol. Revs.*, **15**, 35 (1940).
11. Onslow, M. W., The Anthocyanin Pigments of Plants, 2nd ed.   Cambridge University Press, 1925.   *Nature* **129**, 601 (1932).
12. Beale, G. H., Robinson, G. M., Robinson, R., and Scott-Moncrieff, R., *J. Genetics*, **37**, 375 (1939).
13. Scott-Moncrieff, R., *J. Genetics*, **32**, 117 (1936).
14. Stephens, S. G., *Arch. Biochem.*, **18**, 449 (1948); *Genetics*, **33**, 191 (1948).
15. Everest, A. E., *Proc. Roy. Soc. (London)*, **B87**, 444 (1914).
16. Robinson, R., *Nature*, **137**, 172 (1935).
17. Bate-Smith, E., *Nature*, **161**, 835 (1948).
    Wender, S. H., and Gage, T. B., *Science*, **109**, 287 (1949).

CHAPTER 29

THE PLANT GROWTH SUBSTANCES

Introduction. The term plant growth substance has by common usage come to signify any substance which possesses certain physiological attributes, among which the ability to promote cell elongation is the most widely recognized. Logically speaking the term might be expected to include all substances highly active in influencing plant growth, in which case it would comprise several other important groups of materials in addition to that covered by the term as used in the present narrow and limited sense. The study of the plant growth substances is a field peculiar to plant biochemistry, for these same substances have no recognizable importance in the economy of animals or indeed in the microorganisms with the possible exception of the algae. The type example of a plant growth substance is the compound indoleacetic acid (IAA), which is known to occur widely in the tissues of higher plants and which is physiologically active in the promotion of cell elongation, in the induction of new roots, and in the induction of at least nineteen other morphological or histological changes which will be enumerated below. In the intact plant synthesis of indoleacetic acid takes place most actively in particular growth substance producing centers. Thus the apical bud and young leaves commonly synthesize the growth substance in relatively large amounts and export the material to other lower regions of the plant where it then controls such aspects of growth as cell elongation. Indoleacetic acid is therefore a plant growth hormone, a substance synthesized in one tissue and transported in small amounts to other tissues where it brings about specific physiological effects. There are, in addition to indoleacetic acid, substances with structures more or less resembling that of indoleacetic acid which possess qualitatively similar physiological activity. Thus the application to a plant or plant tissue of α-naphthaleneacetic acid (NAA) induces cell elongation or the formation of new roots very much as does the application of indoleacetic acid itself. Alpha-naphthaleneacetic acid is not known to occur naturally in plants, however. It is then a plant growth substance although it is not a native plant hormone. Many synthetic compounds are now known which can simulate the effects of indoleacetic acid, and so vast is their number and so various their practical uses and applications that we might almost speak of this aspect of the study of plant growth substances as plant pharmacology. Such pharmacological investigations have far outnumbered the purely biochemical studies of the plant growth substances with which this discussion will be chiefly concerned.

Responses to Growth Substances. A simple system for the demonstration of physiological response to plant growth substances is constituted by excised sections of the *Avena* coleoptile.[1] When excised sections from the growing region of this hollow, nearly cylindrical, organ are placed in water they elongate very little. If they are placed instead in a solution of an active growth substance such as indoleacetic acid, they elongate rapidly and may in fact grow as rapidly as they would have if allowed to remain *in situ* in the intact seedling. The growth of such a coleoptile section is essentially exclusively by cell elongation, cell divisions having been completed at an earlier stage, and this growth effect of indoleacetic acid is then an effect upon the process of cell elongation. Relatively low concentrations of indoleacetic acid are needed to bring about the cell elongation response, as little as 0.01 mg. per liter causing a detectable response, with the optimum concentration lying in the range of 1–10 mg. per liter.[2] This simple test for growth substance activity, the *Avena* section test, has been widely used in the analysis of the mechanism of growth substance action. Excised sections of other species respond in a similar manner and sections from the growing regions of pea and other epicotyls have been similarly used.[3]

The assay of native plant growth substances is ordinarily carried out by the *Avena* curvature test.[4] In this method use is made of the fact that the native growth substance of the oat coleoptile is produced in the extreme tip of the organ. Thus elongation in the lower and actively growing portion of the coleoptile is at the expense of growth substances supplied by the tip. When the tip is removed, growth is greatly reduced for a period of approximately two hours, after which the apical portion of the stump assumes the previous growth substance synthesizing function of the tip. The growth of the decapitated coleoptile may, however, be immediately restored by placing upon the apical surface of the stump an agar block containing indoleacetic acid. If the agar block containing indoleacetic acid is placed asymmetrically upon the stump, the indoleacetic acid travels downward on one side of the coleoptile only, causing that side to grow more than the other, and a curvature of the coleoptile results. This curvature is dependent on the concentration of indoleacetic acid in the agar block and is in fact proportional to the concentration of indoleacetic acid supplied, within the limits of approximately 10–50 micrograms of indoleacetic acid per liter. Since only about 0.1 ml. of solution is actually needed to conduct the test with ten test plants, it would theoretically be practicable to assay as little as 1×10^{-3} microgram of indoleacetic acid by this method. In actual practice amounts of the order of 10^{-2} microgram of indoleacetic acid may be readily and quantitatively determined.

In the so-called slit-pea test use is made of sections from the growing region of etiolated pea epicotyls which are divided symmetrically down the middle for a portion of their length.[5] When such sections are placed

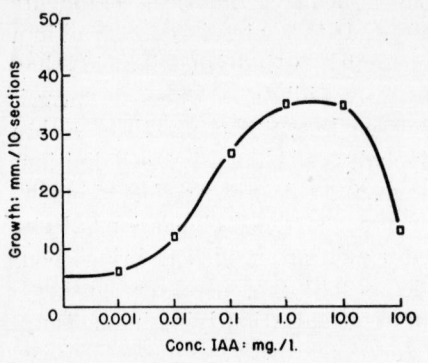 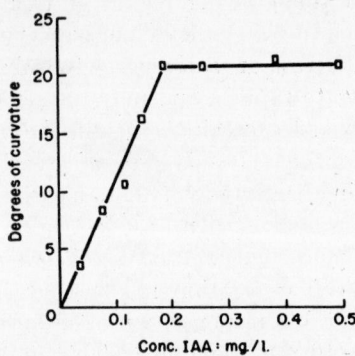

A. Growth of *Avena* coleoptile sections on solution. (After Bonner[52])

B. *Avena* curvature test with decapitated coleoptiles. (After Went and Thimann[4])

FIG. 29-1. Growth responses to applied indoleacetic acid.

in water the slit ends bend outward owing to the tissue tension between epidermis and cortex. In solutions of indoleacetic acid on the other hand, the apical portions of the slit ends come to assume an inward curvature due to the fact that the outer uninjured portions of the tissue grow more than do the inner surfaces. The resultant curvature bears a relation to the log of the concentration of indoleacetic acid applied. The pea test, like the *Avena* curvature test and the *Avena* section test, is a measure of cell elongation. Many other cell elongation responses to applied indoleacetic acid have been noted, including the growth of the stalks of inflorescences,[6] the elongation of veins of leaves,[7] and the asymmetrical elongation of petioles which leads to epinasty of leaves,[8] a phenomenon frequently observed in experiments in which entire plants are treated with growth substances.

Although the plant growth substances bring about cell elongation responses in a host of tissues, still other types of responses are elicited in other tissues. Thus the application of indoleacetic acid promotes the formation of adventitious roots on cuttings or on intact plants of many species.[9] Indoleacetic acid applied to unfertilized ovaries brings about the production of parthenocarpic (seedless) fruits in species such as tomato, tobacco, and others, in which parthenocarpy would not ordinarily occur.[10] Indoleacetic acid applied locally to stems in high concentrations may also produce large growths of callus tissues.[11] Such callus tissue can frequently be grown aseptically in tissue culture as the isolated tissue, provided that nutrient medium containing indoleacetic acid is supplied.[12] Indoleacetic acid causes activation of cambial growth in seedling stems and hypocotyls, as well as in woody shoots.[13] In all these responses indoleacetic acid is associated with the initiation of cell division activity either of a nonspecialized nature (callus) or of a highly specialized type (root initials, cambium, fruit development).

Equally well-known are plant responses in which applied indoleacetic acid brings about inhibition of growth. Thus indoleacetic acid is inhibitory to the elongation of roots in the same range of concentrations which promote growth of coleoptiles, stems, etc.[14] Perhaps the best known growth substance-induced inhibition is that of lateral buds.[15] In the normal shoot the growth of the lateral or axillary buds is suppressed by the presence of an apical bud, a phenomenon known as apical dominance. This suppression is due to the growth substance synthesized in the apical growing point which on its way down the stem not only promotes elongation of the shoot but also in some way inhibits growth of the axillary buds. If the apical bud is removed, axillary buds grow out, since the growth substance-induced inhibition is removed. The application of indoleacetic acid to the decapitated stem results in suppression of lateral bud growth just as does the presence of the apical bud itself. Growth substances are of importance also in the suppression of abscission of leaves and fruits.[16] The normal shedding of both of these types of organs is associated with restriction in the amount of growth substance supplied to the part, reduction of growth substance supply in some manner promoting formation or separation of the abscission layer. The shedding response may, however, be inhibited by application of indoleacetic acid to the organ whose fall is to be delayed.

The plant growth substances are known to have still further effects on plant tissues and Table 29-1 contains a summary of other well-known responses. It is to be anticipated that additional responses will be observed in the future. Discussion of the physiological details of these varied responses will be found in the general references cited at the end of this chapter. The point to be stressed here is merely that the plant growth substances typified by indoleacetic acid elicit many and varied growth responses. It seems that we may in general anticipate that the application of indoleacetic acid to a plant tissue will cause a response. The exact nature of the response to the applied substance differs, however, with the tissue, the previous treatment of the tissue, and the species of plant used.

TABLE 29-1. Physiological responses to plant growth substances

Class of response	Response	Discovery	Reference
Cell elongation	*Avena* curvature test	Went, 1928	4
	Avena section growth	Bonner, 1933	1
	Slit-pea stem curvature	Went, 1934	5
	Flower peduncles	Uyldert, 1927	6
	Epinasty	Uyldert, 1931	6
	Leaf veins	Avery, 1935	7
	Algae	Yin, 1937	17
Cell division	New root formation	Thimann and Went, 1934	9
	Callus growth	Laibach and Fischnich, 1935	11
	Cambial activation	Snow, 1935	13
	Parthenocarpy	Gustafson, 1936	10
Inhibition	Axillary buds	Thimann and Skoog, 1934	15
	Root growth	Kögl *et al.*, 1934	14
	Flower formation	Dostal and Hosek, 1937	18
	Photoperiodic induction	Thurlow and Bonner, 1947	19
	Abscission of petioles	Laibach, 1933	16
Miscellaneous	Promotion of flower initiation	Clark and Kearns, 1942	20
	Ripening of fruit	Mitchell and Marth, 1944	21
	Closing of stomata	Ferri and Lex, 1948	22
	Nonosmotic water uptake	Reinders, 1942	23

Chemistry and Identification of the Native Plant Growth Substances. The *Avena* curvature test was devised by Went in 1928 for the demonstration and estimation of the growth-promoting substance produced in *Avena* coleoptile tips.[4] This particular growth-promoting substance has not yet been isolated in chemically pure form, the difficulties inherent in the obtaining of a sufficient supply of starting material having proved insuperable up to the present time. (We now know that 10,000 tips yield only about 1 microgram of indoleacetic acid.) Urine was early found to be a rich source of material active in the *Avena* curvature test, and in 1934, Kögl, Haagen-Smit, and Erxleben[24] announced the isolation from urine of indoleacetic acid, a compound highly active in the *Avena* curvature test. Another source likewise yielding indoleacetic acid as the active substance was revealed by parallel work on the *Avena* curvature-causing material contained in culture filtrates of the fungus *Rhizopus suinus* (Thimann, 1935).[25] Indoleacetic acid has subsequently been isolated in chemically pure form as a natural product present in the developing corn seed.[26]

Although indoleacetic acid has been directly isolated from only one higher plant material, still evidence other than chemical isolation indicates that indoleacetic acid is a naturally occurring growth substance in many species and in many plant tissues. One such line of evidence is based upon a sensitive chemical and spectrophotometric method for the determination of indoleacetic acid. Indoleacetic acid forms an intensely colored complex with ferric ions in acid solution. Formation of this red complex, which possesses a characteristic absorption curve with a maximum at 5250 Å, is characteristic for indoleacetic acid and may be used as an analytical method for the determination of the substance in purified plant extracts.[27] Although the method has been refined so that as little as 1 microgram of indoleacetic acid may be determined quantitatively, it is nevertheless far less sensitive than the *Avena* curvature technique in which amounts of 10^{-3} to 10^{-2} microgram may be estimated. Application of the spectrophotometric method to determination of the native plant growth substance and comparison of the results obtained with those obtained by the *Avena* curvature method have shown that the extractable growth substance of *Avena* coleoptile tips is largely and perhaps entirely indoleacetic acid.[28] Similarly the native growth substance content of leaves of spinach, cabbage, *Xanthium*, and pineapple, of apical buds of *Xanthium* and tomato, of sunflower tumors, and of other varied tissues can be accounted for on the basis of indoleacetic acid content. Biochemical evidence to be presented below indicates also that enzymes concerned with the biosynthesis and metabolism of indoleacetic acid are widely distributed in many plant tissues. There is then no doubt but that indoleacetic acid is of general importance as a native plant growth substance.

Although indoleacetic acid is the best known of the native plant growth substances, it is nevertheless entirely possible that still other substances of similar physiological activity may also occur in higher plants. This possibility was in fact actually suggested before it became known that indoleacetic acid possessed growth-promoting properties. Kögl and others in 1933[29] announced the isolation from urine of a compound active in the *Avena* curvature test, a compound to which they gave the name auxin *a*, and whose structure as elucidated by Kögl and co-workers is given below:

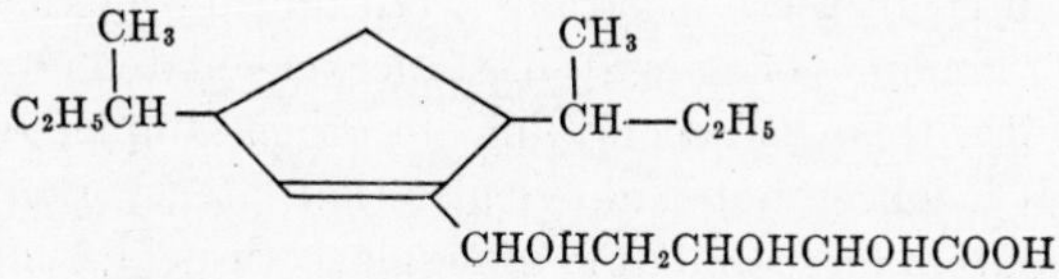

Auxin *a*

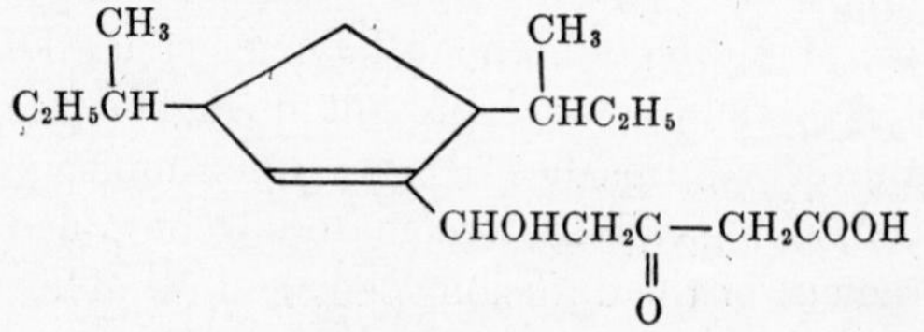

Auxin *b*

The name auxin is frequently applied generically to all substances including indoleacetic acid which are active in the *Avena* curvature test. The designation auxin *a* is, however, properly reserved for the chemical individual of the structure shown above. Shortly after their initial discovery Kögl and his group isolated from corn oil and from malt not only auxin *a* but also a related substance, auxin *b*, which differs from auxin *a* by loss of the elements of one molecule of water. The samples

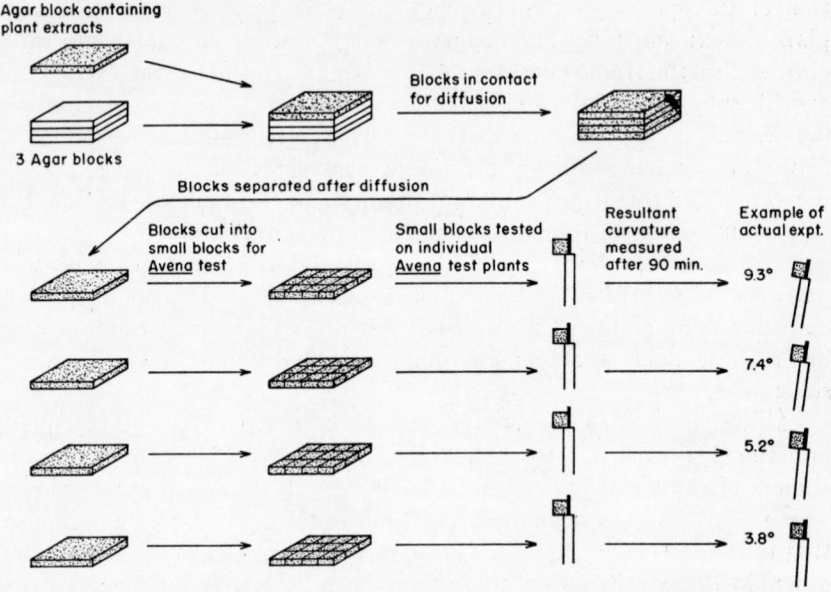

Agar block containing
plant extracts

3 Agar blocks

Blocks in contact
for diffusion

Blocks separated after diffusion

Blocks cut into
small blocks for
<u>Avena</u> test

Small blocks tested
on individual
<u>Avena</u> test plants

Resultant
curvature
measured
after 90 min.

Example of
actual expt.

9.3°

7.4°

5.2°

3.8°

FIG. 29-2. General method for determination of diffusion coefficient of growth substance.

of corn oil and malt used in these isolations were unusual in that they contained much larger amounts of growth substance, calculated on a basis of physiological activity, than has been found in other samples by subsequent workers.[42] It has been noted above also that later workers have found corn seeds to contain large amounts of indoleacetic acid, which was apparently not the case in the sample used by the Dutch investigators.

In addition to the direct methods of chemical isolation or determination several indirect methods have been used for identification of the type of growth substance present in higher plant tissues. The first of these indirect approaches is based upon molecular weight determination of the physiologically active material contained in plant extracts from measurement of the diffusion coefficient.[30] This technique is based upon allowing the extract to diffuse through a set of four agar blocks. The unknown active substance is incorporated in the top block and diffusion is allowed to take place through the stack for a specified period, usually from one to two hours. The four blocks are next separated and each is analyzed for its content of the material under investigation. The method may be used with any substance for which a specific method of quantitative determination is available even though the substance itself may be present only in the form of a crude and impure extract. In the determination of the diffusion coefficient of a plant growth substance, a specific analytical method is available in the *Avena* curvature test. Each of the four blocks from the diffusion is cut into a series of smaller blocks of 10 cu. mm., and these small blocks are tested in the *Avena* curvature test. In this way it is possible to determine the distribution of physiological activity between the four blocks concerned in the original diffusion. From these data, the diffusion coefficient D may be obtained from diffusion data, prepared by Stefan and others, which give the diffusion coefficient as a function of distribution of active material between four equal blocks, based on the assumption that at the beginning of the diffusion all the material was contained in the uppermost block.[31] The distributions

TABLE 29-2. Diffusion table for calculation of diffusion constants from the distribution of a material between four agar blocks after diffusion from a terminal block. (After Heyn[31])

$x = \dfrac{h^2}{4Dt}$	Per cent of total material in block			
	Block 1	Block 2	Block 3	Block 4
0.0784	19.35	22.65	27.35	30.68
0.0900	17.72	21.94	27.99	32.33
0.1024	16.07	21.21	28.66	34.04
0.1156	14.40	20.48	29.34	35.76
0.1296	12.84	19.72	29.94	37.51
0.1444	11.35	18.96	30.49	39.21
0.1606	9.96	18.16	30.97	40.88

are given in terms of the per cent of the total diffused material contained in each block. We now determine for each block the value x associated with the percentage of the total material contained in that block. The quantity x is related to the diffusion coefficient by equation 1

$$x = \frac{h^2}{4Dt} \tag{1}$$

in which h is the thickness of each agar block in centimeters, D is the diffusion coefficient, and t the diffusion time in days. The diffusion coefficient D is related to the molecular weight M by equation 2,

$$\sqrt{M} = \frac{k}{D} \tag{2}$$

in which k is a constant which depends upon temperature and the general nature of the substance whose diffusion is being considered. For a temperature of 26°C and for substances of the molecular weight range and type of the plant growth substances, k assumes the value 8.8,[32] while at a temperature of 20°C, $k = 7.0$.[30] It is wise, however, to determine k experimentally by evaluating it from the diffusion constant of a substance of known molecular weight. The molecular weight values for purified growth substance of *Avena* agree moderately closely to those expected for indoleacetic acid,[28] although values obtained on crude extracts are considerably higher.[4] Evidently factors of an unknown nature present in crude extracts of *Avena* interfere to an appreciable extent with correct estimation by diffusion of the molecular weight of the active growth substance. In other tissues, including for example tomato leaves and buds, pineapple leaves, and seedlings of pea and radish, the values for the molecular weight obtained by diffusion are close to those expected for pure indoleacetic acid.

Still a further qualitative criterion for differentiation of different native plant growth substances is based on relative stability of the physiological activity toward hot acid or base. The behaviors of auxin a, auxin b, and indoleacetic acid are summarized in Table 29-3. Indole-

TABLE 29-3. Stabilities of various growth substances to heating in 1 N HCl or in 1 N NaOH for periods of 1–3 hours at 100°C

Growth substance	Stability in acid	Stability in base
Indoleacetic acid	Destroyed	Relatively stable
Auxin a	Stable	Destroyed
Auxin b	Destroyed	Destroyed

acetic acid is rapidly oxidized and destroyed by heating in acid solution but is more stable toward heating in alkaline solution. Auxin a behaves in the opposite fashion while auxin b is destroyed by heating in acid as well as in base. Application of this method indicates that the growth substance activity of the *Avena* coleoptile behaves as expected for auxin a,[14] a conclusion at variance with the chemical evidence and with the evidence obtained by molecular weight determinations. Many other tissues contain nevertheless growth substances which are destroyed by heating in acid as would be expected of indoleacetic acid. Since the inactivation of indoleacetic acid in hot acid is actually an oxidation, it is only to be expected that the rate and extent of indoleacetic acid destruction by hot acid in crude extracts will depend on the presence or absence of other oxidizable substances as well as on any antioxidants which may be present. It is doubtful therefore that the acid-alkali stability test truly represents a reliable method for determining the nature of native plant growth substances.

Although both molecular weight determinations and acid-base stability leave much to be desired as methods for the diagnosis of the nature of the native plant growth substances, still application of these two methods has by and large yielded results in accordance with those obtained by the spectrophotometric method and is largely in accord with the concept that indoleacetic acid is of widespread occurrence in higher plant tissues.

Biogenesis of Indoleacetic Acid. Discussion of the biogenesis of plant growth substances is necessarily restricted to discussion of the biogenesis of indoleacetic acid itself, the one growth substance definitely established to be a normal plant product. Indoleacetic acid is enzymatically produced in the plant from tryptophane. Thus relatively large amounts of indoleacetic acid are formed in leaves infiltrated with solutions of tryptophane and then allowed to incubate for short periods under aerobic conditions.

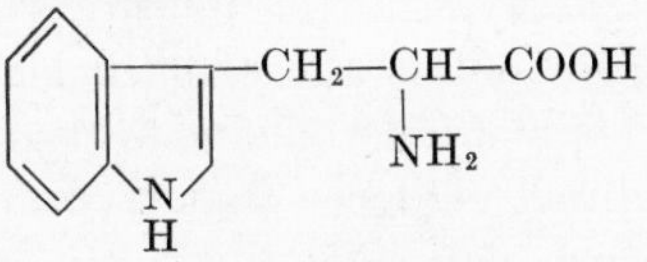

$$\text{Tryptophane} \qquad \text{Indoleacetic acid}$$

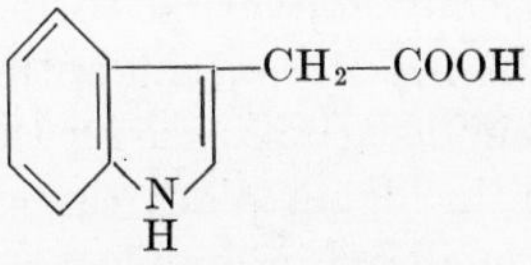

Much more indoleactic acid can be recovered by extraction of the tryptophane treated tissue than from similar non-tryptophane infiltrated tissue.[33] The entire sequence of reactions by which tryptophane is converted to indoleacetic acid can also be caused to take place with enzyme preparations made from various tissues such as leaves, tips of the *Avena* coleoptile, and tumor cultures. Investigation of the mechanism involved has indicated that a neutral substance, possibly indole-3-acetaldehyde, may be an intermediate in the conversion. This neutral substance, which will be referred to for convenience as indole-3-acetaldehyde although its structure has not been confirmed by actual isolation, is a normal constituent of many plant tissues including etiolated plants of pea, *Vicia*, sunflower, and *Brassica*,[30] green leaves of pineapple, and roots such as those of dandelion.[34] When tryptophane is added to enzyme preparations of pineapple leaves, indoleacetaldehyde as well as indoleacetic acid accumulates among the reaction products, indicating that oxidation of indoleacetaldehyde to indoleacetic acid may be a bottleneck in the overall conversion of tryptophane. Added indoleacetaldehyde is, however, oxidized to indoleacetic acid at a moderate rate by enzyme preparations of pineapple leaves[34] and of *Avena* seedlings.[35] The intermediate steps in the production of indoleacetaldehyde from tryptophane are somewhat uncertain, and it is possible that two different pathways may actually be involved. Indolepyruvic acid, the compound which should arise as the product of oxidative deamination of tryptophane, spontaneously decomposes to indoleacetaldehyde, and although indolepyruvic acid is enzymatically converted to indoleacetic acid by spinach and by pineapple leaf preparations, it still is possible that the only enzyme involved may be that for the oxidation of indoleacetaldehyde itself. Direct evidence that

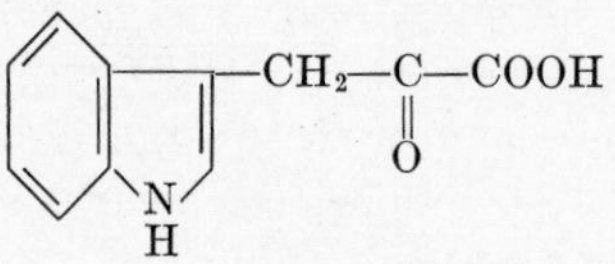

Indolepyruvic acid 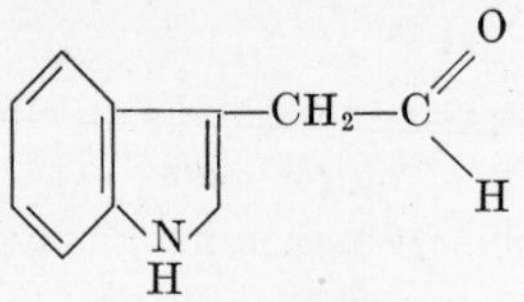   Indoleacetaldehyde

indolepyruvate is formed enzymatically in the plant is also lacking. On the other hand tryptamine, the decarboxylation product of tryptophane is enzymatically converted to indoleacetaldehyde in the pineapple leaf as well as in the *Avena* coleoptile, although not in the spinach leaf. No amino acid decarboxylase for the conversion of tryptophane to tryptamine is known in higher plant tissues, and direct evidence that tryptamine is an intermediate in indoleacetic acid production is then also lacking. Tryptamine has, however, been reported as a natural plant product in occasional cases.[36]

It has been known for some years that zinc deficiency results in a lowered growth substance content of plants, the lowering of the amount of protein-bound growth substance (see below) being particularly marked. The mechanism by which zinc exerts this effect has been investigated by Tsui in the tomato plant.[37] It was found that zinc-deficient tomato plants possess as high an activity of the tryptophane-indoleacetic acid converting enzyme system as plants supplied with the normal amount of zinc. Zinc deficiency was found to affect primarily the tryptophane content of the plant, zinc-deficient plants containing much less tryptophane than plants normally supplied with zinc. Upon the addition of zinc both growth substance and tryptophane levels were rapidly restored (Fig. 29-3). Thus it would appear that zinc deficiency lowers the growth substance level in the plant by affecting the concentration of substrate available to the tryptophane-indoleacetic acid enzyme system.

In summary, indoleacetic acid arises from tryptophane in the higher

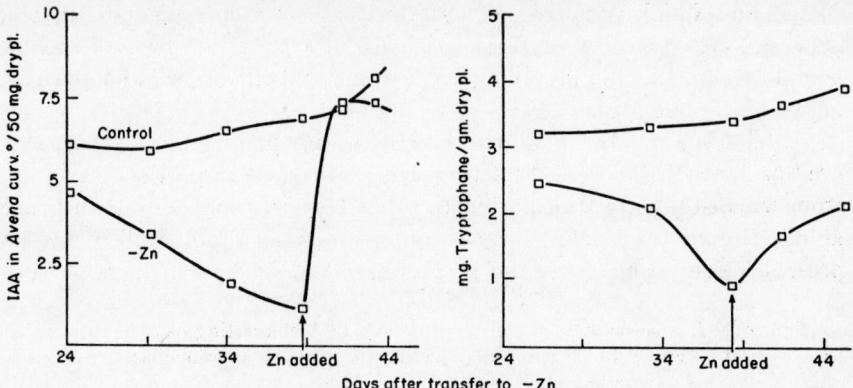

FIG. 29-3. Effect of zinc supply on the tryptophane and protein-bound growth substance content of tomato plants. (After Tsui[37])

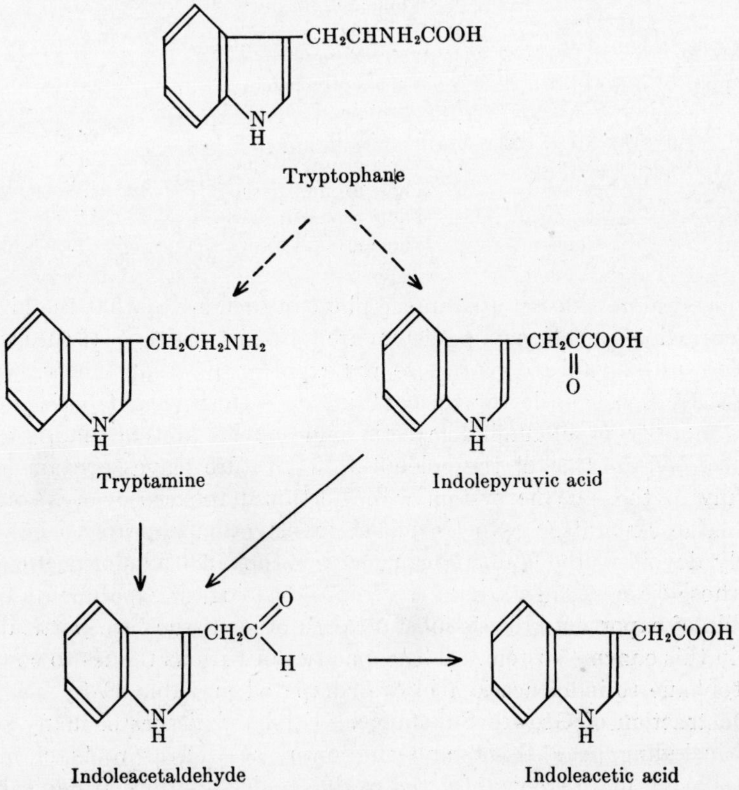

Tryptophane

Tryptamine

Indolepyruvic acid

Indoleacetaldehyde

Indoleacetic acid

FIG. 29-4. Pathways for the conversion of tryptophane to indoleacetic acid.

plant, and indoleacetaldehyde appears to be an intermediate in the process. There are two possible pathways for the production of indole-acetaldehyde, the one involving tryptamine and the other indolepyruvic acid as intermediates.

It has been mentioned above that the native plant growth substances are hormones in the sense that they are synthesized in particular centers from which they are transported to other tissues where they then bring about growth reactions. It is of importance therefore that so far as present information goes, the tryptophane-indoleacetic acid converting

TABLE 29-4. Occurrence of the tryptophane-IAA converting enzyme system in higher plant tissues. In all cases the enzyme preparation was incubated with substrate 3–4 hours at approximately 24°C

Species	Part	Nature of sample	IAA production from tryptophane. $\gamma IAA/gm.$ dry sample
Spinach	Leaves	Whole living leaf	0.39
		Whole cytoplasmic protein	0.30
Tomato	Leaves	Whole living leaf	0.54
Pineapple	Leaves	Whole cytoplasmic protein	0.48
Avena coleoptile	Tip	Whole ground tissue	25.5
Avena coleoptile	Base	Whole ground tissue	4.0
Tobacco	Ovary	Whole ground tissue	940
Sunflower	Stem	Whole ground tissue	14
Carrot	Root	Whole ground tissue	3.8

enzyme system is distributed in the plant in such a way that the highest concentrations of the enzyme system are found in just those tissues which are known to be able to synthesize and export growth substances. Thus in the Avena coleoptile the tip contains on a fresh weight basis several times more tryptophane-indoleacetic acid enzyme system than is found in the lower portion of the coleoptile.[28] Mature leaves contain lesser amounts of the enzyme system while fertilized tobacco ovaries contain very large amounts.[38] No tissue yet investigated appears to be completely devoid of the ability to convert tryptophane to indoleacetic acid, but those tissues which are known from physiological experiments to be capable of exporting growth substances to other tissues are particularly rich in this enzyme system. The capacities of various tissues to convert tryptophane to indoleacetic acid are compared in Table 29-4.

Destruction of Growth Substance. Etiolated tissues of many seedlings including pea, Avena, and sunflower, as well as roots of many species, contain an enzyme for the oxidative degradation of indoleacetic acid, an enzyme which may then be termed indoleacetic acid oxidase.[39]

Indoleacetic acid oxidase attacks indoleacetic acid with an uptake of 1 mol of oxygen and the liberation of 1 mol of CO_2 per molecule of indoleacetic acid destroyed.

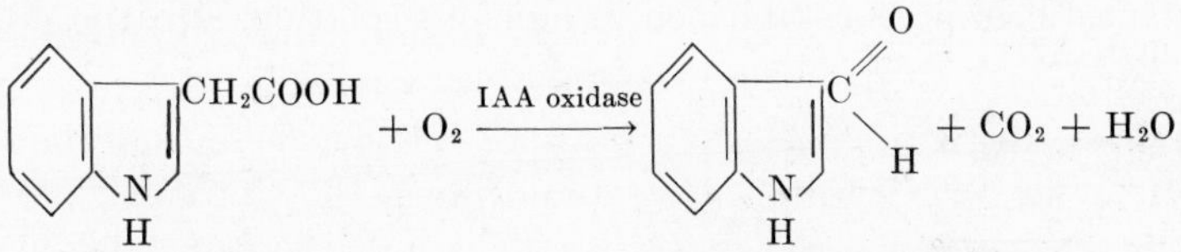

Indoleacetic acid   Indolealdehyde
Attack of indoleacetic acid by indoleacetic acid oxidase

The reaction corresponds thus to oxidative decarboxylation of indoleacetic acid to indole-3-aldehyde, a substance which is not active as a growth substance. Indoleacetic acid oxidase is inhibited by cyanide and other heavy metal poisons and would hence appear to contain a heavy metal as an essential prosthetic group. Interestingly enough, the enzyme is also present in green leaves and stems although these tissues also contain large amounts of a heat stable inhibitor of the enzyme. This inhibitor of unknown structure is also present, although in lower concentrations, in etiolated plants. Indoleacetic acid oxidase is, so far as is known, specific to indoleacetic acid, with the reservation that it may attack the non-naturally occurring indolepropionic and indolebutyric acids, although at a much slower rate. The function of the enzyme may possibly lie in regulation of the hormone level within the plant, a function which is suggested particularly by the presence of the enzyme in roots. It is known, as pointed out above, that the growth of roots is inhibited by high concentrations of indoleacetic acid, and the enzyme may act to prevent the accumulation of excessively high concentrations of the growth substance. It is clear, however, that regulation of hormone level in the plant is a complicated matter, since the indoleacetic acid oxidase, the inhibitor of indoleacetic acid oxidase and enzymes responsible for formation of indoleacetic acid may simultaneously be present in a single tissue.

Green tissues may contain still other systems capable of inactivating indoleacetic acid, although the exact nature of these systems is still unclear. It is known, however, that the destruction of indoleacetic acid in plant extracts is increased when such extracts are placed in bright light, particularly if small amounts of carotene are present in the extract.[30] Indoleacetic acid as well as other compounds containing the indole nucleus are rapidly destroyed also by a riboflavin-catalyzed photooxidation.[40] In this reaction riboflavin, the light-absorbing agent, is reversibly photooxidized, and the oxidized riboflavin is then reduced at the expense of indoleacetic acid.

A great deal has been reported in the literature concerning the light-sensitized destruction of auxin a.[41] An equilibrium between auxin a and auxin a lactone is established in aqueous solution. In the presence

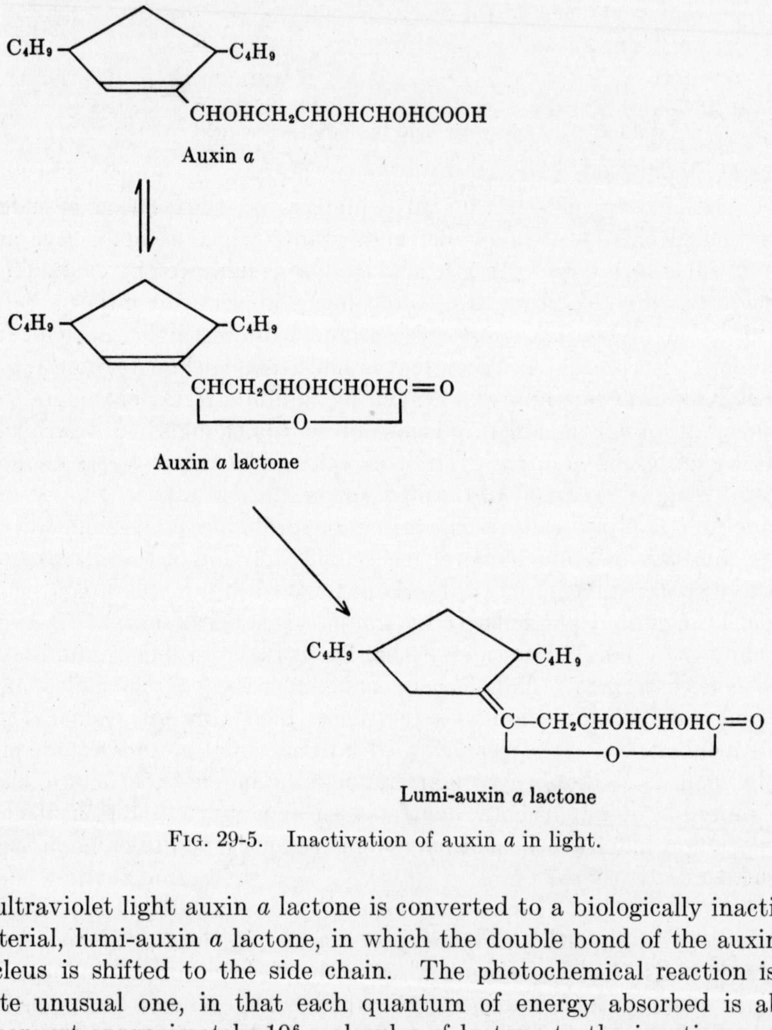

FIG. 29-5. Inactivation of auxin a in light.

of ultraviolet light auxin a lactone is converted to a biologically inactive material, lumi-auxin a lactone, in which the double bond of the auxin a nucleus is shifted to the side chain. The photochemical reaction is a quite unusual one, in that each quantum of energy absorbed is able to convert approximately 10^6 molecules of lactone to the inactive material. In any case the reaction is sensitized by carotenoids, the formation of lumi-auxin a lactone being catalyzed by visible light in the presence of alpha- or beta-carotene. It has been suggested that this mechanism may account for the *in vivo* destruction of auxin *a* in tissues subjected to illumination and may be of some importance therefore in the phototropic growth response.

Growth Substance Complexes. From higher plant sources at least two low-molecular-weight materials are known which are closely related to indoleacetic acid and which yield indoleacetic acid on appropriate chemical treatment. Thus seeds of wheat, corn, rye, and presumably of other cereals, contain in the endosperm a material which while itself inactive as a growth substance, yields indoleacetic acid by incubation at a *p*H of 9–10.5 and at moderate temperature.[42] The process of liberation of indoleacetic acid appears to consist in the hydrolytic release of growth substance from a relatively simple complex (molecular weight less than 500). Corn seeds accumulate this complex in extraordinarily large amounts and may yield over 500 micrograms of indoleacetic acid per gram of whole corn meal upon hydrolysis. The indoleacetic acid-yielding complex is not tryptophane itself nor is it indoleacetaldehyde, since its content of indoleacetic acid is not released by the tryptophane-indoleacetic acid converting enzyme system. This enzyme system is, however, present in the developing corn seed and is presumably responsible for the synthesis of indoleacetic acid, which may be further combined into the inactive endosperm complex. That the endosperm complex is a secondary product synthesized *in vivo* from indoleacetic acid is indicated by the time relations of the accumulation during development of the cereal grain. During this development free indoleacetic acid accumulates to a moderate concentration, only to again disappear. Over the same period the indoleacetic acid complex also accumulates, reaching a maximum after attainment of the free indoleacetic acid maximum. Late in the development of the corn seed, the endosperm complex may itself decrease somewhat in concentration. Although the role of the endosperm indoleacetic acid complex in the corn seedling is not known, it would appear unlikely that it serves as a source of indoleacetic acid for the growing corn seedling. In the corn seedling as in *Avena* the native growth substance of the coleoptile is synthesized in the extreme tip of the organ, and we have seen above that the coleoptile tip is rich in the tryptophane-indoleacetic acid enzyme system.

A second substance in which indoleacetic acid appears to be bound as a complex or derivative is the so-called inhibitor, originally discovered in the radish seedling by Stewart.[44] This substance is a neutral compound, possesses a molecular weight approximately equal to that of indoleacetic acid, and yields indoleacetic acid upon oxidation. Unlike the corn endosperm-indoleacetic acid complex, the radish material is itself physiologically active and inhibits the action of indoleacetic acid in the

Avena curvature test. Thus in the *Avena* curvature test, the radish inhibitor causes positive curvatures, that is, causes curvatures toward rather than away from the applied agar block. When radish inhibitor and indoleacetic acid are applied together in the *Avena* curvature test, the resultant curvature depends upon the ratio between the two substances. In these peculiar physiological properties, the radish inhibitor is unlike any indoleacetic acid derivative of known structure which has been thus far investigated, and it will be of interest to determine the chemical nature of the substance. The role of the radish inhibitor in the economy of the plant is obscure, but it is unlikely that it plays any general role in higher plants, since its distribution is restricted in general to seedlings of the *Cruciferae*. Tissues of other species do, however, occasionally yield on ether extraction materials capable of causing positive curvature in the *Avena* test.

The principal form in which indoleacetic acid is bound in the plant is represented by the indoleacetic acid-protein complexes found in many, perhaps in all, plant tissues. The first indication that indoleacetic acid is, in fact, bound to protein came from the demonstration that growth substances may be liberated from plant tissue by the action of proteolytic enzymes such as chymotrypsin.[45] Subsequent work of Wildman and Gordon showed that this is due to the release of auxin from the cytoplasmic proteins of the tissue.[46] Procedures for the separation of the cytoplasmic proteins of leaves, described in Chapter 17, have made it possible to show that essentially all the protein-bound indoleacetic acid of the leaf is associated with fraction I, the major cytoplasmic protein of the leaf.[47] Indoleacetic acid may be released from this protein either (*a*) by proteolytic hydrolysis or (*b*) by hydrolysis with dilute (0.1 N) KOH at 100°C. Indoleacetic acid is not removed from the protein by prolonged dialysis or by denaturation of the protein with heat or acid. Exact determination of the amount of indoleacetic acid bound to fraction I protein has been made difficult by the fact that all methods for removal of the growth substance result in partial destruction of the indoleacetic acid liberated and probably bring about in addition some conversion of protein-bound tryptophane to indoleacetic acid. The data of Wildman suggest, however, that one to a few molecules of indoleacetic acid are associated with each fraction I protein molecule. Protein-bound indoleacetic acid is found not only in leaves but also in tissues of *Avena* coleoptiles and of pea stems, among others. In no case, however, have these protein-auxin complexes been purified and studied as extensively as those of leaves.

The seed proteins of wheat contain associated growth substances which may be liberated by proteolytic or alkaline hydrolysis.[48] Thus wheat globulin, primarily an embryo protein, contains of the order of 300–500 micrograms of growth substance per kilogram of protein. The endosperm proteins, including the glutelins and gliadins, contain smaller amounts of growth substance. Although these protein-bound growth substances of the seed have been less studied than those of leaves, still it would appear that the globulin-bound growth substance of the embryo may in general resemble the leaf material described above.

Survey of the Known Forms of Growth Substance. We have seen above that the native growth substances of plants include indoleacetic acid and possibly other active compounds and that, in addition, plants contain a variety of substances which are either precursors of indoleacetic acid (tryptophane, indoleacetaldehyde) or secondary products of indoleacetic acid metabolism such as the protein or endosperm complexes. Since each of these various materials may be converted to indoleacetic acid, or perhaps to other active growth substances, a great deal of confusion has arisen as to exactly what form of growth substance is removed or determined by any particular extraction or determination procedure. Table 29-5 gives a summary of the more important forms and sources of growth substances in the plant so far as present information goes. Free

TABLE 29-5. The more important forms and sources of native growth substances in the plant, including precursors and derivatives

Form present in plant	Definition	Chemical nature	Converted to active growth substance by:	Function
Free growth substance	Present in tissue as growth substance molecule or ion	IAA, auxin *a* & *b*	....	Transport form of growth substance
Precursors of growth substance	Chemical intermediate in growth substance synthesis	Tryptophane	Trypt.-IAA convert. enzyme system	Precursor
		Indoleacetaldehyde?	Trypt.-IAA convert. enzyme system	Precursor
Bound growth substance	Growth substance combined in form of derivatives	Endosperm material of corn, wheat (small molecule)	Yields IAA when incubated at pH 10.5	Unknown
		Radish inhibitor	Yields IAA on ox.	Unknown
Protein-bound growth substance	Growth substance combined to native protein	IAA-protein of leaves (fraction I)	Proteolytic enzymes. Hyd. with 0.1 N NaOH at 100°	Functional form of growth substance?
		IAA-protein of wheat embryos	Same as above	Functional form of growth substance?

growth substances are those materials present in the tissue as such, for example as indoleacetic acid present in the form of indoleacetic acid molecules or ions. This form of growth substance may be extracted from acidified tissue with ether or other organic solvents, but care must be taken to avoid generation of growth substances from other sources and in particular the conversion of tryptophane or indoleacetaldehyde to indoleacetic acid during the extraction period. The method of Wildman and Muir is probably the most specific yet devised for extraction from the plant of free growth substance only.[38] In this method the tissue is first frozen rapidly, then dried at low temperature by lyophilization. The dried sample is next extracted with ether at 0°C. If higher extraction temperatures are used, appreciable enzymatic conversion of indole-acetaldehyde and tryptophane to indoleacetic acid may occur. The known precursors of the native growth substances are confined at present to the precursors of indoleacetic acid. Of these, tryptophane is readily separated from indoleacetic acid, since it is not extracted by ether as is the latter. Indoleacetaldehyde is soluble in ether as is indoleacetic acid, but it may be separated from the latter by partition between aqueous bicarbonate and ether. The indoleacetic acid passes into the alkaline water layer while the neutral indoleacetaldehyde remains in the ether layer. This separation depends of course on the fact that indoleacetic acid is an acid while indoleacetaldehyde is a neutral compound. Indole-acetaldehyde may be quantitatively determined after conversion to indoleacetic acid by a suitable mild oxidation. This oxidation may for example be carried out with animal aldehydrase (xanthine oxidase), or possibly with the indoleacetaldehyde-oxidizing enzyme of the *Avena* coleoptile.

The complexes or derivatives in which growth substances are combined in relatively small molecules are not yet known to be of general occurrence and may not then in general constitute complications in the extraction and determination of growth substances. Where such complexes do occur as in seeds, they may frequently be separated from indole-acetic acid on the basis of differing solubilities; this is the case with the corn endosperm complex which is insoluble in ether. Protein-bound growth substance, on the contrary, appears to be of very wide distribution. Growth substance bound to protein is much less readily freed than that bound in the endosperm material, and is also less readily converted to free growth substance than either tryptophane or indole-acetaldehyde. Until better methods for estimation of protein-bound growth substance are devised it would appear to be most satisfactory to determine this material on the basis of active growth substance released by proteolytic or mild alkaline hydrolysis of precipitated or dialyzed protoplasmic protein. In this way all small molecular forms of growth substance as well as growth substance precursors and derivatives are removed from the protein and cannot contribute to the values obtained for protein-bound growth substance.

Mechanism of Growth Substance Action. The action of growth substances in increasing rate of cell elongation has a well-established basis in increased cell wall plasticity (Chapter 8). Applied growth substance measurably increases the plasticity of primary cell walls. The effect is an indirect one which depends on processes mediated by the living tissue, and cell walls of non-living tissue are not affected. It is clear therefore that an explanation of the mechanism of growth substance action must be sought not in the cell wall but in the functioning of the protoplasm. It has been known for many years that the effect of growth substances in bringing about growth responses depends intimately upon respiration. Thus tissues do not respond to growth substance in an anaerobic atmosphere, and certain inhibitors of respiration, such as cyanide, decrease growth substance-induced growth response to an extent proportional to the respiratory inhibition.[1] There is then some general basis for the supposition that the growth substances may exert their effect on growth by influencing a phase of the respiratory process. An attractive extension of this idea is the possibility that the growth substances may exert this same basic effect in all tissues and that the many different manifestations of growth substance activity may be due to differences in the manner in which different plant tissues translate the basic biochemical growth substance effect into growth response.

A great deal of the basic work on the biochemistry of growth substance action has been done with excised sections of *Avena* coleoptiles prepared as described earlier and allowed to float either in solutions lacking growth substance or in solutions to which growth substances have been added. Such sections make a rapid and vigorous growth response to added growth substance, and comparisons of the metabolism of treated and untreated tissue may readily be made. When such sections are supplied with indoleacetic acid in a concentration of 1–10 mg./liter (optimal for growth) the rate of respiration may be increased by 10 to 35%.[1,49] The exact factors which govern the magnitude of the increase are unknown, and increased respiration is not essential to growth response, since in certain experiments growth responses occur without increase in rate of respiration.[50] Nevertheless, such increases in respiration have been frequently found, not only with *Avena* but also with pea sections and other tissues. Analysis of the relation of respiration to growth has been facilitated by the discovery that the action of growth substances in promoting growth and increasing respiration can be suppressed by certain inhibitors which do not influence the basal respiration itself. Iodoacetate is such an inhibitor as are other substances which like iodoacetate possess the power of inactivating sulfhydryl group-containing enzymes.[51] Through the use of iodoacetate it is possible largely to inhibit the effect of indoleacetic acid in inducing growth with only a very small reduction in overall rate of respiration of *Avena* coleoptile tissue. The growth substance response may then involve enzymes more sensitive to sulfhydryl reagents than the respiratory enzymes.

A second type of inhibitor of a wholly different nature is arsenate, whose action has been described in Chapter 15. In the present case, arsenate inhibits both the growth and the increase in respiration induced

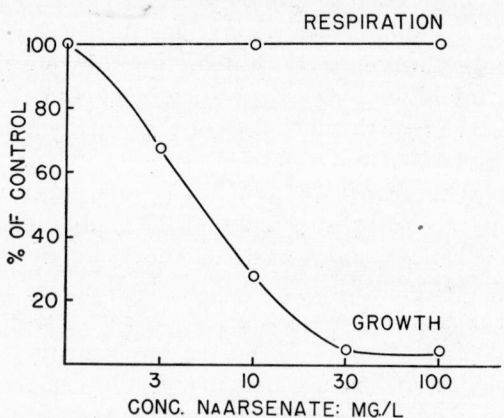

EFFECT OF NaARSENATE ON GROWTH & RESPIRATION

Fig. 29-6. Arsenate inhibition of the indoleacetic acid induced growth response of *Avena* coleoptile sections. The basal respiratory rate of the coleoptile remains unaffected.[43]

by indoleacetic acid, but is without effect on the basal metabolism of *Avena* coleoptile sections, as is shown in Fig. 29-6. We have seen in Chapter 15 that arsenate is capable of replacing inorganic phosphate in oxidative systems for which phosphate uptake is compulsorily coupled with substrate oxidation. That arsenate interferes with the metabolism of phosphate in the *Avena* coleoptile is indicated by the fact that the arsenate inhibition of growth substance action may be relieved by the further addition of phosphate to the tissue. The influence of arsenate in inhibiting the effect of added growth substance is a competitive one and depends upon the ratio between added phosphate and added arsenate. These observations suggest then relationships between growth substance metabolism and the utilization of energy-rich phosphate produced in respiration. That such a relationship may exist is further suggested by the fact that 2,4-dinitrophenol, a substance also discussed in Chapter 15, which is capable of uncoupling the respiratory oxidations from energy-rich phosphate production, is a powerful inhibitor of growth substance action although this same substance actually increases rate of respiratory gas exchange.[52]

In view of the minute quantities of growth substance which suffice to elicit plant growth responses, it has long been assumed that the material must in all probability participate as a component of an enzyme system essential to growth. Confirmation of this hypothesis would then require identification and isolation of the enzyme involved. A direct approach to the problem has been sought in the isolation of the protein-bound growth substance which is contained in plant tissues. It has been shown earlier that all or essentially all of the protein-bound growth substance of vegetative tissues, such as leaves, is found associated with the major protein component, Fraction I. This protein has phosphatase activity associated with it, as has been discussed in Chapter 17. The phosphatase acts *in vitro* to hydrolyze the phosphate linkages in such compounds as ATP and pyrophosphate, as well as the ester phosphate linkages in such compounds as glycerophosphate. It has not as yet proved possible however to demonstrate any direct relation between growth substance content and phosphatase activity of the auxin protein, nor has it proved possible to test directly the possibility that the protein-bound auxin is indeed the form of auxin which is physologically active in bringing about growth responses. The association of growth substance with an enzyme of the phosphate metabolism does however suggest the possibility of some relationship between growth substance and phosphate metabolism.

It will be evident from what has been said that the plant growth substances appear to be involved in a very complex and little understood region of the cellular machinery, namely, in that part which is concerned with making available or directing utilization of respiratory energy. It is little wonder then that application of growth substances to diverse tissues may be attended by such diverse physiological manifestations.

Pharmacology of the Growth Substances.[54] One year after the discovery in 1934 of indoleacetic acid as a native plant growth substance it was shown by several groups of workers that certain other substances related to indoleacetic acid in structure also possess growth substance activity. This is true even of compounds not known to occur in plant tissues and therefore probably to be regarded as synthetic analogs of the naturally occurring substances rather than as possible alternative natural growth substances. These active compounds largely fall into four groups.

1. Indole derivatives other than indoleacetic acid itself.
2. Naphthalene derivatives such as naphthaleneacetic acid.
3. Phenoxyacetic acid derivatives, including 2,4-dichlorophenoxy-acetic acid.
4. The substituted benzoic acids.

This classification is not complete as there are still other compounds, such as phenylacetic acid, which are not included and which show marked growth substance activity. Each of the four groups enumerated above does, however, contain many known active compounds.

A striking feature of the synthetic growth substances is the fact that any given compound does not necessarily possess the same activity relative to indoleacetic acid in bringing about the various growth responses. This is shown in Table 29-6. Naphthaleneacetic acid, for example, is only 2.5% as active as indoleacetic acid in causing curvature in the *Avena* curvature test, but is 15% as effective as indoleacetic acid in the *Avena* section test and as effective or more effective than indoleacetic acid in the split-pea curvature and tomato petiole bending tests. All these methods of estimation of growth substance activity deal with cellular elongation responses. They nevertheless differ in chemical specificity. One reason for this anomaly undoubtedly lies merely in the differing secondary effects which enter into each test. Thus in the *Avena* curvature test it is necessary for the growth substance to be taken up by the cells of the coleoptile stump and then to be passed on from cell to cell through the coleoptile. This transport, which is not a simple diffusion but rather a polar active movement, is itself highly specific, indoleacetic acid being transported more rapidly and in larger quantities than any of the synthetic analogs.[55] The *Avena* curvature test measures therefore the transportability of a substance as well as its activity in causing actual cell elongation. The *Avena* section test and the split-pea stem test measure more nearly the ability of a compound to induce cell elongation, since in these tests the material is applied directly to the tissues which are to undergo elongation.[56] Still another secondary property of a prospective growth substance, a property not directly related to growth substance activity but which may modify apparent activity, is susceptibility to oxidation by indoleacetic acid oxidase or other growth substance destroying enzymes in the tissue. Thus several analogs appear to be more active than indoleacetic acid in the pea curvature test, a tissue which contains much indoleacetic acid oxidase and in which a portion of the applied indoleacetic acid is undoubtedly destroyed without causing growth. In the *Avena* section test, carried out with a tissue which contains a lesser activity of the enzyme, only 2,4-dichlorophenoxyacetic acid has thus far appeared to be actually more active than indoleacetic acid. There are nevertheless unexplained differences between the relative activities of indoleacetic acid and the various synthetic growth substances as compared in the various biological assays. These differences may sometimes be quite spectacular. Thus naphthaleneacetic acid and indolebutyric acid are found to be in general more effective than indoleacetic acid itself in the formation of roots on cuttings, and in fact the optimal rooting of a given species may frequently be obtained only with one of the two above-mentioned compounds. 2,4-Dichlorophenoxy-acetic acid is highly effective in inhibition of abscission of some varieties of apples, but is ineffective with other varieties for which naphthalene-acetic acid may function satisfactorily. The biochemical basis for these differences in activity is unknown, and thus the synthetic growth sub-stance most effective for a given purpose must still be sought on a purely empirical basis.

Our knowledge of physiological activity of substances related to indoleacetic acid constitutes at present, despite the gaps alluded to above, one of the best investigated studies of the relation of chemical structure to physiological activity. So well understood are the general structural requirements for activity in the split-pea curvature test,[57] that it has been possible to lay down fairly well-defined qualifications which a molecule must possess if it is to be effective in causing this cell elongation response. These qualifications are:

1. The molecule must possess a ring system as a nucleus.
2. There must be one or more double bonds in the ring system.
3. The molecule must possess a side chain, in general two or more carbon atoms in length.
4. The side chain must possess a carboxyl group or a group readily convertible to a carboxyl group.
5. The carboxyl group must be capable of occupying a particular spacial relationship to the nucleus.

The possession of these qualifications, while it appears to assure that a substance will be active in the split-pea test, does not guarantee that the substance will be fully active as a growth substance in other tests; thus the structural requirements for transportability in the *Avena* curvature test are apparently much more rigorous than the requirements for growth activity alone.

Metabolic Antagonists of Growth Substances. Among the many substances tested for their ability to simulate the effects of indoleacetic acid as plant growth substances a few have been found which appear to act antagonistically to indoleacetic acid and to inhibit the effect of the

TABLE 29-6.   Relative activities of some synthetic growth substances in eliciting various responses

Activity relative to IAA in per cent

Compound	Structure	Avena curvature test	Avena section test	Pea curvature test	Rooting of cuttings (various sp)	Tomato petiole bending	Bud inhibition (various sp)	Root inhibition (Avena)	Formative effects (tomato)
Indoleacetic acid	—CH₂COOH	100	100	100	100	100	100	100	None
Indolebutyric acid	—(CH₂)₃·COOH	8	9	190	150	6	100+	10	None
Indene-3-acetic acid	—CH₂COOH	1	7	20	100	......	14	.....	......
α-Naphthalene-acetic acid	CH₂COOH	2.5	15	370	150	100	100+	4	None

Compound	Structure								
β-Naphthoxyacetic acid	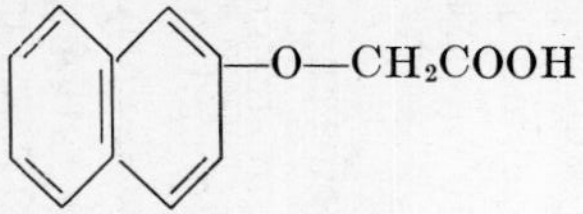 O—CH₂COOH (naphthalene)	0.00	....	....	25	15	ca. 100	.....	Active
Phenylacetic acid	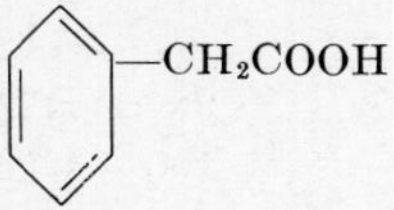 CH₂COOH	0.02	1.0	10	0	0.60	ca. 0	0.3	None
Phenoxyacetic acid	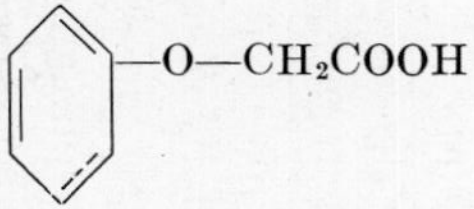 O—CH₂COOH	0.00	0.00	0	0	0.15	....	.....	None
2,4-Dichloro-phenoxyacetic acid	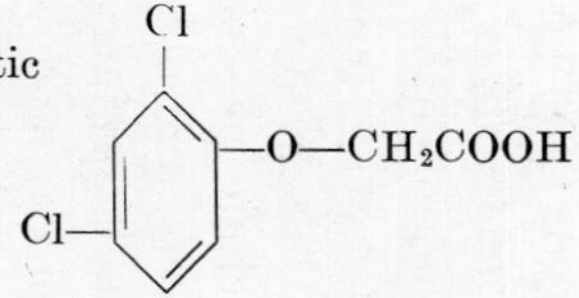 Cl ... O—CH₂COOH ... Cl	0.00	300	200–1200	ca.1500	100	....	30	Active
2,3,5-Triiodobenzoic acid	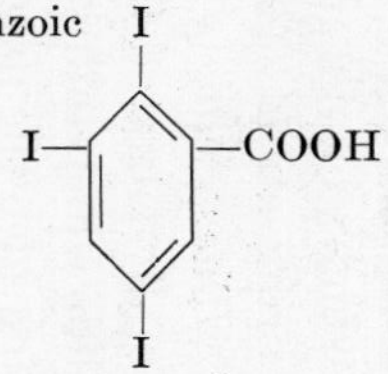 I, I, COOH, I	Inhibits	Inhibits	....	Inhibits	0.0	Promotes growth	.....	Active

latter. Perhaps the first to be specifically noted was γ-phenylbutyric acid which while itself slightly active as a growth substance is neverthe-

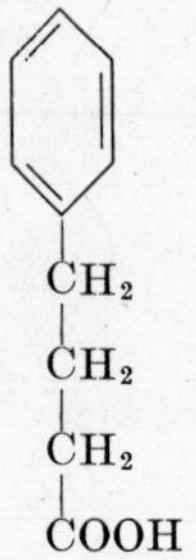

γ-Phenylbutyric acid 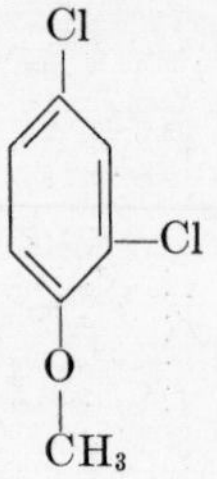   2,4-Dichloro-anisole 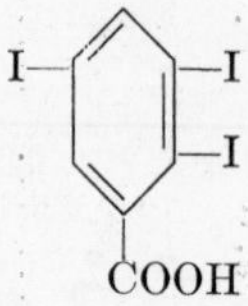   2,3,5-Triiodo-benzoic acid

Structures of compounds which act antagonistically to indoleacetic acid in plant responses

less highly effective in the *Avena* curvature and section tests in inhibiting the effect of simultaneously applied indoleacetic acid.[58] 2,4-Dichloro-anisole, an analog of 2,4-dichlorophenoxyacetic acid, is itself wholly inactive as a growth substance, but is markedly effective in decreasing the response to applied indoleacetic acid.[52] Still a third compound, 2,3,5-triiodobenzoic acid possesses similar properties in that it is effective in reducing response to indoleacetic acid in the *Avena* curvature test, in the section test, and in the rooting of cuttings.[59] Triiodobenzoic acid applied to intact plants yields symptoms which might be interpreted as due to lowering of the effective native growth substance level in the plant. These symptoms include reduction of growth and lessening of apical dominance with outgrowth of lateral buds. Undoubtedly further antagonists of indoleacetic acid remain to be discovered.

Formative Effects. Triiodobenzoic acid together with other compounds which do not necessarily possess any true growth substance activity bring about what has been termed by Zimmerman formative effects.[60] These effects include production of misshapen leaves, early or abnormally situated flowers, and other morphogenetic manifestations. Compounds which induce such formative activities are in general capable of inducing also the parthenocarpic development of fruits on such plants as the tomato. Although triiodobenzoic acid, which is inactive as a growth substance, does possess formative effects, not all compounds which show these effects are necessarily active as antagonists to indole-acetic acid, at least as judged by ability to inhibit growth substance-induced cell elongation. Thus 2,4-dichlorophenoxyacetic acid is highly active in cell elongation and also in induction of formative effects. The property of bringing about formative effects would appear to have a somewhat different chemical specificity than does growth substance activity proper, but the two forms of specificity appear to overlap to an appreciable extent.

General References

Bonner, J., and Wildman, S. G., Sixth Growth Symposium, 1946.   Page 51.
van Overbeek, J., *Ann. Rev. Biochem.*, **13**, 631 (1944).
Skoog, F., *Ann. Rev. Biochem.*, **16**, 529 (1947).
Thimann, K. V., The Hormones. Vol. I.   Academic Press, 1948.   Chapters II and III.
Went, F. W., and Thimann, K. V., Phytohormones.   Macmillan, 1937.
Zimmerman, P. W., and Hitchcock, A. E., *Ann. Rev. Biochem.*, **17**, 601 (1948).

References

1. Bonner, J., *J. Gen. Physiol.*, **17**, 63 (1933).
   Schneider, C. L., *Am. J. Botany*, **25**, 258 (1938).
2. Thimann, K. V., and Bonner, W. D., *Am. J. Botany*, **35**, 271 (1948).
   Bonner, J., *Am. J. Botany*, **36**, 323 (1949).
3. Galston, A. W., and Hand, M. E., *Am. J. Botany*, **36**, 85 (1949).
4. Went, F. W., *Rec. trav. bot. néerland.*, **25**, 1 (1928).
   Went, F. W., and Thimann, K. V., Phytohormones.   Macmillan, 1937.
5. Went, F. W., *Proc. Kon. Akad., Wet. Amsterdam*, **37**, 547 (1934).
   Thimann, K. V., and Schneider, C. L., *Am. J. Botany*, **25**, 270 (1938).
6. Uyldert, I., *Proc. Kon. Akad. Wet. Amsterdam*, **31**, 59 (1927); Thesis, Utrecht, 1931.
7. Avery, G. S., *Bull. Torrey Botan. Club*, **62**, 313 (1935).
8. Hitchcock, A. E., *Contrib. Boyce Thompson Institute*, **7**, 349, 447 (1935).
9. Thimann, K. V., and Went, F. W., *Proc. Kon. Akad. Wet. Amsterdam*, **37**, 456 (1934).
   See review in Thimann, K. V., and Behnke, J., The use of auxins in rooting of woody cuttings.   Harvard Forest, 1947.
10. Gustafson, F. G., *Proc. Natl. Acad. Sci.*, **22**, 628 (1936); *Botan. Rev.*, **8**, 599 (1942).
11. Laibach, F., and Fischnich, O., *Ber.*, **53**, 469 (1935).
    Kraus, E. J., Brown, N. A., and Hamner, K. C., *Botan. Gaz.*, **98**, 370 (1936).
12. Gautheret, R., La culture des tissues.   Gallimard, 1945.   Also 6th Growth Symposium, 1946.
13. Snow, R., *New Phytologist*, **34**, 347 (1935).
    Söding, H., *Jahrb. wiss. Botan.*, **82**, 534 (1936).
14. Kögl, F., Haagen-Smit, A. J., and Erxleben, H., *Z. physiol. Chem.*, **228**, 104 (1934).
    Bonner, J., and Koepfli, J. B., *Am. J. Botany*, **26**, 557 (1939).
15. Thimann, K. V., and Skoog, F., *Proc. Roy. Soc. (London)*, **B114**, 317 (1934).
    Thimann, K. V., *Biol. Revs.*, **14**, 314 (1939).
16. Laibach, F., *Ber.*, **51**, 336 (1933).
    Gardner, F. E., and Cooper, W. C., *Botan. Gaz.*, **105**, 80 (1943).
17. Yin, H. C., *Proc. Natl. Acad. Sci.*, **23**, 174 (1937).
18. Dostál, R., and Hosek, M., *Flora*, **31**, 263 (1937).
19. Thurlow, J., and Bonner, J., *Am. J. Botany*, **34**, 603 (1947).
    Bonner, J., and Thurlow, J., *Botan. Gaz.*, **110**, 613 (1949).
20. Clark, H. E., and Kerns, K. R., *Science*, **95**, 536 (1942).

21. Mitchell, J. W., and Marth, P. C., *Botan. Gaz.*, **106**, 199 (1944).
22. Ferri, M. G., and Lex, A., *Contrib. Boyce Thompson Institute*, **15**, 283 (1948).
23. Reinders, D. E., *Rec. trav. bot. néerland.*, **39**, 1 (1942).
24. Kögl, F., Haagen-Smit, A. J., and Erxleben, H., *Z. physiol. Chem.*, **228**, 90 (1934).
25. Thimann, K. V., *J. Biol. Chem.*, **109**, 279 (1935).
26. Haagen-Smit, A. J., Dandliker, W. B., Wittwer, S. H., and Murneek, A. E., *Am. J. Botany*, **33**, 118 (1946).
27. See review in Tang, Y. W., and Bonner, J., *Arch. Biochem.*, **13**, 11 (1947).
28. Wildman, S. G., and Bonner, J., *Am. J. Botany*, **35**, 740 (1948).
29. Kögl, F., Erxleben, H., and Haagen-Smit, A. J., *Z. physiol. Chem.*, **214**, 241 (1933); **225**, 215 (1934).
30. Full discussion in Larsen, P., *Dansk. Botan. Arkiv.*, **11**, 1 (1944).
31. Stefan, J., *Sitzber. Akad. Wiss. Wien, Math. Nat. Klasse*, **79**, 11, 161 (1879). Oholm, L., *Med. Vet.-Akad., Nobelinst.*, **2**, 1 (1913). Heyn, A., *Proc. Kon. Akad., Wet. Amsterdam*, **38**, 1074 (1935).
32. Kramer, M., and Went, F. W., *Plant Physiol.*, **24**, 207 (1949).
33. Wildman, S. G., Ferri, M. G., and Bonner, J., *Arch. Biochem.*, **13**, 131 (1947).
34. Gordon, S. A., and Sánchez Nieva, F., *Arch. Biochem.*, **20**, 356, 367 (1949).
35. Larsen, P., *Am. J. Botany*, **36**, 32 (1949).
36. White, E. P., *New Zealand J. Sci. Tech.*, **25B**, 137, 1944.
37. Tsui, C., *Am. J. Botany*, **35**, 172 (1948).
38. Wildman, S. G., and Muir, R. M., *Plant Physiol.*, **24**, 84 (1949).
39. Tang, Y. W., and Bonner, J., *Arch. Biochem.*, **13**, 11 (1947); *Am. J. Botany*, **35**, 570 (1948).
40. Galston, A. W., *Proc. Nat. Acad. Sci.*, **35**, 10 (1949).
41. Kögl, F., Koningsberger, C., and Erxleben, H., *Z. physiol. Chem.*, **244**, 266 (1936). Schuringa, G., Thesis, Utrecht, 1941.
42. Avery, G. S., Berger, J., and Shalucha, B., *Am. J. Botany*, **28**, 596 (1941). Berger, J., and Avery, G. S., *Am. J. Botany*, **31**, 199, 203 (1944). Haagen-Smit, A. J., Leech, W. D., and Bergren, W. R., *Am. J. Botany*, **29**, 500 (1942). Hatcher, E. S. J., *Ann. Botany*, **IX**, 235 (1945).
43. Stehsel, M., Thesis, University of California, Berkeley, 1949.
44. Stewart, W. S., *Botan. Gaz.*, **101**, 91 (1939). Also unpublished work of W. Dandliker.
45. Skoog, F., and Thimann, K. V., *Science*, **92**, 64 (1940).
46. Wildman, S. G., and Gordon, S. A., *Proc. Nat. Acad. Sci.*, **28**, 217 (1942).
47. Wildman, S. G., and Bonner, J., *Arch. Biochem.*, **14**, 381 (1947).
48. Gordon, S. A., *Am. J. Botany*, **33**, 160 (1946).
49. Commoner, B., and Thimann, K. V., *J. Gen. Physiol.*, **24**, 279 (1941). Berger, J., Smith, P., and Avery, G. S., *Am. J. Botany*, **33**, 601 (1946). Bonner, J., *Am. J. Botany*, **36**, 429 (1949).
50. Bonner, J., *J. Gen. Physiol.*, **20**, 1 (1936).
51. Thimann, K. V., and Bonner, W. D., *Am. J. Botany*, **36**, 214 (1949).
52. Bonner, J., *Am. J. Botany*, **36**, 323 (1949).
53. Wildman, S. G., Campbell, J. M., and Bonner, J., *Arch. Biochem.*, **24**, 9 (1949).
54. This large field is summarized to some extent in the following papers: Thimann, K. V., and Schneider, C. L., *Am. J. Botany*, **26**, 328 (1939). Thompson, H. E., Swanson, C. P., and Norman, A. G., *Botan. Gaz.*, **107**, 476 (1946).

Zimmerman, P. W., and Hitchcock, A. E., *Contrib. Boyce Thompson Institute,* **12,** 321 (1942).

Haagen-Smit, A. J., and Went, F. W., *Proc. Kon. Akad. Wet. Amsterdam,* **38,** 852 (1935).

Avery, G. S., Berger, J., and Shalucha, B., *Botan. Gaz.,* **104,** 281 (1942).

Veldstra, H., *Enzymologia,* **11,** 97 (1944).

See also references 56, 57, 60.

55. van der Weij, H., *Rec. trav. botan. néerland.,* **29,** 379, 1932; **31,** 810, 1932.

Went, F. W., and White, R., *Botan. Gaz.,* **100,** 465 (1939).

56. Thimann, K. V., *Proc. Kon. Akad. Wet. Amsterdam,* **38,** 896 (1935).

57. Koepfli, J. B., Thimann, K. V., and Went, F. W., *J. Biol. Chem.,* **122,** 763 (1938).

58. Skoog, F., Schneider, C. L., and Malan, P., *Am. J. Botany,* **29,** 568 (1942).

59. Galston, A. W., *Am. J. Botany,* **34,** 356 (1947).

Thimann, K. V., and Bonner, W. D., *Plant Physiol.,* **23,** 158 (1948).

60. Zimmerman, P. W., *Cold Spring Harbor Symposia Qaunt. Biol.,* **10,** 152 (1942).

CHAPTER 30

PHOTOSYNTHESIS

Introduction. The process of photosynthesis is perhaps the most important single activity of the plant and the relative amount of space devoted to the process in this book is in no way proportional to the significance of photosynthesis in comparison with other plant functions here considered. Much of the work which has been done on photosynthesis has been of a clearly physiological rather than of a biochemical nature. Thus, much attention has been paid to the kinetics of the gas exchange and to the influence of external factors on the photosynthetic rate. There is, however, a small but rapidly increasing body of information concerning the biochemistry, the chemical mechanism, of photosynthesis and we will be concerned here only with this phase of the study of the photosynthetic process.

By the term photosynthesis we understand the process in which the energy of light is used by living organisms for the reduction of CO_2 to organic matter, frequently to carbohydrate. The overall reaction of the photosynthetic process as it occurs in green plants is given by the classical equation in which CO_2 and water are taken up by the plant and reduced in the presence of light to carbohydrate with the concomitant evolution of oxygen.

$$CO_2 + H_2O \xrightarrow{\text{light}} (CH_2O) + O_2$$

We will take it as well established that the photosynthesis of green plants is intimately concerned with chlorophyll, and that the light energy which carries out the photosynthetic reaction is largely that absorbed by chlorophyll. Our first question is then the present status of our knowledge of the chemistry of chlorophyll.

Chemistry of Chlorophyll. The isolation of chlorophyll from green leaves by Willstätter[1] established that this pigment consists actually of two major components, chlorophyll a and chlorophyll b. These two components differ from one another in absorption spectra (Fig. 30-1) as well as in solubilities and other chemical properties, chlorophyll a being for example more soluble in organic solvents than chlorophyll b. The proportions of chlorophyll a and b in the leaf are by no means constant but tend nevertheless to approach a value of approximately 2–3 mols of *a* per mol of *b*.   No clear distinction between the activities or roles of the two chlorophylls has as yet emerged, and it is improbable that photosynthesis involves any interconversion of the two forms.   The chemical work of Willstätter early established the presence of magnesium in chlorophyll as well as the empirical formula, $C_{55}H_{72}N_4O_5$ Mg, of chloro-

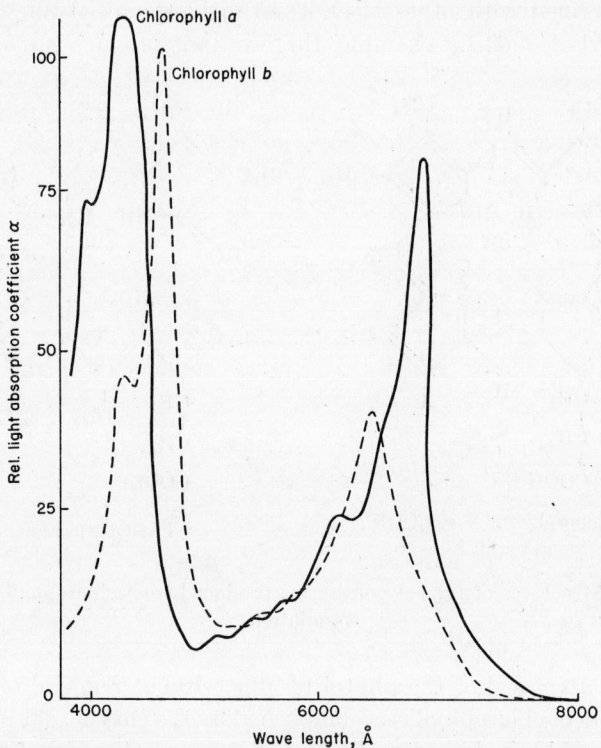

FIG. 30-1.   Absorption spectra of chlorophyll *a* (___) and chlorophyll *b* (_ _).   (From Zscheile[34])

phyll *a*.   In addition Willstätter found that saponification of chlorophyll leads to the release of methyl alcohol and of the reduced diterpene alcohol, phytol (Chapter 26), one molecule of each being found in the chlorophyll molecule.   Early work on the chemistry of chlorophyll also indicated the relationship of this material to the iron-containing hemin of hemoglobin and of catalase (Chapter 15) since both compounds were found to yield porphyrins on alkaline degradation.   The structures of several of the porphyrins thus produced from chlorophyll have been elucidated by the synthetic work of Hans Fischer,[2] firmly establishing chlorophyll as a porphyrin. The action of acid, alkali and certain enzymes also produces well-characterized derivatives of chlorophyll. Thus when chlorophyll a or b is treated with a weak acid, as oxalic, the magnesium is removed with the production of pheophytins. Hydrolysis of pheophytin with stronger acid removes the phytol group to yield pheophorbides a and b, which are both monocarboxylic acids and which can be esterified with other alcohols to yield the alkyl pheophorbides; with methyl alcohol, for example, the resultant product is methyl pheo-

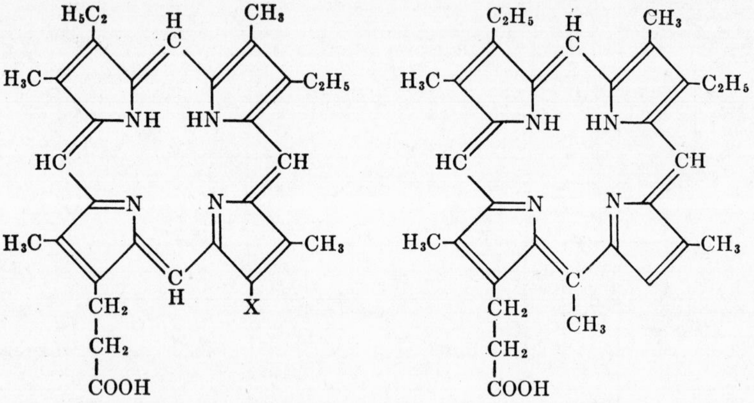

Rhodoporphyrin, X = COOH
Pyrroporphyrin, X = H

Phylloporphyrin

FIG. 30-2. Structures of typical porphyrins produced from chlorophyll by alkaline degradation.

phorbide. Removal of the phytol residue without removal of the magnesium atom of chlorophyll is accomplished by the enzyme chlorophyllase, which is present in leaves and which becomes active when leaves are injured. Thus when leaves are soaked in alcohol, the phytol is removed by the action of chlorophyllase and replaced by the alcohol present in the solution with the production of the corresponding alkyl chlorophyllide. The action of hot alkali on the pheophorbide or ethyl chlorophyllide is to saponify the ester linkages and to produce still another free carboxyl group with the formation of chlorin-e which contains in all three free carboxyl groups. The degradative reactions of chlorophyll a are summarized in Fig. 30-3. Table 30-1 gives in addition a brief survey of the rather involved nomenclature of the chlorophyll derivatives.

The structure of chlorophyll itself has been arrived at through the further studies of Conant, Stoll, and others and especially of Hans Fischer; the details of this work may be found in the reviews cited at the

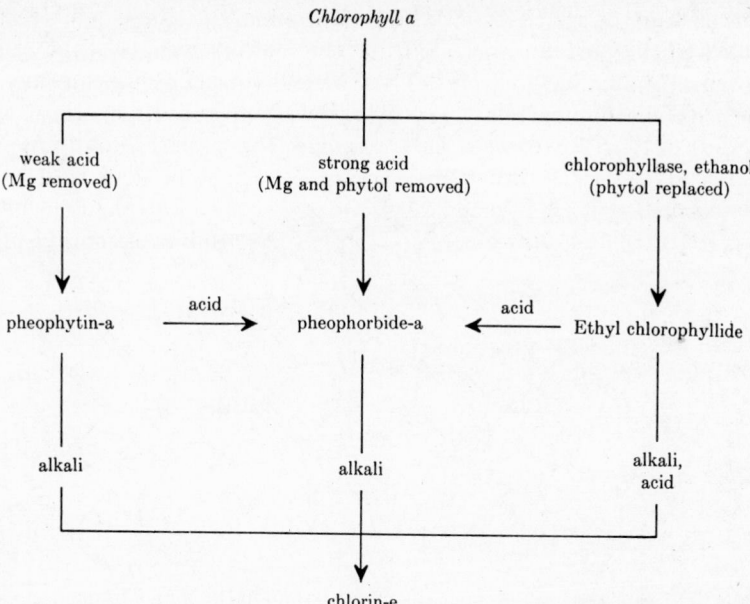

FIG. 30-3. Summary of the action of acid, alkali, and chlorophyllase on chlorophyll.

end of this chapter. The Fischer proposals for the structures of chlorophylls *a* and *b* include, in addition to the tetrapyrrole or porphin ring system, a five-membered carbocyclic or isocyclic ring as well as characteristic substituents. It may be seen that the magnesium atom is

TABLE 30-1. Guide to the nomenclature of chlorophyll derivatives

Name applied	Structural relationship
Porphin	Tetrapyrrole nucleus
Porphyrin	Ring-substituted porphins related to hematin or chlorophyll
Phorbin	Dihydroporphyrin containing carbocyclic ring of chlorophyll
Chlorin	Phorbin with carbocyclic ring opened
Phorbide	Phorbin ester
Pheophorbide	Phorbide derived from chlorophyll, a methyl ester
Phytin	A phytyl phorbide
Pheophytin	Phytin derived from chlorophyll (Mg free)
Phyllin	Mg derivative of phorbide, chlorin, etc.
Chlorophyllin	Phytyl phorbide containing Mg
Chlorophyllide	Pheophorbide containing Mg
Chlorophyll	Phytyl chlorophyllide

present in the center of the phorphin ring and is bound to the four pyrrole nitrogen atoms much as is the iron atom in the iron porphyrins. The conversion of chlorophyll to pheophytin by weak acids involves, then, the replacement of magnesium by two hydrogen atoms. The phytol residue of chlorophyll is esterified to a propionic acid residue which is attached to the porphin nucleus, while the methyl ester group, which is freed only by alkaline hydrolysis, is attached through a carboxyl group of the isocyclic five-carbon ring. The closest approach which has been

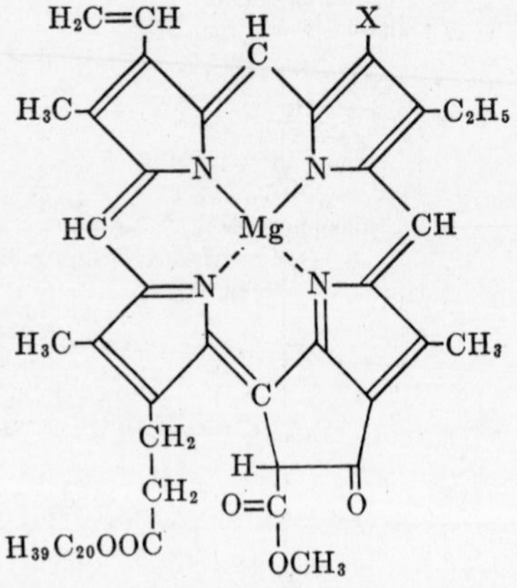

FIG. 30-4. Fischer proposal for structures of chlorophylls a and b. In chlorophyll a X = CH₃, in chlorophyll b X = CHO[3].

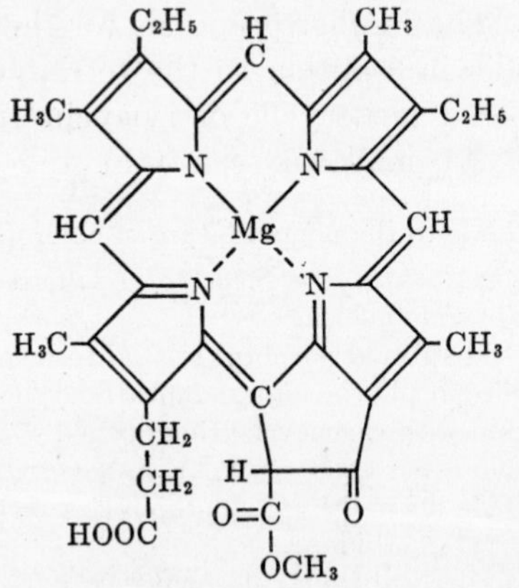

FIG. 30-5. Structure of the phyllin (Mg derivative) of synthetic pheoporphyrin a_5 methyl ester $C_{35}H_{34}N_4O_5Mg$.

made thus far to a total synthesis of chlorophyll has been the synthesis of pheoporphyrin a_5.[3] The phyllin or magnesium-containing derivative of pheoporphyrin a_5 is isomeric with chlorophyllide a and differs from the latter in substitution of an ethyl group for the vinyl group and in the occurrence of one extra double bond in the porphin ring system of pheoporphyrin a_5.[4]

Although the porphyrin of chlorophyll is closely related in structure to those of the iron-containing hemes of plant and animal origin, there are two major innovations in the chlorophyll molecule which are not found in any of the known heme porphyrins. The first of these is the presence of the phorbin ring which is, so far as known, unique to the photosynthetic pigments and which, as will be shown below, is apparently introduced into the structure relatively late in the biogenesis of the chlorophyll molecule. Esterification with the alcohol phytol is likewise unique to the chlorophylls. The presence of magnesium as the coordinated metal in the chlorophyll porphyrin is of course also a property not shared with other pigments.

Chloroplast Structure and Development. The structure of the chloroplast has been considered in Chapter 17. In summary the saucer-shaped chloroplast is made up of smaller particles, the grana, which contain chlorophyll bound in a pigment-protein complex. The grana are in turn composed of numerous layers or laminae. The orientation or arrangement of chlorophyll molecules within the granum is unknown.

Chloroplasts apparently develop only from already existing chloroplast or proplastid bodies, and increase in number in the cell only by self-reproduction of existing chloroplasts. This may be readily observed in cells of algae or of bryophytes in which division of the chloroplasts may precede or accompany cell division, with distribution of the daughter chloroplasts to the daughter cells. The chloroplasts of the higher plant develop from proplastid granules which in maize are $1\ \mu$ or less in diameter and which enlarge to 3–$4\ \mu$ or half of the final diameter before marked pigment production begins to occur.[5] In the case of maize each cell originally contains many more proplastid granules than develop into mature chloroplasts, and as yet unrecognized factors regulate both the reproduction of the proplastid granules and the number of these which develop into chloroplasts.

Chloroplasts may undergo alterations or mutations which are permanent and which are inherited as cytoplasmic rather than as strictly Mendelian characters. In the simplest case, cells which contain no chloroplasts give rise only to chloroplast-free cells, as has been observed directly in *Euglena*.[6] In species such as *Mirabilis* in which no functional proplastids are transmitted to the egg by the pollen tube, chloroplast inheritance is strictly maternal. Flowers which develop on chloroplast-free branches of variegated plants give rise to viable seeds which yield only chloroplast-free seedlings. The iojap character of maize, a characteristic striping of the leaves, is brought about by a gene which induces mutation in the chloroplast. Once the chloroplast mutation has occurred, however, it is inherited as a cytoplasmic factor independent of the presence or absence of the iojap gene.

There are a great many purely genetic effects which are known to influence the development or composition of the chloroplast and in which the chloroplast is affected only when the gene in question is actually present in the genetic complement of the cell. In these cases the effects, unlike that of the iojap character, are strictly reversible and the chloro-

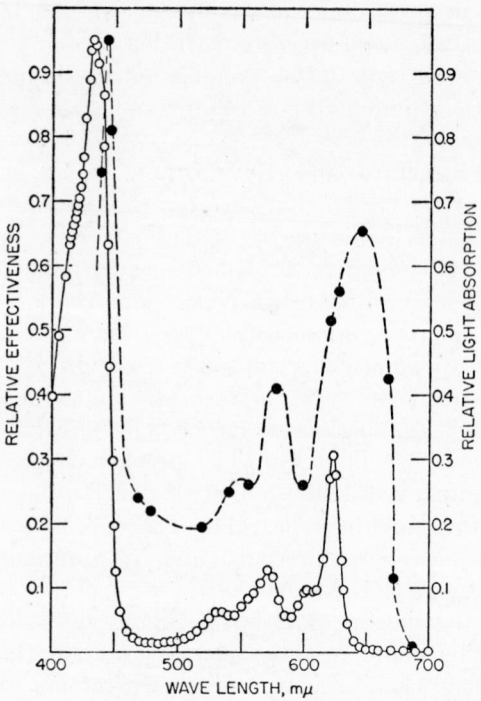

FIG. 30-6. The action (relative effectiveness) curve for chlorophyll formation compared with the relative light absorption curve of protochlorophyll in ether solution. The two curves are adjusted to be identical at the blue peak. (After Smith[8]) Closed circles, action spectrum; open circles, absorption spectrum.

plasts revert to normal in the absence of the particular modifying gene. Of some 65 genes known to influence chloroplast development in maize, thirteen produce albino seedlings in which the proplastids disintegrate without formation of chlorophyll.[7] Other of these genes permit more or less normal plastid development to take place, but chlorophyll is absent or present in reduced amount. In still other cases chlorophyll development is merely delayed or variegated plants are produced.

It is clear then that although the chloroplast possesses a considerable measure of autonomy in the plant and is subject to its own independent type of reproduction and inheritance, still the chloroplast is also subject to a very considerable control by the genetic constitution of the host cell.

Biogenesis of Chlorophyll. In most higher plants the development of chlorophyll is dependent upon light, although chlorophyll formation may proceed to some extent in darkness in certain cases, as in the seedlings of conifers as well as in particular algae and bryophytes. Even in these instances, however, light accelerates rate of chlorophyll formation. The action spectrum of chlorophyll formation, i.e., the relative effectiveness of different wavelengths in bringing about the process, has been

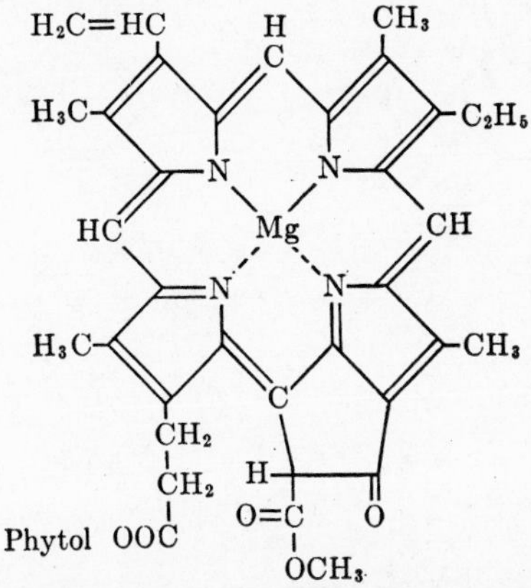

FIG. 30-7.   Structure of protochlorophyll.

accurately determined for barley seedlings[8] and found to coincide closely as to form with the absorption spectrum of a pigment, protochlorophyll, which occurs in etiolated leaves and seedlings (Fig. 30-6). The discrepancies which do exist between the two curves have their origin in secondary matters such as the fact that the position of the absorption maxima of protochlorophyll (as well as of chlorophyll) in the extracted state are shifted slightly to shorter wavelengths as compared to the positions of the same maxima in the living tissue. When etiolated barley seedlings are illuminated under conditions of low temperature ($0-4°$), it is found that the protochlorophyll disappears in stoichiometric proportion to the chlorophyll which appears.[8]   This is due to the fact that under conditions of low temperature the limited amount of protochlorophyll present in the seedling is converted to chlorophyll, but further protochlorophyll synthesis is blocked. Protochlorophyll is, however, regenerated in the dark at higher temperature. It would appear therefore that it is the conversion of protochlorophyll to chlorophyll which requires light. In the greening of etiolated tissue then, protochlorophyll may be produced continuously and immediately transformed photochemically to chlorophyll. In the normal green plant protochlorophyll does not accumulate to a detectable extent and if it is an intermediate in chlorophyll synthesis in this case also, as seems probable, it must be held at a low concentration level by photochemical transformation to chlorophyll. Protochlorophyll contains two less hydrogen atoms than chlorophyll and is in fact a vinyl derivative of pheoporphyrin a_5.[9] The photochemical transformation of protochlorophyll to chlorophyll is therefore a reduction.

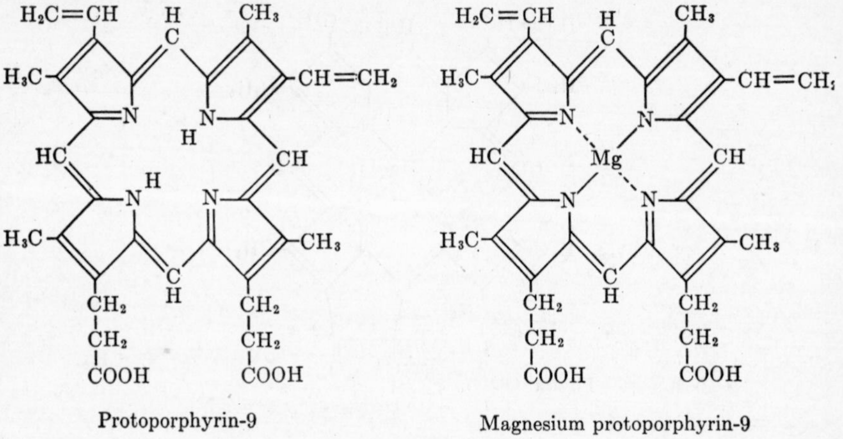

Protoporphyrin-9   Magnesium protoporphyrin-9

FIG. 30-8. Protoporphyrins accumulated by mutant strains of *Chlorella* in which chlorophyll synthesis is blocked by genetic defects.

Thus far our discussion of the biogenesis of chlorophyll has been confined to the terminal step in the process. There is some evidence, however, as to the nature of two earlier steps, evidence based on determination of the porphyrins accumulated by mutant strains of the green alga, *Chlorella*, in which chlorophyll synthesis is blocked by genetic defects.[10] Such chlorophyll-free mutant strains must of course be grown heterotrophically on nutrient medium containing a suitable carbon source such as glucose. Among the mutant strains produced by X-irradiation of normal cells, one was found to lack chlorophyll but to accumulate a brownish pigment, protoporphyrin-9, identical with the protophorphyrin derived from the iron protoporphyrin or hemin of animal hemoglobin. The structure of this porphyrin, which occurs widely in plant tissues in small amounts,[11] resembles in general that of chlorophyll *a*, but differs from the latter in number and arrangement of the double bonds and in the absence of the carbocyclic phorbin ring, as is shown in Fig. 30-8. A second chlorophylless mutant of *Chlorella* was found to accumulate magnesium protoporphyrin. In this case the same protoporphyrin nucleus is formed as in the first mutant but is accumulated as the magnesium-containing material rather than as the free protopor-phyrin. In *Chlorella* then, chlorophyll formation may proceed stepwise through protoporphyrin and magnesium protoporphyrin. The steps in the synthesis of these still relatively complicated compounds remain to be unraveled.

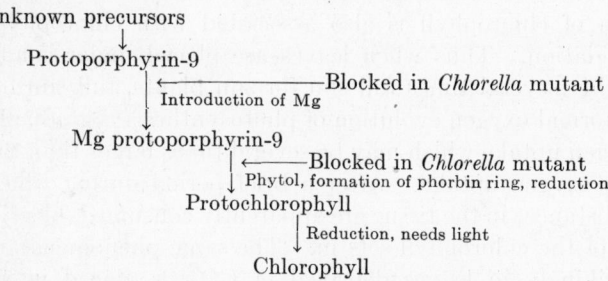

Fig. 30-9. Summary of information concerning the biogenesis of chlorophyll.

A number of nutritional deficiencies interfere strikingly with chlorophyll formation and of these the most important is perhaps iron. When iron is withdrawn from the nutrient or otherwise made unavailable to the plant, the new leaves which develop are deficient in chlorophyll. This deficiency may set in quite suddenly since iron, unlike nitrogen or phosphorus, is not mobilized by the growing tissues from other parts of the plant. It appears from the work of Jacobson, Bennett, and others[12] that the iron which is active in chlorophyll formation is that bound in organic combination actually in the chloroplast. Protein synthesis is also abnormally low in iron-deficient plants and soluble nitrogenous materials may accumulate to several times their normal levels. It would appear possible that the iron requirement for chlorophyll development may be related primarily to the production of chloroplastic protein rather than to the formation of the chlorophyll itself.

Chlorophyll Destruction. There are a variety of conditions under which chlorophyll is destroyed in the plant and attention has been called to one such case in Chapter 20. In leaves undergoing proteolysis, the chlorophyll disappears to the same extent as the chloroplastic protein, leading us to the conclusion that chlorophyll may be stable in the plant only when in its proper association with chloroplastic material. It is known in addition that intense illumination, or illumination when photosynthesis is inhibited by poisons or by lack of CO_2, leads to destruction of chlorophyll. This destruction is dependent on light and is therefore presumably a photochemical reaction.   It is well known that solutions of chlorophyll are unstable and become decolorized in light, the red absorption band in particular decreasing relatively rapidly.   The reaction depends not only on light but also upon the presence of oxygen and is hence a photochemical oxidation in which the chlorophyll itself is the substrate.[13]   The products of this reaction have not been identified. Chlorophyll is also able to sensitize the photochemical oxidation of substrates other than itself; benzidine and allylthiourea are such examples. *In vivo* destruction of chlorophyll is also associated with chlorophyll-sensitized photooxidation.   Thus when leaves are placed under conditions of intense light (several times full sun for sun plants, full sun for shade plants) the normal oxygen evolution of photosynthesis is gradually replaced by an oxygen uptake which may be several times larger than the normal respiratory oxygen uptake.   After a brief period during which other oxidizable substances in the tissue are apparently consumed, bleaching or destruction of the chlorophyll sets in.   The same phenomena are observed more strikingly in leaves deprived of CO_2 or placed in an atmosphere containing higher than atmospheric oxygen pressure.[14]

In summary the destruction or bleaching of chlorophyll both *in vitro* and *in vivo* appears to be the result of an oxidation involving the uptake of oxygen and promoted by the absorption of light by the chlorophyll itself.

Formulation of Photosynthesis.   The photosynthetic reaction carried out by green plants closely approximates that of equation 1 in which 1 mol of oxygen is evolved and 1 mol of carbon reduced to the level of carbohydrate for each mol of CO_2 taken up.   That equation 1 truly

$$CO_2 + H_2O \xrightarrow{\text{light}} (CH_2O) + O_2 \qquad\qquad (1)$$

represents the overall course of the reaction is indicated not only by the equivalence of CO_2 uptake and oxygen evolution in photosynthesis but also by the fact, elegantly demonstrated by Smith,[15] that the carbon assimilated in photosynthesis can be quantitatively accounted for in the various forms of carbohydrate formed, as is shown in Table 30-2.   Over periods of 0.5 to 2.5 hours, the carbon assimilated appears principally as sucrose, starch, and monosaccharides, with sucrose making up over half of the total for the shorter experimental periods.   Despite the fact that equation 1 correctly represents the overall changes in photosynthesis, it does not adequately express the chemical facts, since the oxygen which is evolved in photosynthesis is produced exclusively from water and does not come from the oxygen contained in the CO_2 molecule.   This has been shown[16] by allowing photosynthesis to take place in media in which the oxygen of the water or of the CO_2 was labeled with the oxygen isotope O^{18}. The results of these experiments show that the isotopic composition of the oxygen evolved in photosynthesis corresponds with the isotopic composition of the water but does not correspond with the isotopic com-

TABLE 30-2. Equivalence of carbohydrate formation and carbon dioxide absorption during photosynthesis of sunflower leaves at 20°C. (After Smith[15])

Time of illumination: minutes	27 ± 1	58 ± 4	101 ± 8	146 ± 5
Carbon absorbed: mg.	4.11 ± 0.02	7.77 ± 0.04	15.41 ± 0.10	23.14 ± 0.19
Fraction	*Per cent recovery of carbon absorbed*			
Monosaccharide	4.4	10.0	16.6	22.3
Sucrose	55.3	51.8	45.9	39.9
Unidentified sugar	....	3.1	3.2	1.1
Unidentified polysaccharide	−7.7	1.4	3.3	1.7
Starch	36.5	25.5	27.8	28.7
Total sol. carbohydrate	86.6	91.9	96.7	93.7
Total residue	20.7	6.5	−0.4	−1.5
Total recovery	107.3 ± 4.6	98.4 ± 3.1	96.4 ± 1.5	92.2 ± 1.3

position of the CO_2 utilized. We must therefore depict the photosynthetic reaction as shown in equation 2. The formulation of photo-

$$CO_2 + 2H_2O^* \xrightarrow{\text{Light}} (CH_2O) + O_2^* + H_2O \qquad (2)$$

synthesis given in equation 2 is but a special case of a still more generalized formulation of the process as it occurs in photosynthetic organisms in general and including the green plant. Equation 3, which is due to Van Niel,[17] represents photosynthesis as a process in which some material,

$$CO_2 + 2H_2A \xrightarrow{\text{Light}} (CH_2O) + H_2O + 2A \qquad (3)$$

H_2A, is oxidized to A under the influence of light while the hydrogen liberated in this oxidation is utilized for the reduction of CO_2. In green plant photosynthesis the material oxidized by light happens to be water. In the green sulfur bacteria, on the other hand, the substance oxidized by light is H_2S, and elemental sulfur, rather than oxygen, is evolved. Many other variants on the general formulation of photosynthesis are known in nature, including cases in which H_2A may be represented by organic

$$CO_2 + 2H_2S \xrightarrow{\text{Light}} (CH_2O) + H_2O + 2S \qquad (4)$$

compounds such as secondary alcohols (purple bacteria[18]) or even by molecular hydrogen itself as with certain specially adapted green algae.[19]

In the normal green plant, however, photosynthesis would appear to consist of a photochemical splitting of water in which the hydrogen is used to reduce CO_2 and the oxygen is liberated in the molecular form. It is in fact experimentally possible to separate the photochemical splitting of water from the process of CO_2 reduction by a technique in which hydrogen acceptors other than CO_2 are supplied to the chloroplast; this is discussed in the next section. Experimental separation of the two processes can also be achieved by illumination of green cells in the absence of CO_2. Under these conditions, reductant formed in the light is stored, although in small quantity, and when the cells are subsequently transferred to the dark and CO_2 is admitted, a measurable, although small, fixation of CO_2 takes place over a period of approximately one minute.[32] We may then regard photosynthesis according to the formulation of equation 2 as actually made up of two separate processes, namely the photochemical decomposition of water with the production of some reduced hydrogen acceptor, followed by the non-photochemical reduction of CO_2 at the expense of the reduced intermediate acceptor. This concept is summarized in equations 5 and 6 in which X represents the unknown intermediate acceptor.

$$2H_2O + 2X \rightarrow 2(X \cdot 2H) + O_2 \tag{5}$$
$$CO_2 + 2(X \cdot 2H) \rightarrow (CH_2O) + H_2O \tag{6}$$

Evolution of Oxygen. The photosynthesis of green leaves or cells is extraordinarily susceptible to injury, and even slight mechanical or chemical damage may result in complete cessation of photosynthetic activity. This fact has made it impossible as yet to make *in vitro* preparations capable of carrying out the overall process of photosynthesis. It was however discovered in 1937 by Robin Hill that isolated chloroplasts are capable of evolving oxygen upon illumination, provided only that a suitable hydrogen acceptor is present in the suspension.[20] This reaction, the Hill reaction, may be thought of as representing a partial reaction of photosynthesis, namely reaction 5 above. Hill found ferric salts to be suitable as the oxidant and it has been found furthermore that there is an equivalence between the amount of ferric iron reduced to the ferrous form and the amount of oxygen liberated. For the reduction of ferricyanide to ferrocyanide the reaction was found to be represented by equation 7. That the oxygen liberated during the course of

$$4Fe^{+++} + 2H_2O \xrightarrow[\text{Chloroplasts}]{\text{Light}} 4Fe^{++} + 4H^+ + O_2 \tag{7}$$

the reaction arises in fact from water can be shown[21] for photosynthesis by the use of water containing the oxygen isotope O^{18}. The number of quanta of light energy required for the production of 1 mol of oxygen in the Hill reaction is approximately 10–15 or close to the usual value reported for photosynthesis proper.[22]   The close relationship of the Hill reaction to photosynthesis is also indicated by the fact that intact normal cells of the green alga, *Chlorella*, are able to evolve oxygen upon illumination in the absence of CO_2 provided only that ferric ions or other suitable oxidants are present in the medium.

The oxidant for the Hill reaction need not be ferric ion but may be any one of a number of organic compounds such as p-benzoquinone,[23] 2,6-dichlorophenol-indophenol, or other oxidation-reduction dyes.   In these cases the dye is reduced to the leuco-dye in the light, and estimation of the activity of the chloroplast preparation may be based on colorimetry rather than on measurement of the oxygen evolved.   Naturally occurring oxidation-reduction systems, including dehydroascorbic acid, DPN, glutathione, and cytochrome C, are not reduced by the chloroplast system *in vitro*, and the nature of the native hydrogen acceptor which participates in photosynthesis is as yet obscure.

The structure of the intact chloroplast is not essential to the photochemical liberation of oxygen, and preparations of grana are in fact fully active.   In general, however, treatments which disrupt the grana are attended by losses in the activity, although fragmentation of grana by certain treatments has yielded particles which are smaller than grana and still retain activity.[21]   A successful method for such fragmentation has consisted in forcing grana suspensions through a small orifice under high pressure.[22]   By this treatment grana fragments are produced which are small enough to stay in suspension indefinitely and which retain activity.   The still smaller particles have, however, but little activity. It would appear then that the reaction requires a certain minimum unit of the granum, perhaps a unit comprising a full complement of essential enzymes packed together in an appropriate fashion.   Even in the absence of detailed knowledge of the enzymes and enzymatic reactions involved in the photochemical evolution of oxygen by isolated chloroplasts, it remains nonetheless that the discovery of this reaction has permitted the clear separation of photosynthesis into two component partial reactions and permits of the study of the photochemical reaction in a relatively simple *in vitro* system.

Quantum Yield.   A matter of much concern to students of photosynthesis has been the determination of the efficiency with which light energy is used in the process.   Early attempts to measure the ratio of U, the energy fixed in reduction of CO_2, to E, the light energy absorbed by the plant, were made by Brown and Escomb in 1905 who found that about 0.5% of the light incident on a plant under field conditions is fixed as chemical energy. Their figures are based on incident light energy rather than on energy actually absorbed by the plant and were furthermore carried out at high light intensities where the ratio U/E is now known to be smaller than it is at lower intensities. Particular interest attaches to the determination of the maximum possible value of U/E since knowledge of this value may assist in the formulation of the exact manner in which light energy is transformed to chemical energy. The reduction of CO_2 by water is strongly endothermic and requires of the order of 112,000 cal. for each mol of CO_2 reduced to the level of carbohydrate. Light at a wavelength of 7000 Å and active in photosynthesis supplies only approximately 40,500 cal. per einstein (mol of quanta). At least three

$$CO_2 + H_2O \xrightarrow{\text{nh}\nu} (CH_2O) + O_2 \qquad -\Delta H = -112,000 \text{ cal.} \qquad (8)$$

quanta of red light are therefore required merely to supply the minimum energy needed to reduce one CO_2 molecule, that is, the quantum yield or number of quanta needed to reduce one CO_2 molecule (n in equation 8) could not be less than about three on purely thermodynamic grounds. The first comprehensive attempt to measure accurately the quantum yield in photosynthesis, carried out by Warburg and Negelein in 1922–1923,[24] indicated that with the green alga *Chlorella* four quanta suffice for the reduction of each CO_2 molecule, a number only slightly greater than the minimum required on thermodynamic grounds. In the technique used by Warburg and Negelein and followed in principle by several more recent investigators, the algal cells are allowed to remain in the dark for a period during which the rate of respiratory gas exchange is measured. The algal suspension is then illuminated and rate of gas exchange again measured. The photosynthetic rate can now be calculated on the assumption that gas exchange in the light represents the difference between respiratory gas exchange and photosynthetic gas exchange. It is of course essential that exact measurements be made also of the amount of energy absorbed by the photosynthesizing cells during the period of illumination. In the experiments of Warburg and Negelein, as well as those of Rieke[25] and of Emerson and Lewis,[26] CO_2 and oxygen exchange were followed manometrically by measurements of the pressure changes within the vessel. Other methods of measurement of gas exchange have also been employed in quantum yield determinations as for example the purely chemical methods of oxygen analysis employed by Manning and others.[27]

Exact repetition of the procedures of Warburg and Negelein led to confirmation of their result.[25,26] It was, however, shown by Emerson and Lewis[26] that these high efficiencies are attainable only in experiments in which relatively short alterations of light and dark exposures are combined in a certain arbitrary way. Measurements of gas exchange over longer periods of light and dark lead to a value for the quantum yield of roughly ten quanta absorbed per molecule of oxygen evolved. This inconsistency is apparently due to the fact that the gas exchange of *Chlorella* is composed not only of that due to photosynthesis and to respiration but includes also a third component which is not directly related to photosynthesis. This extraneous gas exchange consists of a rapid burst or gush of CO_2 from the cells immediately on illumination, a response which is essentially completed within five minutes or less and which is accompanied by a brief uptake of CO_2 immediately following return of the cells to darkness. These short duration effects are not of importance in long-term measurements of gas exchange but appear to be of great importance when short periods, e.g., 10–15 minutes, are used as in the early measurements of quantum yield.

An entirely different method of measurement of quantum yield[28,29] has been based on direct calorimetric determination of the energy retained and energy lost as heat by algal suspensions during illumination. This method, which is not open to any of the errors or uncertainties of those based on gas exchange, also indicates that not more than 26–28% of the energy absorbed is converted to chemical energy. This efficiency corresponds to a quantum value of nine-ten quanta absorbed per molecule of CO_2 reduced.

Although there are now a substantial number of cases recorded in which the minimum number of quanta needed to reduce one CO_2 molecule is not less than approximately ten, there are nevertheless still reports which indicate that higher quantum efficiencies may possibly be attained.[30] So far as the mechanism of photosynthesis is concerned, however, it seems clear that the process may be and probably usually is carried on at a relatively low efficiency with the energy of ten or more quanta being in some way combined and concentrated for the reduction of a single CO_2 molecule. In all probability this is achieved by reduction of CO_2 through a series of graded steps, no single one of which involves an energy change greater than the energy contained in a single quantum.

Path of Carbon in Photosynthesis. We have seen that the photochemical reaction of photosynthesis consists essentially of the decomposition of water with the production of oxygen and a reduced hydrogen acceptor. The second portion of the photosynthetic reaction must consist then in the reduction of CO_2 with the aid of the reductant produced in the light reaction. We wish now to ascertain the path by which CO_2 is reduced and to discover the nature of the intermediate compounds involved. We have seen earlier that when photosynthesis is allowed to occur over an extended period, 30 minutes or longer, all the carbon reduced can be accounted for as carbohydrate formed. This means that the intermediates between CO_2 and carbohydrate do not accumulate to any high concentration as compared with the final product. In fact, identification of the intermediates concerned in photosynthetic CO_2 reduction has become possible only with the availability of isotopic carbon and with the development of techniques for the study of the products formed not in photosynthetic periods of thirty minutes or more but formed in periods of thirty seconds or less. Early isotope experiments on the pathway of carbon in photosynthesis were carried out by Ruben, Hassid, and Kamen in 1939–1940.[31] This work established that the course of photosynthesis as followed by rate of uptake of the radioactive carbon isotope parallels the course as measured by the usual manometric gas exchange methods. It was also found by these workers that small amounts of CO_2 are taken up in the dark immediately following an exposure to light and that the product formed is a carboxylic acid or acids. We have seen in Chapter 14 that a great variety of nonphotosynthetic CO_2 fixation mechanisms are found both in heterotrophic lower organisms and in tissues of higher plants and that such dark CO_2 fixation may attain large proportions, particularly in certain leaves as in those of the succulents. It is only natural then to inquire as to whether photosynthetic CO_2 fixation and reduction may not be related to or a variant of these purely heterotrophic CO_2 fixation mechanisms.

Strong support of this view comes from the work of Benson, Calvin, and co-workers on the nature of the intermediates formed during very short periods of photosynthetic CO_2 reduction in the algae *Chlorella* and *Scenedesmus*.[32] The technique used in these experiments consists simply in allowing an algal suspension to photosynthesize in light in the presence of nonradioactive CO_2. After a steady photosynthetic state has been attained, the nonradioactive CO_2 is swept out and rapidly replaced by a charge of CO_2 containing isotopic C^{14}. After a further photosynthetic period of five seconds to five minutes the algal suspension is suddenly ejected into a solution of boiling alcohol which immediately stops all enzymatic activity and extracts the soluble CO_2 fixation products which have been formed. Separation of these products has been based in part on the classical isolation and identification procedures and in part on the methods of paper chromatography. In this latter method the concentrated extract is applied to and dried near one corner of a large sheet of filter paper. One edge of the paper is then placed in a solution of a suitable developing agent, in this case phenol saturated with water. As the solvent moves past the spot containing the mixture of algal products, these substances are carried along, each different substance or class of substances moving a distance characteristic to itself. For further separation of the products, the paper is next dried, rotated through 90° and then dipped in a second developing solvent, in this case a mixture of butanol, propionic acid, and water. At the expiration of this period of development the individual substances of the algal extract are spread out over the surface of the paper, each particular substance occupying an area largely characteristic to and typical of itself. By subjecting known substances or mixtures of known substances to similar chromatographic procedures it is possible to prepare a map indicating the location of each compound in such a chromatogram. Such a guide map is reproduced in Fig. 30-10. Further identification of the substance contained in each spot or area of the chromatogram can be done by elution of the spot and the application of purely chemical methods to the eluate. When substances containing radioactivity have been subjected to chromatographic separation, the location of each substance on the paper may be simply ascertained by allowing the paper to remain in contact with a photographic film. In the radiograph thus obtained, each radioactive spot on the paper will have exposed the film, which on development will then record a dark spot.

When isotopically tagged CO_2 is supplied in the light to *Chlorella* or to *Scenedesmus* for a total photosynthetic period of five seconds the principal product formed is phosphoglyceric acid and in fact approximately 70% of the carbon fixed can be accounted for as this one compound.[32] The radiograms of Fig. 30-11 show this qualitatively and in addition the phosphoglycerate formed has been isolated in pure form and chemically characterized.[33] In the same short photosynthetic period of five seconds small amounts of radioactive carbon appear in triosephosphate, glucose-1-phosphate, and in glucose- and fructose-6-phosphates. As the period of photosynthesis is increased to thirty seconds, isotopic carbon appears in additional compounds, particularly in alanine, malic acid, and aspartic acid. Finally, after photosynthetic periods of ninety seconds or more, the carbon assimilated begins to appear in sucrose, lipids, and other cellular constituents. That phosphoglycerate is also the initial product of photosynthetic CO_2 fixation in the higher plant is shown in the radiograms of Fig. 30-12. Thus when barley leaves are exposed to radioactive CO_2 in the light for only two seconds, essentially all the CO_2 fixed appears in the phosphoglycerate fraction.

The first photosynthetic intermediate to appear in quantity is then the three-carbon compound, phosphoglycerate. This compound would appear to be synthesized by addition of CO_2 to some preexisting two-carbon fragment, since the 2-phosphoglycerate initially contains isotopic carbon only in the carboxyl group. The phosphorylated sugars formed must in turn be formed by condensation of two such terminally labeled

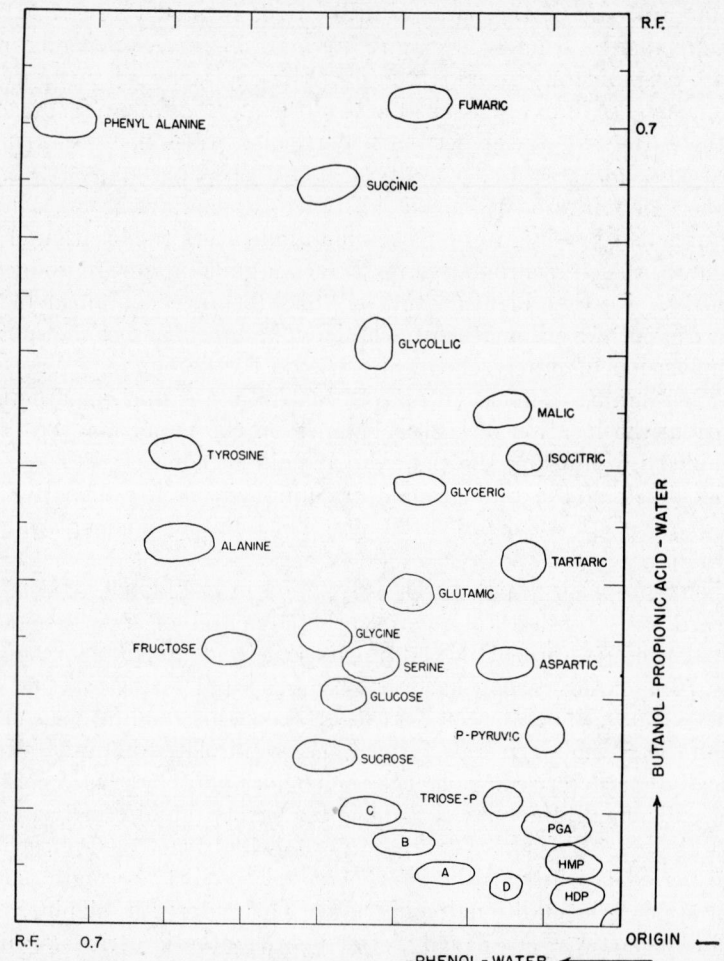

F<small>IG</small>. 30-10. A chromatomap showing the positions occupied by various plant materials after separation by a two-dimensional paper chromatogram. The plant extract is applied to a sheet of filter paper at the point marked *origin*. The materials are next separated by allowing a phenol-water mixture to move from the right-hand edge toward the left. After this separation the paper is dried and the lower edge dipped in a butanol-propionic acid-water mixture and the substances caused to migrate toward the top of the paper. The R.F. values indicated along the two axes represent distance moved by the particular compound in relation to the movement of the solvent front. *A, B,* and *C* are unknown polyglucose compounds. (Courtesy of A. A. Benson and M. Calvin)

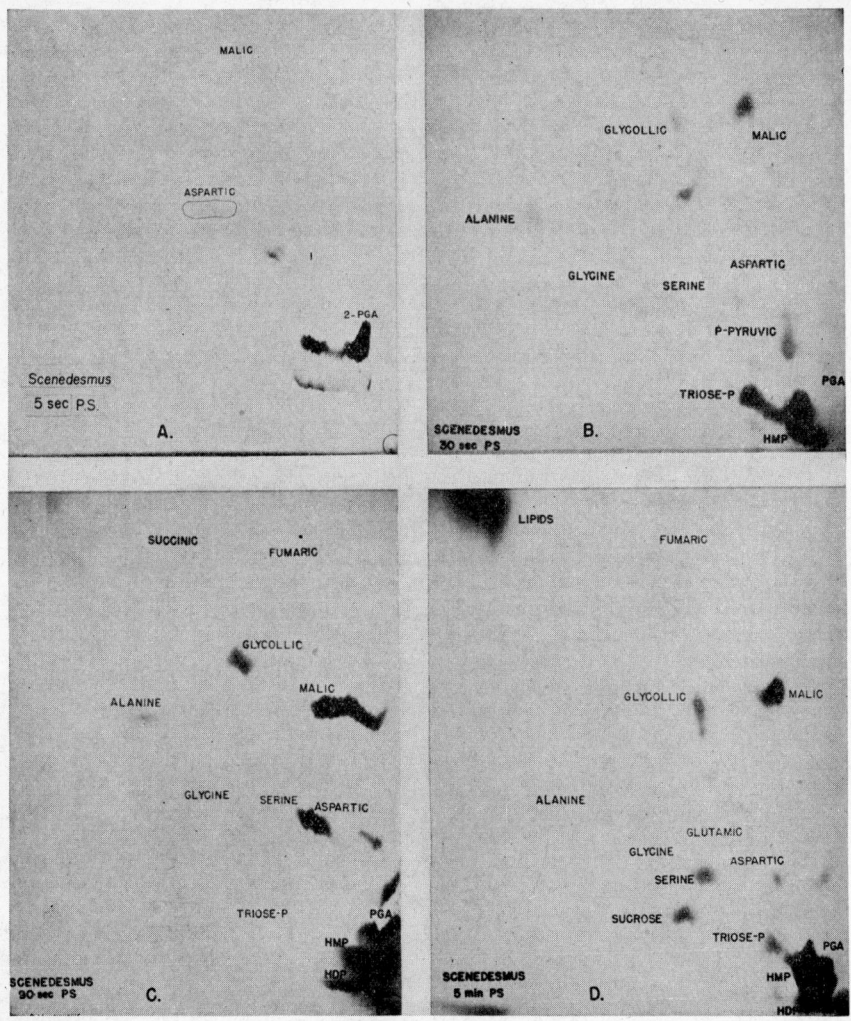

Fig. 30-11. The course of carbon dioxide fixation in the alga *Scenedesmus*. A suspension of the alga is first allowed to photosynthesize until a constant rate is achieved. CO_2 marked with radioactive C^{14} is then added at time $t = 0$. The algae are rapidly harvested after further photosynthetic periods of 5 seconds to 5 minutes. The alcoholic extract of the plant is then subjected to separation by two-dimensional paper chromatography. The dried paper is allowed to expose a photographic film to give the radiograms given here.

A. Photosynthetic products formed in 5 seconds. Phosphoglycerate is the principal product. B. Products formed in 30 seconds. Triose phosphates and hexose phosphates have appeared in addition to phosphoglycerate as major products. C. Products formed in 90 seconds. Major quantities of malic and aspartic acids have appeared. D. Products formed in 5 minutes. Sucrose makes its appearance. However, lipids form the principal reserve of this organism. (Courtesy of A. A. Benson and M. Calvin)

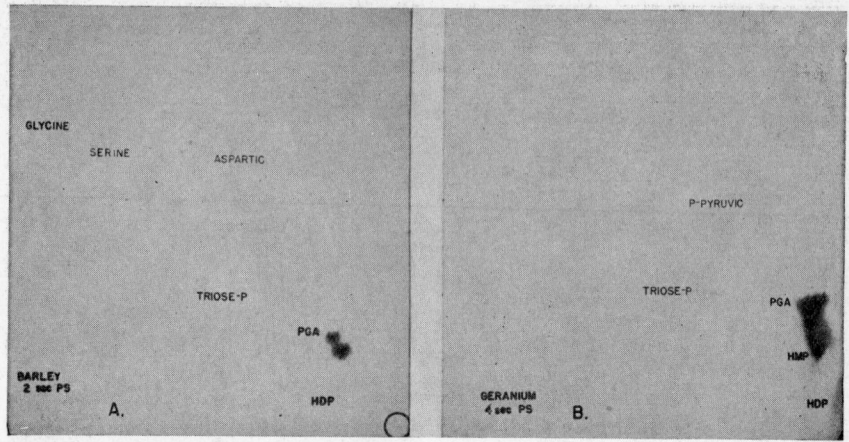

FIG. 30-12. Products formed in leaves of higher plants during extremely short photosynthetic periods in CO_2 containing radioactive C^{14}. The radiograms were prepared as described for Fig. 30-11.

A. Products formed by barley leaves during a 2-second photosynthetic period. Phosphoglycerate is essentially the sole product although a trace of triose phosphate is also found. B. Products formed by geranium leaves during a 4-second photosynthetic period. Phosphoglycerate is the principal product although hexose mono- and diphosphates as well as triose phosphate have also accumulated to some extent. (Courtesy of A. A. Benson and M. Calvin)

three-carbon units since the hexose derivatives initially formed have high radioactivity in the central pair of carbon atoms (3 and 4) and only low radioactivity in the 1,2,5, and 6 carbon atoms at the two ends of the molecule. These facts suggest then that photosynthetic carbon dioxide reduction may consist of the formation of 2-phosphoglycerate followed by reactions which constitute a reversal of the normal glycolytic

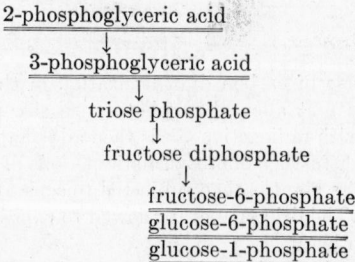

FIG. 30-13. Possible course of carbon in photosynthesis following the initial formation of phosphoglycerate. Doubly underlined compounds are those which accumulate in particularly high concentration during short periods of photosynthesis in *Chlorella* or *Scenedesmus*.

breakdown of hexose and in which two three-carbon units, possibly triose phosphates, are united to form hexose phosphate.

The reactions by which 2-phosphoglycerate is synthesized from CO_2 are evidently cyclical ones which involve firstly the addition of CO_2 to some two-carbon compound and ultimately the reformation of this two-carbon material in such a way that it also is derived from CO_2. This follows from the fact that when photosynthesis is allowed to proceed in the presence of radioactive CO_2 for longer periods, the three-carbon

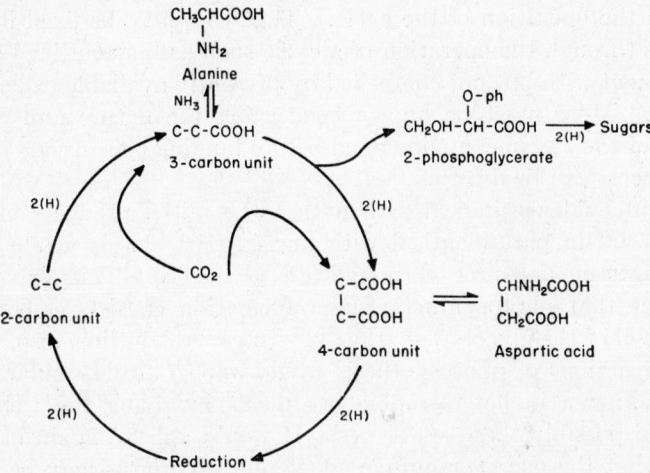

FIG. 30-14. Generalized representation of the cyclical utilization and reformation of a two-carbon CO_2 acceptor in photosynthesis. Modified after Calvin and Benson.

compounds as well as the sugar produced from them tend to attain a uniform distribution of radioactive carbon through all their component carbon atoms. One suggestion as to the possible nature of the reactions involved in formation of 2-phosphoglycerate is presented in Fig. 30-14. In this cycle, CO_2 would first be fixed by addition as a carboxyl group to a preformed two-carbon compound. The three-carbon compound formed could be pyruvic acid or a derivative of pyruvic acid and in this case would constitute the source of the alanine which is observed as an early if minor component of the photosynthetic products. The three-carbon unit could then be transformed to 2-phosphoglycerate, a reaction which would involve the transfer and utilization of one energy-rich phosphate per molecule, and the 2-phosphoglycerate then reduced to the level of hexose through reversal of the glycolytic pathway. On the other hand, the three-carbon atom CO_2 fixation product may alteruatively proceed through reactions leading to the fixation of a second CO_2 molecnle, with the production of a four-carbon atom dicarboxylic acid, possibly oxalo-acetic acid, which would presumably be the immediate source of the aspartic acid observed as an early product of photosynthesis. The further reactions of the four-carbon unit would necessarily constitute its reduction and central cleavage to reconstitute two molecules of the initial two-carbon unit. In each operation of the cycle then, two CO_2 molecules would be taken up and two molecules of the two-carbon unit formed. Of the two molecules of this material formed one would be available for the pathway through phosphoglycerate and one would be needed to maintain the operation of the cycle. That CO_2 may be fixed in photosynthesis through the operation of a cycle such as that outlined above is in agreement with but not compelled by presently available experimental evidence. More must be known concerning the nature and chemical identity of the components involved before binding conclusions as to the exact process can be drawn.

Present evidence indicates then that the initial products of photosynthesis are in part identical with and in part closely related to the normal intermediates of carbohydrate glycolysis. This leaves little doubt but that photosynthetic sugar formation consists essentially in the reversal of the glycolytic pathway. In general outline, also, the CO_2 fixation reactions of photosynthesis would appear to be similar in type to those known in heterotrophic organisms involving the well-known C_2-C_1 and possibly the C_3-C_1 types. It would appear, again in general outline, that the energy requirement of photosynthesis may be twofold and consist in (a) a requirement for reductant to reduce the carboxyl groups formed in CO_2 addition reactions and (b) a requirement for the energy-rich phosphate needed to drive the reverse glycolytic pathway. A critical problem in photosynthesis is then the determination of the exact nature of the carrier system between the photolytic decomposition of water and the energy-using reactions of the CO_2 fixing and reducing system.

General References

Willstätter, R., and Stoll, A. Investigations on Chlorophyll. Science Press, 1928.
Fischer, H., and Stern, A., in Chemie des Pyrrols. Vol. II, 2. Leipzig, 1940.
Rabinowitch, E. I., Photosynthesis. Interscience Publishers, 1945.
Franck, J., and Loomis, W. E., Photosynthesis in Plants. Iowa State College Press, 1949.

References

1. Willstätter, R., Untersuchungen über Chlorophyll. Berlin, 1913.
2. Fischer, H., Berg, H., and Schlormüller, A., Ann., **480,** 109 (1930).
   Fischer, H., and Helberger, H., Ann., **480,** 235 (1930).
3. Fischer, H., Naturwissenschaften, **28,** 401 (1940).
4. Fischer, H., and Kellermann, H., Ann., **524,** 25 (1936).

5. Randolph, L. F., *Botan. Gaz.*, **73**, 337 (1922).
6. Lwoff, A., and Dusi, H., *Compt. rend. soc. biol. Paris*, **119**, 1092 (1935).
7. Demerec, M., *Cold Spring Harbor Symposia Quant. Biol.*, **3**, 80 (1935).
8. Smith, J. H. C., *Arch. Biochem.*, **19**, 449 (1948).
9. Fischer, H., and Oestreicher, A., *Z. physiol. Chem.*, **262**, 243 (1940).
10. Granick, S., *J. Biol. Chem.*, **172**, 717 (1948).
    Granick, S., *J. Biol. Chem.*, **177**, 333 (1948).
11. Granick, S., and Gilder, H., Advances in Enzymology. Interscience Publishers, 1947. Vol. 7, p. 305.
12. Jacobson, L., *Plant Physiol.*, **20**, 233 (1945).
    Bennett, J. P., *Soil Sci.*, **60**, 91 (1945).
13. Rabinowitch, E. I., Photosynthesis. Interscience Publishers, 1945.
14. Franck, J., and French, C. S., *J. Gen. Physiol.*, **25**, 309 (1941).
    Myers, J., and Burr, G. O., *J. Gen. Physiol.*, **24**, 45 (1940).
15. Smith, J. H. C., *Plant Physiol.*, **19**, 394 (1944).
16. Ruben, S., Randall, M., Kamen, M. D., and Hyde, J. L., *J. Am. Chem. Soc.*, **63**, 877 (1941).
    Kamen, M. D., and Barker, H. A., *Proc. Natl. Acad. Sci.*, **31**, 8 (1945).
17. Van Niel, C. B., Advances in Enzymology. Interscience Publishers, 1941. Vol. 1, p. 263.
18. Foster, J. W., *J. Gen. Physiol.*, **24**, 123 (1940); *J. Bact.*, **47**, 355 (1944).
19. Gaffron, H., *J. Gen. Physiol.*, **26**, 195, 241 (1942).
20. Hill, R., *Nature*, **139**, 881 (1937).
    Hill, R., and Scarisbrick, R., *Nature*, **146**, 61 (1940).
21. Holt, A. S., and French, C. S., in Photosynthesis in Plants. Iowa State College Press, 1949.
22. French, C. S., and Rabideau, G. S., *J. Gen. Physiol.*, **28**, 329 (1945).
    Milner, H. W., and Lawrence, N., Abstr., Vancouver meeting. Am. Soc. Plant Physiol., 1949.
23. Warburg, O., and Lüttgens, W., *Naturwissenschaften*, **32**, 161 (1944).
24. Warburg, O., and Negelein, E., *Z. physik. Chem.*, **102**, 235 (1922).
25. Rieke, F. F., in Photosynthesis in Plants. Iowa State College Press, 1949.
26. Emerson, R., and Lewis, C. M., *Am. J. Botany*, **28**, 789 (1941).
27. Manning, W. M., Stauffer, J. F., Duggar, B. M., and Daniels, F., *J. Am. Chem. Soc.*, **60**, 266 (1938).
28. Magee, J. L., DeWitt, T. W., Smith, E. C., and Daniels, F., *J. Am. Chem. Soc.*, **61**, 3529 (1939).
29. Arnold, W., in Photosynthesis in Plants. Iowa State College Press, 1949.
30. Warburg, O., *Am. J. Botany*, **35**, 194 (1948).
31. Ruben, S., Hassid, W. Z., and Kamen, M. D., *J. Am. Chem. Soc.*, **61**, 661 (1939); **62**, 3443, 3450, 3451 (1940).
32. Calvin, M., and Benson, A. A., *Science*, **107**, 476 (1948); **109**, 140 (1949).
    Benson, A. A., and Calvin, M., *Cold Spring Harbor Symposia Quant. Biol.*, **13**, 6 (1948).
    Benson, A. A., *et al.*, in Photosynthesis in Plants. Iowa State College Press, 1949.
33. Calvin, M., *J. Chem. Ed.*, **26**, 639 (1949).
34. Zscheile, F. P., *Cold Spring Harbor Symposia Quant. Biol.*, **3**, 108 (1935).

Author Index*

A

Abegg, F. A., 417, 418, *420*
Abramson, H. A., 247, *280*
Adler, E., 228, 229, 233, *243*
Åkeson, A., 180, *216*, 375, *383*
Albaum, H. G., 208, 210, *218*, 233, *243*
Albers, H., 172, *215*
Aleshin, S., 394, *396*
Allison, F. E., 399, *351*
Allsopp, A., 136, *139*
Alsberg, C. L., 58, *64*
Altman, R. F. A., 414, *420*
Altschul, A. M., 178, *216*
Ambronn, H., 78, 80, *92*
Ambros, O., 288, *292*
Anderson, B., 174, *215*, 228, 243
Anderson, E., 111, *113*, *114*, 116, 119, *121*, 131, 133, 138
Anderson, P., 308, *318*
Angell, H., 164, *166*
Angell, S., 110, 111, *114*, 138, *139*
Annan, E., 193, *217*
Anson, M. L., 245, *280*
Appleman, C. O., *143*
Archibald, R. M., 296, *298*
Arhimo, A. A., 141, *165*
Armstrong, E. F., 16, *25*, *33*
Arnold, W., 483, *491*
Arnon, D., 182, *216*, 260, *281*
Aronovsky, S. I., 90, *92*
Arreguin, B., 34, *47*, 62, *65*, 224, *243*, 418, *420*
Arrhenius, S., 253, *280*
Arthur, J. M., 427, *437*
Asahina, Y., 419, *420*
Astbury, W. T., 245, 271, *280*, *282*
Atchley, W. A., 376, *383*
Averill, F. J., 72, *91*
Avery, G. S., 174, 175, 186, *216*, 229, *243*, 440, 442, 444, 453, 457, 459, *465*, *466*, *467*
Avery, O. T., 30, *33*
Axelrod, A. E., 186, *217*

B

Baas-Becking, L. G. M., 336, *351*
Bach, A., 286, *292*

Bachrach, J., 138, *139*
Baier, W. E., 107, *108*
Bailey, A. E., 254, *280*
Bailey, I. W., 84, 85, 86, *92*
Baker, D., 202, *218*
Baker, G. L., 107, *108*
Baldwin, I. L., 341, *351*
Baldwin, R. R., 54, 55, *64*
Ball, E. G., 178, *216*
Ball, H., 77, *92*
Balls, A. K., 59, *64*, 283, 286, *292*
Bamann, E., *13*
Barker, H. A., 30, *33*, 38, *47*, 152, *166*, 371, *382*, 478, *491*
Barrien, B. S., 256, *281*, 311, *318*
Barron, E. S. G., 197, 201, *218*
Barry, V. C., 120, *121*
Bate-Smith, E., 425, *437*
Batelli, F., 169, *215*
Bates, F. L., 50, *64*
Baumberger, J. P., 207, *218*
Baur, L., 100, *107*
Bawden, F. C., 268, 274, *281*, *282*
Beadle, B. W., 359, *383*
Beadle, G. W., 236, *243*, 331, *334*
Beale, G. H., 432, 434, *437*
Bear, R. S., 54, *64*
Beck, L. V., 190, *217*
Behnke, J., 440, 442, *465*
Behrens, M., 268, *281*
Beijerinck, M., 103, *108*, 338, 341, *351*
Belfanti, S., 380, *383*
Belozersky, A., 268, *281*
Bennet-Clark, T. A., 154, 156, *165*, *166*, 198, *218*
Bennett, E., 111, *114*
Bennett, J. P., 477, *491*
Benson, A. A., 480, 484, 485, *491*
Benz, F., 175, *216*
Béres, T., 409, *410*
Berg, H., 469, *490*
Berger, J., 174, 175, 186, *216*, 229, *243*, 286, 289, *292*, 444, 453, 457, 459, *466*, *467*
Berger, L., 36, *47*, 188, *217*
Bergmann, L., 202, *218*
Bergmann, M., 283, 289, *291*, *292*
Bergren, W. R., 444, 453, *466*

* Numbers in italics indicate pages on which references are listed.

Fosse, R., 320, 322, *327*
Foster, C., 379, 380, *383*
Foster, E. O., 368, *382*
Foster, J. F., 55, 56, 57, *64*
Foster, J. W., 479, *491*
Frampton, V. L., 178, *216*, 249, *280*
Franck, J., 478, *490*, *491*
Franzen, H., 144, *165*
Fraps, G., 408, *410*
Fred, E. B., 341, *351*
Frémy, E., 99, 104, *107*
French, C. S., 478, 480, 481, *491*
French, D., 50, 54, 55, 57, *64*
Freudenberg, K., 122, 123, 124, 125, 126, 127, *129*
Frey-Wyssling, A., 58, *63*, 72, 77, 78, 80, 81, 82, 83, 84, 87, 88, *91*, *92*, 130, *138*, 141, 144, *165*, 258, *282*, 407, *410*, 413, *420*, 427, *437*
Fruton, J. S., 283, 289, *292*
Fuller, W. H., 89, *92*

G

Gaffron, H., 479, *491*
Gale, E. F., 241, *243*
Gallup, W. D., 369, *382*
Galston, A. W., 164, *166*, 331, *334*, 418, *420*, 425, *437*, 439, 451, 464, *465*, *466*
Gaponenkov, T., 394, *396*
Gardner, F. E., 441, *465*
Gary, W., 110, 111, *113*
Gaulis, M., 124, *129*
Gautheret, R., 441, *465*
Geddes, W. F., *63*, 297, *298*
Gee, G., 416, *420*
Gerber, C., 369, 370, *382*
Gert, H., 264, *281*
Gildemeister, E., *396*
Gilder, H., 476, *491*
Gill, R. E., 119, *121*
Gillette, L. A., 119, *121*
Ginter, W., 154, 157, 159, *166*
Gisvold, O., *12*
Glasstone, S., *13*
Glick, D., *12*
Goddard, D. R., 180, 186, 201, 203, 207, *215*, *216*, *217*, *218*
Goddum, L., 106, *108*
Goebel, W. F., 118, *121*
Goepp, R. M., Jr., *25*, *33*, *63*, *91*, *98*
Gollub, M., 148, 149, 151, 153, *165*, 175, 198, *216*
Goodwin, M. W., 107, *108*
Gordon, S. A., 447, 448, 454, *466*
Gorham, P., 407, *410*

Gorin, M. H., 247, *280*
Gortner, R. A., 95, *98*
Goss, M. J., 124, *129*
Gözsy, B., 193, *217*
Graeve, P. de., 320, *327*
Grafflin, A. I., 375, 376, *383*
Gralén, N., 76, *91*, 178, *216*
Gramick, S., 235, *243*, 256, 258, *281*, *282*, 476, *491*
Grassmann, W., 89, *92*
Graubard, M., 182, *216*
Grauer, H., 178, *216*, 230, *243*
Green, A. A., 62, *65*
Green, D. E., 61, 62, *65*, 148, *165*, 178, 183, 185, 207, 211, 212, *215*, *216*, *217*, *218*, 230, 231, *243*, 375, 376, 377, *383*
Greenberg, D. M., 284, 285, 286, *292*
Greenberg, G. R., 322, *327*
Greene, K. A., 368, *382*
Greene, R. D., 254, *280*
Greenhill, A. W., 300, *318*
Greenstein, J. P., 266, 273, 275, *279*, *281*, *282*
Gregory, F. G., 308, 310, 312, *318*
Griese, A., 173, *215*
Griffioen, K., 134, 135, *139*
Guenther, E., *396*
Gullard, J. M., 268, 271, *281*, *282*
Gunsalus, I. C., 236, *243*
Gunther, G., 229, 233, *243*
Gurin, S., 377, *383*
Gustafson, F. G., 441, 442, *465*
Guthrie, J., 59, *64*
Guzman-Barron, E., *46*

H

Haag, A., 125, *129*
Haagen-Smit, A. J., 394, *396*, 417, *420*, 441, 442, 443, 444, 446, 453, 459, *465*, *466*, *467*
Haas, E., 178, 181, *216*
Haas, P., *12*
Hägglund, E., 126, *129*
Halcro-Wardlaw, H. S., 374, *382*
Hale, W. S., 286, *292*
Haley, D. E., 374, *383*
Haller, M., 106, *108*
Halsall, T. G., 48, 49, *64*
Hamner, K. C., 223, *242*, 408, *410*, 441, *465*
Hammarsten, E., 271, *282*
Hammer, H., 308, *318*
Hand, M., 439, *465*
Hanes, C. S., 53, 54, 55, 60, *64*

Lewis, C. M., 407, *410*, 482, *491*
Lewis, S., 183, *217*
Lex, A., 442, *466*
Li, L., 182, 202, *216*, *218*, 260
Little, H., 344, 345, *351*
Linderstrøm-Lang, K., 286, 288, 289, *292*
Ling, A., 99, *107*
Link, K. P., 100, *107*, 164, *166*, *437*
Linser, H., 330, *334*
Lipman, J. G., 337, 346, *351*
Lipmann, F., 152, *166*, 184, 188, 212, 213, 214, 215, *217*, *218*
Lochhead, A. G., 342, *351*
Lockhart, E. E., 177, *216*
Lohmann, K., 35, *47*, 183, 212, *217*, *218*
Long, C., 26, *33*
Longenecker, H. E., 374, *383*
Loomis, W. E., 212, *218*, *490*
Loring, H. S., 271, *282*, 325, *327*
Lovelace, F., 284, *292*
Luchett, S., 100, *107*
Lüdtke, M., 93, *98*
Lunberg, R., 178, *216*
Lüttgens, W., 481, *491*
Lutwak-Mann, C., 187, *217*
Lwoff, A., 473, *491*
Lythgoe, B., 268, *281*

M

Macbeth, A. K., 394, *396*
McCoy, E., 341, *351*
McCready, R. M., 38, 39, *47*, 50, 51, 61, *64*, *65*
MacDonald, J., 48, *64*
McDonald, M. R., 36, *47*, 118, *217*
McGavack, J., 413, *420*
McGettrick, W., *121*
McGilvery, R., 291, *292*
Machemer, H., 48, *64*
McIlroy, R. J., 112, *114*
McKee, H. S., *317*
McKinney, R., 368, *382*
MacLean, H., *382*
McMurdie, H., 412, *420*
McNair, J. N., 355, *382*
MacVicar, R., 229, *243*
Magee, J. L., 483, *491*
Mahdihassan, S., 268, *281*
Malan, P., 464, *467*
Mangelsdorf, P., 408, *410*
Mann, T., 182, 187, 191, *216*, *217*
Manning, W. M., 407, *410*, 482, *491*
Mantell, C. L., *121*
Mark, H., 73, 75, *91*
Markley, K. S., *382*

Marsh, P. B., 201, 203, *218*
Marth, P. C., 442, *466*
Marthaler, H., 222, *242*
Marteny, W. W., 133, *139*
Martius, C., 195, *217*
Marx, E., 169, 183, *215*
Maskell, E. J., 311, *318*
Mason, T. G., 308, 310, 311, 314, *318*
Matthes, E., 374, *382*
Maunfield, J. D., 286, 288, *292*
Mauger, R. P., 112, *114*
Mayer, F., *437*
Mayer, K., 59, *64*
Medes, G., 376, *383*
Medigreceanu, F., 273, *282*
Meek, J. S., 152, *166*
Meisels, A., 392, *396*
Meister, M., 127, *129*
Melnick, J. L., 183, *217*
Memmler, K., *419*
Menke, W., 255, 259, *281*
Mercer, F. V., 306, 307, 309, 312, *318*
Merry, J., 201, 203, *218*
Meyer, K. H., 50, 51, 56, *64*, 73, 74, *91*
Meyerhof, O., 189, 190, 212, *217*, *218*, 336, *351*
Michael, G., 307, *318*
Michaelis, L., 169, 203, *215*, *218*
Michlin, D., 224, *243*
Mieg, W., 406, *410*
Mikhlin, D., 203, *218*
Miller, E., 369, 370, 371, 372, 374, *382*
Miller, N. H. J., 227, *243*
Milner, H. W., 481, *491*
Mirsky, A. E., 245, 266, 272, *280*, *281*, *282*
Misch, L. C., 74, *91*
Misra, P., 136, *139*
Mitchell, H. K., 239, *243*, 249, *283*, 325, 326, *327*, 331, *334*
Mitchell, J. W., 442, *466*
Mitchell, R. L., 90, *92*
Molisch, H., *12*, 102, *107*
Monguillon, P., 225, *243*
Montonna, R. E., 72, *91*
Moog, F., 189, *217*
Moore, D. H., 178, *216*
Morell, S., 100, *107*
Morris, G., 89, *92*
Morris, H. J., 339, *351*
Mothes, K., 296, *298*, 299, 300, 311, *318*, 329, *334*
Moxon, A. C., 301, 306, 307, *318*
Moyer, L. S., 247, *280*, 411, *420*
Muir, R. M., 450, 456, *466*
Murlin, J. R., 373, *382*
Murneek, A. E., 442, *466*

Subject Index

Regulator, growth, see Growth substances
Reseda, flavone in, 429
Resin
  acids, 391, 392
  in crude lipid, 352
  in latex, 413, 414
Respiration, amide formation and, 296
  Bach-Chodat theory of, 169
  definition of, 167
  energetics of, 203, 207
  glycolysis in, 187–192
  growth substances and, 457
  history of, 168, 169
  initial phase, definition of, 167
  metabolism as energy source, 10
  nitrate reduction and, 223
  particulate matter and, 4
  terminal oxidase of, 200
  terminal phase, 167, 192 f.
Respiratory quotient, see R. Q.
Retting, and middle lamella, 103, 104
Rf value, in paper chromatography, 426, 486
rH, definition of, 205
Rhamninose, 32
Rhamnoglucoside, 425
Rhamnose, 23
  in glycosides, 429, 430
  in gum arabic, 115
  in mucilage, 119
  occurrence, 23, 32
Rhamnus
  calcium oxalate in, 144
  rhamninose in, 32
Rhizobium
  amino acid decarboxylase in, 241
  culture, 342
  gums of, 118
  nitrogen fixation by, 342–344
  species of, 341
Rhizopus
  CO₂ fixation in, 151
  pectinase in, 104
Rhodoporphyrin, structure, 470
Rhodotorula, carotenoids biogenesis in, 406
Rhodoviolascin, 402
Rhodoxanthin, 401, 403
Rhubarb
  amides in, 301
  changes in excised leaves, 306
  metabolism of organic acids, 149, 150
  organic acids in, 142
Rhus
  flavonol in, 430
  gallic acid in, 164
  laccase in, 182
Ribes, salicylic acid in, 163
Ribitol, 43, 176
Riboflavin
  in *Azotobacter*, 339
  in flavoproteins, 176
  as growth factor for *Rhizobium*, 342
  IAA destruction and, 451, 452
  structure, 176

Riboflavin phosphate, 175, 176
Ribonuclease, 273
Ribose, 21, 23
  in ATP, 188
  in cytoplasmic proteins, 263
  in flavin-adenine dinucleotide, 176
  in nucleic acids, 267 f.
  in nucleoside synthesis, 325–327
  occurrence of, 22
  in phosphopyridine nucleotides, 172, 173
Ribose-1-phosphate, 274
Ribose-5-phosphate, 37
Rice
  amylose in, 50
  choline phosphatase in, 381
  fatty acids in, 362
  lecithinases in, 380
  seed proteins of, 253
  starch grains of, 57
Ricinoleic acid, 357
  behavior in germination, 373
  in seed fats, 362, 363
Robinia
  flavonol in, 430
  robinose in, 32
Robinose, 32
  as flavonol, 430
  occurrence, 32
Robison ester, structure, 5
Rooting of cuttings, 440, 461–464
Roots, see also individual plants
  anabasine synthesis in, 332
  anthocyanins in, 427
  anthoxanthins in, 428 f.
  atropine in, 332, 333
  CO₂ fixation in, 151
  fats in, 354
  growth substances and, 441, 442, 462, 463
  IAA oxidase in, 450, 451
  nicotine synthesis in, 328, 329
  nitrate reduction in, 222, 223
  nitrogen excretion by, 346
  nodules of, 341 f.
  organic acids in, 143
  protein turnover in, 313
  purines in, 320
Rose
  anthocyanidins in, 425
  emulsin in, 26
  essential oil of, 390
  fats in, 354, 361
  flavonol in, 430
  gene effects on anthocyanins, 433
R. Q.
  definition, 167
  in fat formation, 370
  in nitrate reduction, 223
  of succulents, 157
Rubber
  biosynthesis of, 413, 417 f.
  chemistry of, 414 f.
  distribution of, 411
  function of, 419

particles, electron micrographs of, 412
Rubiaceae
  purines in, 323
  seed fats in, 361
Rubixanthin, 401, 402
Rubus, salicylic acid in, 163
Rutabaga, protopectinase in, 103
Rutaceae, essential oils in, 385
Ruthenium red, 131
Rye
  fructosan in, 68
  glutamine formation in, 300, 301
  IAA complex in, 453
  nucleic acids in, 268
  pyrimidines in, 325
  seed proteins of, 253
  straw, composition, 137
Rye grass, fats in, 354, 360

S

Sabinene, 388
Sabinol, 389
Saccharic acid, 42
Sago, amylose in, 50
Salicylic acid, 162, 163
Salvia, essential oil in, 389
α-Santalene, 391
Sapindaceae
  purines in, 323
  seed fats in, 361
Saponification number of fats, 359
Saponins, 392
Sargassum, alginic acid in, 120
Scanning patterns of proteins, 248, 249, 264, 265
Scarlet H, stain for cuticular substances, 131
Scenedesmus, path of carbon in photosynthesis study, 484 f.
Schinopsis, protocatechuic acid in, 164
Sclerotinia, pectinase in, 104
Schweizer's reagent, 76
Scutellum, proteases in, 286
Scyllitol, 44, 45
Sea buckthorn, fats in, 354
Secondary wall
  composition, see Cell Wall Components
  estimation of pectic substances in, 134, 182
  histology, 84, 85
  micellar structure, 81
  orientation of cellulose in, 85, 87
  structure in flax and ramie, 83
  xylan in, 95
Sedoheptose, see also Altroketoheptose
  structure, 24
  succulent metabolism and, 156
Sedum, as succulent, 154
Seedlings
  amides in, 294 f.
  proteolysis in, 293
Seeds, see also individual plants
  cuticle waxes of, 366
  cis carotenoids in, 404
  cereal proteins of, 251

dispersal of, 407
fats in, 354, 355, 361 f., 463
fat formation in, 368, 369
IAA complex in, 453
mobilization of fats, 372
mucilage of seed coat, 118
phosphatidic acid in, 378, 380
proteases in, 286
proteolysis in germination, 308
purines in, 320, 323
pyrimidines in, 325
Selacholeic acid, structure, 357
Selaginella, trehalose in, 29
β-Selinine, 390, 391
Sequoia
  composition of, 136
  mannan in, 94
  sequoyitol in, 45
Sequoyitol, 44, 45
Serine, 220
  in phospholipids, 379
  relation to indole, 236
  relation to glycine, 239
Sesame, fatty acids in, 362
Sesquiterpenes, 390, 391
Shikimic acid, structure, 45
Sieve tubes, wall structure of, 87, 88
Silage, mannitol in, 42
Silica skeleton of cellulose, 78, 80
Silver, deposition of in fibers, 80
Simarubaceae, seed fats in, 363
Simmondsia, wax in, 368
Sisal, xylan in fiber, 97
Snapdragon, gene effect on anthocyanins, 433, 435
Soaps, preparation of, 353
Solanaceae, seed fats in, 361
Solanain, 286
Solanum
  lycophyll in, 401
  solanain in, 286
Sorbitol, 42
L-Sorbose, 15
  ascorbic acid and, 45
  glucose-1-phosphate and, 40
  as oxidation product of D-sorbitol, 42
Sorghum, seed proteins of, 253
Soybean
  fats in, 354, 355, 363
  lactic dehydrogenase in, 224
  lipoxidase in, 374
  phosphatides in, 379, 380
  proteins of, 254
  Rhizobium of, 341
  stachyose in, 32
  urease in, 235
Spinach
  amino acids in cytoplasmic protein, 262
  cellular constituents of, 256
  chloroplasts of, 258, 259
  cytoplasmic protein, preparation of, 261
  enzymes in chloroplasts, 260
  fats in, 354, 360
  hexokinase in, 36